Trends in Edible Packaging: A sustainable approaches for postharvest management of fruits and vegetables

Nishant Kumar • Pratibha
Ashutosh Upadhyay • Anka Trajkovska Petkoska
Editors

Trends in Edible Packaging: A sustainable approaches for postharvest management of fruits and vegetables

Sustainable Packaging

Springer

Editors
Nishant Kumar
Amity Food and Agriculture Foundation (AFAF)
Amity University
Noida, Uttar Pradesh, India

Pratibha
Centre of Management and Humanities (CMH)
Punjab Engineering College (Deemed to be University)
Chandigarh, India

Ashutosh Upadhyay
Food Science & Technology
NIFTEM-K
Kundli, Haryana, India

Anka Trajkovska Petkoska
Faculty of Technology and Technical Sciences
University St. Kliment Ohridski-Bitola
Veles, North Macedonia

ISBN 978-3-032-13198-0 ISBN 978-3-032-13199-7 (eBook)
https://doi.org/10.1007/978-3-032-13199-7

This Springer imprint is published by the registered company Springer Nature Switzerland AG
The registered company address is: Gewerbestrasse 11, 6330 Cham, Switzerland

Preface

The challenges of managing global hunger have intensified due to rapid population growth and ongoing disruptions and losses within the food supply chain, these challenges further affect the socioeconomic and environmental aspects. About one-third of food is wasted from the total food production around the world. Fruits and vegetables are among the most perishable food commodities due to their high moisture content and metabolic activity, making them highly susceptible to post-harvest losses up to 30-40% throughout the supply chain. These losses are primarily attributed to factors such as microbial spoilage, enzymatic browning, and the degradation of nutritional components. Quality parameters critical to consumer acceptance include sensory attributes such as flavor, color, and texture, as well as the absence of visible decay or physiological disorders in fruits and vegetables. However, the food preservation techniques play a crucial role in minimizing post-harvest losses and maintaining the quality and safety of perishable produce. Each processing step can negatively affect their nutritional and sensory qualities. To maintain quality and extend their shelf life, researchers have developed techniques such as edible coatings, modified and controlled atmosphere packaging, calcium treatments, cold storage, and gamma irradiation. These methods aim to slow down spoilage processes, preserve texture and color, and meet consumer demands for fresh, convenient, and high-quality produce with improved safety and extended marketability. In recent years, there has been a growing consumer demand for minimally processed and health-oriented food products, which has led to increased research interest in edible packaging systems. Edible films and coatings are sustainable packaging and can be used an alternative of plastic- and synthetic-based packaging to overcome the environmental pollutions. The edible coatings can be formulated via natural plant- and animal-derived biopolymers such as polysaccharide, proteins, lipid/wax and their composites to maintain the postharvest shelf life of fruits and vegetables during the storage period by retarded the weight loss and physiological effects. It is served as primary packaging systems that help preserve key sensory characteristics such as flavor, aroma, and visual appeal. These coatings contribute to prolonged shelf life and a slower ripening process, thereby maintaining freshness and improving product stability during storage and distribution.

One of the major highlighting features of this book, *Trends in Edible Packaging: A Sustainable Approach for Postharvest Management of Fruits and Vegetables*, is to draw the attention of food processing industries, researchers, and food scientists for novel techniques and methods of prolonging the shelf life of packed food as well as for a better food quality and safety. The book has unveiled pivotal aspects of edible packaging of significant importance in the sustainable management of post-harvest losses. This book also highlights the use of different types of biopolymers for developing edible coating/films and composite packaging for preservation of fruits and vegetables. In addition, the different types of application methods of coating/films, effects of additives/active agents such as plasticizers, emulsifiers, essential oils, plant extracts, probiotics, nanomaterials, and nutrients and their effect on the properties of edible coating/films are covered in this book. Moreover, the effect of different types of biopolymer-based edible coating and films on the postharvest characteristics of fruits and vegetables also are discussed.

The entire material presented in this work is the collaboration of experts in their distinguished field. This piece of work will cater to food researchers, scientists, academics, environmentalists, material scientists, nutritionists, food process and packaging industries, and people in horticulture and postharvest management research in terms of uniqueness and usefulness. This work will indeed prove to be very informative to the public for efficiently analyzing the significance of biopolymers use for developing eco-friendly, nontoxic, and edible packaging (coating/films), a healthy-friendly solution for preservation of postharvest quality and consumer acceptance of fruits and vegetables.

AFAF, Amity, Noida, Uttar Pradesh, India Nishant Kumar

Contents

Editors and Contributors

About the Editors

Nishant Kumar is working as an Assistant Professor at Amity and Food Foundation, Amity University, Noida, India. He has research experience on synthesis and production of biopolymers from different plant-based sources, development of nanoformulations, edible coating, postharvest management, etc. He has more than 4 years of research experience in this field as a Research Associate, Project Fellow, and Knowledge Management Officer. In addition, he also has 2.5 years of experience in the food processing industry. He did a PhD at the National Institute of Food Technology Entrepreneurship and Management, Kundli, Sonipat (Haryana). During his PhD, he worked on different types of biopolymers, active agents to develop the edible coating, biocomposite, nanocomposite, and their application for shelf-life extension of different fruits and vegetables. In addition, he also studied the structural, biocompatible, and antioxidant potential of natural plant extracts and essential oils as alternatives to chemical fungicides. He has published more than 70 research/review articles in highly reputed journals and has published 1 book, 20+ book chapters in various books, and has been granted patents. He is also serving as an editor for more than 15 journals, and he is also reviewer of more than 50 International/National scientific journals. He also reviews book proposals for various publishers. His significant contribution to scientific research can be reflected in his research profile.

Pratibha is working as an Assistant Professor at the Centre of Management and Humanities (CMH), Punjab Engineering College (Deemed to be University), Chandigarh, India. Her specialization includes food technology, food engineering, food packaging, edible coatings, postharvest management, and sustainable entrepreneurship development. She has published various research/review articles in reputed journals and several book chapters. She is also working as editor and reviewer for various journals.

Ashutosh Upadhyay is a Full Professor at the Department of Food Science and Technology, National Institute of Food Technology Entrepreneurship and Management, Kundli Sonipat, Haryana. His specialization and research focus include healthy foods, food safety, minimal processing and grain processing, novel processing techniques, traceability, grain quality in bulk storage, new product development, edible packaging, and postharvest management. He has published various research/review articles in reputed journals and several book chapters. He is also working as editor and reviewer for various journals.

Anka Trajkovska Petkoska is a Full Professor at the University St. Kliment Ohridski-Bitola (R.N. Macedonia) and an Adjunct Professor at the Department of Materials Science and Engineering at Korea University, Seoul (South Korea). She has been actively involved in the development of numerous novel products, proprietary material formulations, and cutting-edge technologies, like development of micro/nano-polymer composites, food contact materials, antimicrobial formulations, multi-functional thin films and coatings for customized applications, energy efficiency, and utilization of renewable energy sources. Her interests also include sustainable food systems, healthy diets, global syndemic problems, and circular economy. She is a scientific/technical consultant and a collaborator of SMEs in the field of materials science and engineering, novel technologies, and sustainability challenges. Her 20+ years of experience involves working in academia, industrial, and military institutions. She has participated in different project programs starting from TEMPUS, EU FP7, Horizon 2020, CEEPUS, and COST Actions; she is an evaluator for EU programs as well. She is also a reviewer of many scientific journals published by Elsevier, Springer, Wiley, Taylor & Francis, MDPI, and Hindawi and a guest editor of Frontiers journals. She is the author and co-author of numerous peer-reviewed scientific articles, five books, more than 15 book chapters, and training materials and has been granted with three US patents. For her work, she has been awarded prestigious recognitions and awards.

Contributors

Aditya Department of Agriculture and Environmental Sciences, National Institute of Food Technology Entrepreneurship and Management, Kundli (An Institute of National Importance of India, NIFTEM-K), Sonipat, Haryana, India

Waseem Ahmed Department of Horticulture, The University of Haripur, Haripur, Khyber Pakhtunkhwa, Pakistan

Sarra Akermi Laboratory of Microbial and Enzymatic Biotechnologies and Biomolecules (LBMEB), Center of Biotechnology of Sfax (CBS), University of Sfax, Sfax, Tunisia

Aleksandra Anakjioska Faculty of Technology and Technical Sciences, University St. Kliment Ohridski-Bitola, Veles, North Macedonia

Aniqa Anam Department of Botany, The University of Haripur, Haripur, Khyber Pakhtunkhwa, Pakistan

A. R. S. A. Athauda Department of Crop Science, Faculty of Agriculture, University of Peradeniya, Peradeniya, Sri Lanka

Rafia Azmat Department of Chemistry, University of Karachi, Karachi, Pakistan

J. N. Bhatia Department of Plant Pathology, Chaudhary Charan Singh Haryana Agricultural University (CCS HAU), Hisar, Haryana, India

Vikrant Bodana Centre For Sustainable Polymers, Indian Institute of Technology Guwahati, Guwahati, Assam, India

Moufida Chaari Laboratory of Microbial and Enzymatic Biotechnologies and Biomolecules (LBMEB), Center of Biotechnology of Sfax (CBS), University of Sfax, Sfax, Tunisia

Urmila Choudhary Department of Dairy Technology, College of Dairy Science and Technology, Sri Karan Narendra Agricultural University, Jobner, Jaipur, Rajasthan, India

Vijay Dangi Department of Physical Sciences (Chemistry), Baba Mastnath University, Rohtak, Haryana, India

Khaoula Elhadef Laboratory of Microbial and Enzymatic Biotechnologies and Biomolecules (LBMEB), Center of Biotechnology of Sfax (CBS), University of Sfax, Sfax, Tunisia

Mandavi Goswami Centre for Sustainable Polymers, Indian Institute of Technology Guwahati, Guwahati, Assam, India

Atul K. Gupta Central Instrument Facility, Sharda University, Greater Noida, Uttar Pradesh, India

Himani Department of Life Sciences (Botany), Baba Mast Nath University, Rohtak, Haryana, India

Himanshu Department of Life Sciences (Botany), Baba Mast Nath University, Rohtak, Haryana, India

Bouthaina Ben Hadj Hmida Laboratory of Biochemistry and Enzymatic Engineering of Lipases, National Engineering School of Sfax, University of Sfax, Sfax, Tunisia

Rushali Jaiswal National Institute of Food Technology Entrepreneurship and Management, Kundli, Sonipat, India

B. G. U. Janith Department of Crop Science, Faculty of Agriculture, University of Peradeniya, Peradeniya, Sri Lanka

J. M. C. M. Jayasekara Department of Food Science and Technology, Faculty of Agriculture, University of Peradeniya, Peradeniya, Sri Lanka

Jyoti Central Instrument Facility, Sharda University, Greater Noida, Uttar Pradesh, India

Vimal Katiyar Centre for Sustainable Polymers, Indian Institute of Technology Guwahati, Guwahati, Assam, India

Nutan Kaushik Amity Food and Agriculture Foundation (AFAF), Amity University Noida, Noida, Uttar Pradesh, India

Ishu Khangwal Department of Life Sciences (Botany), Baba Mastnath University, Rohtak, Haryana, India

Anand Kishore National Institute of Food Technology Entrepreneurship and Management, Kundli, Sonepat, Haryana, India

Sarita Kodan Department of Life Sciences (Botany), Baba Mastnath University, Rohtak, Haryana, India

Nishant Kumar Amity Food and Agriculture Foundation (AFAF), Amity University Noida, Noida, Uttar Pradesh, India

Khushbu Kumari National Dairy Research Institute, Karnal, Haryana, India

Manoj Kumari Department of Life Sciences (Botany), Baba Mastnath University, Rohtak, Haryana, India

H. M. P. C. Kumarihami Department of Crop Science, Faculty of Agriculture, University of Peradeniya, Peradeniya, Sri Lanka

Muhammad Liaquat Department of Food Science and Technology, The University of Haripur, Haripur, Khyber Pakhtunkhwa, Pakistan

Manisha Department of Life Sciences (Botany), Baba Mastnath University, Rohtak, Haryana, India

Mehnaz Mehboob Department of Botany, The University of Haripur, Haripur, Khyber Pakhtunkhwa, Pakistan

Muskan Department of Physics, Baba Mast Nath University, Rohtak, Haryana, India

P. Nandagopal Department of Dairy Technology, College of Dairy Science and Technology, Kolahalamedu Kerala Veterinary and Animal Sciences University, Kolahalamedu, Kerala, India

Neeraj Department of Agriculture and Environmental Sciences, National Institute of Food Technology Entrepreneurship and Management, Kundli (An Institute of National Importance of India, NIFTEM-K), Sonipat, Haryana, India

Nikita National Institute of Food Technology Entrepreneurship and Management, Kundli, Sonepat, Haryana, India

M. D. A. M. Perera Department of Crop Science, Faculty of Agriculture, University of Peradeniya, Peradeniya, Sri Lanka

Anka Trajkovska Petkoska Faculty of Technology and Technical Sciences, University St. Kliment Ohridski-Bitola, Veles, North Macedonia

Poonam Department of Life Sciences (Botany), Baba Mastnath University, Rohtak, Haryana, India

Amrita Poonia Department of Dairy Science and Food Technology, Institute of Agricultural Sciences, Banaras Hindu University, Varanasi, Uttar Pradesh, India

Pratibha Centre of Management and Humanities (CMH), Punjab Engineering College (Deemed to be University), Chandigarh, India

Abdul Qayyum Department of Agronomy, The University of Haripur, Haripur, Khyber Pakhtunkhwa, Pakistan

Renu Department of Life Sciences (Botany), Baba Mastnath University, Rohtak, Haryana, India

T. M. R. Rusarani Department of Crop Science, Faculty of Agriculture, University of Peradeniya, Peradeniya, Sri Lanka

Himanshi Singh National Institute of Food Technology Entrepreneurship and Management, Kundli, Sonepat, Haryana, India

Slim Smaoui Laboratory of Microbial and Enzymatic Biotechnologies and Biomolecules (LBMEB), Center of Biotechnology of Sfax (CBS), University of Sfax, Sfax, Tunisia

Aneta Stajikj Faculty of Technology and Technical Sciences, University St. Kliment Ohridski-Bitola, Veles, North Macedonia

A. K. A. N. W. M. R. K. Thamarsha Graduate School of Horticulture, Chiba University, Chiba, Japan

Ajar Nath Yadav Department of Biotechnology, Dr. Khem Singh Gill Akal College of Agriculture, Eternal University, Baru Sahib, Sirmour, Himachal Pradesh, India

Ajay Yadav ICAR National Dairy Research Institute, Karnal, Haryana, India

Chapter 1
Introduction to Edible Packaging

Aneta Stajikj, Aleksandra Anakjioska, Nishant Kumar, and Anka Trajkovska Petkoska

Abstract This chapter provides a comprehensive review of edible films and coatings developed from natural biopolymers with incorporated functional (antimicrobial, antioxidant) compounds, primarily derived from food industry by-products and natural extracts. The study emphasises the growing interest in bio-based, biodegradable packaging alternatives to synthetic materials, focusing on compounds such as chitosan, starch, pectin, alginate, and various plant-derived extracts and essential oils. Special attention is given to the dual role of these materials—not only as physical barriers but also as active systems enhancing food safety and shelf life. In addition, this study analyses their physicochemical, mechanical, and barrier properties, as well as their effectiveness against common food-borne pathogens and oxidative processes. It also explores current trends in the application of natural bioactive substances, including polyphenols, essential oils, and nanosized additives, highlighting their mode of action and compatibility with edible matrices, as well as the combination of nonthermal technologies with edible packaging to improve the microbiological status of the product. The commercially available edible packaging is listed here, as well as the importance and challenges of edible packaging in the food-producing sector. The findings support the potential of edible films as sustainable solutions for active food packaging and propose directions for future research in formulation optimisation, regulatory compliance, and industrial scalability.

Keywords Sustainable packaging · Biopolymers · Eco-friendly · Post-harvest management · Shelf life extension · Antimicrobial activity

A. Stajikj · A. Anakjioska · A. Trajkovska Petkoska (✉)
Faculty of Technology and Technical Sciences, University St. Kliment Ohridski-Bitola, Veles, North Macedonia
e-mail: anka.trajkovska@uklo.edu.mk

N. Kumar
Amity Food and Agriculture Foundation (AFAF), Amity University, Noida, Uttar Pradesh, India

N. Kumar et al. (eds.), *Trends in Edible Packaging: A sustainable approaches for postharvest management of fruits and vegetables*,
https://doi.org/10.1007/978-3-032-13199-7_1

Introduction

The evolution of food packaging materials has become increasingly critical due to growing concerns about environmental sustainability and food safety. Traditionally, food packaging has relied heavily on synthetic materials derived from petrochemicals, which are favoured for their mechanical strength, barrier properties, and cost-effectiveness. However, these materials pose significant environmental risks, including the accumulation of non-biodegradable waste and pollution resulting from their production and disposal (Hasanin et al., 2023). As a result, there is a burgeoning interest in alternative materials, especially biopolymers, which offer a more sustainable solution to food packaging needs.

One of the essential aspects of edible packaging is its role in addressing the environmental implications of waste management. Traditional plastics are notorious for their slow degradation rates, which lead to pollution and accumulation in ecosystems, adversely affecting wildlife and human health (Shiam et al., 2024). In contrast, edible films crafted from renewable sources pose no risk to the environment, as they can be consumed along with their contents or biodegrade naturally (Petkoska et al., 2021; Vergel-Alfonso et al., 2025). This shift to biopolymer-based materials, including those made from agricultural by-products, underscores the production systems and waste management practices (Oliveira, 2025; Kumar et al., 2021a; Kumar et al., 2021b; Kumar et al., 2021c).

The history of edible packaging is deeply rooted in various cultural practices, stretching back to the twelfth century, when wax coatings were first employed in China to preserve citrus fruits, such as oranges and lemons. This application was intended to prevent moisture loss and enhance their aesthetic appeal (Nair et al., 2023). This early use of wax exemplifies the concept of preserving food by integrating a layer that is consumable. Additionally, the utilisation of yuba, or soy-milk skin, as a form of edible film in Asian cuisine dates back to the fifteenth century, further demonstrating the longstanding tradition of integrating packaging with consumable materials (Neogi et al., 2022).

One of the most significant advantages of edible packaging is its ability to extend the shelf life of perishable products while simultaneously enhancing their safety and quality. The incorporation of active substances such as essential oils and plant extracts into packaging films and coatings has proven effective in inhibiting microbial growth and oxidative processes (Kumar et al., 2022a; Kumar et al., 2022b; Kumar et al., 2023a; Kumar et al., 2023b; Kumar et al., 2021a, b, c; Petkoska et al., 2021; Trajkovska Petkoska et al., 2021). Natural antimicrobial and/or antioxidant agents, like ginger essential oil and pomegranate peel extract, have shown effectiveness in this domain, offering the dual benefits of extending the shelf life of food products and improving quality standards without the harmful effects often associated with synthetic preservatives (Al-Hilifi et al., 2022a, b; Kumar et al., 2022a, b).

The functional properties of edible films are fundamentally linked to their material composition. Biopolymers such as chitosan, alginate, and pectin not only contribute to the mechanical strength of the films but also impart biodegradability—one of the leading traits sought in contemporary food packaging solutions. Edible coatings have also shown remarkable efficacy in delaying ripening and senescence in fruits by regulating gas exchange and moisture loss. Research indicates that coatings can effectively lower respiration rates, thereby prolonging freshness and enhancing the overall aesthetic value of produce, thus reducing spoilage and food waste (Aboulfotouh et al., 2024; Jamuna et al., 2024; Nunes et al., 2023). For instance, chitosan-based coatings have been widely studied for their potential to create barriers that restrict moisture loss and microbial invasion, providing a sustainable alternative to conventional plastic wrappers (Pratiwi et al., 2024; Ramkumar et al., 2024).

Moreover, the integration of technology in developing edible coatings—such as nanomodifications—has been a vital innovation aimed at enhancing the performance characteristics of these materials (Aboulfotouh et al., 2024; Ma et al., 2024; Trajkovska Petkoska et al., 2025). This approach can improve mechanical properties and barrier functionality while utilising eco-friendly ingredients to ensure the sustainability of the overall life cycle of food packaging (Nunes et al., 2023; Sun et al., 2024). Nanotechnology can also be employed to improve the release profiles of encapsulated active compounds, thereby making edible coatings more effective (Antonino, 2023; Ma et al., 2024; Trajkovska Petkoska et al., 2025).

Despite the tremendous potential of edible packaging, several challenges remain, particularly regarding production scalability, economic feasibility, and consumer acceptance. The durability and mechanical properties of edible films need continual improvement to meet the demands of food storage and transportation while ensuring cost-effectiveness. Furthermore, regulatory hurdles should be addressed to standardise the use of active ingredients in edible packaging to ensure consumer safety and product efficacy (Bremenkamp, 2025; Rout & Pradhan, 2024; Rodríguez-Parada et al., 2025).

Edible Coatings and Edible Films

Edible films and edible coatings are both innovative approaches in food packaging that utilise edible materials; however, they fundamentally differ in terms of their formation, application, and functional behaviour. Edible films are free-standing, preformed sheets that are manufactured separately from the food product through processes such as casting or extrusion. In contrast, edible coatings are formulated as liquid solutions that are directly applied to the food surface using techniques such as dipping, spraying, fluidised bed processing, or panning, as demonstrated in Fig. 1.1 (Kumar et al., 2022a, b; Suhag et al., 2020). The methodology of film formation is central to distinguishing the two. Films are produced independently and dried to create a continuous substrate that can be subsequently used to wrap or layer food products. This enables the films to be engineered with specific mechanical and barrier properties prior to their application. For instance, films produced by extrusion or casting can be tailored for high tensile strength or specific permeability

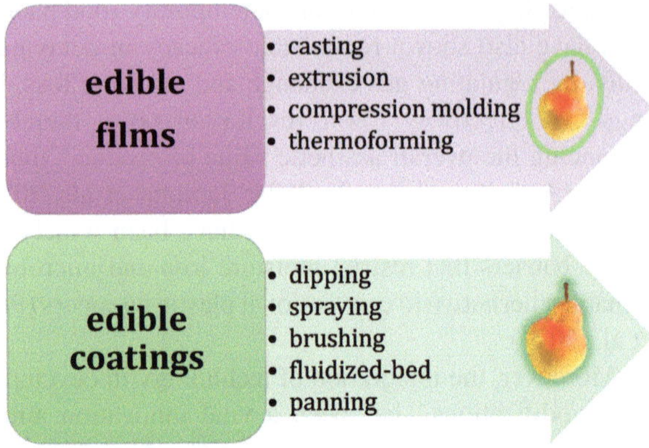

Fig. 1.1 Deposition methods for edible films and coatings

characteristics, making them useful for applications where a robust barrier is required (Kumar et al., 2022a, b; Suhag et al., 2020). In contrast, coatings are applied directly to the food, creating an in situ barrier that conforms to the food's surface. This method allows coatings to provide adaptive coverage that minimises gas exchange and moisture loss, effectively extending shelf life while being sensitive to the shape and size of the produce (Liyanapathiranage et al., 2023).

The differences in processing also lead to differences in physical and mechanical characteristics. Edible films are typically characterised by their uniformity and strength, as they are preformed and optimised under controlled conditions. They can be handled and stored as separate packaging components before being applied to food products. On the other hand, edible coatings are applied in a continuous process over the food, where their final properties (mechanical strength, barrier performance, and surface morphology) can be influenced by the conditions of application, such as temperature, viscosity of the coating solution, and drying environment (Liyanapathiranage et al., 2023). This inherent adaptability makes edible coatings particularly suitable for direct application to fresh produce that may have irregular surfaces or require immediate post-harvest treatment (Kumar et al., 2022a, b). The main ingredients for creating edible packaging typically fall into two primary categories—biopolymers and functional additives—but solvents are also sometimes necessary.

Biopolymers

The foundation of edible films and coatings is predominantly formed by biopolymers, which are derived from natural sources. These include proteins, polysaccharides, and lipids. Common proteins utilised in edible films include casein, whey, collagen, gelatin, and various plant proteins such as soy and wheat gluten. These proteins provide films with mechanical strength and biocompatibility, thus making

Table 1.1 The commonly used substances for edible packaging

Biopolymers	Substances	References
Proteins	Casein, whey, collagen, gelatin, soy, wheat gluten	Kumar et al. (2023a, b), Pires et al. (2024a, b), Daniloski et al. (2021)
Polysaccharides	Chitosan, alginate, pectin, starch	Petkoska et al. (2021), Kumar et al. (2021a, b, c), Ramkumar et al. (2024)
Lipids	Beeswax, vegetable oils	Pires et al. (2024a, b)
Composite materials	Combinations of the above-mentioned substances	Kumar et al. (2021a, b, c), Al-Hilifi et al. (2022a, b), Liyanapathiranage et al. (2023)

them suitable for food-contact applications (Kumar et al., 2023a, b; Pires et al., 2024a, b; Daniloski et al., 2021). Polysaccharides such as chitosan, alginate, pectin, starch, and others (Table 1.1) are widely used due to their gelling and film-forming properties, allowing for enhanced moisture retention and barrier functions against gases (Kumar et al., 2021a, b, c; Ramkumar et al., 2024). For instance, chitosan is recognised for its antimicrobial properties and ability to form stable edible coatings that can also improve the shelf life of products such as fruits and vegetables (Petkoska et al., 2021). Lipids, including those from beeswax or vegetable oils, can be incorporated to improve moisture resistance and create a barrier against oxygen and lipids, further enhancing the shelf life of packaged foods (Pires et al., 2024a, b). In addition, a combination of polysaccharides, proteins, and/or lipids is possible in the form of composite materials used for edible packaging.

Functional Additives

To enhance the mechanical, physical, and barrier properties of biopolymer films, various functional additives are often incorporated. These can include essential oils, plant extracts, and antimicrobial agents. Essential oils such as ginger and various plant extracts have shown promising effects as antimicrobial agents within edible films (Table 1.2), helping mitigate food-borne pathogens and extend shelf life (Kumar et al., 2021a, b, c; Al-Hilifi et al., 2022a, b). Additionally, bioactive compounds derived from fruit and vegetable by-products (e.g., pomegranate peel extract) are added not only for their antimicrobial properties but also for their ability to enrich the nutritional profile of the packaging (Ma et al., 2024; Oliveira, 2025). Materials derived from fruit and vegetable purees, juices, and residues not only contribute to the film's structure but also align with the movement towards sustainability by promoting the use of food waste in packaging technology (Antonino et al., 2023). Functional ingredients assist in enhancing antioxidant activity and improving the overall sensory characteristics of the food products contained within the packaging (Al-Hilifi et al., 2022a, b; Aboulfotouh et al., 2024).

An essential component of edible films is the plasticiser, a low-molecular-weight substance that intercalates within the polymer matrix to reduce intermolecular attractions, thereby enhancing chain mobility, reducing brittleness, and improving

Table 1.2 Functional additives for edible film

Functional additives	Examples	References
Essential oils	Ginger essential oil, cinnamon essential oil, clove essential oil	Kumar et al. (2021a, b, c), Al-Hilifi et al. (2022a, b), Antonino (2023)
Plant materials	Materials derived from fruit and vegetable purees, juices, residues	Ma et al. (2024), Oliveira (2025)
Antimicrobial agents	Bioactive compounds derived from fruit and vegetable by-products	Antonino (2023)
Plasticiser	Glycerol, sorbitol, other polyols	Song et al. (2021), Dong et al. (2023)
Probiotics and prebiotics	Various probiotic cultures, fibres	Ramazanidoroh et al. (2024)

film flexibility. The selection, type, and concentration of plasticisers are crucial for tailoring the physical and mechanical properties of edible films, as evidenced by studies examining tensile strength, elongation at break, and other mechanical attributes. Plasticisers such as glycerol, sorbitol, and other polyols are commonly employed due to their favourable characteristics, such as low toxicity, low volatility, and efficiency in reducing moisture absorption (Song et al., 2021; Dong et al., 2023). Moreover, the inclusion of probiotics and prebiotics in edible films represents an emerging trend whereby these beneficial microorganisms can contribute to food safety and quality maintenance. Such films can act as functional carriers, delivering probiotics to food products and enhancing their health benefits (Ramazanidoroh et al., 2024). This trend reflects an ongoing shift towards integrating health-orientated solutions within the food packaging sector, illustrating a dual focus on sustainability and functional performance (Gupta et al., 2022).

Incorporation of Compounds from Biowaste and Food By-Products in Edible Packaging

The versatility of edible coatings extends beyond the by-products from fruits and vegetables. For example, recent innovations have highlighted the use of edible films derived from whey proteins to create biodegradable packaging for a range of food products, including meat (Shiam et al., 2024; Pires et al., 2024a, b; Kumar et al., 2023a, b). This reflects a significant shift in the food industry towards sustainable practices, with edible films capable not only of extending shelf life but also of contributing to reducing food waste generated along the supply chain (Nunes et al., 2023). The unique properties of these materials often enable tailored features, such as antimicrobial effects, while enhancing the aesthetic appeal of the packaged product (Al-Hilifi et al., 2022a, b). Previously, several researchers have reported the use of different types of by-product compounds in the formation of edible coatings and films (Table 1.3).

Table 1.3 Biowaste/by-product compounds used in edible packaging materials

Biowaste/by-products	Reference
Mandarin peels	Yun and Liu (2024)
Grape	Mileti et al. (2023)
Sweet potato	Real et al. (2023)
Beetroot peel extract	Chaari et al. (2022)
Coffee grounds	Rizal et al. (2023), Papadaki et al. (2023)
Apple pomace/peel	Pakulska et al. (2023)
Pomegranate peel	Bodana et al. (2023), Kumar et al. (2023a, b)
Pineapple peel	Jatav et al. (2022)
Grapefruit seed extract	Roy et al. (2021)
Orange peel	Ghafoor et al. (2022), Azizkhani et al. (2023)
Almond shell	Valdés et al. (2021)
Tomato peel	Umeohia and Olapade (2024)
Onion peel extract	Amran et al. (2024)
Citrus limetta pomace extract	Singhi et al. (2023)
Avocado seed starch	Suhartini et al. (2023)
Mango kernel seed starch	Yadav et al. (2022a, b)
Oilcakes	Šuput et al. (2023), Petraru and Amariei (2023)
Vegetables	da Silva Rios et al. (2022)

Methods for Obtaining Edible Films and Coatings

The development of edible films can be achieved through various methodologies, broadly categorised into dry and wet methods. Each of these approaches relies on distinct physicochemical principles and offers unique advantages suited for different applications in food preservation and packaging. Dry methods for obtaining edible films primarily involve solvent-evaporation techniques. In these methods, the selected biopolymers, often polysaccharides or proteins, are first dissolved in an appropriate solvent, such as water or organic solvents. The resultant viscous solution is then cast onto a surface and subjected to drying, allowing moisture removal and solidification of the film (Ma et al., 2024). Techniques such as casting and air-drying are commonly employed, where the film's thickness and mechanical integrity can be controlled by varying the casting conditions and the concentration of the polymer solution.

Conversely, wet methods for preparing edible films often involve the formation of emulsions or gels. In these approaches, biopolymers are typically blended with plasticisers in a solvent to create a homogeneous mixture, which can then be coated directly onto the food surface or into moulds to set (Akhavan-Mahdavi et al., 2023). Methods such as spray drying or dip coating may be utilised to create thin films directly on food products. These methods often promote better adhesion to substrates, effectively enhancing moisture retention and reducing spoilage (Ramazanidoroh et al., 2024).

Edible coatings are thin layers of food-grade materials that are applied directly to food surfaces to provide protection, extend shelf life, and maintain quality without forming self-standing films. The initial step in obtaining an edible coating involves selecting appropriate natural polymers based on the desired barrier and protective effects. These polymers are usually water-soluble; therefore, dissolving them in a solvent (commonly water) forms a viscous dispersion. Heating may assist dissolution, and subsequent cooling is necessary to avoid degradation of thermally labile compounds. During formulation, additives such as glycerol may be incorporated to adjust viscosity and improve the coating's adherence to the food surface (Kusnadi et al., 2022; Nakum et al., 2024). The final film-forming solution must be optimised for viscosity and concentration, since these physical parameters significantly influence the thickness and uniformity of the resulting coating layer, thereby affecting its performance (Tufan et al., 2021; Pedreiro et al., 2021).

Edible Films and Nanotechnologies

The integration of nanotechnology into edible packaging has garnered significant attention due to its potential to enhance food preservation, quality, and safety while addressing environmental concerns. The use of nanomaterials offers innovative solutions that improve the functionality of edible packaging materials, which are vital for mitigating the environmental impact associated with conventional plastic packaging (Trajkovska Petkoska et al., 2025). The use of nanomaterials in edible coatings helps optimise the transfer and release of bioactive compounds, as well as boost their protective qualities by increasing surface area and interaction points with microbes. This trend aligns well with the overarching goal of ensuring food safety through innovative packaging solutions that maintain product integrity without compromising environmental sustainability (Oliveira, 2025).

Nanotechnology enhances the barrier properties of edible films, making them more effective in preventing moisture and gas transfer, which is crucial for prolonging shelf life. Recent studies highlight the development of nanocomposites that incorporate nanoparticles, microcapsules, and bioactive compounds into edible films. These composites not only improve mechanical properties but also exhibit antimicrobial and antioxidant features that can help inhibit microbial growth and oxidation in food products (Dholariya et al., 2021; Mei & Wang, 2020; Sousa et al., 2023). For instance, edible films reinforced with nanocellulose and starch have shown increased durability and enhanced barrier performance, thus promoting their use in food packaging applications (Dmitrenko, 2025; Dmitrenko et al., 2023; Basavegowda & Baek, 2021).

Moreover, the incorporation of natural extracts, such as essential oils, with nano-sized carriers into edible coatings demonstrates significant benefits. These nano-emulsions not only improve the sensory attributes of food products but also act as active components that extend shelf life by providing antimicrobial protection (Shafiq et al., 2020; Kong et al., 2022). The use of polysaccharide-based materials, which are renewable and generally recognised as safe (GRAS), further aligns with sustainability goals in food packaging (Zhao et al., 2021a, b; Atta et al., 2021). Furthermore, the combination of edible packaging with intelligent systems that can monitor food quality (e.g., colour change indicators or sensors) represents a forward-thinking approach in the food industry, optimising both food safety and consumer experience (Guo et al., 2024; Pudake et al., 2024). This intelligent packaging leverages nanotechnology to provide real-time responses to changes in food conditions, enhancing consumer trust and reducing food waste.

The application of ionising radiation for nanomodification has emerged as an innovative method for developing active edible packaging (Aboulfotouh et al., 2024). Such methods have demonstrated promising results in enhancing the thermal and mechanical properties of edible films made from materials such as carboxymethyl cellulose and starch, which are particularly useful for prolonging the shelf life of perishable items such as fruits (Kumari et al., 2024a, b; Ramkumar et al., 2024).

Integration of Edible Packaging with Nonthermal Technologies

The discussions around edible coatings are further enriched by exploring their integration with joint nonthermal technologies. Employing methods such as pulsed light and ozone treatments combined with edible coatings represents a forward-thinking approach to enhancing food preservation while retaining sensory and nutritional quality (Ma et al., 2024). The synergy between edible packaging and nonthermal technologies, like high hydrostatic pressure (HHP), is particularly noteworthy. HHP techniques can enhance the physical and antimicrobial properties of edible films, thus improving their performance in food preservation (Evrendilek et al., 2022). By applying HHP, researchers have demonstrated the ability to fortify edible films from agricultural by-products, thereby providing an additional layer of sustainability to food packaging solutions (Bayram et al., 2021). This method improves the safety and quality of food products while also revalorising food industry waste, a critical component in addressing the global plastic crisis (Bayram et al., 2021). Edible packaging and cold plasma technologies represent distinct yet interconnected segments within the food preservation and packaging domain. The utilisation of cold plasma technology has gained traction as a nonthermal sterilisation method that effectively enhances food safety and extends shelf life (Table 1.4). Cold plasma involves the treatment of food surfaces with ionised gases to inactivate microorganisms, thus preventing spoilage without significantly altering the food's nutritional and sensory properties (Birania et al., 2022). This innovative approach has been critically assessed for its applications in conjunction with conventional packaging methods, highlighting its potential to revolutionise the food storage sector (Alaguthevar et al., 2024). For example, recent studies indicate that in-package cold plasma treatment combined with nanotechnology can yield superior results in food preservation and shelf life extension (Karthik et al., 2024). Research has

Table 1.4 Examined combination of nonthermal technologies and edible packaging

Title	Effect	Reference
Edible Coating for Fresh-Cut Fruit and Vegetable Preservation: Biomaterials, Functional Ingredients, and Joint Nonthermal Technology	This paper gives a review of recent advances in fresh-cut fruit and vegetable preservation from the perspective of biomacromolecule-based edible coatings combined with some joint nonthermal technologies such as pulsed light, short-wave ultraviolet, modified atmosphere packaging, ozonation, and γ-irradiation	Ma et al. (2024)
Assessing Edible Composite Film Polymer from Potato Industry Effluent under High Hydrostatic Pressure and Its Antimicrobial Properties	This study investigates the development of an edible film from potato industry effluent having antimicrobial properties against *Salmonella enteritidis* and *Escherichia coli* by the addition of *Citrus sinensis* volatile oil and the changes in its textural properties under high hydrostatic pressure	Evrendilek et al. (2022)
Cold Plasma Technology- Based Eco-Friendly Food Packaging Biomaterials	In this review, the focus is on recent cold plasma advances in various food matrices derived from plants and animals, with the aim of highlighting potential applications, ongoing research, and market trends	Karthik et al. (2024)
Current Status of Bio-based and Biodegradable Food Packaging Materials: Impact on Food Quality and Effect of Innovative Processing Technologies	This review provides insight into the connection between bio-based packaging materials and innovative technologies such as high pressure, cold plasma, microwave, ultrasound, and ultraviolet light	Nilsen-Nygaard et al. (2021)
Combined Effect of Chitosan Coating and Modified Atmosphere Packaging on Fresh-cut Cucumber	In this paper, it is shown that the combined chitosan coating with MA packaging maintained quality and microbial safety and extended the shelf life of fresh-cut cucumber	Olawuyi et al. (2019)
Cold Plasma Technology: Applications in Improving Edible Films and Food Packaging	This paper demonstrates cold plasma's effectiveness in modifying edible films and enhancing packaging performance	Sani et al. (2023)
The Effects of Novel Thermal and Nonthermal Technologies on the Properties of Edible Food Packaging	This study states the definitions and mechanisms of novel technologies. The experimental conditions, the packaging matrix, and the results pertaining to the effects of these technologies on various types of edible packaging are also discussed	Beikzadeh et al. (2020)
Natural Polymers in Biodegradable/Edible Film: A Review on Environmental Concerns, Cold Plasma Technology and Nanotechnology Application on Food Packaging- A Recent Trends	This paper evaluates the mutual use of cold plasma technology and nanotechnology application with natural polymer-based biodegradable packaging materials	Gupta et al. (2022)
The Effect of Chitosan Coating Combined with Cold Plasma on the Quality and Safety of Pistachios	This paper investigates combined chitosan and cold plasma treatment for nut preservation	Akhavan-Mahdavi et al. (2023)

illustrated that edible films not only serve as physical barriers to oxygen and moisture but also can be enhanced with functional additives such as antimicrobial agents and antioxidants (Moghadas et al., 2024). These combinations can further improve the effectiveness of both the packaging and cold plasma technology in preserving food quality (Iversen et al., 2022).

As the food packaging industry transitions towards greener solutions, it is evident that the fusion of nonthermal technologies and edible packaging yields significant benefits across multiple dimensions. The dual emphasis on functionality and sustainability drives the demand for materials that can effectively address food safety while minimising environmental footprints. Furthermore, the ongoing evolution of edible packaging, driven by advances in materials science and consumer demand for environmentally conscious alternatives, sets the stage for innovative applications across various food sectors, including dairy, meat, and produce (Meshram et al., 2023; Song et al., 2021). Ultimately, this integration of edible packaging with nonthermal technologies represents a critical advancement in food science, where researchers and industry practitioners are forging new paths toward achieving sustainable food preservation. By leveraging the inherent properties of biopolymers, enhancing them with advanced nonthermal techniques, and focusing on usability and biodegradability, edible films are poised to play integral roles in the future of food packaging ecosystems. As research continues to uncover novel materials and methods, the potential for widespread adoption of these technologies in both consumer and industrial settings appears increasingly viable.

Importance of Edible Materials in the Food Processing Sector

Nowadays, the food industry has a duty to offer the consumer fresh, pleasant, good-quality food with beneficial properties for health. However, the challenge of meeting consumers' needs is very demanding, since there is no food that can remain in optimal conditions and maintain its properties permanently due to the natural deterioration generated by chemical and biochemical reactions and physical changes (Diaz & Castro, 2021). Currently, edible coatings and films have obtained significant attention in the food industry because of their advantages over non-edible, non-biodegradable films in maintaining food quality. The surge in interest in edible films and coatings can be attributed to consumers' growing awareness of the importance of making healthy food choices and their worries about the negative environmental effects of synthetic, non-biodegradable packaging materials (Perez-Vazquez, 2023). Benefits of edible coatings include minimising packaging waste, extending the shelf life of fresh and minimally processed products, and protecting them from hazardous environmental effects by preserving oxygen, carbon dioxide, moisture, scent, and taste-compound transmission in a food system (Iversen, 2022). Figure 1.2 summarises the main properties of edible packaging.

Edible packaging (coatings and films) is an excellent, eco-friendly option to extend the shelf life of food products for a longer period. The extension of shelf life

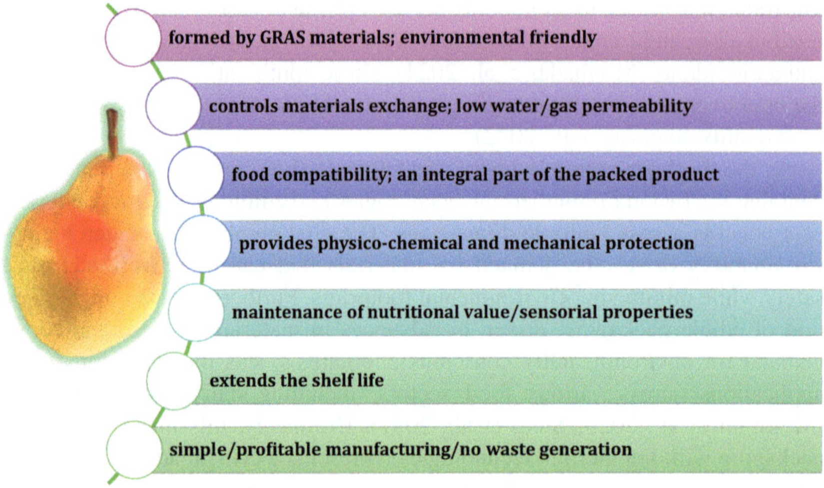

Fig. 1.2 Main properties of edible packaging

in food products for at least a few days represents significant economic profit for food companies (Vaishnavi, 2023). Numerous studies have investigated the use of edible coatings in combination with traditional synthetic primary packaging, with a focus on increasing product shelf life and reducing food waste by reducing degradation processes such as microbiological growth, but also increasing the product's robustness against temperature changes during the supply chain. Studies focusing on the coating performance without an external layer have shown that the coating alone can protect and extend product shelf life. This could indicate that an edible coating in combination with external packaging provides protection even when the integrity of the external packaging is broken at the consumer stage. This approach can support the use of larger purchasing units, reducing packaging waste and providing convenience to the consumer by reducing shopping frequency and offering more flexibility in consuming the product (Bremenkamp & Sousa Gallagher, 2025).

The food industry currently confronts significant challenges in its efforts to nourish the global population. By 2060, the global population is projected to reach 10 billion, resulting in an anticipated increase in total food demand of 35–56% from 2010 to 2050 (Worldometers, n.d.). These trends indicate that market demand for food will persistently increase, thereby impacting the development of the food industry. Although this may enhance dietary diversity for certain demographic groups, it can also promote nutritionally imbalanced diets and sedentary behaviours, leading to overnutrition and associated noncommunicable diseases, as well as undernutrition, including micronutrient deficiencies (WHO, 2025). Essential nutrition is provided by consuming nutritious food, yet some of the good ingredients in foods may not be well absorbed. Low or non-solubility, chemical instability, and disagreeable taste are some of the drawbacks of nutraceuticals, including curcumin, carotenoids, and omega-3 fatty acids. These substances may also be degraded by digestive enzymes and intestinal walls (Gupta et al., 2024). The incorporation of nutraceuticals into food coatings can potentially compensate for the loss of nutrients that usually occurs during food processing or increase the nutrient composition of the newly coated food products (Petkoska et al., 2021; Trajkovska Petkoska et al., 2021). The main influence of the nutraceuticals, besides being a source of nutrients and energy, is to provide potential beneficial health effects. Many authors have reported that some nutraceutical compounds have a positive influence on human health, such as ascorbic acid (antioxidant), pectin (cardiovascular support), casein phosphopeptides (anti-atherosclerotic), omega fatty acids (anti-inflammatory), polyphenols (antioxidant and many other benefits), capsaicin (anticarcinogenic), beta-carotene (enhances the immune system, protects from age-related macular degeneration), etc. (Petkoska et al., 2021; Trajkovska Petkoska et al., 2021).

Additional values of edible coatings are customisation and personalisation potential. Consumers are seeking more personalised and customised products, and edible coatings can provide a platform for product differentiation. For example, coatings can be customised with different flavours or colours to enhance the sensory experience of the consumer, and they can also be used as carriers for nutraceuticals such as vitamins or minerals, allowing product tailoring to meet the specific demands of consumer groups such as elderly people (Bremenkamp & Sousa Gallagher, 2025). Finally, as consumers increasingly seek natural and minimally processed food products, edible coatings enhance product appeal while meeting these demands (Vaishnavi, 2023).

Challenges and Sustainable Approach of Edible Packaging in the Food Processing Sector

Nowadays, it is crucial to adopt an integrated approach to improve the sustainability of the food system that covers the whole supply chain, from primary production to packaging and distribution. To meet the world's growing need for sustainable and high-quality food supply chains, two major challenges need to be addressed: food loss and waste reduction and packaging waste reduction, which are target issues of the Sustainable Development Goals (SDGs) defined by the United Nations (2015). Recently, with the advancement in sustainable packaging, countries are moving from the linear economy to the circular economy. The European Union (EU) in 2020 adopted the Circular Economy Action Plan that aims at reducing traditional packaging waste and promoting the use of biodegradable polymers for safeguarding the environment. Additionally, consumer demands such as mildly preserved, fresh, nutritious, safe, and convenient food products with enhanced shelf life and longer retailing distances have driven the need for novel and innovative food packaging. Studies have identified edible films and coatings as a natural, inexpensive, and renewable method for preserving food from a variety of harmful substances (Hamed et al., 2021). One of the key challenges in the field of edible films is ensuring sustainability throughout the entire life cycle, from sourcing raw materials to disposal.

Edible packaging research is a constantly evolving field, exploring new materials, active packaging, and nanotechnology. Moreover, advances in 3D printing technology have opened new possibilities for the fabrication of edible films with intricate structures and customised designs. Recent advancements in 3D printing have allowed for precise control over the deposition of plant-based materials, enabling the creation of complex geometries and customised shapes (Mittal et al., 2023). This technology enables the production of personalised edible films with tailored properties, such as thickness, texture, and release profiles. Additionally, the development of food-grade 3D printers and edible ink formulations has facilitated the direct printing of edible films. According to Olawade et al. (2024), the potential for personalised nutrition, on-demand production, and novel sensory experiences makes 3D printing a promising technique in the field of edible film. While 3D printing of edible films offers exciting opportunities, challenges remain. Optimising the rheological properties of the printing materials, ensuring their food safety and functionality, and achieving high printing resolution are ongoing areas of research. Despite all the advantages, like all other newly developed technologies, edible films also face some limitations that need to be overcome to make them a commercial success. The main benefits and limitations of edible packaging are displayed in Fig. 1.3.

Consumer acceptance is an important and critical factor, as edible packaging must match or exceed the performance of conventional plastics to gain trust and preference. Consumers are concerned about the coating material type, the coating material's safety, the sensory characteristics of food products, the shelf life of food products, and the price of packaged food items. Freshness, texture, and nutritive value are important parameters considered by consumers at the time of purchasing. In a study, Bucher et al. (2023) found that edible coatings were removed from apples due to consumer demands and suggested that Australian customers rejected the coated apples because they lacked an understanding of the food coating and its benefits. Therefore, combining information about the food product and its edible coating on the label, together with the coating ingredients, may reduce the risk perception and thus improve consumer acceptance (Gupta et al., 2024). Consumer awareness programmes and advertisements of edible films can also increase their acceptability (Kumar et al., 2022a, b).

Despite the challenges in material properties, production techniques, cost competitiveness, and consumer acceptance, ongoing research and innovation hold the potential to overcome these obstacles. Achieving the commercialisation of edible films on a global scale will depend on advancements in material formulations, scalable processing technologies, and cost-effective production methods. Establishing regulatory frameworks and industry standards will ensure the safety and sustainability of these products. By addressing these challenges through collaborative efforts, it is possible to bring environmentally friendly edible materials from the laboratory to the market, contributing to a more sustainable future in food packaging (Weng et al., 2025).

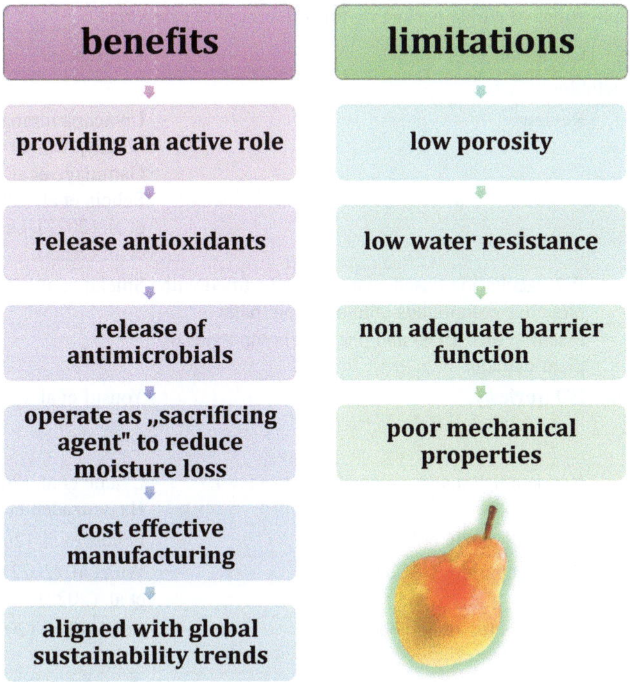

Fig. 1.3 Benefits and limitations of edible packaging

Applications of Edible Packaging

There are many products, including cereals, spices, meats, drinking water, chocolate and confectionery, fruits and vegetables, marine products, and many more, that can be effectively and economically packaged by edible films and coatings. There are numerous examples of edible films currently in use, including wax coatings for fruit, chocolate coatings for confectionery, lipid films for meat products, and the use of soy-milk-based lipoprotein films to improve the appearance and preservation of various food products (Trajkovska Petkoska et al., 2021). Table 1.5 presents recent cases of edible packaging for a variety of food categories.

Commercially Available Edible Packaging

Commercially available edible coatings and films are integral to food preservation, enhancing shelf life while maintaining the quality of fresh produce. Several types of commercially available materials are utilised in the production of these coatings, including proteins, polysaccharides, and lipids or their combination (composite materials). In Table 1.6, commercially available edible films and coatings are listed,

Table 1.5 Applications of edible packaging for different food categories

Food category	Composition of edible	References
Fruits and vegetables		
Fruits/ vegetables	Review	Liyanapathiranage et al. (2023), Ammar et al. (2025), Gammage et al. (2025), Felicia et al. (2024), Khalid et al. (2022), Ungureanu et al. (2023).
Fresh fruits	Polysaccharide-based coatings are fortified with bioactive compounds obtained from plant essential oils (EOs) and phenolic compounds of plant extracts	Shiekh et al. (2022)
Fresh/fresh-cut fruit and vegetables	EO (review)	Yousuf et al. (2021)
Fruits and vegetables	Nanoemulsion and nanoparticles (review)	Cvanic et al. (2023), Hassanzadeh et al. (2025)
Fruits	Essential oils in polysaccharides	Felicia et al. (2022)
Fruits/ vegetables	Polysaccharides (review)	Wu et al. (2022), Felicia et al. (2022)
Fresh/fresh-cut fruits	Citrus pectin, oregano EO	Kostic et al. (2023)
Fresh-cut watermelon	Review	Quandoh and Albornoz (2025)
Mango	Amaranthus flour (AF), montmorillonite, three essential oils (clove, muña, matico).	Ramos et al. (2025)
Mango	Hydrophobic deep eutectic oil-in-water nanoemulsion	Gidado et al. (2024)
Mango	Aloe-pectin	Haneef et al. (2022)
Citrus	Plant extract from pomegranate peels (PPE), biosynthesised selenium nanoparticles with PPE (PPE/SeNPs) and chitosan nanoparticles (NCT)	Salem et al. (2022)
Guava	Carboxymethyl cellulose (CMC) and gum arabic (GA)-based active edible films activated with rose leaf (RL) and marigold petal (MP) extracts	Lekshmi et al. (2025)
Guava	The mango kernel seed starch-based combined with different concentrations of lemongrass EO	Yadav et al. (2022a, b)
Kiwi	Alginate enriched with eugenol and citral	Guerreiro et al. (2025)
Kiwi	Thymol-halloysite nanohybrids (in chitosan/ poly-vinylalcohol films)	Salmas et al. (2022)
Fresh dates	Chitosan, chitosan nanoparticle, $CaCl_2$	El-Gioushy et al. (2022)
Dates	Gum Arabic	Alqahtani et al. (2025)
Strawberries	Chitosan and chitosan infused with apple pomace extract	Maslov Bandić et al. (2025)

(continued)

Table 1.5 (continued)

Food category	Composition of edible	References
Strawberries (frozen)	Edible coating (RP-7) with carvacrol and citral	Pié-Amill et al. (2024)
Apples (fresh cut)	Whey protein-based film	Xin et al. (2023)
Apples	Arabinoxylan (AX) and stearic acid ester of β-glucan (SABG) coating	Singh et al. (2025)
Pomegranate	Coating—carnauba wax (0.5%, 1%, 1.5%) and carboxymethyl cellulose (0.5%, 1%, 1.5%).	Nazoori et al. (2023)
Pomegranate/ arils	Gum arabic (GA) (1.5% w/v) alone or enriched with lemongrass oil (LM) (0.1% v/v) and/or pomegranate peel extract (PP) (1% w/v)	Kawhena et al. (2022)
White grape	Pomegranate peel extract of varying concentrations (0.02, 0.04, 0.06, 0.08, 0.1 g/ml) incorporated into jackfruit seed starch (5% w/w)-based films and coatings.	Bodana et al. (2023)
Grape	Various levels of EOs (0.5–5%, v/v) from clove, cinnamon, thyme, peppermint, lemon, bergamot, ginger, spearmint, lemongrass	Luesuwan et al. (2021)
Tomato	Beeswax	Rudiyanto and Ramisah (2023)
Tomato	Coating of tomato peel fibre and moringa leaf extract	Umeohia and Olapade (2024)
Cherry tomato	Chitosan (CS) enriched with green-synthesised zinc oxide nanoparticles (zno-nps)	Buthelezi (2025)
Plum	Gum arabic (GA) with pomegranate seed oil, GA5% with ascorbic acid	Fawole et al. (2024)
Plum	Hydroxypropyl methylcellulose (HPMC) and gum arabic (GA) coatings amended with 0.2% geraniol (GE)	Asgarian et al. (2023)
Pear (fresh cut)	A 2% sodium alginate solution with 0.5% ginger oil as a herbal antimicrobial agent was used as coating material and a 2% calcium chloride	Lamani and Ramaswamy (2023)
Pear	Chitosan (CH) coatings alone (CH 1.0–2.0%) and incorporated with 2.0 mm salicylic acid	Sinha et al. (2021)
Mandarin	Corn starch (PS), with a cellulose nanofibre (CF) Pickering emulsifier, and is enriched with basil EO	Wigati et al. (2023a, b)
Mandarin	Corn starch (PCS), basil essential oil (BEO), and cellulose nanofibers (CNF)	Wigati et al. (2023a, b)
Pineapple (fresh cut)	Chitosan (CH), pullulan (PU), linseed (LM), nopal cactus (NM), aloe mucilage (AM), (lbl)	Treviño-Garza et al. (2017)

(continued)

Table 1.5 (continued)

Food category	Composition of edible	References
Figs	Four different coatings—Gelatin, guar gum, guar gum—gelatin bilayer coating and gelatin—guar gum bilayer coating incorporated with grapefruit seed extract (GSE)	Chitra et al. (2022)
Figs	A mucilage solution of Opuntia ficus-indica cladode	Allegra et al. (2017)
Strawberries	The presence/absence of essential oil/extract in the coating, the polymer type (polysaccharide, protein, combination of both)	Taban et al. (2024)
Strawberries	Flaxseed mucilage (Mu) coatings supplemented with postbiotics (P) obtained from *Lactobacillus acidophilus* LA-5	Sharafi et al. (2024)
Blackberries	Alginate and limonene-liposomes	Joshi et al. (2021)
Blackberries	Different chitosan concentrations prepared with either lactic or acetic acid	Vilaplana et al. (2020)
Blueberries (frozen)	Coating (RP-7) with carvacrol and citral	Pié-Amill et al. (2024)
Blueberries	Clove essential oil (CO) in the composite film based on sodium alginate (SA) and aloe vera gel (AG)	Cui et al. (2023)
Red raspberry	Chitosan (Ch) with the incorporation of green tea (GTE) and rosemary (RSME) ethanolic extracts as natural antifungal agents	Vieira et al. (2022)
Banana (fresh cut)	Active packaging using chitosan (CS) and esterified chitin nanofibers (CF) combined with different contents (1, 2, 4 wt% on CS basis) of scallion flower extract (SFE)	Xie et al. (2023)
Lime	Coating formulated from xanthan gum, enriched with *Spirulina platensis* (Sp) and pomegranate seed oil (PSO)	Mohammadi et al. (2024)
Sweet orange	Muskmelon seeds' flour with oregano EO	Mehra et al. (2024)
Mushrooms	Orange peel essential oils (eos) in aloe vera gel-based coating	Shenbagam et al. (2023)
Mushrooms	A combination of apple peel powder and CMC in distilled water	Thakur et al. (2020)
Cucumber	2% chitosan and different concentrations of olive cake extract (OCE) and orange peel extract (OPE)	Ghafoor et al. (2022)
Chilli	Pure chitosan, cassava starch, gum arabic, and gelatin coatings	Kumari et al. (2024a, b)
Red chilli	Gelatin–pectin composite coating (75:25, 50:50, 25:75), incorporated with garlic essential oil (2–3%)	Heristika et al. (2023)

(continued)

Table 1.5 (continued)

Food category	Composition of edible	References
Okra	Composite of guar gum: almond gum (75:25 *V/V*) prepared with addition of different concentrations (0.05, 0.1, 0.15%) (*V/V*) of oregano essential oils	Shinde et al. (2024)
Fresh collard greens	Chitosan (CH)- based films were developed enriched with thyme essential oil (TEO) incorporated with different additives, including zinc oxide (ZnO), polyethylene glycol (PEG), nano clay (NC), calcium chloride ($CaCl_2$)	Zehra et al. (2022)
Sweet potato (fresh cut)	Chitosan (Ch) and sodium alginate (SA) solutions. Their gel coating (chcsa) is formed by layer-by-layer (LbL) electrostatic deposition using calcium chloride	Chit et al. (2022)
Parsley (fresh cut)	Whey-protein-isolate-based coatings with the addition of jojoba oil at concentrations of 1–2%	Galus et al. (2022)
Avocado	Poly(lactic acid) (PLA)/gelatin/PLA multilayer films supplemented with cellulose nanocrystals and antioxidant extract from almond shell (AS)	Valdés et al. (2021)
Carrot (fresh cut)	Chitosan containing free thyme essential oil (TO) and microencapsulated TO in β-cyclodextrin (β-CD)	Viacava et al. (2022)
Carrots	Starch and wheat gluten, containing cinnamon and turmeric essential oils	Rivera Leiva et al. (2022)
Red bell pepper	Chitosan, gelatin and chitosan/gelatin Layer-by-layer (LbL) coating	Kehila et al. (2021)
Dried vegetables	Cassia gum and ethyl cellulose	Li et al. (2022)
Meat, fish, and seafood		
Eggs	Chitosan and shellac coatings	SAEED et al. (2017)
Meat	Review	Prakoso et al. (2023)
Meat	Gums	Gao et al. (2024)
Meat	Active films and coatings incorporated with natural extracts, essential oils, natural polymers, protein hydrolysates, enzymes, and nanocomponents (materials and emulsions)	Umaraw et al. (2020)
Meat-based products	Seaweed polysaccharides	Chaudhary (2023)
Fish	Review (bio-based coatings)	Eranda et al. (2024)
Fish	Active films and coatings incorporated with natural extracts, essential oils, natural polymers, protein hydrolysates, enzymes, and nanocomponents (materials and emulsions)	Umaraw et al. (2020)
Tuna fish	Chitosan-gelatin-based edible films/coatings.	Eranda et al. (2024)
Sardine fillets	Hydrogels, oleogels, and bigel coatings.	Kanelaki et al. (2022)
Sea bream fillets	Films containing cinnamon-perilla essential oil (C-PEO)	Zhao et al. (2022)

(continued)

Table 1.5 (continued)

Food category	Composition of edible	References
Salmon fillets	Postbiotics generated by a pre-fermentation of *Lactiplantibacillus plantarum* 299 V in the coating together with the probiotic bacteria cells	Hua et al. (2022)
Chicken breast	Carboxymethyl cellulose (CMC) coating with various concentrations (0, 1, 2, 3, 4%) of an ethanolic propolis extract (EPE)	El Sheikha et al. (2022)
Chicken fillet	Coatings based on gelatin (Gel), chitosan (Ch), and modified starch (MS) incorporated with bitter orange peel extract (BOE)	Azizkhani et al. (2023)
Burger meat	Chitosan aerogels reinforced with lignocellulose and cellulose micro/nanofibers from bay tree pruning waste (BTPW)	Rincón et al. (2023)
Beef burger patties	Film based on onion (*Allium cepa* l.)	Soares et al. (2021)
Pork	Coatings based on chitosan (CH), gelatin (GL), *aloe vera* gel (AVG), nanoemulsified carvacrol (CNE)	Kalita et al. (2025)
Pork	Copper-modified zinc oxide nanoparticles (ZnO@Cu) in gelatin/agar-based multifunctional films loaded with clove essential oil Pickering emulsion (PEC)	Roy et al. (2021)
Sausage	Bovine collagen gels	Šupová et al. (2023)
Lamb meat	Film based on cassava starch and pomegranate peel powder (PPP)	Esfahani et al. (2022)
Bakery and confectionery products		
Small loaf	*Malva sylvestris* L. (mallow) flower extract as an antimicrobial agent in edible coatings of three types of polysaccharides, pectin(P), xanthan(X), and carboxymethylcellulose(C)	Alexieva et al. (2022)
Bread (wheat)	Aqueous-ethanol dispersions of zein and zein-sunflower oil (10–25% dry basis) emulsions.	Mouzakitis et al. (2022)
Bread	Bacterial cellulose nanofibers (BCNF) and various levels (0.5%, 1%, 2%) of silver nanoparticles (AgNPs) in chitosan (CS) film	Yazdi et al. (2024)
Mini buns	Pectin, alginate and whey protein	Chakravartula et al. (2019)
Muffins	Triticale (× Triticosecale Wittmack) is the hybrid between wheat (Triticum ssp.) and rye (Secale ssp.)-based coatings	Bartolozzo et al. (2016)
Cookies	Chitosan and gum arabic (GA) enriched with grape seed extract (GSE)	Molnar et al. (2023)
Spirulina snack bar	Coated with 6% maltodextrin and 1% gelatin	Dewanti et al. (2024)
Dairy products		

(continued)

Table 1.5 (continued)

Food category	Composition of edible	References
Cheese	Cheese whey (SCW) and oregano (Origanum compactum) and clary sage (Salvia sclarea) EOs	Pires et al. (2024a, b)
Cheese	Whey protein concentrate (WPC) coating enriched With 1–2% of cumin essential oil (CEO)	Nemati et al. (2023)
Cheese	Film from whey protein isolate (WPI), potato peel extract (PPE), and mango peel extract (MPE)	Elfadaly et al. (2023)
Semi-ripened cheese	Coatings based on blends of potato starch (2 g/100 g) and carvacrol (0.1 g/100 g)	López-Córdoba (2021)
Artisanal acid-curd Cheese	Liquid acid whey protein concentrate coating with entrapped indigenous antimicrobial *Lactobacillus helveticus* MI-LH13	Vasiliauskaite et al. (2022)
Nuts		
Brazil nuts	Coating composed of carboxymethyl cellulose and sorbitol, physically reinforced with nanocellulose, and chemically fortified with tocopherol	Nascimento et al. (2024)
Pistachio	Different concentrations of chitosan as an edible coating (0.5–1.5% w/v)	Akhavan-Mahdavi et al. (2023)
Nuts	Two polysaccharides of sodium alginate (SA) and xanthan gum (XG) are compounded in proportions of 100:0, 75:25, 50:50, 25:75, 0:100	Wen et al. (2024)
Walnuts	Coatings based on carboxymethylcellulose (CMC) or hydroxypropyilmethylcellulose (HPMC)	Caballero et al. (2021)
Walnuts	Walnut flour (WF)	Grosso et al. (2020)

including their base material, manufacturer, regulatory status, and applications for which they are meant to serve.

The market potential for edible coatings and films is vast, particularly in sectors dealing with fruits, vegetables, meats, and dairy products, where traditional plastic packaging typically falls short in terms of sustainability and safety (Prakash & Mishra, 2023). Ongoing research is focused not only on optimising material properties for broader application but also on ensuring the economic viability of scalable production methods, which include extrusion processes and solvent-casting techniques (Ochoa-Yepes et al., 2022; Liu et al., 2021). The growing interest in edible coatings and films underscores a pivotal shift towards sustainable food packaging solutions that prioritise both consumer health and environmental integrity. As this field continues to evolve, further research will be essential in addressing the current limitations and in expanding the applicability of these innovative materials.

Table 1.6 Edible films and coatings—worldwide commercial products

Product name	Film/ coating	Base material	Manufacturer	Regulatory status[a]	Applications
ZeinCoat™	Coating	Zein (corn protein)	Flo Chemical Corp (USA)	FDA GRAS, EU Approved	Nuts, confectionery
NatureSeal®	Coating	Plant-based polysaccharides	NatureWorks LLC (USA)	FDA GRAS, EFSA Approved	Fresh produce
Solanyl®	Film	Potato/corn starch	Rodenburg Biopolymers (NL)	EU EN 13432 Certified	Snack packaging
ChitoPack™	Film/ coating	Chitosan	Kitozyme (Belgium)	FDA GRAS, EU Novel Food	Meat, seafood
Pullulan Film	Film	Pullulan	Hayashibara Co. (Japan)	FDA GRAS, JFS Compliant	Supplements, strips
MonoFlex™	Coating	Lipid-protein composite	Mantrose-Haeuser (USA)	FDA GRAS, Canada Approved	Cheese, baked goods
AquaCoat®	Coating	Cellulose derivatives	Ashland Inc. (USA)	FDA GRAS, REACH Compliant	Pharmaceuticals
Alginate Coat	Coating	Sodium alginate	FMC BioPolymer (Global)	FDA GRAS, EFSA Approved	Meat, fish, produce
WheyLayer™	Film/ coating	Whey protein	Lactalis Ingredients (FR)	FDA GRAS, EU Compliant	Protein bars, frozen foods
EcoFilm®	Film	Pectin + starch	DuPont (USA)	FDA GRAS, EU Certified	Dry food packaging

[a]Key to Regulatory Abbreviations: *FDA GRAS* Generally Recognised as Safe (USA), *EFSA* European Food Safety Authority, *JFS* Japan Food Safety Standards, *REACH* EU Regulation on chemicals, *EN 13432* EU compostability standard

Regulations About Biopolymers and Edible Packaging

Biopolymers and edible packaging have become increasingly significant in the context of sustainable materials and environmental regulations. Various regulatory bodies are recognising the need for guidelines and standards that govern the use and production of biodegradable materials, particularly in food packaging applications. Biopolymers, derived from renewable biological sources, including polysaccharides and proteins as well as lipids, are often used as alternatives to conventional plastics. The regulation of these materials is crucial, especially considering their environmental impacts and safety for food contact. For instance, the European Bioplastics Organisation defines *bioplastics* as materials that are either bio-based, biodegradable, or possess both properties, setting a foundation for regulatory definitions and classifications (Nilsen-Nygaard et al., 2021). Such definitions play a pivotal role in ensuring consumer safety and environmental sustainability.

The Food and Drug Administration (FDA) and the European Food Safety Authority (EFSA) are key regulatory bodies recognised worldwide that establish guidelines for materials that come into contact with food. In the European Union, plastic materials intended for food contact must comply with Commission Regulation (EU) No. 10/2011, which includes specifications for the use of biopolymers (Lafraya et al., 2021). These regulatory frameworks ensure that biopolymers used in food packaging meet safety and environmental standards, addressing concerns about migration levels of substances from packaging into food (Westlake et al., 2023). Furthermore, studies highlight the importance of understanding the interactions between food simulants and biopolymer properties to guarantee food safety (Bumbudsanpharoke et al., 2023). The development of edible coatings using biopolymers is being encouraged as a method to enhance the shelf life and safety of food products. Recent research showcases how these coatings can incorporate antimicrobial agents to further extend shelf life while being safe for consumption (Zhao et al., 2023; Simović, 2023). Such innovations not only fulfil consumer demands for natural additives but also align with regulatory objectives of reducing synthetic chemical use in food preservation.

Legislation is evolving to recognise the potential of biodegradable packaging as a strategy to tackle the plastic-waste crisis. The reduction of reliance on fossil fuels for packaging materials is echoed in regulations aimed at promoting life cycle sustainability (Mtibe et al., 2021). Current practices in the food industry include the formation of biocomposite films, which not only utilise biopolymers but also incorporate enhancing agents to improve mechanical and barrier properties, thereby fulfilling regulatory requirements (Pasquier et al., 2022; Ponnusamy et al., 2022; Califano et al., 2024).

Conclusion

Rising consumer environmental concern, driven by the widespread use of plastic packaging, has stimulated the development of new environmentally friendly food packaging materials. Edible packaging has attracted considerable interest in recent years, as it offers a sustainable and versatile solution to various environmental problems and benefits human health, thanks to its biodegradable and readily available characteristics. It uses edible coatings or wrappings made of biodegradable materials to release chemicals that preserve food quality and extend its shelf life, such as antioxidants and antimicrobials. Nanotechnology could add bioactives, vitamins, and nutrients to edible packaging, improving functionality and retaining the freshness of fruits and vegetables. It also helps to minimise moisture loss, stop deterioration, and avoid spoilage of food products. The nutritional value and health benefits of edible packaging are enhanced when nutraceuticals and bioactive compounds are included.

The development of edible packaging involves the careful selection of biopolymers, functional additives, and natural components. The synergistic effects of these ingredients not only contribute to the mechanical and barrier properties of the packaging but also play a vital role in prolonging the shelf life and safety of food products, thus addressing both consumer demands and environmental concerns surrounding food packaging.

Edible packaging has enormous potential and benefits in the food industry. In addition to increasing the shelf life of food, this packaging also promotes the protection of the environment by reusing materials that contribute to the circular economy, reducing waste, and meeting the increased consumer demand for more natural and environmentally friendly products. Ongoing research and development efforts are focused on addressing challenges related to sustainability, scalability, regulation, and consumer acceptance. Advancements in processing techniques, such as nanotechnology and 3D printing, offer exciting possibilities for tailoring materials and creating customised packaging solutions. Additionally, the functionalisation of plant-based materials with active and intelligent components enhances food safety, preservation, and quality.

The regulatory landscape surrounding biopolymers and edible packaging is complex and continuously evolving, driven by both scientific advancements and growing environmental concerns. Regulatory bodies like the FDA and EFSA are critical in establishing guidelines that ensure food safety, while innovative applications of biopolymers present viable solutions to meet the challenges of contemporary food packaging needs. In conclusion, edible coatings show promise for the food industry, offering benefits for both products and the environment. With ongoing research, technological advancements, and increased consumer awareness, these materials can contribute to a more environmentally friendly and sustainable future.

References

Aboulfotouh, M. E., Ali, H. E., & Mohamed, M. R. (2024). Nano-modifications of edible materials using ionizing radiation for potential application in active food safety. *Functional Composite Materials, 5*(1). https://doi.org/10.1186/s42252-024-00056-4

Akhavan-Mahdavi, S., Mirzazadeh, M., Alam, Z., & Solaimanimehr, S. (2023). The effect of chitosan coating combined with cold plasma on the quality and safety of pistachio during storage. *Food Science & Nutrition, 11*(7), 4296–4307. https://doi.org/10.1002/fsn3.3355

Alaguthevar, R., Packialakshmi, J. S., Balakrishnan, M., Rhim, J., & Umamaheswari, T. (2024). In-package cold plasma treatment to extend the shelf life of food. *Comprehensive Reviews in Food Science and Food Safety, 23*(2). https://doi.org/10.1111/1541-4337.13318

Alexieva, I., Baeva, M., Popova, A., Fidan, H., Goranova, Z., & Milkova-Tomova, I. (2022). Development and application of edible coatings with Malva sylvestris L. extract to extend shelf-life of small loaf. *Foods, 11*, 3831. https://doi.org/10.3390/foods11233831

Al-Hilifi, S. A., Al-Ali, R. M., & Petkoska, A. T. (2022a). Ginger essential oil as an active addition to composite chitosan films: Development and characterization. *Gels, 8*(6), 327. https://doi.org/10.3390/gels8060327

Al-Hilifi, S. A., Al-Ali, R. M., Al-Ibresam, O. T., Kumar, N., Paidari, S., Petkoska, A. T., & Agarwal, V. (2022b). Physicochemical, morphological, and functional characterization of edible anthocyanin-enriched aloevera coatings on fresh figs (Ficus Carica L.). *Gels, 8*(10), 645. https://doi.org/10.3390/gels8100645

Alqahtani, N. K., Alkhamis, B., Alnemr, T. M., & Mohammed, M. (2025). Combined influences of edible coating and storage conditions on the quality of fresh dates: An investigation and predictive analysis using artificial neural networks. *Heliyon*. https://doi.org/10.1016/j.heliyon.2025.e42373

Ammar, E. E., Zou, X., Ghosh, S., Helen, O., Elmasry, S. A., Alkeay, A. M., AL-Farga, A., Rady, H. A., EL-Shershaby, N. A., & Sallam, A. S. (2025). Fresh futures: Cutting-edge eco-friendly coating techniques for fruits. *Journal of Food Processing and Preservation, 2025*, 5201632, 19 pages. https://doi.org/10.1155/jfpp/5201632

Amran, F. I. S., Zamri, A. I., Arifin, H. R., & Shafrina, A.-H. (2024). Active food packaging from alginate-based film containing onion peel extract (Allium cepa). *Journal of Biochemistry, Microbiology and Biotechnology, 12*(SP1), 5–8. https://doi.org/10.54987/jobimb.v12iSP1.923

Antonino, C. D. (2023). Effect of edible coatings and films enriched with plant extracts and essential oils on the preservation of animal-derived foods. *Journal of Food Science, 89*, 748–772. https://doi.org/10.1111/1750-3841.16894

Armghan Khalid, M., Niaz, B., Saeed, F., Afzaal, M., Islam, F., Hussain, M., et al. (2022). Edible coatings for enhancing safety and quality attributes of fresh produce: A comprehensive review. *International Journal of Food Properties, 25*(1), 1817–1847. https://doi.org/10.1080/10942912.2022.2107005

Asgarian, Z. S., Palou, L., de Souza, R. F. L., Quintanilla, P. G., Taberner, V., Karimi, R., & Pérez-Gago, M. B. (2023). Hydroxypropyl methylcellulose and gum arabic composite edible coatings amended with geraniol to control postharvest brown rot and maintain quality of cold-stored plums. *Foods, 12*, 2978. https://doi.org/10.3390/foods12152978

Atta, O. M., Manan, S., Ahmed, A. A. Q., Awad, M. F., Ul-Islam, M., Subhan, F., Ullah, M. W., & Yang, G. (2021). Development and characterization of yeast-incorporated antimicrobial cellulose biofilms for edible food packaging application. *Polymers, 13*(14), 2310. https://doi.org/10.3390/polym13142310

Azizkhani, M., Kavosi, S., & Partovi, R. (2023). Improving the quality of the chicken fillet using chitosan, gelatin, and starch coatings incorporated with bitter orange peel extract during refrigeration. *Food Science & Nutrition, 11*, 4700–4712. https://doi.org/10.1002/fsn3.3432

Bartolozzo, J., Borneo, R., & Aguirre, A. (2016). Effect of triticale-based edible coating on muffin quality maintenance during storage. *Food Measure, 10*, 88–95. https://doi.org/10.1007/s11694-015-9280-1

Basavegowda, N., & Baek, K. (2021). Advances in functional biopolymer-based nanocomposites for active food packaging applications. *Polymers, 13*(23), 4198. https://doi.org/10.3390/polym13234198

Bayram, B., Özkan, G., Kostka, T., Çapanoğlu, E., & Esatbeyoglu, T. (2021). Valorization and application of fruit and vegetable wastes and by-products for food packaging materials. *Molecules, 26*(13), 4031. https://doi.org/10.3390/molecules26134031

Beikzadeh, S., Ghorbani, M., Shahbazi, N., Izadi, F., Pilevar, Z., & Mortazavian, A. M. (2020). The effects of novel thermal and nonthermal technologies on the properties of edible food packaging. *Food Engineering Reviews, 12*(3), 333–345. https://doi.org/10.1007/s12393-020-09227-y

Birania, S., Attkan, A. K., Kumar, S., Kumar, N., & Singh, V. K. (2022). Cold plasma in food processing and preservation: A review. *Journal of Food Process Engineering, 45*(9). https://doi.org/10.1111/jfpe.14110

Bodana, V., Swer, T. L., Kumar, N., et al. (2023). Development and characterization of pomegranate peel extract-functionalized jackfruit seed starch-based edible films and coatings for prolonging the shelf life of white grapes. *International Journal of Biological Macromolecules*. https://doi.org/10.1016/j.ijbiomac.2023.127234

Bremenkamp, I., & Sousa Gallagher, M. (2025). Edible coatings for ready-to-eat products: Critical review of recent studies. *Sustainable Packaging Perspectives, Challenges and Emerging Trends*. https://doi.org/10.3390/polym17030376

Bucher, T., Jaala, M., Mukhopadhyay, S. P., Vuong, Q., & Beckett, E. (2023). Consumer acceptance of edible coatings on apples: The role of food technology neophobia and information about purpose. *Food Quality and Preference, 112*. https://doi.org/10.1016/j.foodqual.2023.105024

Bumbudsanpharoke, N., Chongcharoenyanon, B., Harnkarnsujarit, N., Kwon, S., & Ko, S. (2023). Effect of food simulants on stability of copper oxide in bionanocomposite food packaging film. *Packaging Technology and Science, 36*(8), 657–665. https://doi.org/10.1002/pts.2734

Buthelezi, N. M. D. (2025). Chitosan–Zinc oxide nanoparticle composite coatings preserve postharvest quality of 'rosada' cherry tomato. *HortScience, 60*(4), 613–623. https://doi.org/10.21273/HORTSCI18426-24

Caballero, E., Bernal, D., & Wilckens, J. (2021). Edible coatings and lipid oxidation data in walnuts. *Data in Brief, 107295*. https://doi.org/10.1016/j.dib.2021.107295

Califano, G., Lombardi, A., Giudice, T. D., Caracciolo, F., & Cembalo, L. (2024). Bioplastics in the basket of italians: A hybrid framework for understanding the adoption of bioplastic food packaging. *Australian Journal of Agricultural and Resource Economics, 68*(4), 826–846. https://doi.org/10.1111/1467-8489.12578

Chaari, M., Elhadef, K., Akermi, S., Ben Akacha, B., Fourati, M., Chakchouk Mtibaa, A., Ennouri, M., Sarkar, T., Shariati, M. A., Rebezov, M., et al. (2022). Novel active food packaging films based on gelatin-sodium alginate containing beetroot peel extract. *Antioxidants, 2095*, 11. https://doi.org/10.3390/antiox11112095

Chakravartula, S. S. N., Cevoli, C., Balestra, F., Fabbri, A., & Dalla Rosa, M. (2019). Evaluation of the effect of edible coating on mini-buns during storage by using NIR spectroscopy. *Journal of Food Engineering, 263*, 46–52. https://doi.org/10.1016/j.jfoodeng.2019.05.035

Chaudhary, S. (2023). Seaweed polysaccharide coatings/films for meat based foods. *Food and Humanity, 1*, 777–792. https://doi.org/10.1016/j.foohum.2023.07.029

Chitra, K., Rasheeda Shameem, M., & Arivoli, S. (2022). Bilayer edible coating of figs (Ficus carica) using gelatin and guar gum incorporated with grapefruit seed extract. *The Pharma Innovation Journal, SP-11*(9), 1904–1911.

Cui, Z., Li, Y., Feng, X., & Hu, Z. (2023). Development, RSM-based optimization, and characterization of a unique edible composite film by incorporating clove essential oil based on sodium alginate and aloe vera gel to preserve the quality of blueberries. *Journal of Food Processing and Preservation, 2023*, 3578799. https://doi.org/10.1155/2023/3578799

Cvanic, T., Šovljanski, O., Popovic, S., Erceg, T., Vulić, J., Canadanović-Brunet, J., Cetković, G., & Travicic, V. (2023). Progress in fruit and vegetable preservation: Plant-based nanoemulsion coatings and their evolving trends. *Coatings, 13*, 1835. https://doi.org/10.3390/coatings13111835

da Rios, D. A. S., Masako Nakamoto, M., Cavalcante Braga, A. R., & da Cunha Silva, E. M. (2022). Food coating using vegetable sources: Importance and industrial potential, gaps of knowledge, current application, and future trends. *Applied Food Research, 2*, 100073. https://doi.org/10.1016/j.afres.2022.100073

Daniloski, D., Petkoska, A. T., Lee, N. A., Bekhit, A. E. A., Carne, A., Vaskoska, R., & Vasiljevic, T. (2021). Active edible packaging based on milk proteins: A route to carry and deliver nutraceuticals. *Trends in Food Science & Technology, 111*, 688–705. https://doi.org/10.1016/j.tifs.2021.03.024

de Sousa, M. S., Schlogl, A. E., Estanislau, F. R., Souza, V. G. L., dos Reis Coimbra, J. S., & Santos, I. J. B. (2023). Nanotechnology in packaging for food industry: Past, present, and future. *Coatings, 13*(8), 1411. https://doi.org/10.3390/coatings13081411

Dewanti, E., Lee, W.-C., Tarman, K., Santoso, J., Nurjanah, N., & Hardiningtyas, S. (2024). Total phenolic content and antioxidant properties of hydrophobic compounds edible coating Spirulina snack bar. *BIO Web of Conferences, 106*, 05003. https://doi.org/10.1051/bioconf/202410605003

Dholariya, P. K., Borkar, S., & Borah, A. (2021). Prospect of nanotechnology in food and edible packaging: A review. *The Pharma Innovation, 10*(5), 197–203. https://doi.org/10.22271/tpi.2021.v10.i5c.6202

Díaz-Montes, E., & Castro-Muñoz, R. (2021). Edible films and coatings as food-quality preservers: An overview. *Foods, 10*(2), 249. https://doi.org/10.3390/foods10020249

Dmitrenko, M. (2025). Development and study of sustainable edible coating from carrageenan/starch/nanocellulose for enhancing fruit/vegetable shelf life and preservation. *Acs Omega, 10*(13), 13555–13569. https://doi.org/10.1021/acsomega.5c00480

Dmitrenko, M., Kuzminova, A., Cherian, R. M., Joshy, K. S., Pasquini, D., John, M. J., Hato, M. J., Thomas, S., & Penkova, A. (2023). Edible carrageenan films reinforced with starch and nanocellulose: Development and characterization. *Sustainability, 15*(22), 15817. https://doi.org/10.3390/su152215817

Dong, Y., Li, Y., Ma, Z., Rao, Z., Zheng, X., Tang, K., & Liu, J. (2023). Effect of polyol plasticizers on properties and microstructure of soluble soybean polysaccharide edible films. *Food Packaging and Shelf Life, 35*, 101023. https://doi.org/10.1016/j.fpsl.2022.101023

Elfadaly, S., Mattar, A., Sorour, M., Karam-Allah, A., & Soliman, T. (2023). Preparation and characterization of whey protein edible coating with potato and mango peels: Application in processed cheese. *Egyptian Journal of Chemistry*. https://doi.org/10.21608/ejchem.2023.191365.7552

El-Gioushy, S. F., El-Masry, A. M., Fikry, M., El-Kholy, M. F., Shaban, A. E., Sami, R., Algarni, E., Alshehry, G., Aljumayi, H., Benajiba, N., et al. (2022). Utilization of active edible films (Chitosan, Chitosan Nanoparticle, and CaCl2) for enhancing the quality properties and the shelf life of date palm fruits (Barhi Cultivar) during cold storage. *Coatings, 12*, 255. https://doi.org/10.3390/coatings12020255

Esfahani, A., Mohammadi Nafchi, A., Baghaei, H., & Nouri, L. (2022). Fabrication and characterization of a smart film based on cassava starch and pomegranate peel powder for monitoring lamb meat freshness. *Food Science & Nutrition, 10*, 3293–3301. https://doi.org/10.1002/fsn3.2918

Evrendilek, G. A., Bulut, N., & Uzuner, S. (2022). Assessing edible composite film polymer from potato industry effluent under high hydrostatic pressure and its antimicrobial properties. *CLEAN – Soil, Air, Water, 50*(11). https://doi.org/10.1002/clen.202100406

Fawole, O. A., Riva, S., Silue, Y., & Opara, U. L. (2024). Evaluating commercial viability of gum Arabic-based edible coatings for enhancing shelf life of 'African Delight™' plum under simulated packhouse conditions. *South African Journal of Botany, 174*, 902–915. https://doi.org/10.1016/j.sajb.2024.10.005

Galus, S., Mikus, M., Ciurzyńska, A., & Janowicz, M. (2022). Whey protein isolate-based edible coatings incorporated with jojoba oil as a novel approach for improving the quality of fresh-cut root parsley during refrigerated storage. *Applied Science, 12*, 9023. https://doi.org/10.3390/app12189023

Gammage, S., & Marangoni, A. G. (2025). Safety of edible coatings on fruits and vegetables. *Comprehensive Reviews in Food Science and Food Safety, 24*, e70108. https://doi.org/10.1111/1541-4337.70108

Gao, X., Pourramezan, H., Ramezan, Y., et al. (2024). Application of gums as techno-functional hydrocolloids in meat processing and preservation: A review. *International Journal of Biological Macromolecules*. https://doi.org/10.1016/j.ijbiomac.2024.131614

Ghafoor, K., Al-Juhaimi, F. Y., Mohamed Ahmed, I. A., Babiker, E. E., Shahzad, S. A., & Alsawmahi, O. N. (2022). Effects of functional coatings containing chitosan, orange peel and olive cake extracts on the quality attributes of cucumber during cold storage. *Plants, 11*, 1895. https://doi.org/10.3390/plants11141895

Gidado, M. J., Gunny, A. A. N., Gopinath, S. C. B., Wongs- Aree, C., Aifaa Yusoff, N. H., Ibrahim, R., Laboh, R., & Ali, A. (2024). Effect of hydrophobic deep eutectic oil-in-water nano coating on the quality preservation of postharvest 'Harumanis' mango. *Colloids and Surfaces A: Physicochemical and Engineering Aspects, 691*, 133818. https://doi.org/10.1016/j.colsurfa.2024.133818

Grosso, A. L., Asensio, C. M., Grosso, N. R., & Nepote, V. (2020). Increase of walnuts' shelf life using a walnut flour protein-based edible coating. *Lwt - Food Science and Technology, 118*, 108712.

Guerreiro, A. C., Gago, C., Passos, D., Martins, J., Cruz, S., Veloso, F., Guerra, R., & Antunes, M. D. (2025). Edible coatings enhance storability and preserve quality of Kiwiberry (Actinidia arguta L.) cv. Ken's Red. *Horticulturae, 11*, 105. https://doi.org/10.3390/horticulturae11010105

Guo, Y., Qiu, Y., & Chen, Z. C. (2024). Integrating nanocomposites and biosensors: Towards an intelligent food packaging system. *Bio Web of Conferences, 111*, 01005. https://doi.org/10.1051/bioconf/202411101005

Gupta, R. K., Guha, P., & Srivastav, P. P. (2022). Natural polymers in bio-degradable/edible film: A review on environmental concerns, cold plasma technology and nanotechnology application on food packaging- A recent trends. *Food Chemistry Advances, 1*, 100135. https://doi.org/10.1016/j.focha.2022.100135

Gupta, D., Lall, A., Kumar, S. S. A., Patil, T. D., & Gaikwad, K. K. (2024). Plant-based edible films and coatings for food-packaging applications: Recent advances, applications, and trends. *Sustainable Food Technology, 2*(5), 1428–1455. https://doi.org/10.1039/d4fb00110a

Hamed, I., Jakobsen, A. N., & Lerfall, J. (2021). Sustainable edible packaging systems based on active compounds from food processing byproducts: A review. *Comprehensive Reviews in Food Science and Food Safety*. https://doi.org/10.1111/1541-4337.12870

Haneef, N., Garièpy, Y., Raghavan, V., Kurian, J. K., Hanif, N., & Hanif, T. (2022). Effects of Aloe-pectin coatings and osmotic dehydration on storage stability of mango slices. *Brazilian Journal of Food Technology, 25*, e2022028. https://doi.org/10.1590/1981-6723.02822

Hasanin, M. S., Kader, A. H. A., El-Sayed, E. S. A., & Kamel, S. (2023). Green chitosan-flaxseed gum film loaded with ZnO for packaging applications. *Starch - Stärke, 75*(5–6). https://doi.org/10.1002/star.202200132

Hassanzadeh, H., Ahmed, S. A., & Qadir, N. S. H. (2025). Application of the active edible film reinforced with nanoparticles and nanoemulsions as the coating systems to improve the quality and shelf life of fruits and vegetables. *Journal of Nanotechnology, 2025*, 7036931. https://doi.org/10.1155/jnt/7036931

Heristika, W., Ningrum, A., Supriyadi, Helimatul Munawaroh, H. S., & Show, P. L. (2023). Development of composite edible coating from gelatin-pectin incorporated garlic essential oil on physicochemical characteristics of red chili (Capsicum annnum L.). *Gels, 9*, 49. https://doi.org/10.3390/gels9010049

Hua, Q. H., Wong, C., & Li, D. (2022). Postbiotics enhance the functionality of a probiotic edible coating for salmon fillets and the probiotic stability during simulated digestion. *Food Packaging and Shelf Life, 34*, 100954. https://doi.org/10.1016/j.fpsl.2022.100954

Iversen, L. J. L., Rovina, K., Vonnie, J. M., Matanjun, P., Erna, K. H., Aqilah, N. M. N., et al. (2022). The emergence of edible and food-application coatings for food packaging: A review. *Molecules, 27*(17), 5604. https://doi.org/10.3390/molecules27175604

Jamuna, P., Rathnasamy, V. K., Devi, A. N., Jegadeeswari, V., Kumar, R. S., Geetha, K. A., Lakshmi, I., Maheswari, T. U., & Thiyagarajan, A. (2024). Edible coatings: Classification, applications and innovations in food preservation. *Plant Science Today, 11*(sp4). https://doi.org/10.14719/pst.5263

Jatav, J., Tarafdar, A., Saravanan, C., & Bhattacharya, B. (2022). Assessment of antioxidant and antimicrobial property of polyphenol-rich chitosan-pineapple peel film. *Journal of Food Quality, 2022*, 8064114, 10 pages. https://doi.org/10.1155/2022/8064114

Kalita, S., Kumar, S., Dutta, J., & Mukherjee, A. (2025). Chitosan and gelatin based antimicrobial composite coatings functionalized with natural active agents for preservation of pork meat. *Future Foods, 11*. https://doi.org/10.1016/j.fufo.2025.100607

Kanelaki, A., Zampouni, K., Mourtzinos, I., & Katsanidis, E. (2022). Hydrogels, oleogels and bigels as edible coatings of sardine fillets and delivery systems of rosemary extract. *Gels, 8*, 660. https://doi.org/10.3390/gels8100660

Karthik, C., Mavelil-Sam, R., Thomas, S., & Thomas, V. (2024). Cold plasma technology based eco-friendly food packaging biomaterials. *Polymers, 16*(2), 230. https://doi.org/10.3390/polym16020230

Kawhena, T. G., Opara, U. L., & Fawole, O. A. (2022). Effects of gum arabic coatings enriched with lemongrass essential oil and pomegranate peel extract on quality maintenance of pomegranate whole fruit and arils. *Foods, 11*, 593. https://doi.org/10.3390/foods11040593

Kehila, S., Alkalai-Tuvia, S., Chalupowicz, D., Poverenov, E., & Fallik, E. (2021). Can edible coatings maintain sweet pepper quality after prolonged storage at sub-optimal temperatures? *Horticulturae, 7*, 387. https://doi.org/10.3390/horticulturae7100387

Kong, I., Degraeve, P., & Pui, L. P. (2022). Polysaccharide-based edible films incorporated with essential oil nanoemulsions: Physico-chemical, mechanical properties and its application in food preservation—a review. *Foods, 11*(4), 555. https://doi.org/10.3390/foods11040555

Kostić, M., Bajac, B., Janjušević, L., Bajac, J., & Antov, M. (2023). Edible coatings based on plant components for active packaging of fresh/fresh-cut fruits. *South African Journal of Botany, 161*, 395–403. https://doi.org/10.1016/j.sajb.2023.08.039

Kumar, N., Phogat, N., Pratibha, N. A., & Petkoska, A. T. (2021a). Improved shelf life and quality of tomato (Solanum Lycopersicum L.) by using chitosan-pullulan composite edible coating enriched with pomegranate peel extract. *ACS Food Science & Technology, 1*(4), 500–510. https://doi.org/10.1021/acsfoodscitech.0c00076

Kumar, N., Pratibha, N. A., Petkoska, A. T., Khojah, E., Sami, R., & Al-Mushhin, A. A. M. (2021b). Chitosan edible films enhanced with pomegranate peel extract: Study on physical, biological, thermal, and barrier properties. *Materials, 14*(12), 3305. https://doi.org/10.3390/ma14123305

Kumar, N., Pratibha, N. A., Phogat, N., Petkoska, A. T., Al-Hilifi, S. A., & Fawole, O. A. (2021c). Effect of chitosan–pullulan composite edible coating functionalized with pomegranate peel extract on the shelf life of Mango (Mangifera Indica). *Coatings, 11*(7), 764. https://doi.org/10.3390/coatings11070764

Kumar, N., Daniloski, D., Pratibha, N. A., Phogat, N., D'Cunha, N. M., Naumovski, N., & Petkoska, A. T. (2022a). Pomegranate peel extract – a natural bioactive addition to novel active edible packaging. *Food Research International, 156*, 111378. https://doi.org/10.1016/j.foodres.2022.111378

Kumar, A., Muzaffar, H., Shukadev, M., Pravitha, M., & Deepak, V. K. (2022b). Trends in edible packaging films and its prospective future in food: A review. *Applied Food Research.* https://doi.org/10.1016/j.afres.2022.100118

Kumar, N., Pratibha, N. A., Prasad, J., Yadav, A., Upadhyay, A., Phogat, N., Shukla, S., Petkoska, A. T., Heena, Suri, S., Gniewosz, M., & Kieliszek, M. (2023a). Recent trends in edible packaging for food applications—perspective for the future. *Food Engineering Reviews, 15*(4), 718–747. https://doi.org/10.1007/s12393-023-09358-y

Kumar, N., Tripathi, S., Pratibha, Mehra, M., Heena, & Trajkovska Petkoska, A. (2023b). Whey protein based edible coatings: Recent trends. In A. Poonia & A. Trajkovska Petkoska (Eds.), *Whey valorization* (pp. 187–209). Springer Nature Singapore. https://doi.org/10.1007/978-981-99-5459-9_10

Kumari, S., Debbarma, R., Nasrin, N., Khan, T., Taj, S., & Bhuyan, T. (2024a). Recent advances in packaging materials for food products. *Food Bioengineering, 3*(2), 236–249. https://doi.org/10.1002/fbe2.12096

Kumari, K., Riram, K., Maibam, P., & Srikanth, P. (2024b). Effect of novel edible coating formulations on post-harvest quality of Chilli (Capsicum annuum var. Pusa Jwala): A review. *BIO Web of Conferences, 110.* https://doi.org/10.1051/bioconf/202411002007

Kusnadi, J., Mahatmanto, T., Marsheli, N., Fawzia, N., Rahmawani, D. E., & Alexander, K. (2022). Development of low-cost edible coatings based on polysaccharides with active lactic acid bacteria for the protection of fresh produce modeled using fresh cut apples. *Food Science and Technology International, 29*(4), 287–298. https://doi.org/10.1177/10820132221075117

Lafraya, Á., Prieto, C., Pardo-Figuerez, M., Chiva, A., & Lagarón, J. M. (2021). Super-repellent paper coated with electrospun biopolymers and electrosprayed silica of interest in food packaging applications. *Nanomaterials, 11*(12), 3354. https://doi.org/10.3390/nano11123354

Lamani, N. A., & Ramaswamy, H. S. (2023). Composite alginate–ginger oil edible coating for fresh-cut pears. *Journal of Composites Sciences, 7*, 245. https://doi.org/10.3390/jcs7060245

Lekshmi, S. G., Sethi, S., Asrey, R., Singh, K. P., Kumar, R., Sindhu, P. M., Singh, A. K., Gunjan, P., & Goswami, A. K. (2025). Comprehensive characterization of biodegradable edible films activated with rose and marigold extracts and application of active edible coatings to extend the postharvest storage life of guava. *Food Research International*. https://doi.org/10.1016/j.foodres.2025.115895

Li, T., Meng, F., Chi, W., Xu, S., & Wang, L. (2022). An edible and quick-dissolving film from cassia gum and ethyl cellulose with improved moisture barrier for packaging dried vegetables. *Polymers, 14*, 4035. https://doi.org/10.3390/polym14194035

Liu, L., Xiang, H., Geng, F., & Huang, Q. (2021). Optimization of preparation process of egg white protein/ κ -carrageenan composite film. *Journal of Food Processing and Preservation, 46*(9). https://doi.org/10.1111/jfpp.16167

Liyanapathiranage, A., Dassanayake, R. S., Gamage, A., Karri, R. R., Manamperi, A., Evon, P., Jayakodi, Y., Madhujith, T., & Merah, O. (2023). Recent developments in edible films and coatings for fruits and vegetables. *Coatings, 13*(7), 1177. https://doi.org/10.3390/coatings13071177

López-Córdoba, A. (2021). Feasibility of using carvacrol/starch edible coatings to improve the quality of paipa cheese. *Polymers, 13*, 2516. https://doi.org/10.3390/polym13152516

Luesuwan, S., Naradisorn, M., Shiekh, K. A., Rachtanapun, P., & Tongdeesoontorn, W. (2021). Effect of active packaging material fortified with clove essential oil on fungal growth and post-harvest quality changes in table grape during cold storage. *Polymers, 13*, 3445. https://doi.org/10.3390/polym13193445

Ma, M., Liu, Y., Zhang, S., & Yuan, Y. (2024). Edible coating for fresh-cut fruit and vegetable preservation: Biomaterials, functional ingredients, and joint non-thermal technology. *Foods, 13*(23), 3937. https://doi.org/10.3390/foods13233937

Maslov Bandić, L., Vuković, M., Gavranić, D., Tanuwidjaja, I., Mrkonjić Fuka, M., Duralija, B., & Jurić, S. (2025). Elevating the bioactive potential of strawberries with apple pomace extract-infused chitosan edible coating. *Polish Journal of Food and Nutrition Sciences, 75*(1), 24–36. https://doi.org/10.31883/pjfns/199628

Mehra, M., Kumar, N., Ojha, A., et al. (2024). Extending the postharvest shelf life of sweet orange (Citrus sinensis L.) fruits using muskmelon seeds flour based edible coatings functionalized with oregano essential oil. *Food Measure, 18*, 8472–8488. https://doi.org/10.1007/s11694-024-02815-0

Mei, L., & Wang, Q. (2020). Advances in using nanotechnology structuring approaches for improving food packaging. *Annual Review of Food Science and Technology, 11*(1), 339–364. https://doi.org/10.1146/annurev-food-032519-051804

Meshram, B. D., Lule, V. K., Vyawahare, S., & Rani, R. (2023). Application of edible packaging in dairy and food industry. *Food Processing and Packaging Technologies - Recent Advances*. https://doi.org/10.5772/intechopen.107850

Mileti, O., Baldino, N., Filice, F., Lupi, F. R., Sinicropi, M. S., & Gabriele, D. (2023). Formulation study on edible film fromwaste grape and red cabbage. *Foods, 12*, 2804. https://doi.org/10.3390/foods12142804

Mittal, S., Bhuiyan, H. R., & Ngadi, M. O. (2023). Challenges and prospects of plant-protein-based 3D printing. *Foods, 2023*. https://doi.org/10.3390/foods12244490

Moghadas, H. C., Chauhan, R., & Smith, J. S. (2024). Application of plant oils as functional additives in edible films and coatings for food packaging: A review. *Foods, 13*(7), 997. https://doi.org/10.3390/foods13070997

Mohammadi, M., Rastegar, S., & Rohani, A. (2024). Enhancing Mexican lime (Citrus aurantifolia cv.) shelf life with innovative edible coatings: Xanthan gum edible coating enriched with Spirulina platensis and pomegranate seed oils. *BMC Plant Biology, 24*, 906. https://doi.org/10.1186/s12870-024-05606-3

Molnar, D., Novotni, D., Kurek, M., Galić, K., Iveković, D., Bionda, H., & Ščetar, M. (2023). Characteristics of edible films enriched with fruit by-products and their application on cookies. *Food Hydrocolloids, 135*, 108191. https://doi.org/10.1016/j.foodhyd.2022.108191

Mtibe, A., Motloung, M. P., Bandyopadhyay, J., & Ray, S. S. (2021). Synthetic biopolymers and their composites: Advantages and limitations—an overview. *Macromolecular Rapid Communications, 42*(15). https://doi.org/10.1002/marc.202100130

Nair, S. S., Trafiałek, J., & Kolanowski, W. (2023). Edible food packaging- an overview. *Chemistry and Materials Science*. https://doi.org/10.20944/preprints202306.0827.v1

Nakum, K. R., Chudasama, B., & Bhola, D. (2024). Development of edible coating and its preservation effect on peeled and deveined brown shrimp (metapenaeus affinis) during refrigerated storage. *International Journal of Advanced Biochemistry Research, 8*(7), 292–299. https://doi.org/10.33545/26174693.2024.v8.i7d.1487

Nascimento, D. R., Mesquita, J., da Silva, T., Hernandes, T., Lengowski, E. C., & Takeuchi, K. (2024). A novel approach to protect Brazil nuts from lipid oxidation: Efficacy of nanocellulose–tocopherol edible coatings. *Coatings, 14*, 1182. https://doi.org/10.3390/coatings14091182

Nazoori, F., Mollai, S., Sobhani, F., Mirdehghan, S., & Sahhafi, S. (2023). Carboxymethyl cellulose and carnauba wax treatments kept the pomegranate fruit (Punica granatum L.) quality during cold storage via improving enzymatic defense system and bioactive compounds. *Scientia Horticulturae, 309*, 111645. https://doi.org/10.1016/j.scienta.2022.111645

Nemati, V., Hashempour-Baltork, F., Sadat Gharavi-Nakhjavani, M., Feizollahi, E., Marangoni Júnior, L., & Mirza Alizadeh, A. (2023). Application of a whey protein edible film incorporated with cumin essential oil in cheese preservation. *Coatings, 13*, 1470. https://doi.org/10.3390/coatings13081470

Neogi, A. G., Upadhyaya, A., Manokaran, S., & Reddy, M. (2022). Edible packaging: Food for thought and food for the future. *ECS Transactions, 107*(1), 13757–13771. https://doi.org/10.1149/10701.13757ecst

Nilsen-Nygaard, J., Fernández, E. N., Radusin, T., Rotabakk, B. T., Sarfraz, J., Sharmin, N., Sivertsvik, M., Sone, I., & Pettersen, M. K. (2021). Current status of biobased and biodegradable food packaging materials: Impact on food quality and effect of innovative processing technologies. *Comprehensive Reviews in Food Science and Food Safety, 20*(2), 1333–1380. https://doi.org/10.1111/1541-4337.12715

Nunes, C., Silva, M., Farinha, D., Sales, H., Pontes, R., & Nunes, J. (2023). Edible coatings and future trends in active food packaging–fruits' and traditional sausages' shelf life increasing. *Foods, 12*(17), 3308. https://doi.org/10.3390/foods12173308

Ochoa-Yepes, O., Ceballos, R. L., & Famá, L. (2022). Effect of wheat and oat bran on the physicochemical properties of edible starch-based films obtained by extrusion for food packaging applications. *Starch - Stärke, 75*(1-2). https://doi.org/10.1002/star.202200087

Olawade, D. B., Wada, O. Z., & Ige, A. O. (2024). Advances and recent trends in plant-based materials and edible films: A mini-review. *Frontiers in Chemistry, 12*, 1441650. https://doi.org/10.3389/fchem.2024.1441650

Olawuyi, I. F., Park, J. J., Lee, J. J., & Lee, W. Y. (2019). Combined effect of chitosan coating and modified atmosphere packaging on fresh-cut cucumber. *Food Science & Nutrition, 7*(3), 1043–1052. https://doi.org/10.1002/fsn3.937

Oliveira, I. (2025). Sustainability in bio-based edible films, coatings, and packaging for small fruits. *Applied Sciences, 15*(3), 1462. https://doi.org/10.3390/app15031462

Pakulska, A., Bartosiewicz, E., & Galus, S. (2023). The potential of apple and blackcurrant pomace powders as the components of pectin packaging films. *Coatings, 13*, 1409. https://doi.org/10.3390/coatings13081409

Papadaki, A., Alexandri, M., Kachrimanidou, V., Tzora, P., & Kopsahelis, N. (2023). Enhancing hydrophobic and antioxidant properties of whey protein films through the inclusion of oil and oleogel sourced from spent coffee grounds. *Sustainable Chemistry and Pharmacy, 36*. https://doi.org/10.1016/j.scp.2023.101293

Pasquier, E., Mattos, B. D., Koivula, H., Khakalo, A., Belgacem, M. N., Rojas, O. J., & Bras, J. (2022). Multilayers of renewable nanostructured materials with high oxygen and water vapor barriers for food packaging. *Acs Applied Materials & Interfaces, 14*(26), 30236–30245. https://doi.org/10.1021/acsami.2c07579

Pedreiro, S., Figueirinha, A., Sanches-Silva, A., & Ramos, F. (2021). Bioactive edible films and coatings based in gums and starch: Phenolic enrichment and foods application. *Coatings, 11*(11), 1393. https://doi.org/10.3390/coatings11111393

Perez-Vazquez, A. P. (2023). Edible coatings as a natural packaging system to improve fruit and vegetable shelf life and quality. *Foods, 12*, 3570. https://doi.org/10.3390/foods12193570

Petkoska, A. T., Daniloski, D., D'Cunha, N. M., Naumovski, N., & Trajkovska-Broach, A. (2021). Edible packaging: Sustainable solutions and novel trends in food packaging. *Food Research International, 140*, 109981. https://doi.org/10.1016/j.foodres.2020.109981

Petraru, A., & Amariei, S. (2023). A novel approach about edible packaging materials based on oilcakes—a review. *Polymers, 15*, 3431. https://doi.org/10.3390/polym15163431

Pié-Amill, A., Colás-Medà, P., Vinas, I., Falcó, I., & Alegre, I. (2024). Efficacy of an edible coating with carvacrol and citral in frozen strawberries and blueberries to control foodborne pathogens. *Foods, 13*, 3167. https://doi.org/10.3390/foods13193167

Pires, A. F., Rubio, O. D., Cobos, Á., & Pereira, C. (2024a). A review of recent developments in edible films and coatings-focus on whey-based materials. *Foods, 13*(16), 2638. https://doi.org/10.3390/foods13162638

Pires, A., Pietruszka, H., Bożek, A., Szkolnicka, K., Gomes, D., Díaz, O., Cobos, A., & Pereira, C. (2024b). Sheep's second cheese whey edible coatings with oregano and clary sage essential oils used as sustainable packaging material in cheese. *Foods, 13*, 674. https://doi.org/10.3390/foods13050674

Ponnusamy, P. G., Sharma, S., & Mani, S. (2022). Cotton noil based cellulose microfibers reinforced polylactic acid composite films for improved water vapor and ultraviolet light barrier properties. *Journal of Applied Polymer Science, 139*(24). https://doi.org/10.1002/app.52329

Prakash, G., & Mishra, R. K. (2023). Edible and bio-based food packaging: A review. *The Pharma Innovation, 12*(7S), 1497–1505. https://doi.org/10.22271/tpi.2023.v12.i7sr.21623

Prakoso, F. A. H., Indiarto, R., & Utama, G. L. (2023). Edible film casting techniques and materials and their utilization for meat-based product packaging. *Polymers, 15*, 2800. https://doi.org/10.3390/polym15132800

Pratiwi, D., Saputri, R., & Setiyawan, A. (2024). Whey edible film enhanced with ginger essential oil: Characterization and antimicrobial activity. *Iop Conference Series Earth and Environmental Science, 1377*(1), 012036. https://doi.org/10.1088/1755-1315/1377/1/012036

Pudake, R. N., Mohanta, T. K., & Mahato, N. (2024). Editorial: Opportunities and challenges for nanotechnology in sustainable agri-food production. *Frontiers in Nanotechnology, 6*. https://doi.org/10.3389/fnano.2024.1420192

Quandoh, E., & Albornoz, K. (2025). Fresh-cut watermelon: Postharvest physiology, technology, and opportunities for quality improvement. *Frontiers in Genetics, 16*, 1523240. https://doi.org/10.3389/fgene.2025.1523240

Ramazanidoroh, F., Hosseininezhad, M., Shahrampour, D., & Wu, X. (2024). Edible packaging as a functional carrier of prebiotics, probiotics, and postbiotics to boost food safety, quality, and shelf life. *Probiotics and Antimicrobial Proteins, 16*(4), 1327–1347. https://doi.org/10.1007/s12602-023-10110-5

Ramkumar, V., Kumar, S. M. S., Gopiraman, M., Rathinam, B., & Kim, S. (2024). A current trend in efficient biopolymer coatings for edible fruits to enhance shelf life. *Polymers, 16*(18), 2639. https://doi.org/10.3390/polym16182639

Ramos, E., Pacco-Huamani, M., Pizato, S., Pinedo, R., Cortez-Vega, W., & Delgado, G. (2025). Coating based on Montmorillonite, essential oils, and amaranth to preserve mango. *Polímeros, 35*. https://doi.org/10.1590/0104-1428.20240099

Real, A., Raquel, G., Medeiros, R., Pais, V., Bessa, J., Cardoso, S., Fernando, S., & Raul Fangueiro, C. (2023). Starch-based edible protective coating obtained from sweet potato waste. *MET Management Review - MMR, 10*(1). https://doi.org/10.34047/MMR.2020.10102

Rincón, E., Espinosa, E., Pinillos, M., & Serrano, L. (2023). Bioactive absorbent chitosan aerogels reinforced with bay tree pruning waste nanocellulose with antioxidant properties for burger meat preservation. *Polymers, 15*, 866. https://doi.org/10.3390/polym15040866

Rivera Leiva, A. F., Hernández-Fernández, J., & Ortega Toro, R. (2022). Active films based on starch and Wheat Gluten (Triticum vulgare) for shelf-life extension of carrots. *Polymers, 14*, 5077. https://doi.org/10.3390/polym14235077

Rizal, S., Abdul Khalil, H., Hamid, S. A., Ikramullah, I., Kurniawan, R., Hazwan, C. M., Muksin, U., Aprilia, S., & Alfatah, T. (2023). CoffeeWaste macro-particle enhancement in biopolymer materials for edible packaging. *Polymers, 15*, 365. https://doi.org/10.3390/polym15020365

Rodríguez-Parada, L., de la Rosa, S., Salado, J. S., Desmet, P., & Vicente, M. Á. P. (2025). Edible innovations: Testing the WOW impact of 3D printed chocolate packaging. *Food Quality and Preference, 123*, 105337. https://doi.org/10.1016/j.foodqual.2024.105337

Rout, S. S., & Pradhan, K. C. (2024). A review on antimicrobial nano-based edible packaging: Sustainable applications and emerging trends in food industry. *Food Control, 163*, 110470. https://doi.org/10.1016/j.foodcont.2024.110470

Roy, S., Kim, H. C., Panicker, P. S., Rhim, J.-W., & Kim, J. (2021). Cellulose nanofiber-based nanocomposite films reinforced with zinc oxide nanorods and grapefruit seed extract. *Nanomaterials, 11*, 877. https://doi.org/10.3390/nano11040877

Rudiyanto, W. N. S. H. W. I., & Ramisah, M. S. (2023). Effect of edible beeswax coating on tomato (Solanum lycopersicum) postharvest quality. *E3S Web of Conferences, 444*(04010). https://doi.org/10.1051/e3sconf/202344404010

Saeed, F., Javaid, A., Ahmed, N., Nadeem, M. T., Arshad, M. S., Imran, A., Sohaib, M., & Khan, A. U. (2017). Influence of edible coating techniques on quality characteristics of eggs. *Journal of Food Processing and Preservation, 41*, e12815. https://doi.org/10.1111/jfpp.12815

Salem, M. F., Abd-Elraoof, W. A., Tayel, A. A., et al. (2022). Antifungal application of biosynthesized selenium nanoparticles with pomegranate peels and nanochitosan as edible coatings for citrus green mold protection. *Journal of Nanobiotechnology, 20*, 182. https://doi.org/10.1186/s12951-022-01393-x

Salmas, C. E., Giannakas, A. E., Moschovas, D., Kollia, E., Georgopoulos, S., Gioti, C., Leontiou, A., Avgeropoulos, A., Kopsacheili, A., Avdylaj, L., et al. (2022). Kiwi fruits preservation using novel edible active coatings based on rich thymol halloysite nanostructures and chitosan/polyvinyl alcohol gels. *Gels, 8*, 823. https://doi.org/10.3390/gels8120823

Sani, I. K., Aminoleslami, L., Mirtalebi, S. S., Sani, M. A., Mansouri, E., Eghbaljoo, H., Jalil, A. T., Thanoon, R. D., Khodaei, S. M., Mohammadi, F., & Kazemzadeh, B. (2023). Cold plasma technology: Applications in improving edible films and food packaging. *Food Packaging and Shelf Life, 37*, 101087. https://doi.org/10.1016/j.fpsl.2023.101087

Shafiq, M., Anjum, S., Hano, C., Anjum, I., & Abbasi, B. H. (2020). An overview of the applications of nanomaterials and nanodevices in the food industry. *Foods, 9*(2), 148. https://doi.org/10.3390/foods9020148

Sharafi, H., Alirezalu, A., Liu, S. Q., Karami, A., & Moradi, M. (2024). Postbiotics-enriched flaxseed mucilage coating: A solution to improving postharvest quality and shelf life of strawberry. *International Journal of Biological Macromolecules, 265*(Pt 2), 131398. https://doi.org/10.1016/j.ijbiomac.2024.131398

Shiam, M. A. H., Alam, M. A., Biswas, M., Alam, M., Misha, M. H., Ahmed, S., & Hasan, M. K. (2024). A review on biodegradable films from banana peel. *Asian Food Science Journal, 23*(12), 33–46. https://doi.org/10.9734/afsj/2024/v23i12756

Shiekh, K. A., Ngiwngam, K., & Tongdeesoontorn, W. (2022). Polysaccharide-based active coatings incorporated with bioactive compounds for reducing postharvest losses of fresh fruits. *Coatings, 12*, 8. https://doi.org/10.3390/coatings12010008

Shinde, M. M., Malik, M., Kaur, K., Gahlawat, V. K., Kumar, N., Chiraang, P., & Upadhyay, A. (2024). Formulization and characterization of guar gum and almond gum based composite coating and their application for shelf-life extension of okra (Hibiscus esculentus). *International Journal of Biological Macromolecules, 262*(Pt 1), 129630. https://doi.org/10.1016/j.ijbiomac.2024.129630

Simović, N. (2023). Biodegradable biopolymers with controlled release of antimicrobial agents. *Biljni Lekar, 51*(4), 613–621. https://doi.org/10.5937/biljlek2304613s

Singh, A., Gumber, S., Sohel, A., Rani, R., Saini, S., Nayak, M. K., Patel, M., & Mazumder, K. (2025). Effect of conventional dipping and twin-nozzle electrostatic spray coating methods on the shelf-life quality parameters of apples (rich red): A comparative study. *Applied Food Research, 5*, 100829. https://doi.org/10.1016/j.afres.2025.100829

Singhi, H., Kumar, L., Sarkar, P., et al. (2023). Chitosan based antioxidant biofilm with waste Citrus limetta pomace extract and impregnated with halloysite nanotubes for food packaging. *Food Measure, 17*, 2741–2754. https://doi.org/10.1007/s11694-023-01825-8

Sinha, A., Pps, G., Jawandha, D., Kaur, P., & Grewal, S. (2021). Chitosan-enriched salicylic acid coatings preserves antioxidant properties and alleviates internal browning of pear fruit under cold storage and supermarket conditions. *Postharvest Biology and Technology, 182*, 111721. https://doi.org/10.1016/j.postharvbio.2021.111721

Soares, K. S., Souza, M. P., Silva-Filho, E. C., Barud, H. S., Ribeiro, C. A., Santos, D. D., Rocha, K. N. S., de Moura, J. F. P., Oliveira, R. L., & Bezerra, L. R. (2021). Effect of Edible Onion (Allium cepa L.) film on quality, sensory properties and shelf life of beef burger patties. *Molecules, 26*, 7202. https://doi.org/10.3390/molecules26237202

Song, D., Hoa, V. B., Kim, H. W., Khang, S. M., Cho, S., Ham, J., et al. (2021). Edible films on meat and meat products. *Coatings, 11*(11), 1344. https://doi.org/10.3390/coatings11111344

Suhag, R., Kumar, N., Petkoska, A. T., & Upadhyay, A. (2020). Film formation and deposition methods of edible coating on food products: A review. *Food Research International, 136*, 109582. https://doi.org/10.1016/j.foodres.2020.109582

Suhartini, I. S., Foliatini, S. R. S., Nurdiani, L. S., & Muhammad, F. K. (2023). Synthesis and characterization of nano chitosan-avocado seed starch as edible films. *Jurnal Kimia Riset, 8*(1).

Sun, Y., Tao, R., Yang, J., Duan, J., Xie, Q., Fan, B., & Wang, F. (2024). Natural active products in fruit postharvest preservation: A review. *Food Frontiers, 5*(5), 2043–2083. https://doi.org/10.1002/fft2.454

Šupová, M., Suchý, T., Chlup, H., Štípek, J., Žitný, R., Landfeld, A., Skočilas, J., Žaloudková, M., Rýglová, Š., Braun, M., Štancl, J., & Houška, M. (2023). The comprehensive evaluation of two collagen gels used for sausage casing extrusion purposes: The role of the structural and mechanical properties. *Journal of Food Engineering*. https://doi.org/10.1016/j.jfoodeng.2022.111387

Šuput, D., Pezo, L., Loncar, B., Popovic, S., Tepi'c Horecki, A., Danicic, T., Cvetkovic, D., Ranitovic, A., Hromiš, N., & Ugarkovic, J. (2023). The influence of biopolymer coating based on pumpkin oil cake activated with Mentha piperita essential oil on the quality and shelf-life of grape. *Coatings, 13*, 299. https://doi.org/10.3390/coatings13020299

Taban, A., Haghighi, T. M., Mousavi, S. S., & Sadeghi, H. (2024). Are edible coatings (with or without essential oil/extract) game changers for maintaining the postharvest quality of strawberries? A meta-analysis. *Postharvest Biology and Technology, 216*, 113082. https://doi.org/10.1016/j.postharvbio.2024.113082

Thakur, R., Shahi, N., Mangaraj, S., Lohani, U., & Chand, K. (2020). Effect of apple peel based edible coating material on physicochemical properties of button mushrooms (Agaricus bisporus) under ambient condition. *International Journal of Chemical Studies, 8*, 2362–2370. https://doi.org/10.22271/chemi.2020.v8.i1aj.8622

Trajkovska Petkoska, A., Daniloski, D., Kumar, N., Pratibha, & Broach, A. T. (2021). Active edible packaging: A sustainable way to deliver functional bioactive compounds and nutraceuticals. In S. S. Muthu (Ed.), *Sustainable packaging* (pp. 225–264). Springer Singapore. https://doi.org/10.1007/978-981-16-4609-6_9

Trajkovska Petkoska, A. et al. (2025) Bionanocomposites in food packaging and preservation. In: Sustainable Materials for Food Packaging and Preservation - Food Security and Sustainability. (Eds: T. Ghosh, R. Priyadarshi, S. Roy). Elsevier ISBN: 9780443135675

Treviño-Garza, M., García, S., Heredia, N., Alanís-Guzmán, M., & Arevalo, K. (2017). Layer-by-layer edible coatings based on mucilages, pullulan and chitosan and its effect on quality and preservation of fresh-cut pineapple (Ananas comosus). *Postharvest Biology and Technology, 128*, 63–75. https://doi.org/10.1016/j.postharvbio.2017.01.007

Tufan, E. G., Borazan, A. A., & Koçkar, Ö. M. (2021). A review on edible film and coating applications for fresh and dried fruits and vegetables. *Bilecik Şeyh Edebali Üniversitesi Fen Bilimleri Dergisi, 8*(2), 1073–1085. https://doi.org/10.35193/bseufbd.996827

Umaraw, P., Munekata, P., Verma, A., Barba, F., Singh, V. P., Kumar, P., & Lorenzo, J. M. (2020). Edible films/coating with tailored properties for active packaging of meat, fish and derived products. *Trends in Food Science & Technology, 98*. https://doi.org/10.1016/j.tifs.2020.01.032

Umeohia, U. E., & Olapade, A. A. (2024). Influence of tomato peel fiber and moringa leaf extract bioactive coatings on the quality, shelf life, and sensory properties of fresh tomatoes. *Food Chemistry: X*. https://doi.org/10.1016/j.fochx.2025.102396

Ungureanu, C., Tihan, G., Zgârian, R., & Pandelea, G. (2023). Bio-coatings for preservation of fresh fruits and vegetables. *Coatings, 13*, 1420. https://doi.org/10.3390/coatings13081420

Vaishnavi, P., & Shams, R. (2023). Techno-functional characteristics, and potential applications of edible coatings: A comprehensive review. *Journal of Agriculture and Food Research*. https://doi.org/10.1016/j.jafr.2023.100886

Valdés, A., Martínez, C., Garrigos, M. C., & Jimenez, A. (2021). Multilayer films based on Poly(lactic acid)/gelatin supplemented with cellulose nanocrystals and antioxidant extract from almond shell by-product and its application on hass avocado preservation. *Polymers, 13*, 3615. https://doi.org/10.3390/polym13213615

Vasiliauskaite, A., Mileriene, J., Songisepp, E., Rud, I., Muizniece-Brasava, S., Ciprovica, I., Axelsson, L., Lutter, L., Aleksandrovas, E., Tammsaar, E., et al. (2022). Application of edible coating based on liquid acid whey protein concentrate with indigenous Lactobacillus helveticus for acid-curd cheese quality improvement. *Foods, 11*, 3353. https://doi.org/10.3390/foods11213353

Vergel-Alfonso, A. A., Arias-Avelenda, R., Casariego-Año, A., Giménez, M. J., Ruíz-Cruz, S., López-Corona, B. E., Del-Toro-Sánchez, C. L., Gonzalez-Bravo, A. L., Plascencia-Jatomea, M., Menchaca-Armenta, M., Canizales-Rodríguez, D. F., Ramos-Enríquez, J. R., Tapia-Hernández, J. A., & Rodríguez-Félix, F. (2025). Development and characterization of pectin and beeswax-based coatings enhanced with anthocyanins and its antioxidant and antifungal properties. *Processes, 13*(2), 542. https://doi.org/10.3390/pr13020542

Viacava, G. E., Cenci, M. P., & Ansorena, M. R. (2022). Effect of chitosan edible coatings incorporated with free or microencapsulated thyme essential oil on quality characteristics of fresh-cut carrot slices. *Food and Bioprocess Technology, 15*, 768–784. https://doi.org/10.1007/s11947-022-02783-7

Vieira, T. M., Alves, V. D., & Moldão Martins, M. (2022). Application of an eco-friendly antifungal active package to extend the shelf life of fresh red raspberry (Rubus idaeus L. cv. 'Kweli'). *Foods, 11*, 1805. https://doi.org/10.3390/foods11121805

Vilaplana, R., Guerrero, K., Guevara, J., & Valencia-Chamorro, S. (2020). Chitosan coatings to control soft mold on fresh blackberries (Rubus glaucus Benth.) during postharvest period. *Scientia Horticulturae, 262*, 109049. https://doi.org/10.1016/j.scienta.2019.109049

Wen, Q., Wang, X., Liu, B., Lu, L., Zhang, X., Swing, C. J., & Xia, S. (2022). Effect of synergism between sodium alginate and xanthan gum on characteristics of composite film and gloss of areca nut coating. *Food Bioscience*. https://doi.org/10.1016/j.fbio.2022.102113

Weng, S., Marcet, I., Rendueles, M., & Diaz, M. (2025). Edible films from the laboratory to industry: A review of the diferent production methods. *Food and Bioprocess Technology*. https://doi.org/10.1007/s11947-024-03641-4

Westlake, J. R., Tran, M. W., Jiang, Y., Zhang, X., Burrows, A. D., & Xie, M. (2023). Biodegradable biopolymers for active packaging: Demand, development and directions. *Sustainable Food Technology, 1*(1), 50–72. https://doi.org/10.1039/d2fb00004k

WHO. Diet, nutrition and the prevention of chronic diseases: Report of a joint WHO/FAO expert consultation. Retrieved 9 April 2025., from https://www.who.int/publications/i/item/924120916X

Wigati, L., Wardana, A., Jothi, J., Leonard, S., Van, T., Xirui, Y., Tanaka, F., & Tanaka, F. (2023a). Preserving mandarin quality during ambient storage using edible coatings of pregelatinized corn starch Pickering emulsions and essential oil. *Food Bioscience, 53*, 102710. https://doi.org/10.1016/j.fbio.2023.102710

Wigati, L., Wardana, A., Tanaka, F., & Tanaka, F. (2023b). Application of pregelatinized corn starch and basil essential oil edible coating with cellulose nanofiber as Pickering emulsion agent to prevent quality-quantity loss of mandarin orange. *Food Packaging and Shelf Life, 35*, 101010. https://doi.org/10.1016/j.fpsl.2022.101010

Worldometers. (n.d.). World population: Past, present, and future. Retrieved 10 April 2025, from https://www.worldometers.info/world-population/

Xie, C., Wang, F., He, Z., Tang, H., Li, H., Hou, J., Liu, Y., & Jiang, L. (2023). Development and characterization of active packaging based on chitosan/chitin nanofibers incorporated with scallion flower extract and its preservation in fresh-cut bananas. *International Journal of Biological Macromolecules, 242*(Pt 3), 125045. https://doi.org/10.1016/j.ijbiomac.2023.125045

Xin, Y., Yang, C., Zhang, J., & Xiong, L. (2023). Application of whey protein-based emulsion coating treatment in fresh-cut apple preservation. *Foods, 12*, 1140. https://doi.org/10.3390/foods12061140

Yadav, A., Kumar, N., Upadhyay, A., Singh, A., Kumar Anurag, R., & Pandiselvam, R. (2022a). Effect of mango kernel seed starch-based active edible coating functionalized with lemongrass essential oil on the shelf-life of guava fruit. *Quality Assurance and Safety of Crops & Foods, 14*(3), 103–115. https://doi.org/10.15586/qas.v14i3.1094

Yadav, A., Kumar, N., Upadhyay, A., Sethi, S., & Singh, A. (2022b). Edible coating as postharvest management strategy for shelf-life extension of fresh tomato (Solanum lycopersicum L.): An overview. *Food Science, 87*(6), 2256–2290. https://doi.org/10.1111/1750-3841.16145

Yazdi, J. S., Salari, M., Ehrampoush, M. H., & Bakouei, M. (2024). Development of active chitosan film containing bacterial cellulose nanofibers and silver nanoparticles for bread packaging. *Food Science & Nutrition, 12*, 8186–8199. https://doi.org/10.1002/fsn3.4424

Yousuf, B., Wu, S., & Wasim Siddiqui, M. (2021). Incorporating essential oils or compounds derived thereof into edible coatings: Effect on quality and shelf life of fresh/fresh-cut produce. *Trends in Food Science & Technology, 108*, 245–257. https://doi.org/10.1016/j.tifs.2021.01.016

Yun, D., & Liu, J. (2024). Preparation, characterization and application of active food packaging films based on sodium alginate and twelve varieties of mandarin peel powder. *Foods, 13*, 1174. https://doi.org/10.3390/foods13081174

Zhao, R., Guan, W., Zheng, P., Tian, F., Zhang, Z., Sun, Z., & Cai, L. (2021a). Development of edible composite film based on chitosan nanoparticles and their application in packaging of fresh red sea bream fillets. *Food Control, 132*, 108545. https://doi.org/10.1016/j.foodcont.2021.108545

Zhao, Y., Li, B., Li, C., Xu, Y., Luo, Y., Liang, D., & Huang, C. (2021b). Comprehensive review of polysaccharide-based materials in edible packaging: A sustainable approach. *Foods, 10*(8), 1845. https://doi.org/10.3390/foods10081845

Zhao, Y., Li, B., Zhang, W., Zhang, L., Zhao, H., Wang, S., & Huang, C. (2023). Recent advances in sustainable antimicrobial food packaging: Insights into release mechanisms, design strategies, and applications in the food industry. *Journal of Agricultural and Food Chemistry, 71*(31), 11806–11833. https://doi.org/10.1021/acs.jafc.3c02608

Chapter 2
Polysaccharides: Biopolymers in Edible Coatings and Films

Jyoti and Atul K. Gupta

Abstract Polysaccharides, a heterogeneous group of biopolymers, hold the key to the creation of sustainable and active functional materials for food packaging. This chapter is directed toward understanding the intrinsic properties of polysaccharides, emphasizing their structural variability and functionality. Among them, starch, cellulose, chitosan (CS), gums, alginate (Alg), and pectin are prominent because they are biodegradable, form films, and have the ability to extend the shelf life of food. Starch, which is widely available and relatively low-cost, is discussed for its film-forming potential, with cellulose and its derivatives offering mechanical strength and barrier characteristics. Chitosan, by its antimicrobial and antioxidant properties, is a highly useful material in active packaging. Natural gums, alginate, and pectin add to packaging by enhancing film flexibility, water barrier properties, and bioactivity. Each polysaccharide is investigated based on its application in active packaging, considering its interaction with food products, moisture transfer control ability, and the possibility of the inclusion of bioactive compounds. The chapter also highlights recent developments and uses of polysaccharide-based materials in food packaging, overcoming challenges like mechanical restraint and moisture sensitivity. In general, polysaccharides present a promising and sustainable alternative to traditional packaging materials in terms of food safety, quality preservation, and environmental sustainability.

Keywords Edible packaging · Biopolymers · Polysaccharide · Chitosan · Starch · Cellulose · Barrier properties · Food applications

Jyoti (✉) · A. K. Gupta
Central Instrument Facility, Sharda University, Greater Noida, Uttar Pradesh, India
e-mail: jyoti.10@sharda.ac.in

N. Kumar et al. (eds.), *Trends in Edible Packaging: A sustainable approaches for postharvest management of fruits and vegetables*,
https://doi.org/10.1007/978-3-032-13199-7_2

Introduction

The availability, renewable nature, biocompatibility, and environmental friendliness of biopolymers have made them a viable and sustainable alternative to plastic packaging materials (Jabeen et al., 2015; Santagata et al., 2018). Developing biopolymer packaging from food-grade ingredients presents a challenge to controlling the change of tastes, scents, oxygen, carbon dioxide, and moisture among the food's component elements and their surroundings. It is also used to enhance the microbial burden, shelf life, and quality of food (Janjarasskul & Krochta, 2010). Films made of biopolymers can inherently control the mass transfer between the environment and food components. Food quality is enhanced, and shelf life is prolonged as a result. The ability to function as a water and gas barrier; exceptional mechanical strength and adhesion; notable physicochemical and microbiological stability; effective antioxidant, antimicrobial, and color carriers; nutritional additives; human consumption safety; acceptable sensory qualities; affordable raw materials; and straightforward production technology are just a few of the specific requirements that biopolymer films should meet, in addition to the characteristics mentioned earlier (Deshmukh & Gaikwad, 2022). The building pieces for creating biopolymer films are basic materials (lipids, proteins, and polysaccharides), plasticizers, additives, and solvents. Although many natural biopolymers have been investigated thus far, most of these studies have focused on proteins and polysaccharides. Figure 2.1 shows the classification of bio-based polymers. Biopolymers based on polysaccharides and proteins have been demonstrated to have poor water barrier function due to their hydrophilic nature despite being efficient oxygen barriers at low-to-intermediate relative humidity (Jabeen et al., 2015; Bhargava et al., 2020). Table 2.1 depicts the functional properties of biopolymer polysaccharides in food packaging applications.

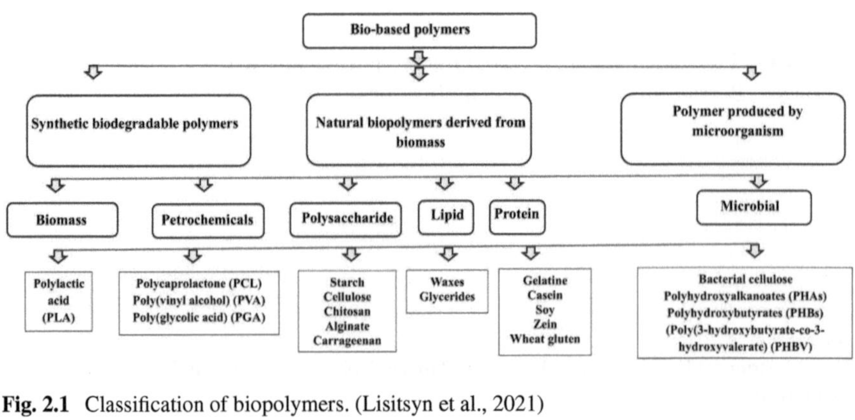

Fig. 2.1 Classification of biopolymers. (Lisitsyn et al., 2021)

Table 2.1 Functional properties of polysaccharide biopolymers in food packaging

Biopolymer	Structure	Positive characteristics	Negative characteristics	References
Starch	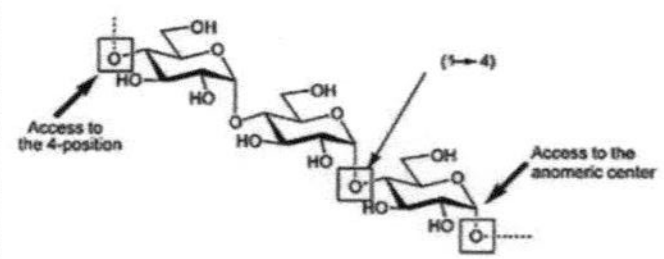	Nontoxic, renewable, biodegradable, inexpensive, plentiful, transparent, good lipids, oxygen, UV barrier qualities, colorless, flavorless, and tasteless Excellent ability to create films and low permeability to water vapor	Low mechanical qualities, poor water resistance, and limited processability Low heat characteristics, hydrophilic, and brittleness	Molavi et al. (2015)
Cellulose	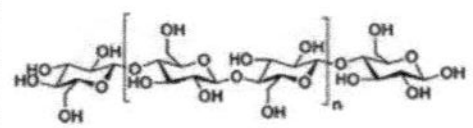	Nontoxic, renewable, and biodegradable Little use of energy Excellent oxygen and hydrocarbon barrier qualities, large surface area, strong mechanical properties, high permeability to water vapor, inexpensive, high specificity, low density, biocompatibility, tasteless, odorless, and stability of chemicals	Opacity, increased color value, hydrophilia, low mechanical properties, and poor water vapor barrier qualities	Xu et al., 2016; Wang et al. (2016)
Chitin and chitosan (CS)	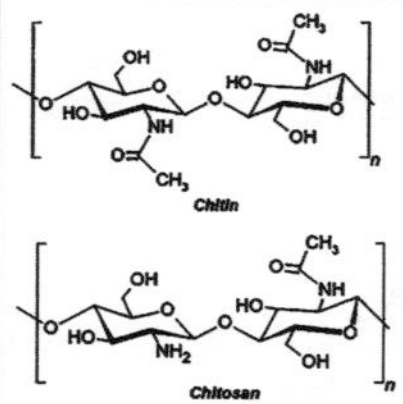	Nontoxic, renewable, and biodegradable Improved absorption characteristics High levels of antibacterial activity Excellent biocompatibility Low cost of manufacture Good gas, fragrance, UV, water-insoluble, wettability, oil barrier, and antioxidant qualities Strong capacity to make films, favorable optical characteristics Flexible and transparent	Low water vapor barrier qualities, high hydrophobicity, and low mechanical qualities Brittleness Insufficient elasticity	Van Den Broek et al. (2015); Lekjing (2016)
Xanthan gum	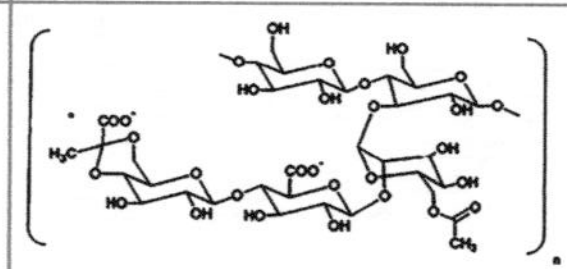	Nontoxic, renewable, and biodegradable Low cytotoxicity, biocompatibility, and controlled viscosity	Low mechanical, rheological, and barrier properties, as well as high production costs	Sharma and Rao (2015)

(continued)

Table 2.1 (continued)

Biopolymer	Structure	Positive characteristics	Negative characteristics	References
Gellan gum	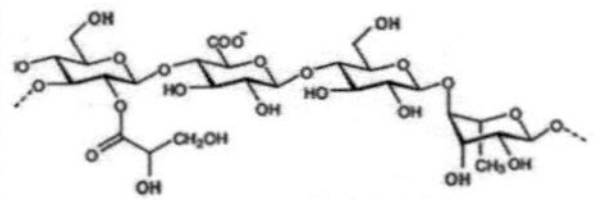	Nontoxic, renewable, and biodegradable Low cytotoxicity, biocompatibility, and controlled viscosity	Low mechanical, rheological, and barrier properties, as well as high production costs	Wu et al. (2021)
Pullulan	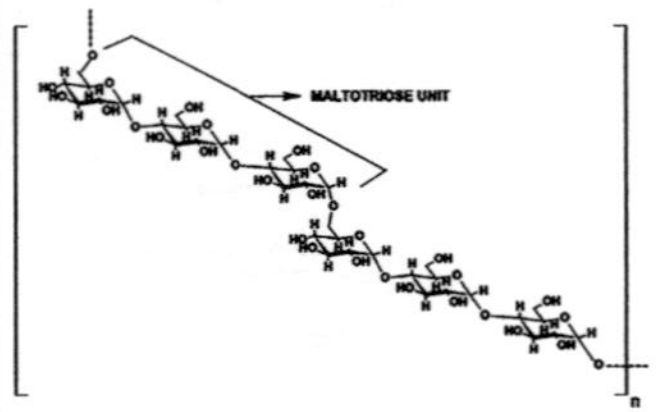	Nontoxic, renewable, and biodegradable Tasteless, colorless, odorless, flexible, transparent, and heat-resistant Excellent oil with oxygen barrier qualities and biocompatibility Water-permeable and heat-sealable	Brittleness, low water resistance, low mechanical qualities, and moisture sensitivity	Falsafi et al. (2023)
Alginate (Alg)	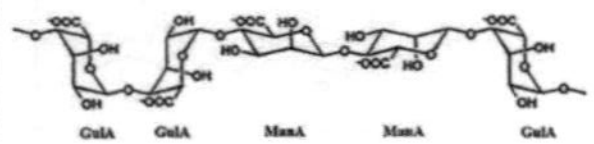	Nontoxic, renewable, and biodegradable Regulates the qualities of swelling Good mechanical qualities, chemical stability, low cost, and biocompatibility Strong water-repellent qualities, favorable mechanical characteristics, and rigidity Reduces fat oxidation and preserves taste	Brittleness, inadequate water resistance, inadequate moisture barriers, high permeability to water vapor, and elevated hydrophilicity	Benavides et al. (2012)
Carrageenan	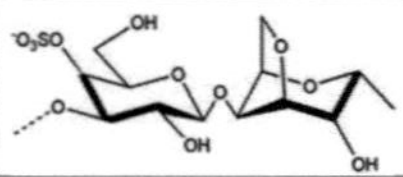	Nontoxic, renewable, and biodegradable, Excellent moisture and gas barrier qualities, Stability of antimicrobial qualities		S. Falsafi et al. (2023); Fani et al. (2022)
Pectin		Good oil, scent, gas barrier, mechanical qualities, good rheological qualities, sustainability, non-toxicity, affordability, and good film-forming ability	Ineffective at preventing the transfer of moisture Inadequate mechanical characteristics	S. Falsafi et al. (2023)

Polysaccharides

The earliest and most researched biopolymers for food packaging are polysaccharides. Polysaccharides from a variety of sources are effective building blocks for bio-based films, according to numerous studies (Gao et al., 2017). They are natural, biodegradable, renewable, nontoxic, and easily available. Numerous polysaccharides and their derivatives, such as chitosan, cellulose, starch, pectin, alginate, and carrageenan, can be used to create biodegradable or edible films (Santagata et al., 2018). Biopolymers known as polysaccharides are composed of saccharide units connected by glycosidic linkages. The hydrophilic nature of the polysaccharides, which allows them to form films and connect with other macromolecules, is attributed to their hydroxyl groups (Idumah et al., 2019). Polysaccharides are a possible substitute for synthetic polymers due to their accessibility, affordability, average mechanical properties, and barrier features. This is especially true for food packaging materials that are bioactive and biodegradable for a brief period of time. Since polysaccharides are made from the biomass of plants or marine life, they are the raw material most often used. Structural changes could enhance polysaccharide natural qualities, depending on the needs (Tajeddin & Arabkhedri, 2020).

Starch

One of the most plentiful sources of regenerative feedstock is starch, a fully natural, biodegradable polysaccharide produced by a variety of plants, like corn, wheat, potatoes, rice, and peas. Additionally, it is a major energy source for the kingdom of plants (Samsudin & Hani, 2019). Starch is considered the most promising biologically derived polymer because of its high biodegradability, wide availability, and affordability. Its easy accessibility and yearly renewal make it the perfect basis material for broad-spectrum applications. The two polysaccharides that make up natural starch are branching amylopectin, which has (1-6)-D-glucopyranosyl side units, and linear amylose, which has (1-4)-D-glucopyranosyl units. Figure 2.2 represents the structure of starch (Acosta et al., 2015). The ratio of amylose to amylopectin is crucial when making starch films. High amylose content in starch produces strong, flexible films, as opposed to high amylopectin content, which takes longer to crystallize due to its high molecular weight and hyperbranched structure (Bangar et al., 2021). Starch does not melt like conventional plastics, degrades at a tWtemperature lower than the melting point, and is insoluble in cold water. Granules of starch undergo irreversible loss of their semicrystalline characteristic and transform into a continuous matrix when they are heated, pressed, or subjected to plasticizers. Films made of starch are transparent, tasteless, and odorless (Chiumarelli & Hubinger, 2012). Furthermore, starch's wide range of applications and versatility, relative affordability, and exceptional ability to create films with exceptional barrier qualities to O_2 and CO_2—as opposed to its poor barrier capabilities to water because

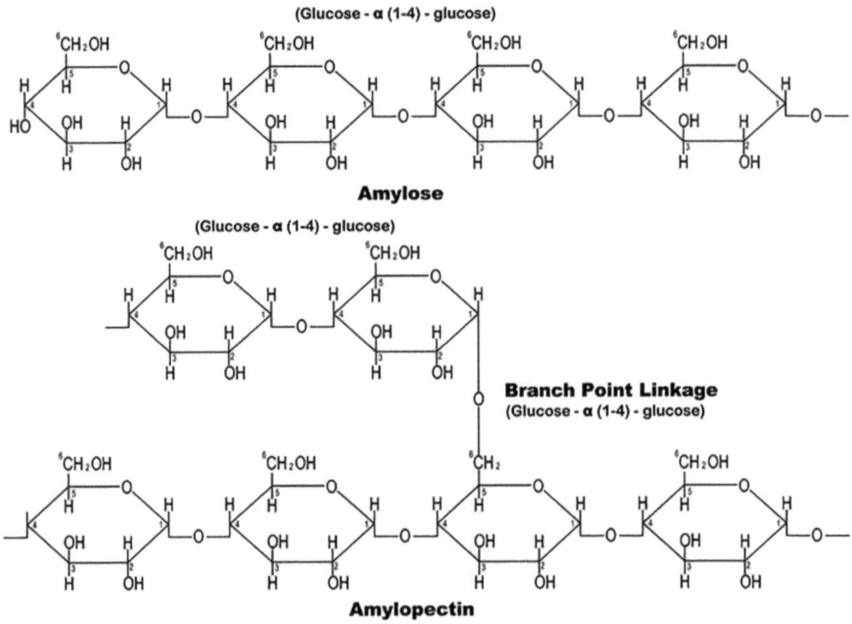

Fig. 2.2 Structure of amylose and amylopectin

of its high hydrophilicity—make it a valuable ingredient for food preservation (Acosta et al., 2015; Idumah et al., 2019). Also, a starch-based film demonstrated that sorbitol was essential for retaining mechanical firmness and limiting water loss in starch-based formulations (Nawab et al., 2017). Fruits and vegetables are examples of perishable goods that are commonly the targets of microbial attacks during storage, mostly due to pathogenic microorganisms. By incorporating antimicrobial substances like organic acids, various plant extracts, and essential oils (EOs) into the starch-based coating, a natural defense against bacterial flora is produced. This also extends the postharvest shelf life of fruits and vegetables (Sapper & Chiralt, 2018). Table 2.2 shows the applications of starch in food packaging. The addition of bioactive essential oils, including carvacrol, thyme, cinnamon, eugenol, and lemon essential oils, increases the antibacterial activity and decreases the starch's hygroscopic inclination (Marín et al., 2017). Nonetheless, meat products, batters, and confections commonly use starch films (Galus et al., 2020).

Cellulose

Cellulose, the most common biopolymer found in nature, is a resource that is both renewable and biodegradable and is essential for both plants and microorganisms (Pandey et al., 2012). Cellobiose, which is composed of two anhydrous glucose

Table 2.2 Applications of starch in food packaging

Packaging material	Bioactive compound	Method	Inference	References
Potato starch (PS)/gelatin	Resveratrol (RES)	Thermal extrusion, followed by compression molding	Enhanced tensile strength and thickness Improved water and oxygen barrier properties High roughness observed in Scanning Electron Microscope (SEM) Improved antioxidant activity Improved anti-permeability for oil RES film significantly reduced oxidative rancidity stored in soybean oil (SO) by 20% Suitable for the packaging of oil-based food products prone to oxidation	Wu et al. (2023)
Starch (corn)	Citric acid, tartaric acid, oxalic acid	Cross-linking	Citric acid showed maximum cross-linking efficiency by having additional carboxyl groups, which improved water resistance, tensile strength, and stability of dimensions Tartaric acid was superior to oxalic acid because of extra hydroxyl groups. Carbonyl groups were more critical in cross-linking efficiency	Gebresas et al. (2023)
Starch-based blend film	Starch-based graphite oxide carbon (S-GOCs), nano-$CaCO_3$, acetyl tributyl citrate (ATC)	Hydrophobic modification	Enhanced hydrophobicity with increased S-GOCs Decreased moisture content and water vapor permeability (WVP) and enhanced flexibility Nano-$CaCO_3$ and ATC enhance water-proof properties Shows potential for water-resistant packaging due to enhanced hydrophobicity and mechanical stability	Geng et al. (2023)
Yam starch (YS)	Gallic acid Chinese yam starch Chitosan (GA/YS/CS)	Gallic acid/Chinese yam starch/Chitosan (GA/YS/CS) composite film	CS lowered viscosity, resulting in thinner films with improved light transmittance Because of hydrogen-bonding interactions, GA and CS enhanced the film's tensile strength and film compatibility GA/YS/CS films improved pork storage quality compared to polyethylene films. The films demonstrated antibacterial activity, oxidation resistance, and mechanical strength and can be used in food preservation, as well as curbing environmental pollution	Rong et al. (2023)

(continued)

Table 2.2 (continued)

Packaging material	Bioactive compound	Method	Inference	References
Cornstarch	Graphene, polydopamine, copper ions	Mussel-inspired polydopamine modification of graphene	Enhanced stretching of graphene within the starch matrix led to an improvement in the film's performance Modification on the surface of graphene was performed successfully Nanocomposite films exhibited improved strength, barrier properties, and thermal stability Films showed antibacterial activities against pathogenic microorganisms for food Added functional properties improved the active properties of the film	Xu et al. (2023)
Konjac glucomannan (KGM)/chitosan (CS) antibacterial film	Microcapsules	Solvent casting	Overall antibacterial properties improved with increased microcapsule content, with the greatest effect observed at 10% The 10% microcapsule film froze pork the longest and was the best at stopping flavor degradation Pork packed with 10% microcapsule film showed maximal inhibition of flavor deterioration	Ma et al. (2023)
Cornstarch	Oxidized debranched starch/chitosan nanoparticles (SC-NPs)	Casting method Schiff base cross-linking	SC-NPs enhanced the structural integrity of the film Increased water contact angle, significantly improving hydrophobicity Improved UV and visible light barrier properties Decreased water permeability, solubility, and swelling ratio, which increases the durability of the film	Q. Liu et al. (2023)
Sweet potato starch–based film	Sugarcane cellulose, cornstalk cellulose, wheat/bamboo cellulose	Extrusion, casting method	Sugarcane cellulose showed the strongest hydrogen bond interaction, tensile strength, and best compatibility Cornstalk cellulose showed the highest thermal degradation Wheat/bamboo cellulose showed better UV resistance Bamboo and cornstalk cellulose showed enhanced moisture barrier properties	Zhang et al. (2023)

High-amylose cornstarch	Quinoa cellulose nanocrystals (QCNCs)	Casting	The starch film exhibited enhanced mechanical strength and improved water resistance High crystallinity, with good water dispersibility QCNCs ensure durability in film applications	Ren et al. (2023)
Banana peel starch	Nanocellulose	Film casting	Banana peel starch has high amylopectin content and thermal stability Enhanced antioxidant activity Nanocellulose enhanced the tensile strength and reduced the permeability Films blocked 98% UV rays and extended bread shelf life by 10 days	Chandrasekar et al. (2023)
Cornstarch	Arabica coffee oil (ACO) Robusta coffee oil (RCO)	Solvent casting method	Green coffee oil acts as a biologically active emulsifier Coffee oil improves the thickness of the film, UV shielding, and transparency of the film Arabica coffee oil improves water barrier properties Arabica coffee oil shows inhibition against *S. aureus* and acts as a natural surfactant	Wang et al. (2022)
Mango kernel starch (MKS) film	Chemical modification of mango kernel starch	Oxidation and benzylation	Surmounted restrictions of native MKS like low solubility, retrogradation, and difficulty in gelatinization Preserved A-type crystal structure akin to that of endosperm starch Integration of functional groups was confirmed with the use of oxidation and benzylation The lightness (L*) of the modified starch films was found to be higher, thus making them lighter in color Tensile strength increased more than twofold in comparison to native starch films Oxidation slowed down recrystallization, thus inhibiting retrogradation Water vapor permeability was reduced by 58%, and UV absorption was increased by 80% due to benzylation	Vellaisamy Singaram et al. (2021)

(continued)

Table 2.2 (continued)

Packaging material	Bioactive compound	Method	Inference	References
Cornstarch	–	Solvent casting	Tackled the physical fragility of starch films without chemical alteration or use of plasticizers Films formulated at pH 11 had high flexibility with no physical defects Films prepared under strong acidic/alkaline conditions were far more hydrophobic Alkaline pH films had their water sorption reduced by almost 50% Their improved elongation and flexibility allow for food packaging applications This method serves multiple bioproducts in an economically priced biodegradable manner	Sakkara et al. (2020)
Corn/octenyl succinate starch (C/OS) films	Soybean oil (SO)	Solvent casting	Films with SO exhibited molecular interactions through hydrogen bonding Low SO concentrations increased starch intermolecular bonding More irregular and coarser surface structures were produced by increasing SO Increased SO content enhanced water repellency (max contact angle: 76.14°) C/OS-1.0% SO exhibited maximum tensile strength (6.54 MPa); C/OS-1.5% SO exhibited maximum elongation (71.84%) Increased hydrophobicity and mechanical hardness make these films applicable in food packaging	Gao et al. (2020)
Potato starch	Phenolic extract and cellulose fibers from sunflower hulls	Thermoplastic processing	Enhanced antioxidant activity Citric acid improves the integrity of the film through hydrolysis and cross-linking of starch polymers Debranching of isoamylase confirmed structural changes in starch polymer	Menzel (2020)

Cassava starch (CaS)	Grape stalks (GSs)	Thermal processing	SEM images did not show the agglomerations and cell structures that are frequently observed in films or foams created by thermal expansion The structures that developed were suitable for low-moisture food packaging The mechanical characteristics of foams made using petroleum-based polymers (expanded polystyrene (EPS)) were similar for short-term or single-use applications CaS films and films based on CaS–GS demonstrated rapid biodegradability	Engel et al. (2019)
Cassava starch	Yerba mate extract (YME)	Solvent casting	Higher YME concentration reduced tensile strength but raised Young's modulus and elongation Enhanced antioxidant activity because of YME availability More moisture absorption was recorded in YME-containing films (5–15%) than in the control	Knapp et al. (2019)
Rye starch	Rosehip extract (RHE)	Solvent casting	Increased rosehip extract concentration reduces tensile strength Improves film flexibility by enhancing elongation at break Films block UV light entirely by reducing light transmittance The highest antioxidant activity is shown by 1% of the rosehip extract films Chicken breast packaged with 1% RHE film inhibits lipid peroxidation	Go and Song (2019)
Eggplant flour and cornstarch flour		Solvent casting	Films with higher cornstarch content enhanced the mechanical strength Substitution of cornstarch with eggplant flour increased solubility, water vapor permeability (WVP), moisture content, and swelling index Eggplant flour incorporation enhanced film biodegradation compared to the pure starch films	Mahfoudhi and Hamdi (2015)

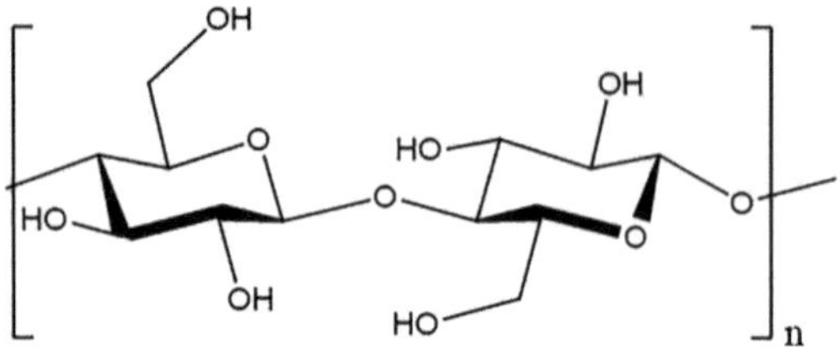

Fig. 2.3 Structure of cellulose

rings connected by a β-1,4 glycosidic link, repeats to produce this linear polymer. The structure of cellulose is displayed in Fig. 2.3 (Siqueira et al., 2010). Cellulose and its derivatives are edible and biodegradable and can be added to food to increase its nutritional content. It plays a significant role in determining the color, appearance, flavor, smell, and taste of cellulose, as well as its overall organoleptic properties. It helps reduce waste and synthetic material packaging (Pooresmaeil et al., 2019).

The hydrophilic nature, water-blocking qualities, low film-forming capacity, and high crystallinity of cellulose make it unsuitable for use in its natural condition. Since its derivatives are biodegradable, flavorless, and odorless, they are mostly used to make biologically edible films (Erginkaya et al., 2014). Methylcellulose (MC), carboxymethyl cellulose (CMC), and hydroxypropyl methylcellulose (HPMC) are the three cellulose derivatives that are most frequently utilized (von Schantz et al., 2014). The common sources of cellulose and its derivatives are shown in Fig. 2.4, along with how they are used as a food packaging material to preserve a range of foods.

Reports state that heat gelatinization and a water-soluble polymer give CMC its remarkable film-forming capabilities (Almasi et al., 2010). Since the turn of the twentieth century, food packaging has made use of cellulose derivatives. The barrier qualities of polymers can be mechanically strengthened by cellulose and its derivatives. Furthermore, cellulose derivatives are more resilient to enzymatic cleavage and microbiological attacks than native cellulose. However, due to their hydrophilic nature, films made from cellulose offer inadequate protection against water vapor (Tabari, 2017).

Cellulose in Food Packaging

The benefits of using cellulose-based materials in food packaging include biodegradability, sustainability, and effective moisture and oxygen barriers. Additionally, cellulose fibers can improve packaging materials' mechanical strength and moisture-repellent qualities. They are in great demand in the international market and are readily accessible at affordable costs. However, some challenges, such as poor interfacial adhesion and high hydrophilicity, need to be surmounted. The hydrophilicity

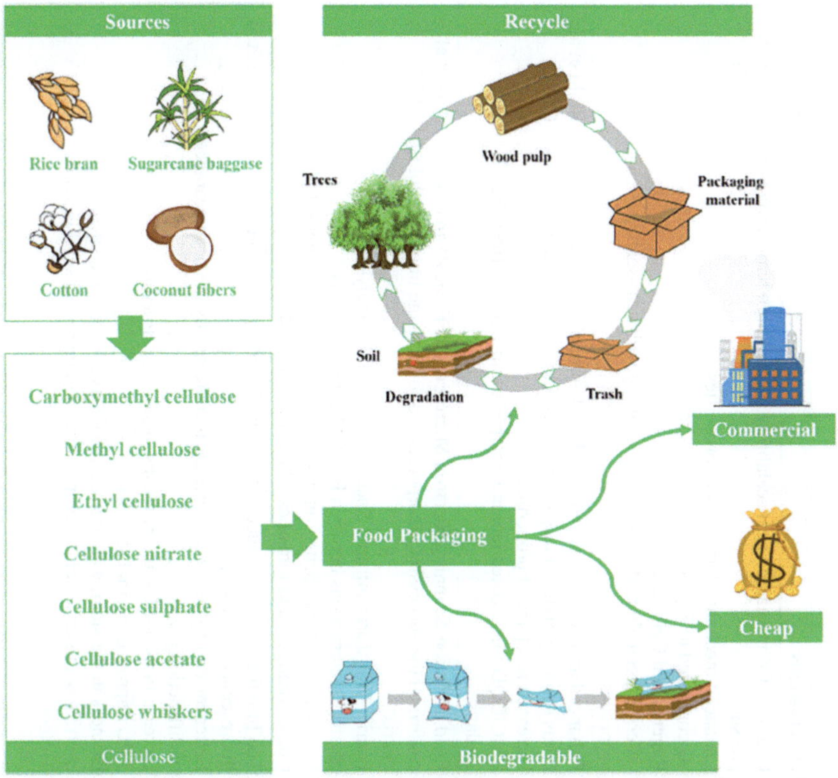

Fig. 2.4 Common cellulose derivative sources and their applications in food packaging. (Liu et al., 2021)

of cellulose is influenced by the polar nanocellulose dispersion problems caused by the many hydroxyl groups on the surface of the nanofibers in nonpolar polymers or nonaqueous liquids. Additionally, most polymers are incompatible with cellulose fibers. These issues can be fixed by modifying cellulose fibers to improve their qualities for use in food packaging (Liu et al., 2021). Table 2.3 shows the applications of cellulose in food packaging.

Poor interfacial adhesion and susceptibility to both gaseous and liquid water are two of the issues with cellulose and its derivatives. As a result, cellulose fibers can be chemically or physically modified to increase their heat stability, decrease their hydrophilicity, and improve their compatibility. This improves the materials' mechanical strength, dispersion in polymer matrices, and barrier qualities. These changes also aid in the creation of biodegradable and sustainable materials, which lessens the amount of plastic trash that ends up in the environment (Liu et al., 2021).

Table 2.3 Applications of cellulose derivatives in food packaging

Biopolymer	Additional material	Method used	Observations	References
Cellulose	Zein	–	Coating of zein enabled heat sealability without high temperature or pressure Enhanced mechanical and barrier properties As compared to Ziploc®, blueberries remained fresh for 12 days in Regenerated Cellulose (RC)/zein bags Offers a green alternative to food packaging with heat-sealing properties	Chu et al. (2023)
Carboxymethyl cellulose (CMC)	Chitosan-based carbon quantum dots (CQDs)		Enhanced antioxidant, antimicrobial, and antifungal activities Films showing highly transparent, UV-blocking functional properties Increased mechanical properties CMC/CQD film–coated lemon remained mold-free and fresh for 21 days	Riahi et al. (2022)
Bacterial cellulose (BC)–based films	Bacterial cellulose (BC), plant oils (POs) (olive oil and ginger oil), glycerol (Gly)	Olive oil and ginger oil	Field Emission-Scanning Electron Microscope (FE-SEM) analysis confirmed the successful incorporation of oil into the fibrous cellulose matrix Fourier Transform Infrared Spectroscopy (FTIR) analysis shows chemical interactions between BC, CMC, Gly, and oils Olive oil film shows better antimicrobial activity than ginger oil films Coated oranges' and tomatoes' shelf life increased to up to 9 weeks	Atta et al. (2022)
Octenyl succinate cassava starch (OSCS) ester–based films	Chitosan (CS), nano–zinc oxide (ZnO), ε-polylysine (ε-PL)	Solvent casting	The film formation was successfully confirmed by FTIR and fluorescence microscopy Better miscibility between materials is indicated by the disappearance of diffraction peaks Increase in the ε-PL content results in a denser structure ε-PL improved thermal stability Dryness in the stem, weight loss, and decay reduced in coated cherries with surface color maintenance and increased soluble solids content	Zhang et al. (2022)

Cellulose natural fibers	Polylactic acid (PLA)	Solvent casting	Natural fiber is optimally incorporated into a PLA matrix Mechanical and water absorption properties	Vanitha and Kavitha (2021)
Bacterial cellulose (BC)–based edible films	Yeast (*Meyerozyma guilliermondii*), carboxymethyl cellulose (CMC), glycerol (Gly)		FE-SEM showed that Gly and yeast were successfully incorporated into the BC matrix Chemical interaction confirmed by FTIR spectroscopy between BC, CMC, Gly, and yeast Enhanced antimicrobial activity Shelf life of oranges and tomatoes was enhanced by up to 2 weeks at storage temperatures of **6 °C, 30 °C, and 40 °C**	Atta et al. (2021)
Nanofibrilated cellulose (NFC)		High-pressure homogenization	NFC improved the rheological properties of paper coatings Enhanced barrier property, with an increase in NFC concentration Improved mechanical properties of coated paper	Jin et al. (2021)
Cellulose nanocrystals (CNCs) extracted from pea hull and carboxymethyl cellulose	–	–	CNC shows a needle-like structure having high crystallinity Enhanced mechanical, barrier, and thermal properties Red chilies' vitamin C content was preserved while their weight loss was decreased	
Hydroxyethyl cellulose–gelatin electrospun fibers and pressure-sensitive adhesive (PSA)	Allyl isothiocyanate (AIC)	Electrospinning	Gelatin fibers loaded with AIC were successfully fabricated using electrospinning Two percent AIC composite shows significant antimicrobial properties Ten percent of AIC on headspace extended the shelf life of cheese from 4 to 8 weeks	Al-Moghazy et al. (2020)

(continued)

Table 2.3 (continued)

Biopolymer	Additional material	Method used	Observations	References
Cellulose	Synthetic polymers (poly(ε-caprolactone), poly(butylene succinate), polylactic acid, poly(hydroxyl alkanoate), polyamide 6)	Solution–precipitation process with hot-pressing using ionic liquids	Transparent films were produced with homogeneous morphology	Müller and Zollfrank (2020)
Cellulose nanocrystal	Silver (Ag)/alginate	Hydrothermal process	Barrier properties enhanced by CNC/Ag/Alg films Tensile strength of the formulated films was enhanced as compared to the control Alg films Enhanced moisture resistance by decreasing water vapor permeability Enhanced mechanical and barrier properties	Yadav et al. (2019b)
Sodium Carboxymethyl cellulose	Sodium alginate (SA)/chitosan, glycerol, calcium chloride ($CaCl_2$)	Solvent casting	Film shows improved mechanical and barrier properties Enhanced antimicrobial properties	Lan et al. (2018)

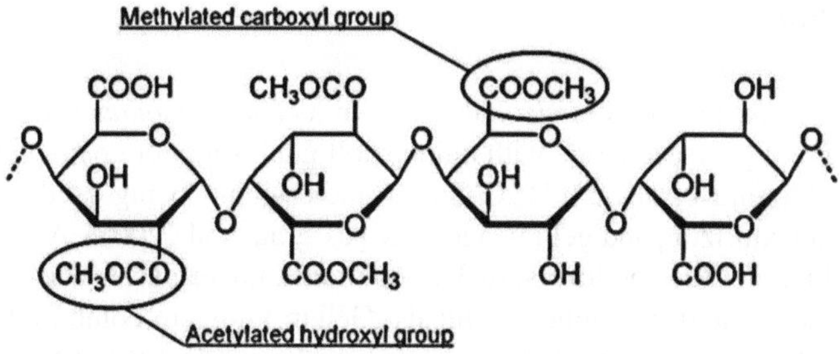

Fig. 2.5 Structure of pectin

Pectin

Among other fruits and vegetables, pectin is a family of polysaccharides that can be found in apple pomace and citrus peel (Dhanapal et al., 2012). As shown in Fig. 2.5, it is an anionic polymer with a (1–4)-linked d-galacturonic acid unit as its structural core. In foods like yogurt, milk, ice cream, and jams, it serves as a thickening, gelling, and stabilizing agent (Espitia et al., 2014). Pectin-based edible films have a high initial modulus, great resistance to oil and odor, and excellent mechanical qualities; nevertheless, they are also somewhat brittle, have limited elongations, and are not very resilient to moisture; the addition of a plasticizer increases their flexibility (Giancone et al., 2011). Polyvalent cations, such as calcium, are used to cross-link pectin-made films with moderate mechanical properties (Sucheta et al., 2019). Pectin films and gels work well to protect foods with little moisture (Liu et al., 2007). Fresh fruits and vegetables, such as carrots, apples, apricots, avocados, berries, guavas, chestnuts, melons, peaches, and walnuts, have all been packaged using pectin-formulated films (Valdés et al., 2015).

Gum Arabic

Gum arabic is the most utilized polysaccharide in the industry because of its unique emulsification and film-forming properties. It is derived from the stem portion of many *Acacia* species (Ali et al., 2013). Gum arabic is made up of glucuronic acid, arabinose, rhamnose, and galactose (Anderson et al., 1991) and is employed by the food, cosmetics, and pharmaceutical industries, in addition to the confectionery, flavoring, and bread industries (Maqbool et al., 2011). Additionally, it has been applied to bananas, tomatoes, and papayas to enhance their quality and prolong their shelf life (Ali et al., 2010; Maqbool et al., 2011). Fruits coated with gum arabic or almond gum showed a considerable decrease in respiration rate and ethylene production (Mahfoudhi & Hamdi, 2015).

Gellan Gum

The bacteria *Sphingomonas elodea,* also known as *Pseudomonas elodea,* generate gellan, a type of polysaccharide with unique colloidal, superior coating and gelling properties (Moreira et al., 2015). Gellan is increasingly being used in the food industry as a texturizing and gelling agent (Rojas-grau et al., 2008). Additionally, it acts as a vehicle for food additives, such as colorants, flavorings, nutraceuticals, and antimicrobial and anti-browning chemicals. Gellan films are composed of brittle, stiff gels. Gellan gum–based edible coatings have been effectively used to increase the freshness and shelf life of freshly chopped vegetables. These polymers gel when cations such as potassium, sodium, magnesium, and calcium ions are present (Rojas-grau et al., 2008; Oms-oliu et al., 2010; Danalache et al., 2016).

Xanthan Gum

Xanthan gum is an exopolysaccharide that is produced by the bacteria *Xanthomonas campestris*. When utilized as a thickening, emulsifying, and food-stabilizing agent, it is typically thought to be a safe substance. It creates a persistent viscous solution in hot or cold water at various pH levels and temperatures and is impervious to enzymatic degradation (Sharma & Rao, 2015). A trisaccharide subunit and alternating 1,4-linked d-glucose residues make up its structure (Zambrano-Zaragoza et al., 2013). Recently, a consumable coating based on xanthan gum has been used to enhance the freshness, extend the shelf life, and transfer the bioactive components of fresh-cut fruits. The ripening of acerola fruit coated with edible xanthan gum was delayed (Freitas et al., 2013; Quoc et al., 2015). Table 2.4 shows the applications of pectin and gum in food packaging.

Alginates

Brown marine algae (*Phaeophyceae*) are the source of alginates, which are naturally occurring polysaccharides. Alginates are composed of units of a-L-guluronate (G) and β-D-mannuronate (M) in different ratios and orders in the (1-4) chain, depending on the age of the plant and the source of the alginate (Erginkaya et al., 2014). Stronger gels are associated with higher G unit concentrations because calcium and sodium ions are involved in the gel formation process (Mohamed et al., 2020). Figure 2.6 represents the common sodium alginate structure. Alginate-based coatings can prolong the food's shelf life and quality by improving water barrier, maintaining flavor, and postponing fat oxidation (Song et al., 2011).

Table 2.4 Applications of pectin and gum in food packaging

Source of pectin + other components	Bioactive compound/nanofiller	Inference	References
Gellan gum, citrus pectin	Natamycin, essential clove oil, Tween 80	Enhanced microbial property Improved mechanical and barrier properties Extends the shelf life of corn tortillas	Callejas-Quijada et al. (2023)
Citrus pectin and chitosan	Noni fruit extract (NFE)	Enhanced tensile strength, barrier properties, and opacity Reduced thickness, moisture content, and water solubility Enhancement of antioxidant and antibacterial properties	Lin et al. (2023)
Waste pitaya peel pectin	Pitaya peel betacyanins , montmorillonite	The film reduced moisture sensitivity and showed colorimetric response to pH/ammonia Smart packaging indication by changing color from red to reddish brown and monitoring the freshness of shrimp	Jiang et al. (2023)
Mango peel waste pectin	–	Enhanced tensile strength, hydrophobicity, and thermal stability	Zhang et al. (2023)
Citrus pectin and chitosan	Quercetin-loaded nanoliposomes (QNLs)	Enhanced tensile strength, Young's modulus, and thermal stability Reduced opacity and moisture content Enhanced antimicrobial activity Preserve meat from spoilage bacteria	Afroz Ali et al. (2023)
Citrus pectin, sodium alginate, and xanthan gum	Blueberry anthocyanin extract (BAE)	Smart monitoring of blueberry freshness with color change from purple to light pink during storage and transportation	Li et al. (2023)
Citrus pectin	Hibiscus extract with inulin and maltodextrin	Enhanced mechanical and barrier properties Enhanced antioxidant activity Smart packaging indication with pH-dependent color change	Mendes et al. (2022)
Lemon peel waste pectin	Purple sugarcane peel anthocyanin extracts (PSPAEs)	Improved barrier, mechanical, and thermal stability Improved antioxidant activity Smart freshness detection of shrimp and beef	Jiang et al. (2022)

(continued)

Table 2.4 (continued)

Source of pectin + other components	Bioactive compound/nanofiller	Inference	References
Citrus pectin and gelatin	Sulfur-functionalized turmeric-derived carbon dots	Enhanced mechanical strength, barrier properties, hydrophobicity, and UV protection Enhanced antimicrobial and antioxidant activity	Ezati et al. (2022)
Gum arabic/ carboxymethyl chitosan	*Cinnamomum camphora* seed extract	Enhanced mechanical and barrier properties Enhanced antioxidant and antibacterial properties Extend oil stability and biodegradability	Alnadari et al. (2022)
Lysozyme (LZ), iota carrageenan (IC), gum arabic	–	Enhanced barrier properties Enhanced antimicrobial property with high film density	Koca and Bayramoğlu (2022)
Lysozyme (LZ), iota carrageenan (IC), gum arabic	–	Enhanced barrier properties Enhanced antimicrobial property with high film density	Koca and Bayramoğlu (2022)
Sodium alginate/ gum arabic	*Syzygium cumini* seed extract	Enhanced solubility and water vapor permeability Antioxidant protection against oil oxidation	Abdin et al. (2022)
Gum arabic	*Alpinia officinarum* phenolic extract	Enhanced dark yellow color Reduced moisture content, swelling, solubility, and water vapor permeability Enhanced thermal property Maintained the firmness of *Agaricus bisporus*	Gomaa et al. (2022)
Citrus pectin/agar	Melanin nanoparticles (MNPs) and grapefruit seed extract (GFSE)	Enhanced mechanical and barrier properties Enhanced antioxidant and antimicrobial properties	Roy and Rhim (2021)
Watermelon peel pectin	Beet extract (BE)	Improve mechanical and thermal properties Reduced oxygen and water vapor permeability A change in pH color from pink to brown indicates beef spoilage and shows the smart property of films	Guo et al. (2021)
Citrus pectin	Mulberry (*Morus alba*) leaf extract, deoxynojirimycin, chlorogenic acid	Enhanced mechanical, thermal, and barrier properties and soil degradability Enhanced antimicrobial and antioxidant properties Enhanced shelf life of coated capsicum by up to 12 days	Shivangi et al. (2021)

Gum arabic	Cellulose nanocrystals (CNCs)	Enhanced mechanical, barrier, and thermal stability Reduced water vapor and oxygen permeability (OP) Bio-nanocomposite films enhanced the shelf life of strawberries by delaying weight loss and quality deterioration	Kang et al. (2021)
Gum arabic and cellulose	Iron hydroxide nanoparticles	Enhanced thermal activity and antimicrobial activity Enhanced shelf life of paneer and dairy products	Chawla et al. (2021)
Citrus pectin	Nanoemulsions (NEs) and pickering emulsions (PEs) of marjoram EO (*Origanum majorana* L.)	PE increased the mechanical and barrier properties NE increased the antioxidant properties Essential oil releasing rate is slower in PE films	Almasi et al. (2020)
Gum arabic	–	Reduced weight loss Suppressed enzymatic degradation Enhanced antioxidant activity Enhanced shelf life of persimmons	Saleem et al. (2020)
Chitosan, gum arabic, and polyethylene glycol (PEG)	Black pepper essential oil/ginger essential oil	Enhanced mechanical and thermal stability Enhanced antimicrobial activity	
Citrus pectin/fish gelatin/beeswax	Olive antioxidants (hydroxytyrosol (HT) + 3,4-dihydroxyphenylglycol (DHPG))	HT (0.5%) with beeswax reduces lipid oxidation by lowering the Thiobarbituric Acid Reactive Substances (TBARS) value and extending freshness and meat preservation	Bermúdez-Oria et al. (2019)

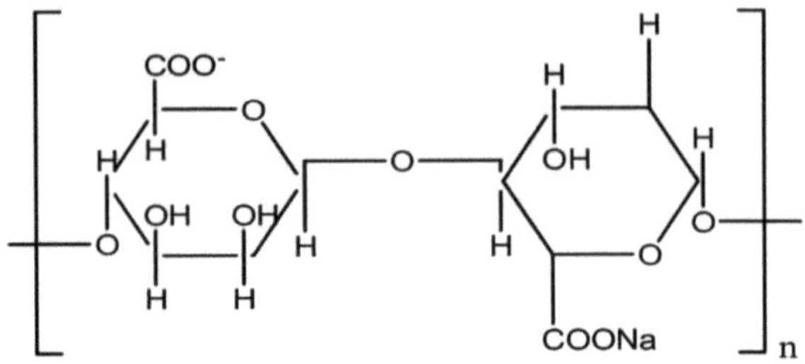

Fig. 2.6 Structure of sodium alginate

Calcium decreases the permeability to water vapor, rendering alginate films insoluble in water despite their poor water vapor barriers. In the presence of calcium cations or polyvalent metals, alginate can react to form a potent, insoluble or gel polymer. This calcium alginate gel is found in processed foods, such as crabsticks, onion rings, poultry products, and cocktail berries. But when subjected to calcium metal ions, alginate gels appear to do so quickly that it could hinder casting for the formation of films.

Carrageenan

Carrageenan, which is derived from a variety of red seaweeds in the Rhodophyceae family, is another polymer with marine origins. A sulfated water-soluble polymer, carrageenan, is used in the food, dairy, and pharmaceutical industries as an emulsifier, gel, and stabilizer (Pulp et al., 2021). Edible carrageenan films were utilized in meat, fish, poultry, sausage casings, fatty foods, and parched solid foods to reduce surface dehydration. The process of making carrageenan films starts with gelation during drying at a reasonable temperature and culminates in a solid film composed of double helices of polysaccharides. Carrageenan is made up of units of 3-linked and 4-linked D-galactopyranose that alternate (Karbowiak et al., 2006). Table 2.5 shows the applications of alginate and carrageenan in food packaging.

Pullulan

Aureobasidium pullulans uses starch to generate pullulan, a microbiological polysaccharide. It consists of maltotriose units joined by glycosidic units α-1,6 and α-1-4. Pullulan is very soluble in water and can be used to make edible films. The films are transparent, water-permeable, thermally permeable, colorless, odorless, with decreased oxygen and oil permeability, and tasteless (Kanmani & Lim, 2013).

Table 2.5 Applications of alginate and carrageenan in food packaging

Polymers	Added fillers	Method used	Inference	References
Sodium alginate	Carboxymethyl cellulose (CMC) Potato starch	Solvent casting	Increasing PS concentration enhanced the barrier properties, whereas moisture content and elongation decreased Enhanced mechanical and thermal properties with uniform thickness Black and green grapes can be stored for up to 16 days after harvest	Ramakrishnan et al. (2023b)
Alginate/CMC/potato starch	Grapefruit seed extract	Solvent casting	A homogeneous film was formulated by using grapefruit seed extract Thermal and UV barrier qualities without compromising the film's transparency Enhanced antioxidant and antimicrobial properties of the film The shelf life of green chili was extended by 25 days by using 0.5 g of concentration	Ramakrishnan et al. (2023a)
Alginate films	Melanin from watermelon seeds, silver nanoparticles (AgNPs), zinc oxide nanoparticles (ZnONPs)	Solvent casting	Enhanced antioxidant and antimicrobial properties Tensile strength enhanced by AgNP ZnONP improved barrier properties by maintaining light transmittance, whereas AgNP completely reduced light transmittance In cell viability in cytotoxicity screening, no adverse effects were observed	Łopusiewicz et al. (2022)
Alginate	Aloe vera (AV) and garlic oil (GO)	Solvent casting	Incorporation of AV and GO improved the thermal and mechanical properties of the films Reduced water vapor permeability Enhanced antimicrobial property Extended shelf life of coated tomatoes	Abdel Aziz and Salama (2021)
Pectin and alginate films	Halloysites (Hals) derived from Dragon and Dunino mines, loaded with salicylic acid	Adsorption and solvent casting	Alginate films showed stronger antimicrobial activity against four bacterial strains, whereas pectin films inhibited two bacterial strains Better control release was shown by alginate films	Kurczewska et al. (2021)
Calcium cross-linked alginate films	Sulfur nanoparticles (SNPs)	Nanoparticle embedding	Tensile strength and water vapor barrier by 12% and 41%, respectively, with 2 wt% SNPs Enhanced barrier properties Improved antimicrobial activity	Roy and Rhim (2020)

(continued)

Table 2.5 (continued)

Polymers	Added fillers	Method used	Inference	References
Alginate-based nanocomposite films	Copper sulfide nanoparticles (CuSNPs)	Solvent casting	Incorporation of 0.5 wt% enhanced barrier, mechanical, and antimicrobial properties	Roy and Rhim (2020)
Alginate	Cottonseed protein hydrolysates (CPHs)	Enzymatic hydrolysis	No effect on moisture content, biodegradability, solubility, and oil barrier properties by the incorporation of CPH CPH incorporation increased film thickness, water vapor permeability, and barrier properties Enhanced antioxidant and antimicrobial properties In an aqueous medium, within 30 min, there is a 60% peptide release	
Alginate	Functionalized halloysite nanotubes and zinc oxide nanoparticles (ZnONPs)	–	Change in surface charge increased mechanical and barrier properties Enhanced antimicrobial property	Shankar et al. (2018)
Carrageenan	Curcumin, zein, and epigallocatechin gallate (EGCG), carrageenan (CZEC)	Layer-by-layer nanoparticle fabrication	CZEC nanoparticles enhanced barrier, mechanical, and thermal properties Enhanced antioxidant activity Extended the freshness of packaged fish	Han et al. (2023)
Carrageenan	Silver ion loaded with (3-aminopropyl) trimethoxysilane (APTMS)–modified halloysite nanotubes		Increase in silver ion adsorption improved barrier and antimicrobial properties	Saedi et al. (2021)
Carrageenan	Jaboticaba peel extract	Casting	Enhanced antimicrobial and antioxidant properties Increased thickness and barrier properties	Avila et al. (2020)
Sodium alginate (SA) and cellulose nano-whisker (CNW)	Copper nanoparticles (NPs)		Enhanced barrier properties Enhanced antioxidant and antimicrobial properties Prevented microbial contamination in fresh-cut pepper	Saravanakumar et al. (2020)

Chitosan

A structural biopolymer present in fungal cell walls is called chitin (Portes et al., 2009), and certain types of algae, in addition to the exoskeletons of arthropods such as crabs (No et al., 2007; Al-Sagheer et al., 2014), arachnids, and insects (García et al., 2010; Alishahi & Aïder, 2012), are the sources of chitosan, a biodegradable and edible polysaccharide (Lizardi-mendoza et al., 2016). The exoskeletal shells of crustaceans, including shrimp, crabs, crayfish, and krill, are the only source of commercial chitosan due to their high chitin concentration (15–40%) and accessibility as a cheap food processing by-product (Chen et al., 2016). Chitin comes in three different types: α-chitin, β-chitin, and γ-chitin. The three chitin polymorphs' schematic differences are shown in Fig. 2.7. Chitin from crabs and shrimp is mostly utilized in the business and belongs to the α-chitin group. Stable polymeric chains have a nonparallel structure and strong hydrogen bonds (Ioelovich, 2017; Kumari & Kishor, 2020). β-chitin can be found in the cuttlebones of cuttlefish and mollusks like squid. Commercially accessible on the market, the polymeric strands are stacked in a parallel fashion and resemble α-chitin (Chen et al., 2016). The third kind, called γ-chitin, is a polysaccharide molecule that is present in both parallel and nonparallel strands in fungi (Ioelovich, 2017; Kaya et al., 2017). However, both β-chitin and γ-chitin forms are structurally exceedingly unstable, very susceptible to swelling, and hardly ever found (Annu et al., 2017).

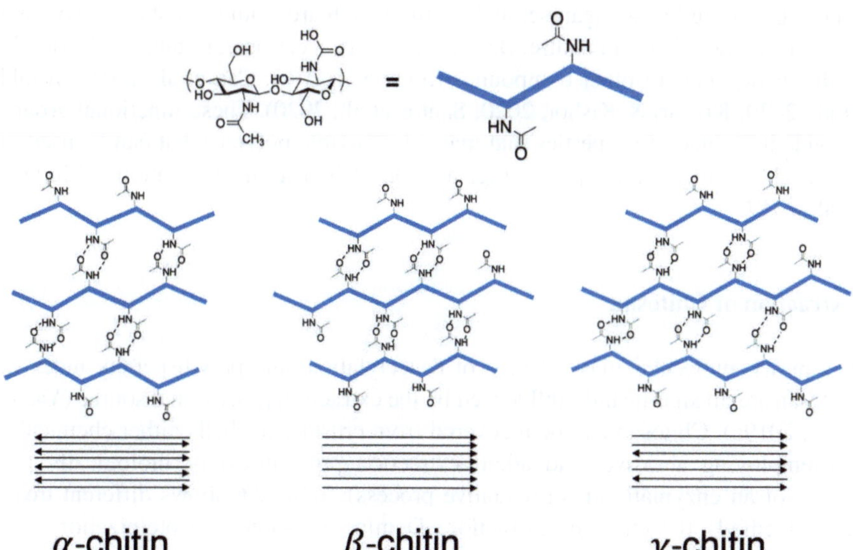

Fig. 2.7 The figure shows a schematic of the variations among the three chitin polymorphs. The lower part shows how the chains' orientation changes as they are stacked, while the upper part shows how the chains are stacked via hydrogen-bonding interactions. (Horiz et al., 2021)

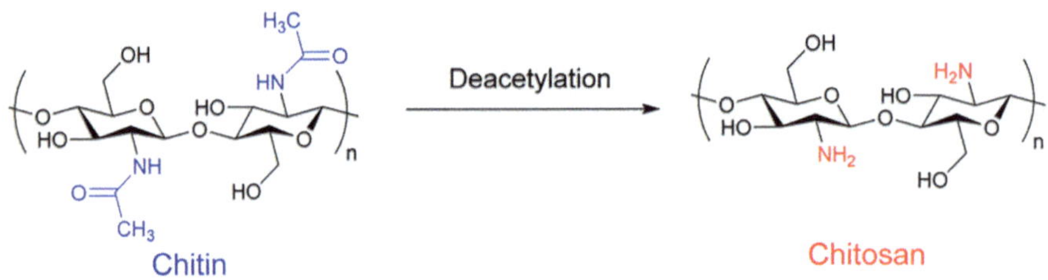

Fig. 2.8 Chemical structure of chitin and chitosan. (Horiz et al., 2021; Pakizeh et al., 2021)

Chitin is a homopolymer made up of N-acetyl-D-glucosamine (GlcNAc) monomers connected linearly by (1–4) glycosidic linkages. The subunits of D-glucosamine (GlcN) and GlcNAc, the latter of which is more prevalent in the polymer chain, combine to form its random copolymer in nature. Figure 2.8 displays the chemical composition of chitin and chitosan (Arnold et al., 2020; Kumari & Kishor, 2020). The hydrophobic acetyl groups and strong intra- and intermolecular hydrogen bonds in chitosan make it insoluble in water and most organic solvents and extremely resistant to physical and chemical degradation (Lizardi-mendoza et al., 2016; Annu et al., 2017; Santos et al., 2020). The molecule gains a net positive charge in water and significantly lessens its hydrophobic properties as the amount of N-acetylation (DA%) decreases because more hydrophilic free amine groups (NH_2) become available in the structure (Kumari & Kishor, 2020). Partially deacetylated chitin with a DA of less than 30–40% that dissolves well in diluted acidic aqueous solutions is known as chitosan (Khouri, 2019). Unlike many other naturally occurring polymers, such as cellulose, agarose, and pectin, which are acidic in nature, chitin and chitosan are both basic in nature. Because chitosan is cationic, it interacts favorably with a wide range of other compounds (Ioelovich, 2017; Cao et al., 2018; Arnold et al., 2020; Kumari & Kishor, 2020; Santos et al., 2020). These functional groups provide it a range of properties that make it a flexible polymer that can be used in textile, biomedical, food packaging, and other industrial applications (Mujtaba et al., 2019).

Extraction of Chitosan

The purity, molecular mass, degree of deacetylation, and polydispersity index of chitosan are all significantly influenced by the extraction process and source (Yadav et al., 2019a). Chitosan can be recovered from crustacean shells either chemically (by employing abrasive acid–alkali extraction procedures) or biologically (by means of an enzymatic or fermentative process). Table 2.6 shows different treatment methods for chitosan extraction. Demineralization, deproteinization, and deacetylation are the three fundamental processes in the primary extraction

Table 2.6 Treatment methods for chitosan extraction

Treatment	Advantages	Disadvantages	References
Chemical treatment (harsh acid–alkali extraction techniques)	Has high extraction efficiencies Short processing times	Unsustainable and unprofitable (because of the significant freshwater and energy requirements) Use of corrosive chemicals Negatively affects the physicochemical properties of chitosan	Hamed et al. (2016); Kumari and Kishor (2020)
Biological treatment			
Enzymatic treatment	Reduces solvent consumption Reduces the demand for freshwater Environmentally friendlier	High cost of purified proteases, crude enzymes Economically viable only for potential industrial-scale applications	Younes and Rinaudo (2015)
Fermentation treatment	Saves the cost for the management of effluents Recovers the solubilized proteins and minerals It can produce homogeneous and high-quality products	High processing time Limited to laboratory-scale studies	Kaur and Dhillon (2015); Yadav et al. (2019a)

procedure used to create chitosan. Along with chitins, proteins, and minerals, crustacean shells can also include carotenoid pigments. To get rid of undesirable pigments like beta-carotene and astaxanthin, a process known as the decolorization technique (acetone, hydrogen peroxide, and sodium hypochlorite) can be applied using a range of inorganic and organic solvents. Figure 2.9 represents the steps for chitosan extraction (Kerton et al., 2013; Annu et al., 2017; Knidri et al., 2018; Kumari & Kishor, 2020). While chemical extraction methods hurt the physicochemical properties of chitosan and are not regarded as sustainable or profitable due to their use of hazardous and caustic chemicals and high energy and freshwater requirements, biological processes yield a reproducible product with the desired qualities while being environmentally sustainable and energy-sustainable (Hamed et al., 2016; Yadav et al., 2019a). They are slow and inefficient and generate only small amounts of via biological means, which restricts their use in small-scale laboratory tests. Consequently, the majority of commercially accessible chitosan manufacture relies on chemical extraction processes, which raises the cost of chitosan procurement, even if its raw material (crustacean shell waste) is inexpensive and widely available (Kaur & Dhillon, 2015; Younes & Rinaudo, 2015; Kumari & Kishor, 2020).

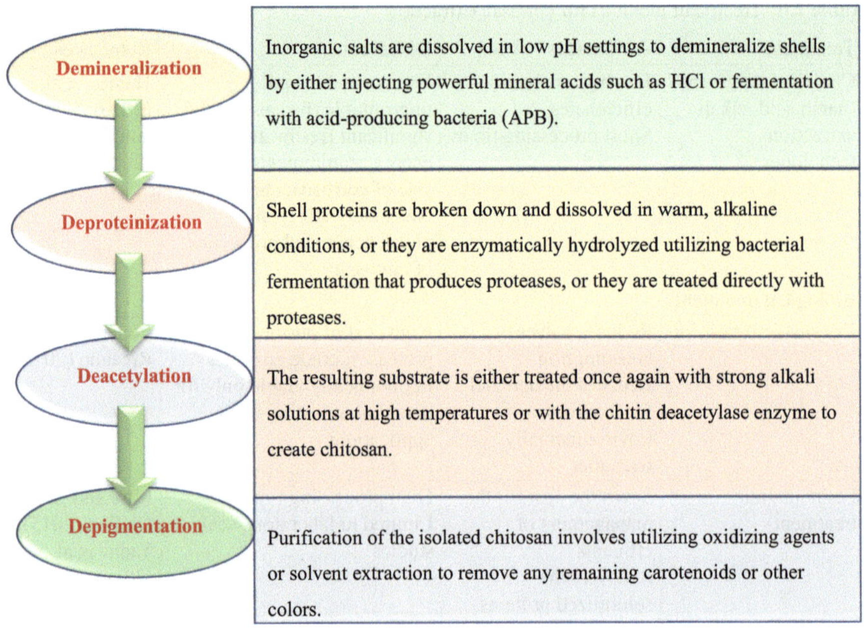

Fig. 2.9 Steps for chitosan extraction

Antioxidant and Antimicrobial Activity of Chitosan

Chitosan and its derivatives have garnered significant interest from both academia and industry, particularly in the food, biotechnology, pharmaceutical, and biomedical industries, due to their many beneficial properties, including their potent antibacterial and antioxidant properties (Knidri et al., 2018; Shanmuganathan et al., 2019; Abd El-Hack et al., 2020). Chitosan's ability to scavenge free radicals is thought to be the outcome of the H-abstraction reaction. Furthermore, the $-NH_2$ groups in chitosan can change into ammonium groups (NH_3+) by absorbing an H+ ion from the solution. These groups can then join forces with the -OH radical to form stable macromolecules (Xie et al., 2001; Tamer et al., 2016). However, many studies have shown that because chitosan lacks the H+ ion necessary to function as a chain-breaking antioxidant, its antioxidant activity is reduced at higher pH values (Tamer et al., 2016). It is difficult to separate the -OH and $-NH_2$ groups and have them react with hydroxyl-free radicals because chitosan molecules have extremely strong intra- and intermolecular hydrogen interactions (high bond dissociation energy) (Xie et al., 2001). The scavenging ability of chitosan can also be impacted by modifications to the Molecular Weight (MW) and Degree of Deacetylation (DDA), as well as the addition of a natural agent (plant extract or essential oil) (Xing et al., 2005). The lack of hydrogen atoms in chitosan and the presence of free amino groups at the C-2 position of chitosan chains, which quickly supply an appropriate antioxidant to act as a good radical scavenger, can be used to explain the incredibly low level of radical scavenging activity of pure chitosan–based films (Schreiber et al., 2013; Genskowsky et al., 2015).

Research on the antibacterial qualities of chitosan has revealed that these qualities may be influenced by both external variables, such as the kinds of microbes present and the pH of the surrounding environment, and intrinsic parameters, such as the molecular weight and the degree of deacetylation (Bagheri-Khoulenjani et al., 2009). These studies indicate that chitosan has a broad spectrum of antibacterial activity against fungi, bacteria, and algae. According to certain research, chitosan exhibits more potent antibacterial activity against gram-positive bacteria than gram-negative ones (Costa et al., 2012). Although there is disagreement about whether chitosan is more efficient against gram-positive or gram-negative bacteria, researchers concur that it has an antibacterial impact (Rhim et al., 2009; Torlak & Sert, 2013; Vardaka et al., 2016). According to the most widely accepted explanation for chitosan's antimicrobial properties, electrostatic interactions between positively charged chitosan and negatively charged microbial cell wall components enhanced the permeability of the cell membrane, allowing intracellular components to seep out and, ultimately, resulting in cell death (Hu & Gänzle, 2019). This interaction has a detrimental impact on the cell's nutrition supply, which results in cell death (Bonilla et al., 2013), On the other hand, No et al. (2007) and Yılmaz Atay and Çelik (2017) assert that this contact results in a leak in the cell membrane, causing intracellular chemicals to leak out of the cell. Although Wang et al. (2011) do not specify the precise mechanism, they concur that the aforementioned interaction brought on by charges results in the microbial cell dying. According to Martínez-Camacho et al. (2010), there are three possible processes: (1) The positively charged chitosan kills the bacterial cell by causing intracellular chemical leakage or by lowering the bacterial food supply; (2) chitosan functions as a chelating agent, limiting the availability of the metals that an organism needs; and (3) the low-molecular-weight chitosan can enter the cell and disrupt Messenger Ribonucleic Acid (mRNA) synthesis, impair protein synthesis, and inhibit enzymes.

The antibacterial effectiveness of chitosan-based films was influenced by a number of factors, including temperature, pH level, food ingredients, chitosan content, molecular weight, and source, as well as the kind and stage of development of microorganisms (Li & Zhuang, 2020; Sohrabi et al., 2020). Additionally, because chitosan is thought to work better in a solution than a film, its physical state is stated to be significant (Zivanovic et al., 2005; Leceta et al., 2013). This is because, in contrast to a rigid film, a solution facilitates simpler access to the chitosan molecules (Abdollahi et al., 2012). Despite chitosan's inherent antibacterial properties, chitosan films did not reveal any inhibitory zones in the targeted microorganisms. This might be due to the active compounds' low mobility. In the agar diffusion test method, chitosan can only prevent the growth of organisms that come into direct contact with the active sites since it does not permeate the surrounding agar material (Coma et al., 2002).

Another reason for the reduced antibacterial activity is the curled-up chitosan films on the agar surface, which reduce the surface and contact areas between the film and the culture on the agar plate, preventing diffusion (Hartley, 2013; Atarés &

Chiralt, 2016). Therefore, in order to decrease bacterial populations more effectively, chitosan needs to be combined with other antibacterial substances, such as plant extracts and essential oils (Foster & Butt, 2011).

Additionally, it has been found that chitosan film does not show a zone of inhibition when combined with propolis extract at a concentration of 0.75%; nevertheless, it does when the concentration is increased to 5% (Siripatrawan & Vitchayakitti, 2016).

Chitosan as a Food Packaging Material

Chitosan's strong adsorption capacity, antibacterial qualities, and applications in food packaging are important areas of research. It has several favorable characteristics for active substances in packaging, including non-toxicity, biodegradability, and biocompatibility. Its enhanced solubility is another factor that contributes to improved processability and functional performance (Dutta et al., 2012; Bonilla et al., 2013; Albertos et al., 2015; Chen et al., 2016; Hahn et al., 2020). The functional, optical, barrier, mechanical (strength, flexibility, and elasticity), and physical (transparency and color) characteristics of a food film are frequently what decide if it is suitable as a material for food packaging. The form and concentrations of acid solutions and plasticizing agents, helpful additives, blending polymer compounds, processing and post-processing factors (drying temperature, storage conditions, film neutralization, heat curing, cross-linking, etc.), and the properties of chitosan (purity, MW, and %DD (Percentage Degree of Deacetylation)) all have a significant impact on these properties (Khouri, 2019; Flórez et al., 2022).

The most popular method for producing pure chitosan films is solvent casting. However, a variety of processing techniques, including solvent blending, extrusion blending, and reactive extrusion blending, are employed to create chitosan-composite films (Van Den Broek et al., 2015).

To make chitosan films, acidified water is required since chitosan is insoluble in both water and the majority of organic solvents. Due to their superior resistance to gas and lipid molecule permeability (apart from vaporized water diffusion because of their water-friendly nature), indistinguishable transparent form, inherent antimicrobial and dual mineral-binding capacity, and high rigidity and tensile strength, solvent-cast chitosan films have generally gained popularity for food packaging and preservation purposes (Alishahi & Aïder, 2012; Khouri, 2019; Flórez et al., 2022). Using more chitosan powder can make the film-forming fluid very viscous and difficult to work with. Consequently, in the majority of studies, the highest chitosan concentration used in the formation of films is 2%. A pure chitosan film is composed of a matrix of chitosan molecules joined by numerous intramolecular and intermolecular hydrogen bonds (Riaz et al., 2021). The chitosan chains immobilize within the matrix due to the relatively dense structure of the chitosan film, making it stiffer and having less flexibility (Reyes-chaparro et al., 2015).

Since solvent-cast films are often very inflexible and have high levels of oxygen and water vapor permeability, it has been observed that these films, in particular, have poor mechanical and barrier characteristics (Rubilar et al., 2013; Riaz et al., 2021).

Plasticizers are substances that are added to polymers to enhance their mechanical characteristics, such as their flexibility. These substances include sugars (glucose and sucrose), polyols (glycerol, sorbitol, and polyethylene glycol), and lipids. Three theories have been put out to explain how plasticizers improve polymers' mechanical qualities. Plasticizers are mostly employed as internal lubricants that decrease frictional forces to enhance the mobility of polymer chains. By lowering the glass transition temperature of polymers and weakening the intermolecular hydrogen and ionic interactions, plasticizers can also increase flexibility. According to the third theory, the polymer film matrix's free volume increases when plasticizers are used (Bujang et al., 2013). Although a wide range of compounds have been studied as possible plasticizers for chitosan films, polyols are receiving the most attention (Bilck et al., 2015). For example, by decreasing the ionic and hydrogen connections between the chitosan molecules and enhancing the flexibility of the polymer chains, plasticizers are known to reduce the rigidity of chitosan films (Yoshida & Franco, 2009).

Glycerol has the right physical and chemical properties and is a viscous liquid. It is colorless, nontoxic, and odorless and has a sweet taste. It is an oleochemical made from vegetable and/or animal fats and oils (Salimon et al., 2012). The melting and boiling points of glycerol are 18.2 and 290 °C, respectively. Glycerol has a density of 1.261 g/cm^3, which makes it denser than water, and it is completely miscible with it. Because glycerol is composed of three hydroxyl groups, it reacts with water. Glycerol's high boiling point and high viscosity are caused by the hydrogen bonds that hold the majority of its molecules together. Its ability to form hydrogen bonds both within and between molecules makes it a very versatile chemical. To make polymeric films more flexible and mobile, glycerol is employed as a plasticizer. Numerous investigations have demonstrated that the development of flexible chitosan films is made possible by the participation of glycerol (Oma, 2005; Rivero et al., 2016; Deband et al., 2016).

Different quantities of propylene glycol (PG), ethylene glycol (EG), polyethylene glycol (PEG), and glycerol (Gly) were assessed as plasticizers in a prior study. In terms of plasticization efficiency and storage stability, it was found that Gly and PEG are superior choices for chitosan at a 20% concentration compared to EG and PG (Oma, 2005). Glycerol and polyethylene glycol 400 were compared for the surface and structural investigations of chitosan films. It was noticed that glycerol would be a better plasticizer for chitosan film because it could create H-bonds with chitosan more effectively (Bajdik et al., 2009).

Four varying amounts of glycerol were added to the chitosan solution: 0.25, 0.5, 0.75, and 1%. According to the results of the mechanical strength test, the chitosan films devoid of glycerol exhibited the highest tensile strength, which declined as more glycerol was added. Glycerol was added to the chitosan solutions at concentrations of 0.75 and 1% to produce more stretchable films. When glycerol was added to the solution, the water vapor permeability values significantly altered, increasing the permeation value (Rivero et al., 2016).

In addition, polyols, including glycerol, sorbitol, and polyethylene glycol, were examined by Srinivasa et al. (2007) to see how they impacted the chitosan films' mechanical and barrier properties. Reports state that when polyols were combined, the Elongation at Break (EAB) increased, but the Tensile Strength (TS) decreased. Similarly, when glycerol was added, the films' WVP dropped; however, when sorbitol and polyethylene glycol were added, it rose. Films produced with polyethylene glycol were stiffer than those produced with the other three polyol plasticizers. Yoshida and Franco (2009) examined the effects of four different glycerol concentrations and found that higher concentrations of glycerol increased the EAB but decreased the TS. This is accounted for by the breaking of molecular bonds within the chitosan film matrix, which enhances polymer chain mobility within the structure. Similar results were reported by Leceta et al. (2013) when they added glycerol to chitosan at 15% and 30%. They found that the high glycerol content increases the EAB and OP but decreases the TS while not affecting the WVP.

The TS of the film dropped while the EAB and WVP rose when 20% and 40% of glycerol and sorbitol, respectively, were added to chitosan. The investigation's findings demonstrated that films with 20% glycerol and chitosan had superior mechanical qualities over the other films. The films became softer and more flexible as a result, and their TS was reduced. However, when intermolecular interactions weaken, the films' WVP rises (Rodríguez-Núñez et al., 2014). It is commonly recognized that polyols alter the physicochemical characteristics of chitosan by weakening the bonds that hold its molecules together, which increases the mobility of the polymer chain.

Other than polyols, a wide range of chemicals were examined for their potential use as plasticizers in chitosan films. To give chitosan films their unique qualities, spermidine and glycerol were added. Films containing spermidine were more extensible than those containing glycerol. The most efficient ratio for creating a flexible film that satisfied industrial standards was found to be 20 mM glycerol to 2 mM spermidine, based on the mechanical properties. Furthermore, it has been demonstrated that spermidine preserves the uniform surface of chitosan films, in contrast to those that contain glycerol, which produce a rough surface (Sabbah et al., 2019). According to a different study, adding a surfactant (like Tween 20) to chitosan film-forming solutions (FFSs) lowered their surface tension (Ziani et al., 2008). Glycerol, gelatin, and resinous materials like propolis, which honeybees gather from a variety of plant sources, can be added to chitosan films to improve their water vapor barrier qualities (Leceta et al., 2013; Siripatrawan & Vitchayakitti, 2016). It was discovered that adding natural bioactive ingredients to the chitosan matrix enhanced the film's mechanical qualities as well as their antibacterial and antioxidant capabilities. The combination of chitosan and banana peel extract increased the film's antioxidant capacity while also lowering its moisture content and water solubility (Zhang et al., 2020a). It has been demonstrated that mango leaves increase chitosan films' antioxidant and tensile strength while lowering their moisture content and water solubility (Rambabu et al., 2019). The chitosan film solution's barrier qualities and antioxidant capacity were further improved by the addition of pine nutshell, peanut shell, and jujube leaf extract (Zhang et al., 2020b).

In a previous study, the addition of polyphenols from apple peels boosted the swelling ratio while decreasing the chitosan film's moisture content and barrier qualities. Additionally, apple peel polyphenols enhance the antibacterial and antioxidant capabilities of chitosan films (Riaz et al., 2021). The antioxidant and antibacterial qualities of chitosan films significantly increased with the addition of different amounts of *Eucalyptus globulus* essential oil. The addition of *Eucalyptus globulus* essential oil reduced the chitosan film's moisture content and water solubility while simultaneously enhancing its functionality (Hafsa et al., 2016). Additionally, propolis extract was utilized to decrease the barrier property of chitosan films while increasing their antibacterial and antioxidant efficacy (Siripatrawan & Vitchayakitti, 2016). Table 2.7 summarizes recent studies on the plant extract components in the chitosan matrix and their impacts on the physicochemical and biological properties.

Table 2.7 Effects of the incorporation of plant extracts and their components in the chitosan matrix

Film extract	Plant extract	Major findings	References
Chitosan	Grapefruit seed extract (GFSE)	TS reduced EAB values increased Antifungal action on bread	Tan et al. (2014)
Chitosan–starch	Polyphenol-rich extract of murta leaf extract	Improved elasticity of chitosan film-forming solution (FFS) CS films: brittle CS–starch films: reduced TS and EAB	Silva-weiss et al. (2013)
Chitosan	Phenolic extract of *Nigella sativa* seedcake	TS decreased EAB improved Water vapor permeability decreased Improved antioxidant activity	Kadam et al. (2018)
Chitosan–polyvinyl alcohol	Aqueous mint extract/pomegranate peel extract	Increased TS and opacity Improved antioxidant and antimicrobial activities	Kanatt et al. (2012)
Chitosan	Maqui berry	Films showed higher total phenolic content (TPC) and total flavonoid content Improved antioxidant and antimicrobial activities	Genskowsky et al. (2015)
Chitosan	Polyphenolic propolis extract	Increased TS and EAB Enhanced TPC, hence antioxidant activity Improved antimicrobial activity	Siripatrawan and Vitchayakitti (2016)
Chitosan	Carvacrol, grape seed extract (GSE)	Films developed with carvacrol reduced WVP, TS, and EAB Films developed with GSE enhanced barrier property (WVP and oxygen permeability (OP)) and reduced TS	Rubilar et al. (2013)
Chitosan	Young apple polyphenol	Improved antioxidant and antimicrobial activities Lowered WVP, TS, and EAB	Sun et al. (2017)

(continued)

Table 2.7 (continued)

Film extract	Plant extract	Major findings	References
Chitosan, chitosan–pea starch blend	Thyme extract (TE)	TE-incorporated CS-based films showed potential antioxidant activity despite the lower polyphenol release in different solvents	Talón et al. (2017)
Chitosan	Apple peel polyphenol	Significantly improved antioxidant and antimicrobial activities Reduced WVP, thermal stability, TS, and EAB	Riaz et al. (2021)
Chitosan	*Pistacia terebinthus*	Increased antimicrobial and antioxidant activity	
Chitosan–starch	Peanut shell and peanut skin extract	Improved antioxidant activity Lower WVP Peanut skin extracts increased TS, whereas peanut shell extract decreased TS	Meng et al. (2020)
Chitosan–starch	Pine nutshell, peanut shell, and jujube leaf extract	Increased antioxidant activity Decreased TS and EAB Increased WVP	Zhang et al. (2020b)
Chitosan	Purple-fleshed sweet potato extract	Improved antioxidant activity Decreased barrier and thermal stability Decreased EAB	Yong et al. (2019)
Chitosan	Black soybean seed coat extract	Improved antioxidant activity Improved WVP and mechanical property	Wang et al. (2019)
Chitosan	Banana peel extract	Improved antioxidant activity Decreased WVP	Zhang et al. (2020a)
Chitosan	Mango leaf extract	Improved antioxidant activity Improved TS, decreased EAB 56% higher oxidation resistance of cashew nuts over 28 days compared to commercial polyamide/polyethylene film	Rambabu et al. (2019)
Chitosan	*Zataria multiflora* and grape seed extract	Improved antioxidant activity Increased surface wettability	Moradi et al. (2012)
Chitosan	Thinned young apple polyphenol	Increased antimicrobial and antioxidant capacity of the film	Sun et al. (2017)
Chitosan	Apricot kernel essential oil (EO)	Enhanced antimicrobial and antioxidant activity Inhibited fungal growth on bread slices	Priyadarshi et al. (2018)
Chitosan	*Thymus moroderi* and *Thymus piperella* EO	Decreased antimicrobial activity (mesophilic and lactobacillus count in cooked, cured ham) Decreased lipid peroxidation	

(continued)

Table 2.7 (continued)

Film extract	Plant extract	Major findings	References
Chitosan	Lemon and cinnamon EO	A combination of both the EOs improved the antimicrobial property of the film	Peng and Li (2014)
Chitosan	Chitosan–pine needle extract	Improved the antioxidant property of the film	Kadam et al. (2021)
Chitosan	Turmeric extract	Enhanced mechanical strength Improved antimicrobial activity	Aytac et al. (2017)
Chitosan	Tea polyphenol	Improved the antioxidant activity of the film	Wang et al. (2013)

Conclusion

Polysaccharides are new-generation biopolymers used for active packaging, providing sustainable, biodegradable, and functional alternatives to traditional synthetic materials. This chapter has underlined the versatile packaging applications of starch, cellulose, chitosan, gums, alginate, and pectin and their functions to prolong food shelf life, increase mechanical properties, and provide barrier properties. Their capacity to include bioactive components, like antimicrobial and antioxidant substances, further enhances their prospect for active packaging applications. Nevertheless, limitations associated with moisture sensitivity, mechanical properties, and cost-effectiveness must be overcome to maximize their commercial application.

In the future, research efforts should be directed toward enhancing the mechanical and barrier behavior of polysaccharide films by applying nanotechnology, biopolymer blends, and cross-linking methods. The adoption of polysaccharide-based food packaging can be further improved by the integration of intelligent packaging technologies, including sensors and indicators for monitoring food quality. Progress in scalable process techniques and low-cost manufacturing will also be key to industrial uptake. Encouragement of collaboration among researchers, industries, and policymakers is the key to enhancing the innovations in and commercialization of polysaccharide-based food packaging. As they continue to evolve, these biopolymers have the capability to transform the packaging sector, helping pave the way for a greener, more sustainable tomorrow.

References

Abd El-Hack, M. E., El-Saadony, M. T., Shafi, M. E., et al. (2020). Antimicrobial and antioxidant properties of chitosan and its derivatives and their applications: A review. *International Journal of Biological Macromolecules, 164*, 2726–2744. https://doi.org/10.1016/j.ijbiomac.2020.08.153

Abdel Aziz, M. S., & Salama, H. E. (2021). Developing multifunctional edible coatings based on alginate for active food packaging. *International Journal of Biological Macromolecules, 190*, 837–844. https://doi.org/10.1016/J.IJBIOMAC.2021.09.031

Abdin, M., El-Beltagy, A. E., El-sayed, M. E., & Naeem, M. A. (2022). Production and characterization of sodium alginate/gum Arabic based films enriched with Syzygium cumini seeds extracts for food application. *Journal of Polymers and the Environment, 30*, 1615–1626. https://doi.org/10.1007/S10924-021-02306-Z/TABLES/3

Abdollahi, M., Rezaei, M., & Farzi, G. (2012). Improvement of active chitosan film properties with rosemary essential oil for food packaging. *International Journal of Food Science and Technology, 47*, 847–853. https://doi.org/10.1111/j.1365-2621.2011.02917.x

Acosta, S., Jiménez, A., Cháfer, M., et al. (2015). Physical properties and stability of starch-gelatin based films as affected by the addition of esters of fatty acids. *Food Hydrocolloids, 49*, 135–143. https://doi.org/10.1016/j.foodhyd.2015.03.015

Afroz Ali, S. M., Niaz, T., Munir, A., et al. (2023). Potential of pectin-chitosan based composite films embedded with quercetin-loaded nanofillers to control meat associated spoilage bacteria. *Food Bioscience, 53*, 102547. https://doi.org/10.1016/J.FBIO.2023.102547

Albertos, I., Rico, D., Diez, A. M., et al. (2015). Effect of edible chitosan/clove oil films and high-pressure processing on the microbiological shelf life of trout fillets. *Journal of the Science of Food and Agriculture, 95*, 2858–2865. https://doi.org/10.1002/jsfa.7026

Ali, A., Maqbool, M., Alderson, P. G., & Zahid, N. (2013). Effect of gum arabic as an edible coating on antioxidant capacity of tomato (Solanum lycopersicum L.) fruit during storage. *Postharvest Biology and Technology, 76*, 119–124. https://doi.org/10.1016/j.postharvbio.2012.09.011

Ali, A., Maqbool, M., Ramachandran, S., & Alderson, P. G. (2010). Gum arabic as a novel edible coating for enhancing shelf-life and improving postharvest quality of tomato (Solanum lycopersicum L.) fruit. *Postharvest Biology and Technology, 58*, 42–47. https://doi.org/10.1016/j.postharvbio.2010.05.005

Alishahi, A., & Aïder, M. (2012). Applications of chitosan in the seafood industry and aquaculture: A review. *Food and Bioprocess Technology, 5*, 817–830. https://doi.org/10.1007/s11947-011-0664-x

Almasi, H., Azizi, S., & Amjadi, S. (2020). Development and characterization of pectin films activated by nanoemulsion and Pickering emulsion stabilized marjoram (Origanum majorana L.) essential oil. *Food Hydrocolloids, 99*, 105338. https://doi.org/10.1016/J.FOODHYD.2019.105338

Almasi, H., Ghanbarzadeh, B., & Entezami, A. A. (2010). Physicochemical properties of starch-CMC-nanoclay biodegradable films. *International Journal of Biological Macromolecules, 46*, 1–5. https://doi.org/10.1016/j.ijbiomac.2009.10.001

Al-Moghazy, M., Mahmoud, M., & Nada, A. A. (2020). Fabrication of cellulose-based adhesive composite as an active packaging material to extend the shelf life of cheese. *International Journal of Biological Macromolecules, 160*, 264–275.

Alnadari, F., Bassey, A. P., Abdin, M., et al. (2022). Development of hybrid film based on carboxymethyl chitosan-gum Arabic incorporated citric acid and polyphenols from Cinnamomum camphora seeds for active food packaging. *Journal of Polymers and the Environment, 30*, 3582–3597. https://doi.org/10.1007/S10924-022-02453-X/FIGURES/6

Al-Sagheer, F. A., Ibrahim, E. I., & Khalil, K. D. (2014). Crystallinity, antimicrobial activity and dyeing properties of chitosan-g-poly(N-acryloyl morpholine) copolymer. *European Polymer Journal, 58*, 164–172. https://doi.org/10.1016/j.eurpolymj.2014.06.027

Anderson, D. M. W., Millar, J. R. A., & Weiping, W. (1991). Gum arabic (acacia Senegal): Unambiguous identification by 13C-NMR spectroscopy as an adjunct to the revised jecfa specification, and the application of 13C-NMR spectra for regulatory/legislative purposes. *Food Additives and Contaminants, 8*, 405–421. https://doi.org/10.1080/02652039109373991

Annu, A. S., Ahmed, S., & Ikram, S. (2017). *Chitin and chitosan: History, composition and properties* (pp. 1–24). Chitosan. https://doi.org/10.1002/9781119364849.ch1

Arnold, N. D., Brück, W. M., Garbe, D., & Brück, T. B. (2020). Enzymatic modification of native chitin and conversion to specialty chemical products. *Marine Drugs, 18*, 1–26. https://doi.org/10.3390/md18020093

Atarés, L., & Chiralt, A. (2016). Essential oils as additives in biodegradable films and coatings for active food packaging. *Trends in Food Science and Technology, 48*, 51–62. https://doi.org/10.1016/j.tifs.2015.12.001

Atta, O. M., Manan, S., Ahmed, A. A. Q., et al. (2021). Development and characterization of yeast-incorporated antimicrobial cellulose biofilms for edible food packaging application. *Polymers, 13*, 2310. https://doi.org/10.3390/POLYM13142310

Atta, O. M., Manan, S., Ul-Islam, M., et al. (2022). Development and characterization of plant oil-incorporated carboxymethyl cellulose/bacterial cellulose/glycerol-based antimicrobial edible films for food packaging applications. *Advanced Composites and Hybrid Materials, 5*, 973–990. https://doi.org/10.1007/S42114-021-00408-9/TABLES/4

Avila, L. B., Barreto, E. R. C., de Souza, P. K., et al. (2020). Carrageenan-based films incorporated with Jaboticaba peel extract: An innovative material for active food packaging. *Molecules, 25*, 5563. https://doi.org/10.3390/MOLECULES25235563

Aytac, Z., Ipek, S., Durgun, E., et al. (2017). Antibacterial electrospun zein nanofibrous web encapsulating thymol/cyclodextrin-inclusion complex for food packaging. *Food Chemistry, 233*, 117–124. https://doi.org/10.1016/j.foodchem.2017.04.095

Bagheri-Khoulenjani, S., Taghizadeh, S. M., & Mirzadeh, H. (2009). An investigation on the short-term biodegradability of chitosan with various molecular weights and degrees of deacetylation. *Carbohydrate Polymers, 78*, 773–778. https://doi.org/10.1016/j.carbpol.2009.06.020

Bajdik, J., Marciello, M., Caramella, C., et al. (2009). Evaluation of surface and microstructure of differently plasticized chitosan films. *Journal of Pharmaceutical and Biomedical Analysis, 49*, 655–659. https://doi.org/10.1016/j.jpba.2008.12.020

Bangar, S. P., Purewal, S. S., Trif, M., Maqsood, S., Kumar, M., Manjunatha, V., & Rusu, A. V. (2021). Functionality and applicability of starch-based films: An eco-friendly approach. *Food, 10*(9), 2181.

Benavides, S., Villalobos-Carvajal, R., & Reyes, J. E. (2012). Physical, mechanical and antibacterial properties of alginate film: Effect of the crosslinking degree and oregano essential oil concentration. *Journal of Food Engineering, 110*, 232–239. https://doi.org/10.1016/J.JFOODENG.2011.05.023

Bermúdez-Oria, A., Rodríguez-Gutiérrez, G., Rubio-Senent, F., et al. (2019). Effect of edible pectin-fish gelatin films containing the olive antioxidants hydroxytyrosol and 3,4-dihydroxyphenylglycol on beef meat during refrigerated storage. *Meat Science, 148*, 213–218. https://doi.org/10.1016/J.MEATSCI.2018.07.003

Bhargava, N., Sharanagat, V. S., Mor, R. S., & Kumar, K. (2020). Active and intelligent biodegradable packaging films using food and food waste-derived bioactive compounds: A review. *Trends in Food Science and Technology, 105*, 385–401. https://doi.org/10.1016/j.tifs.2020.09.015

Bilck, A. P., Maria, C., Müller, O., et al. (2015). Using glycerol produced from biodiesel as a plasticiser in extruded biodegradable films. *Polímeros, 25*, 331–335.

Bonilla, J., Fortunati, E., Vargas, M., et al. (2013). Effects of chitosan on the physicochemical and antimicrobial properties of PLA films. *Journal of Food Engineering, 119*, 236–243. https://doi.org/10.1016/j.jfoodeng.2013.05.026

Bujang, A., Adila, S. N., & Suyatma, N. E. (2013). Physical properties of chitosan films as affected by concentration of lactic acid and glycerol. *International Proceedings of Chemical Biological and Environmental Engineering., 58*, 27–31. https://doi.org/10.7763/IPCBEE

Callejas-Quijada, G., Chavarría-Hernández, N., López-Cuellar, M. d. R., et al. (2023). Films of biopolymers, pectin and gellan, enriched with natamycin and clove essential oils for the packaging of Corn tortilla: Protection against Staphylococcus aureus and Candida parapsilosis. *Food Microbiology, 110*, 104156. https://doi.org/10.1016/J.FM.2022.104156

Cao, Q., Zhang, Y., Chen, W., et al. (2018). Hydrophobicity and physicochemical properties of agarose film as affected by chitosan addition. *International Journal of Biological Macromolecules, 106*, 1307–1313. https://doi.org/10.1016/j.ijbiomac.2017.08.134

Chandrasekar, C. M., Krishnamachari, H., Farris, S., & Romano, D. (2023). Development and characterization of starch-based bioactive thermoplastic packaging films derived from banana peels. *Carbohydrate Polymer Technologies and Applications, 5*, 100328. https://doi.org/10.1016/J.CARPTA.2023.100328

Chawla, P., Najda, A., Bains, A., et al. (2021). Potential of gum Arabic functionalized iron hydroxide nanoparticles embedded cellulose paper for packaging of paneer. *Nanomaterials, 11*, 1308. https://doi.org/10.3390/NANO11051308

Chen, X., Yang, H., & Yan, N. (2016). Shell biorefinery: Dream or reality? *Chemistry - A European Journal, 22*, 13402–13421. https://doi.org/10.1002/chem.201602389

Chiumarelli, M., & Hubinger, M. D. (2012). Stability, solubility, mechanical and barrier properties of cassava starch e Carnauba wax edible coatings to preserve fresh-cut apples. *Food Hydrocolloids, 28*, 59–67. https://doi.org/10.1016/j.foodhyd.2011.12.006

Chu, Y., Popovich, C., & Wang, Y. (2023). Heat sealable regenerated cellulose films enabled by zein coating for sustainable food packaging. *Composites Part C: Open Access, 12*, 100390. https://doi.org/10.1016/J.JCOMC.2023.100390

Coma, V., Martial-Gros, A., Garreau, S., et al. (2002). Edible antimicrobial films based on chitosan matrix. *Journal of Food Science, 67*, 1162–1169. https://doi.org/10.1111/j.1365-2621.2002.tb09470.x

Costa, E. M., Silva, S., Pina, C., et al. (2012). Evaluation and insights into chitosan antimicrobial activity against anaerobic oral pathogens. *Anaerobe, 18*, 305–309. https://doi.org/10.1016/j.anaerobe.2012.04.009

Danalache, F., Carvalho, C. Y., Alves, V. D., et al. (2016). Optimisation of gellan gum edible coating for ready-to-eat mango (Mangifera indica L.) bars. *International Journal of Biological Macromolecules, 84*, 43–53. https://doi.org/10.1016/j.ijbiomac.2015.11.079

Deband, M. V., Bernal, C., & Francois, N. J. (2016). Development of biodegradable films based on chitosan/glycerol blends suitable for biomedical applications. *Tissue Science & Engineering, 7*, 3. https://doi.org/10.4172/2157-7552.1000187

Deshmukh, R. K., & Gaikwad, K. K. (2022). Natural antimicrobial and antioxidant compounds for active food packaging applications. *Biomass Conversion and Biorefinery, 2022*, 1–22. https://doi.org/10.1007/S13399-022-02623-W

Dhanapal, A., Rajamani, L., & Banu, M. (2012). Edible films from polysaccharides. *Food Science and Quality, 3*, 9–18.

Dutta, J., Tripathi, S., & Dutta, P. K. (2012). Progress in antimicrobial activities of chitin, chitosan and its oligosaccharides: A systematic study needs for food applications. *Food Science and Technology International, 18*(1), 3–34.

Knidri, H. E., Belaabed, R., Addaou, A., et al. (2018). Extraction, chemical modification and characterization of chitin and chitosan. *International Journal of Biological Macromolecules, 120*, 1181–1189. https://doi.org/10.1016/j.ijbiomac.2018.08.139

Engel, J. B., Ambrosi, A., & Tessaro, I. C. (2019). Development of biodegradable starch-based foams incorporated with grape stalks for food packaging. *Carbohydrate Polymers, 225*, 115234. https://doi.org/10.1016/J.CARBPOL.2019.115234

Erginkaya, Z., Kalkan, S., & Ünal, E. (2014). Use of antimicrobial edible films and coatings as packaging materials for food safety. In *Food processing: Strategies for quality assessment.*

Espitia, P. J. P., Du, W. X., Avena-Bustillos, R. d. J., et al. (2014). Edible films from pectin: Physical-mechanical and antimicrobial properties - A review. *Food Hydrocolloids, 35*, 287–296. https://doi.org/10.1016/j.foodhyd.2013.06.005

Ezati, P., Roy, S., & Rhim, J. W. (2022). Pectin/gelatin-based bioactive composite films reinforced with sulfur functionalized carbon dots. *Colloids and Surfaces A: Physicochemical and Engineering Aspects, 636*, 128123. https://doi.org/10.1016/J.COLSURFA.2021.128123

Falsafi, S. R., Topuz, F., & Rostamabadi, H. (2023). Dialdehyde carbohydrates—Advanced functional materials for biomedical applications. *Carbohydrate Polymers, 321*, 121276. https://doi.org/10.1016/J.CARBPOL.2023.121276

Fani, N., Enayati, M. H., Rostamabadi, H., & Falsafi, S. R. (2022). Encapsulation of bioactives within electrosprayed κ-carrageenan nanoparticles. *Carbohydrate Polymers, 294*, 119761. https://doi.org/10.1016/J.CARBPOL.2022.119761

Flórez, M., Guerra-Rodríguez, E., Cazón, P., & Vázquez, M. (2022). Chitosan for food packaging: Recent advances in active and intelligent films. *Food Hydrocolloids, 124*. https://doi.org/10.1016/j.foodhyd.2021.107328

Foster, L. J. R., & Butt, J. (2011). Chitosan films are NOT antimicrobial. *Biotechnology Letters, 33*, 417–421. https://doi.org/10.1007/s10529-010-0435-1

Freitas, I. R., Cortez-vega, W. R., Pizato, S., et al. (2013). Xanthan gum as a carrier of preservative agents and calcium chloride applied on fresh-cut apple. *Journal of Food Science, 33*, 229–238. https://doi.org/10.1111/jfs.12044

Galus, S., Aytunga, E., Kibar, A., et al. (2020). Novel materials in the preparation of edible films and coatings — A review. *Coatings, 10*(7), 674.

Gao, C., Pollet, E., & Avérous, L. (2017). Properties of glycerol-plasticized alginate films obtained by thermo-mechanical mixing. *Food Hydrocolloids, 63*, 414–420. https://doi.org/10.1016/j.foodhyd.2016.09.023

Gao, W., Wu, W., Liu, P., et al. (2020). Preparation and evaluation of hydrophobic biodegradable films made from corn/octenylsuccinated starch incorporated with different concentrations of soybean oil. *International Journal of Biological Macromolecules, 142*, 376–383. https://doi.org/10.1016/J.IJBIOMAC.2019.09.108

García, M., Díaz, R., Puerta, F., et al. (2010). Influence of chitosan addition on quality properties of vacuum-packaged pork sausages. *Ciência e Tecnologia de Alimentos, 30*, 560–564. https://doi.org/10.1590/s0101-20612010000200041

Gebresas, G. A., Szabó, T., & Marossy, K. (2023). A comparative study of carboxylic acids on the cross-linking potential of corn starch films. *Journal of Molecular Structure, 1277*, 134886. https://doi.org/10.1016/J.MOLSTRUC.2022.134886

Geng, C., Zhang, Z., Lin, R., et al. (2023). Simulation of hydrophobic lotus leaves for preparing hydrophobic starch blending films based on starch-based graphite oxide carbon. *Industrial Crops and Products, 195*, 116404. https://doi.org/10.1016/J.INDCROP.2023.116404

Genskowsky, E., Puente, L. A., Pérez-Álvarez, J. A., et al. (2015). Assessment of antibacterial and antioxidant properties of chitosan edible films incorporated with maqui berry (Aristotelia chilensis). *LWT - Food Science and Technology, 64*, 1057–1062. https://doi.org/10.1016/j.lwt.2015.07.026

Giancone, T., Torrieri, E., Di, P. P., et al. (2011). Effect of surface density on the engineering properties of high methoxyl pectin-based edible films. *Food and Bioprocess Technology, 4*, 1228–1236. https://doi.org/10.1007/s11947-009-0208-9

Go, E. J., & Song, K. B. (2019). Antioxidant properties of rye starch films containing rosehip extract and their application in packaging of chicken breast. *Starch - Stärke, 71*, 1900116. https://doi.org/10.1002/STAR.201900116

Gomaa, M. M., El, F. E., Salama, M. A., & Abdin, M. (2022). Production of bio-composite films from gum Arabic and galangal extract to prolong the shelf life of Agaricus bisporus. *Journal of Polymers and the Environment, 30*, 4787–4799. https://doi.org/10.1007/S10924-022-02551-W/FIGURES/7

Guo, Z., Ge, X., Li, W., et al. (2021). Active-intelligent film based on pectin from watermelon peel containing beetroot extract to monitor the freshness of packaged chilled beef. *Food Hydrocolloids, 119*, 106751. https://doi.org/10.1016/J.FOODHYD.2021.106751

Hafsa, J., Smach, M. a., Ben Khedher, M. R., et al. (2016). Physical, antioxidant and antimicrobial properties of chitosan films containing Eucalyptus globulus essential oil. *LWT - Food Science and Technology, 68*, 356–364. https://doi.org/10.1016/j.lwt.2015.12.050

Hahn, T., Tafi, E., Paul, A., et al. (2020). Current state of chitin purification and chitosan production from insects. *Journal of Chemical Technology and Biotechnology, 95*, 2775–2795. https://doi.org/10.1002/jctb.6533

Hamed, I., Özogul, F., & Regenstein, J. M. (2016). Industrial applications of crustacean by-products (chitin, chitosan, and chitooligosaccharides): A review. *Trends in Food Science and Technology, 48*, 40–50. https://doi.org/10.1016/j.tifs.2015.11.007

Han, Y., Zhou, M., D, M. C., et al. (2023). Investigation of a novel smart and active packaging materials: Nanoparticle-filled carrageenan-based composite films. *Carbohydrate Polymers, 301*, 120331.

Hartley, S. (2013). *Comparison of methods to evaluate chitosan as an*. All Theses 8.

Horiz, N., Jin, T., Jin, T., & Liu, T. (2021). Chitin and chitosan on the nanoscale. *Nanoscale Horizons, 6*, 505–542. https://doi.org/10.1039/d0nh00696c

Hu, Z., & Gänzle, M. G. (2019). Challenges and opportunities related to the use of chitosan as a food preservative. *Journal of Applied Microbiology, 126*, 1318–1331. https://doi.org/10.1111/jam.14131

Idumah, C. I., Hassan, A., & Ihuoma, D. E. (2019). Recently emerging trends in polymer nanocomposites packaging materials. *Polymer - Plastics Technology and Engineering, 58*, 1054–1109. https://doi.org/10.1080/03602559.2018.1542718

Ioelovich, M. (2017). *Nitrogenated polysaccharides—Chitin and chitosan, characterization and application* (pp. 25–70). Chitosan. https://doi.org/10.1002/9781119364849.ch2

Jabeen, N., Majid, I., & Nayik, G. A. (2015). Bioplastics and food packaging: A review. *Cogent Food & Agriculture, 1*(1). https://doi.org/10.1080/23311932.2015.1117749

Janjarasskul, T., & Krochta, J. M. (2010). Edible packaging materials. *Annual Review of Food Science and Technology, 1*, 415–448. https://doi.org/10.1146/annurev.food.080708.100836

Jiang, H., Zhang, W., Cao, J., & Jiang, W. (2022). Effect of purple sugarcane peel extracts on properties of films based on lemon peel waste pectin and the application in the visible detection of food freshness. *Food Hydrocolloids, 133*, 107982. https://doi.org/10.1016/J.FOODHYD.2022.107982

Jiang, H., Zhang, W., Pu, Y., et al. (2023). Development and characterization of a novel active and intelligent film based on pectin and betacyanins from peel waste of pitaya (Hylocereus undatus). *Food Chemistry, 404*, 134444. https://doi.org/10.1016/J.FOODCHEM.2022.134444

Jin, K., Tang, Y., Liu, J., et al. (2021). Nanofibrillated cellulose as coating agent for food packaging paper. *International Journal of Biological Macromolecules, 168*, 331–338. https://doi.org/10.1016/J.IJBIOMAC.2020.12.066

Kadam, A. A., Singh, S., & Gaikwad, K. K. (2021). Chitosan based antioxidant films incorporated with pine needles (Cedrus deodara) extract for active food packaging applications. *Food Control, 107877*. https://doi.org/10.1016/j.foodcont.2021.107877

Kadam, D., Shah, N., Palamthodi, S., & Lele, S. S. (2018). An investigation on the effect of polyphenolic extracts of Nigella sativa seedcake on physicochemical properties of chitosan-based films. *Carbohydrate Polymers, 192*, 347–355. https://doi.org/10.1016/j.carbpol.2018.03.052

Kanatt, S. R., Rao, M. S., Chawla, S. P., & Sharma, A. (2012). Active chitosan e polyvinyl alcohol fi lms with natural extracts. *Food Hydrocolloids, 29*, 290–297. https://doi.org/10.1016/j.foodhyd.2012.03.005

Kang, S., Xiao, Y., Guo, X., et al. (2021). Development of gum arabic-based nanocomposite films reinforced with cellulose nanocrystals for strawberry preservation. *Food Chemistry, 350*, 129199. https://doi.org/10.1016/J.FOODCHEM.2021.129199

Kanmani, P., & Lim, S. T. (2013). Development and characterization of novel probiotic-residing pullulan/starch edible films. *Food Chemistry, 141*, 1041–1049. https://doi.org/10.1016/j.foodchem.2013.03.103

Karbowiak, T., Debeaufort, F., Champion, D., & Voilley, A. (2006). Wetting properties at the surface of iota-carrageenan-based edible films. *Journal of Colloid and Interface Science, 294*, 400–410. https://doi.org/10.1016/j.jcis.2005.07.030

Kaur, S., & Dhillon, G. S. (2015). Recent trends in biological extraction of chitin from marine shell wastes: a review. *Critical Reviews in Biotechnology, 35*, –44. https://doi.org/10.3109/07388551.2013.798256

Kaya, M., Mujtaba, M., Ehrlich, H., et al. (2017). On chemistry of γ-chitin. *Carbohydrate Polymers, 176*, 177–186. https://doi.org/10.1016/j.carbpol.2017.08.076

Kerton, F. M., Liu, Y., Omari, K. W., & Hawboldt, K. (2013). Green chemistry and the ocean-based biorefinery. *Green Chemistry, 15*, 860–871. https://doi.org/10.1039/c3gc36994c

Khouri J (2019) Chitosan Edible Films Crosslinked by Citric Acid. 188.

Knapp, M. A., dos Santos, D. F., Pilatti-Riccio, D., et al. (2019). Yerba mate extract in active starch films: Mechanical and antioxidant properties. *Journal of Food Processing & Preservation, 43*, e13897. https://doi.org/10.1111/JFPP.13897

Koca, N., & Bayramoğlu, B. (2022). Layer-by-layer assembly of lysozyme with iota-carrageenan and gum Arabic for surface modification of food packaging materials with improved barrier properties. *Colloids and Surfaces A: Physicochemical and Engineering Aspects, 639*, 128391. https://doi.org/10.1016/J.COLSURFA.2022.128391

Kumari, S., & Kishor, R. (2020). *Chitin and chitosan: Origin, properties, and applications*. INC.

Kurczewska, J., Ratajczak, M., & Gajecka, M. (2021). Alginate and pectin films covering halloysite with encapsulated salicylic acid as food packaging components. *Applied Clay Science, 214*, 106270. https://doi.org/10.1016/J.CLAY.2021.106270

Lan, W., He, L., & Liu, Y. (2018). Preparation and properties of sodium carboxymethyl cellulose/sodium alginate/chitosan composite Film. *Coatings, 8*, 291. https://doi.org/10.3390/COATINGS8080291

Leceta, I., Guerrero, P., & De La Caba, K. (2013). Functional properties of chitosan-based films. *Carbohydrate Polymers, 93*, 339–346. https://doi.org/10.1016/j.carbpol.2012.04.031

Lekjing, S. (2016). A chitosan-based coating with or without clove oil extends the shelf life of cooked pork sausages in refrigerated storage. *Meat Science, 111*, 192–197. https://doi.org/10.1016/J.MEATSCI.2015.10.003

Li, J., & Zhuang, S. (2020). Antibacterial activity of chitosan and its derivatives and their interaction mechanism with bacteria: Current state and perspectives. *European Polymer Journal, 138*, 109984. https://doi.org/10.1016/j.eurpolymj.2020.109984

Li, Y., Hu, Z., Huo, R., & Cui, Z. (2023). Preparation of an indicator film based on pectin, sodium alginate, and xanthan gum containing blueberry anthocyanin extract and its application in blueberry freshness monitoring. *Heliyon, 9*, e14421. https://doi.org/10.1016/J.HELIYON.2023.E14421/ASSET/FBAA0764-66DF-4E0F-B69B-D1554C40117C/MAIN.ASSETS/FX36_LRG.JPG

Lin, X., Chen, S., Wang, R., et al. (2023). Fabrication, characterization and biological properties of pectin and/or chitosan-based films incorporated with noni (Morinda citrifolia) fruit extract. *Food Hydrocolloids, 134*, 108025. https://doi.org/10.1016/J.FOODHYD.2022.108025

Lisitsyn, A., Semenova, A., Nasonova, V., et al. (2021). Approaches in animal proteins and natural polysaccharides application for food packaging: Edible film production and quality estimation. *Polymers (Basel), 13*, 1–59. https://doi.org/10.3390/polym13101592

Liu, L. S., Liu, C. K., Fishman, M. L., & Hicks, K. B. (2007). Composite films from pectin and fish skin gelatin or soybean flour protein. *Journal of Agricultural and Food Chemistry, 55*, 2349–2355. https://doi.org/10.1021/jf062612u

Liu, Q., Gao, L., Qin, Y., et al. (2023). *Incorporation of oxidized debranched starch/chitosan nanoparticles for enhanced hydrophobicity of corn starch films*. Elsevier.

Liu, Y., Ahmed, S., Sameen, D. E., et al. (2021). A review of cellulose and its derivatives in biopolymer-based for food packaging application. *Trends in Food Science and Technology, 112*, 532–546. https://doi.org/10.1016/j.tifs.2021.04.016

Lizardi-mendoza, J., Monal, W. M. A., & Valencia, F. M. G. (2016). *2016-Chemical Characteristics and.pdf*. Elsevier.

Łopusiewicz, Ł., Macieja, S., Śliwiński, M., et al. (2022). Alginate biofunctional films modified with melanin from watermelon seeds and zinc oxide/silver nanoparticles. *Materials, 15*, 2381. https://doi.org/10.3390/MA15072381/S1

Ma, Y., Ye, K., Liu, P., et al. (2023). Effect of a Konjac glucomannan/chitosan antibacterial composite membrane microencapsulated with oregano essential oil on the quality of chilled pork. *Applied Food Research, 3*, 100249. https://doi.org/10.1016/J.AFRES.2022.100249

Mahfoudhi, N., & Hamdi, S. (2015). Use of almond gum and gum Arabic as novel edible coating to delay postharvest ripening and to maintain sweet cherry (Prunus avium) quality during storage. *Journal of Food Processing & Preservation, 39*, 1499–1508. https://doi.org/10.1111/JFPP.12369

Maqbool, M., Ali, A., Alderson, P. G., et al. (2011). Postharvest application of gum arabic and essential oils for controlling anthracnose and quality of banana and papaya during cold storage. *Postharvest Biology and Technology, 62*, 71–76. https://doi.org/10.1016/j.postharvbio.2011.04.002

Marín, A., Atarés, L., & Chiralt, A. (2017). Improving function of biocontrol agents incorporated in antifungal fruit coatings: A review. *Biocontrol Science and Technology, 27*, 3157. https://doi.org/10.1080/09583157.2017.1390068

Martínez-Camacho, A. P., Cortez-Rocha, M. O., Ezquerra-Brauer, J. M., et al. (2010). Chitosan composite films: Thermal, structural, mechanical and antifungal properties. *Carbohydrate Polymers, 82*, 305–315. https://doi.org/10.1016/j.carbpol.2010.04.069

Mendes, J. F., Norcino, L. B., Manrich, A., et al. (2022). Pectin-based color indicator films incorporated with spray-dried Hibiscus extract microparticles. *Food Research International, 162*, 111914. https://doi.org/10.1016/J.FOODRES.2022.111914

Meng, W., Shi, J., Zhang, X., et al. (2020). Effects of peanut shell and skin extracts on the antioxidant ability, physical and structure properties of starch-chitosan active packaging films. *International Journal of Biological Macromolecules, 152*, 137–146. https://doi.org/10.1016/j.ijbiomac.2020.02.235

Menzel, C. (2020). Improvement of starch films for food packaging through a three-principle approach: Antioxidants, cross-linking and reinforcement. *Carbohydrate Polymers, 250*, 116828. https://doi.org/10.1016/J.CARBPOL.2020.116828

Mohamed, S. A. A., El-sakhawy, M., & El-sakhawy, M. A. (2020). Polysaccharides, protein and lipid -based natural edible films in food packaging: A review. *Carbohydrate Polymers, 116178*. https://doi.org/10.1016/j.carbpol.2020.116178

Molavi, H., Behfar, S., Shariati, M., et al. (2015). A review on biodegradable starch based film. *The Journal of Microbiology, Biotechnology and Food Sciences, 4*, 456.

Moradi, M., Tajik, H., Razavi Rohani, S. M., et al. (2012). Characterization of antioxidant chitosan film incorporated with Zataria multiflora Boiss essential oil and grape seed extract. *LWT - Food Science and Technology, 46*, 477–484. https://doi.org/10.1016/j.lwt.2011.11.020

Moreira, M. R., Cassani, L., Martín-belloso, O., & Soliva-fortuny, R. (2015). Effects of polysaccharide-based edible coatings enriched with dietary fiber on quality attributes of fresh-cut apples. *Journal of Food Science and Technology, 52*, 7795–7805. https://doi.org/10.1007/s13197-015-1907-z

Mujtaba, M., Morsi, R. E., Kerch, G., et al. (2019). Current advancements in chitosan-based film production for food technology; A review. *International Journal of Biological Macromolecules, 121*, 889–904.

Müller, K., & Zollfrank, C. (2020). Ionic liquid aided solution-precipitation method to prepare polymer blends from cellulose with polyesters or polyamide. *European Polymer Journal, 133*, 109743. https://doi.org/10.1016/J.EURPOLYMJ.2020.109743

Nawab, A., Alam, F., & Hasnain, A. (2017). Mango kernel starch as a novel edible coating for enhancing shelf- life of tomato (Solanum lycopersicum) fruit. *International Journal of Biological Macromolecules*. https://doi.org/10.1016/j.ijbiomac.2017.05.057

No, H. K., Meyers, S. P., Prinyawiwatkul, W., & Xu, Z. (2007). Applications of chitosan for improvement of quality and shelf life of foods: A review. *Journal of Food Science, 72*. https://doi.org/10.1111/j.1750-3841.2007.00383.x

Oma, Ä. R. C. (2005). Effects of hydrophilic plasticizers on mechanical, thermal, and surface properties of chitosan films. *Journal of Agricultural and Food Chemistry, 53*, 3950–3957.

Oms-oliu, G., Martín-belloso, O., & Soliva-fortuny, R. (2010). Pulsed light treatments for food preservation. A review. *Food and Bioprocess Technology, 3*, 13–23. https://doi.org/10.1007/s11947-008-0147-x

Pakizeh, M., Moradi, A., & Ghassemi, T. (2021). Chemical extraction and modification of chitin and chitosan from shrimp shells. *European Polymer Journal, 159*, 110709. https://doi.org/10.1016/j.eurpolymj.2021.110709

Pandey, J. K., Takagi, H., Nakagaito, A. N., et al. (2012). An overview on the cellulose based conducting composites. *Composites Part B: Engineering, 43*, 2822–2826. https://doi.org/10.1016/j.compositesb.2012.04.045

Peng, Y., & Li, Y. (2014). Combined effects of two kinds of essential oils on physical, mechanical and structural properties of chitosan films. *Food Hydrocolloids, 36*, 287–293. https://doi.org/10.1016/j.foodhyd.2013.10.013

Pooresmaeil, M., Behzadi Nia, S., & Namazi, H. (2019). Green encapsulation of LDH(Zn/Al)-5-Fu with carboxymethyl cellulose biopolymer; new nanovehicle for oral colorectal cancer treatment. *International Journal of Biological Macromolecules, 139*, 994–1001. https://doi.org/10.1016/j.ijbiomac.2019.08.060

Portes, E., Gardrat, C., Castellan, A., & Coma, V. (2009). Environmentally friendly films based on chitosan and tetrahydrocurcuminoid derivatives exhibiting antibacterial and antioxidative properties. *Carbohydrate Polymers, 76*, 578–584. https://doi.org/10.1016/j.carbpol.2008.11.031

Priyadarshi, R., Kumar, B., Deeba, F., & Kulshreshtha, A. (2018). Chitosan films incorporated with Apricot (Prunus armeniaca) kernel essential oil as active food packaging material. *Food Hydrocolloids, 85*, 158–166. https://doi.org/10.1016/j.foodhyd.2018.07.003

Pulp, G., Film, E., Thi, T., et al. (2021). Impact of various essential oils and plant extracts on the characterization of the composite seaweed hydrocolloid and Gac Pulp (*Momordica cochinchinensis*) edible Film. *PRO, 9*(11), -2038.

Quoc, L. P. T., Hoa, D. P., Ngoc, H. T. B., & Phi, T. T. Y. (2015). Effect of xanthan gum solution on the preservation of acerola (Malpighia glabra L.). *Cercetari Agronomice in Moldova (Agronomic Research in Moldavia), 48*, 89–97. https://doi.org/10.1515/cerce-2015-0045

Ramakrishnan, R., Kulandhaivelu, S. V., & Roy, S. (2023a). Alginate/carboxymethyl cellulose/starch-based active coating with grapefruit seed extract to extend the shelf life of green chilli. *Industrial Crops and Products, 199*, 116752. https://doi.org/10.1016/J.INDCROP.2023.116752

Ramakrishnan, R., Kulandhaivelu, S. V., Roy, S., & Viswanathan, V. P. (2023b). Characterisation of ternary blend film of alginate/carboxymethyl cellulose/starch for packaging applications. *Industrial Crops and Products, 193*, 116114. https://doi.org/10.1016/J.INDCROP.2022.116114

Rambabu, K., Bharath, G., Banat, F., et al. (2019). Mango leaf extract incorporated chitosan antioxidant film for active food packaging. *International Journal of Biological Macromolecules, 126*, 1234–1243. https://doi.org/10.1016/j.ijbiomac.2018.12.196

Ren, H., Xu, Z., Du, C., et al. (2023). Preparation and characterization of starch-based composite films reinforced by quinoa (Chenopodium quinoa Willd.) straw cellulose nanocrystals. *International Journal of Biological Macromolecules, 242*, 124938. https://doi.org/10.1016/J.IJBIOMAC.2023.124938

Reyes-chaparro, P., Gutierrez-mendez, N., Salas-muñoz, E., et al. (2015). Effect of the addition of essential oils and functional extracts of clove on physicochemical properties of chitosan-based films. *International Journal of Polymer Science, 2015*(10).

Rhim, J. W., Hong, S. I., & Ha, C. S. (2009). Tensile, water vapor barrier and antimicrobial properties of PLA/nanoclay composite films. *Lwt, 42*, 612–617. https://doi.org/10.1016/j.lwt.2008.02.015

Riahi, Z., Rhim, J. W., Bagheri, R., et al. (2022). Carboxymethyl cellulose-based functional film integrated with chitosan-based carbon quantum dots for active food packaging applications. *Progress in Organic Coating, 166*, 106794. https://doi.org/10.1016/J.PORGCOAT.2022.106794

Riaz, A., Aadil, R. M., Amoussa, A. M. O., et al. (2021). Application of chitosan-based apple peel polyphenols edible coating on the preservation of strawberry (Fragaria ananassa cv Hongyan) fruit. *Journal of Food Processing & Preservation, 45*, 1–10. https://doi.org/10.1111/jfpp.15018

Rivero, S., Damonte, L., García, M. A., & Pinotti, A. (2016). An insight into the role of glycerol in chitosan films. *Food Biophysics, 11*(2). https://doi.org/10.1007/s11483-015-9421-4

Rodríguez-Núñez, J. R., Madera-Santana, T. J., Sánchez-Machado, D. I., et al. (2014). Chitosan/hydrophilic plasticizer-based films: Preparation, physicochemical and antimicrobial properties. *Journal of Polymers and the Environment, 22*, 41–51. https://doi.org/10.1007/s10924-013-0621-z

Rojas-grau, M. A., Tapia, M. S., & Martı, O. (2008). Using polysaccharide-based edible coatings to maintain quality of fresh-cut Fuji apples. *LWT, 41*, 139–147. https://doi.org/10.1016/j.lwt.2007.01.009

Rong, L., Ji, X., Shen, M., et al. (2023). Characterization of gallic acid-Chinese yam starch biodegradable film incorporated with chitosan for potential use in pork preservation. *Food Research International, 164*, 112331. https://doi.org/10.1016/J.FOODRES.2022.112331

Roy, S., & Rhim, J. W. (2020). Effect of CuS reinforcement on the mechanical, water vapor barrier, UV-light barrier, and antibacterial properties of alginate-based composite films. *International Journal of Biological Macromolecules, 164*, 37–44. https://doi.org/10.1016/J.IJBIOMAC.2020.07.092

Roy, S., & Rhim, J. W. (2021). Fabrication of pectin/agar blended functional film: Effect of reinforcement of melanin nanoparticles and grapefruit seed extract. *Food Hydrocolloids, 118*, 106823. https://doi.org/10.1016/J.FOODHYD.2021.106823

Rubilar, J. F., Cruz, R. M. S., Silva, H. D., et al. (2013). Physico-mechanical properties of chitosan films with carvacrol and grape seed extract. *Journal of Food Engineering, 115*, 466–474. https://doi.org/10.1016/j.jfoodeng.2012.07.009

Sabbah, M., Di, P., Cammarota, M., et al. (2019). Development and properties of new chitosan-based films plasticized with spermidine and/or glycerol. *Food Hydrocolloids, 87*, 245–252. https://doi.org/10.1016/j.foodhyd.2018.08.008

Saedi, S., Shokri, M., Roy, S., & Rhim, J. W. (2021). Silver loaded aminosilane modified halloysite for the preparation of carrageenan-based functional films. *Applied Clay Science, 211*, 106170. https://doi.org/10.1016/J.CLAY.2021.106170

Sakkara, S., Nataraj, D., Venkatesh, K., et al. (2020). Effect of pH on the physicochemical properties of starch films. *Journal of Applied Polymer Science, 137*, 48563. https://doi.org/10.1002/APP.48563

Saleem, M. S., Ejaz, S., Anjum, M. A., et al. (2020). Postharvest application of gum arabic edible coating delays ripening and maintains quality of persimmon fruits during storage. *Journal of Food Processing & Preservation, 44*, e14583. https://doi.org/10.1111/JFPP.14583

Salimon, J., Salih, N., & Yousif, E. (2012). Industrial development and applications of plant oils and their biobased oleochemicals. *Arabian Journal of Chemistry, 5*, 135–145. https://doi.org/10.1016/j.arabjc.2010.08.007

Samsudin, H., & Hani, N. M. (2019). *Use of starch in food packaging*. Elsevier.

Santagata, G., Mallardo, S., Fasulo, G., et al. (2018). Pectin-honey coating as novel dehydrating bioactive agent for cut fruit: Enhancement of the functional properties of coated dried fruits. *Food Chemistry, 258*, 104–110. https://doi.org/10.1016/j.foodchem.2018.03.064

Santos, V. P., Marques, N. S. S., Maia, P. C. S. V., et al. (2020). Seafood waste as attractive source of chitin and chitosan production and their applications. *International Journal of Molecular Sciences, 21*, 1–17. https://doi.org/10.3390/ijms21124290

Sapper, M., & Chiralt, A. (2018). Starch-based coatings for preservation of fruits and vegetables. *Coatings, 8*(5), 152. https://doi.org/10.3390/coatings8050152

Saravanakumar, K., Sathiyaseelan, A., Mariadoss, A. V. A., et al. (2020). Physical and bioactivities of biopolymeric films incorporated with cellulose, sodium alginate and copper oxide nanoparticles for food packaging application. *International Journal of Biological Macromolecules, 153*, 207–214. https://doi.org/10.1016/J.IJBIOMAC.2020.02.250

Schreiber, S. B., Bozell, J. J., Hayes, D. G., & Zivanovic, S. (2013). Introduction of primary antioxidant activity to chitosan for application as a multifunctional food packaging material. *Food Hydrocolloids, 33*, 207–214. https://doi.org/10.1016/j.foodhyd.2013.03.006

Shankar, S., Kasapis, S., & Rhim, J. W. (2018). Alginate-based nanocomposite films reinforced with halloysite nanotubes functionalized by alkali treatment and zinc oxide nanoparticles. *International Journal of Biological Macromolecules, 118*, 1824–1832. https://doi.org/10.1016/J.IJBIOMAC.2018.07.026

Shanmuganathan, R., Edison, T. N. J. I., LewisOscar, F., et al. (2019). Chitosan nanopolymers: An overview of drug delivery against cancer. *International Journal of Biological Macromolecules, 130*, 727–736. https://doi.org/10.1016/j.ijbiomac.2019.02.060

Sharma, S., & Rao, T. V. R. (2015). Xanthan gum based edible coating enriched with cinnamic acid prevents browning and extends the shelf-life of fresh-cut pears. *LWT - Food Science and Technology, 62*, 791–800. https://doi.org/10.1016/j.lwt.2014.11.050

Shivangi, S., Dorairaj, D., Negi, P. S., & Shetty, N. P. (2021). Development and characterisation of a pectin-based edible film that contains mulberry leaf extract and its bio-active components. *Food Hydrocolloids, 121*, 107046. https://doi.org/10.1016/J.FOODHYD.2021.107046

Silva-weiss, A., Bifani, V., Ihl, M., et al. (2013). Structural properties of films and rheology of film-forming solutions based on chitosan and chitosan-starch blend enriched with murta leaf extract. *Food Hydrocolloids, 31*, 458–466. https://doi.org/10.1016/j.foodhyd.2012.11.028

Siqueira, G., Bras, J., & Dufresne, A. (2010). Cellulosic bionanocomposites: A review of preparation, properties and applications. *Polymers (Basel), 2*, 728–765. https://doi.org/10.3390/polym2040728

Siripatrawan, U., & Vitchayakitti, W. (2016). Improving functional properties of chitosan films as active food packaging by incorporating with propolis. *Food Hydrocolloids, 61*, 695–702. https://doi.org/10.1016/j.foodhyd.2016.06.001

Sohrabi, C., Alsafi, Z., Neill, N. O., et al. (2020). World Health Organization declares global emergency: A review of the 2019 novel coronavirus (COVID-19). *International Journal of Surgery, 76*, 71–76. https://doi.org/10.1016/j.ijsu.2020.02.034

Song, Y., Liu, L., Shen, H., et al. (2011). Effect of sodium alginate-based edible coating containing different anti-oxidants on quality and shelf life of refrigerated bream (Megalobrama amblycephala). *Food Control, 22*, 608–615. https://doi.org/10.1016/j.foodcont.2010.10.012

Srinivasa, P. C., Ramesh, M. N., & Tharanathan, R. N. (2007). Effect of plasticizers and fatty acids on mechanical and permeability characteristics of chitosan films. *Food Hydrocolloids, 21*, 1113–1122. https://doi.org/10.1016/j.foodhyd.2006.08.005

Sucheta, C. K., Sharma, N., & Yadav, S. K. (2019). Composite edible coatings from commercial pectin, corn flour and beetroot powder minimize post-harvest decay, reduces ripening and improves sensory liking of tomatoes. *International Journal of Biological Macromolecules, 133*, 284–293. https://doi.org/10.1016/j.ijbiomac.2019.04.132

Sun, L., Sun, J., Chen, L., et al. (2017). Preparation and characterization of chitosan film incorporated with thinned young apple polyphenols as an active packaging material. *Carbohydrate Polymers, 163*, 81–91. https://doi.org/10.1016/j.carbpol.2017.01.016

Tabari, M. (2017). Investigation of carboxymethyl cellulose (Cmc) on mechanical properties of cold water fish gelatin biodegradable edible films. *Food, 6*, 1–7. https://doi.org/10.3390/foods6060041

Tajeddin, B., & Arabkhedri, M. (2020). *Chapter 16. Polymers and food packaging*. INC.

Talón, E., Trifkovic, K. T., Vargas, M., et al. (2017). Release of polyphenols from starch-chitosan based films containing thyme extract. *Carbohydrate Polymers, 175*, 122–130. https://doi.org/10.1016/j.carbpol.2017.07.067

Tamer, T. M., Valachová, K., Mohyeldin, M. S., & Soltes, L. (2016). Free radical scavenger activity of chitosan and its aminated derivative. *Journal of Applied Pharmaceutical Science, 6*, 195–201. https://doi.org/10.7324/JAPS.2016.60428

Tan, Y. M., Lim, S. H., Tay, B. Y., et al. (2014). Functional chitosan-based grapefruit seed extract composite films for applications in food packaging technology. *Materials Research Bulletin, 69*, 142–146. https://doi.org/10.1016/j.materresbull.2014.11.041

Torlak, E., & Sert, D. (2013). Antibacterial effectiveness of chitosan-propolis coated polypropylene films against foodborne pathogens. *International Journal of Biological Macromolecules, 60*, 52–55. https://doi.org/10.1016/j.ijbiomac.2013.05.013

Valdés, A., Burgos, N., Jiménez, A., & Garrigós, M. C. (2015). Natural pectin polysaccharides as edible coatings. *Coatings, 5*, 865–886. https://doi.org/10.3390/coatings5040865

Van Den Broek, L. A. M., Knoop, R. J. I., Kappen, F. H. J., & Boeriu, C. G. (2015). Chitosan films and blends for packaging material. *Carbohydrate Polymers, 116*, 237–242. https://doi.org/10.1016/j.carbpol.2014.07.039

Vanitha, R., & Kavitha, C. (2021). Development of natural cellulose fiber and its food packaging application. *Materials Today: Proceedings, 36*, 903–906. https://doi.org/10.1016/J.MATPR.2020.07.029

Vardaka, V. D., Yehia, H. M., & Savvaidis, I. N. (2016). Effects of Citrox and chitosan on the survival of Escherichia coli O157: H7 and Salmonella enterica in vacuum-packaged Turkey meat. *Food Microbiology, 58*, 128–134. https://doi.org/10.1016/j.fm.2016.04.003

Vellaisamy Singaram, A. J., Guruchandran, S., Bakshi, A., et al. (2021). Study on enhanced mechanical, barrier and optical properties of chemically modified mango kernel starch films. *Packaging Technology and Science, 34*, 485–495. https://doi.org/10.1002/PTS.2574

von Schantz, L., Schagerlöf, H., Karlsson, E. N., & Ohlin, M. (2014). Characterization of the substitution pattern of cellulose derivatives using carbohydrate-binding modules. *BMC Biotechnology, 14*. https://doi.org/10.1186/s12896-014-0113-9

Wang, L., Dong, Y., Men, H., et al. (2013). Preparation and characterization of active fi lms based on chitosan incorporated tea polyphenols. *Food Hydrocolloids, 32*, 35–41. https://doi.org/10.1016/j.foodhyd.2012.11.034

Wang, L., Liu, F., Jiang, Y., et al. (2011). Synergistic antimicrobial activities of natural essential oils with chitosan films. *Journal of Agricultural and Food Chemistry, 59*, 12411–12419. https://doi.org/10.1021/jf203165k

Wang, S., Lu, A., & Zhang, L. (2016). Recent advances in regenerated cellulose materials. *Progress in Polymer Science, 53*, 169–206. https://doi.org/10.1016/J.PROGPOLYMSCI.2015.07.003

Wang, X., Yong, H., Gao, L., et al. (2019). Preparation and characterization of antioxidant and pH-sensitive films based on chitosan and black soybean seed coat extract. *Food Hydrocolloids, 89*, 56–66. https://doi.org/10.1016/j.foodhyd.2018.10.019

Wang, Y., Wang, X., Hu, G., et al. (2022). Effect of green coffee oil as a natural active emulsifying agent on the properties of corn starch-based films. *LWT, 170*, 114087. https://doi.org/10.1016/J.LWT.2022.114087

Wu, H., Li, T., Peng, L., et al. (2023). Development and characterization of antioxidant composite films based on starch and gelatin incorporating resveratrol fabricated by extrusion compression moulding. *Food Hydrocolloids, 139*, 108509. https://doi.org/10.1016/J.FOODHYD.2023.108509

Wu, L. T., Tsai, I. L., Ho, Y. C., et al. (2021). Active and intelligent gellan gum-based packaging films for controlling anthocyanins release and monitoring food freshness. *Carbohydrate Polymers, 254*, 117410. https://doi.org/10.1016/J.CARBPOL.2020.117410

Xie, W., Xu, P., & Liu, Q. (2001). Antioxidant activity of water-soluble chitosan derivatives. *Bioorganic & Medicinal Chemistry Letters, 11*, 1699–1701. https://doi.org/10.1016/S0960-894X(01)00285-2

Xing, R., Liu, S., Guo, Z., et al. (2005). Relevance of molecular weight of chitosan and its derivatives and their antioxidant activities in vitro. *Bioorganic & Medicinal Chemistry, 13*, 1573–1577. https://doi.org/10.1016/j.bmc.2004.12.022

Xu, H., Chen, L., McClements, D. J., et al. (2023). Development and characterization of active starch-based films incorporating graphene/polydopamine/Cu2+ nanocomposite fillers. *Carbohydrate Polymers, 305*, 120498. https://doi.org/10.1016/J.CARBPOL.2022.120498

Xu, Q., Chen, C., Rosswurm, K., et al. (2016). A facile route to prepare cellulose-based films. *Carbohydrate Polymers, 149*, 274–281.

Yadav, M., Goswami, P., Paritosh, K., et al. (2019a). Seafood waste: a source for preparation of commercially employable chitin/chitosan materials. *Bioresources and Bioprocessing, 6*. https://doi.org/10.1186/s40643-019-0243-y

Yadav, M., Liu, Y. K., & Chiu, F. C. (2019b). Fabrication of cellulose nanocrystal/silver/alginate bionanocomposite films with enhanced mechanical and barrier properties for food packaging application. *Nanomaterials, 9*, 1523. https://doi.org/10.3390/NANO9111523

Yılmaz Atay, H., & Çelik, E. (2017). Investigations of antibacterial activity of chitosan in the polymeric composite coatings. *Progress in Organic Coating, 102*, 194–200. https://doi.org/10.1016/j.porgcoat.2016.10.013

Yong, H., Wang, X., Bai, R., et al. (2019). Development of antioxidant and intelligent pH-sensing packaging films by incorporating purple-fleshed sweet potato extract into chitosan matrix. *Food Hydrocolloids, 90*, 216–224. https://doi.org/10.1016/j.foodhyd.2018.12.015

Yoshida, B. C. M. P., & Franco, T. T. (2009). Chitosan tailor-made films: The effects of additives on barrier and mechanical properties and. *Science, 22*, 161–170.

Younes, I., & Rinaudo, M. (2015). Chitin and chitosan preparation from marine sources. Structure, properties and applications. *Marine Drugs, 13*, 1133–1174. https://doi.org/10.3390/md13031133

Zambrano-Zaragoza, M. L., Mercado-Silva, E., Ramirez-Zamorano, P., et al. (2013). Use of solid lipid nanoparticles (SLNs) in edible coatings to increase guava (Psidium guajava L.) shelf-life. *Food Research International, 51*, 946–953. https://doi.org/10.1016/j.foodres.2013.02.012

Zhang, S., Pan, X., Zhao, J., et al. (2023). Characterization of ionically crosslinked mango peel pectin-based films: Effect of different cations on the improved properties of film. *Food Packaging and Shelf Life, 38*, 101131. https://doi.org/10.1016/J.FPSL.2023.101131

Zhang, W., Li, X., & Jiang, W. (2020a). Development of antioxidant chitosan film with banana peels extract and its application as coating in maintaining the storage quality of apple. *International Journal of Biological Macromolecules, 154*, 1205–1214. https://doi.org/10.1016/j.ijbiomac.2019.10.275

Zhang, X., Guo, H., Luo, W., et al. (2022). Development of functional hydroxyethyl cellulose-based composite films for food packaging applications. *Frontiers in Bioengineering and Biotechnology, 10*, 989893. https://doi.org/10.3389/FBIOE.2022.989893/BIBTEX

Zhang, X., Lian, H., Shi, J., et al. (2020b). Plant extracts such as pine nut shell, peanut shell and jujube leaf improved the antioxidant ability and gas permeability of chitosan films. *International Journal of Biological Macromolecules, 148*, 1242–1250. https://doi.org/10.1016/j.ijbiomac.2019.11.108

Ziani, K., Oses, J., Coma, V., & Mate, J. I. (2008). Effect of the presence of glycerol and Tween 20 on the chemical and physical properties of films based on chitosan with different degree of deacetylation. *LWT - Food Science and Technology, 41*, 2159–2165. https://doi.org/10.1016/j.lwt.2007.11.023

Zivanovic, S., Chi, S., & Draughon a. (2005). Antimicrobial activity of chitosan. *Science (1979), 70*, 45–51.

Chapter 3
Protein: Biopolymers in Edible Coating and Films

Manisha, Renu, Poonam, Vijay Dangi, and Ishu Khangwal

Abstract Biopolymers derived from living organisms, such as bacteria, plants, algae, and animals, represent a varied group of naturally occurring polymers with a wide range of applications. These include polysaccharides, polypeptides, and polynucleotides, which are frequently derived from biological sources or waste using chemical methods. While their hierarchical structures provide them with unique physiochemical features, such as great mechanical strength and superhydrophobicity, creating synthetic replicas of these features is difficult. Biopolymers are highly suitable for biomedical applications, especially in drug delivery and tissue engineering, due to their low toxicity, biodegradability, and biocompatibility. Methods such as electrospinning use proteins like collagen, gelatin, keratin, and fibroin to create hydrogels, nanofibers, and 3D-printed scaffolds. Despite their versatility, extracted biopolymers often demonstrate brittleness and inadequate mechanical performance. Functional polymers, synthesized through techniques like living anionic polymerization, demonstrate enhanced chemical reactivity, structural adaptability, and self-assembly capabilities, making them indispensable for applications in electronics, catalysis, and medical devices. Films produced from proteins such as casein, soy, zein, gelatin, and gluten are gaining popularity as sustainable food packaging alternatives due to their biodegradability and functional properties, such as emulsification and oxygen barrier capabilities. In the food industry, proteins improve texture, stability, and nutritional value across dairy, baking, and meat processing sectors. In the fields of cosmetics and pharmaceuticals, collagen and gelatin are widely used, especially in hydrogels for drug delivery and wound healing purposes. Moreover, proteins can be structurally classified into fibrous, globular, and intermediate forms, with each type serving essential functions in providing structural support, enabling enzymatic activity, and preserving cell integrity. The ongoing advancement of

Manisha · Renu · Poonam · I. Khangwal (✉)
Department of Life Sciences (Botany), Baba Mastnath University, Rohtak, Haryana, India

V. Dangi
Department of Physical Sciences (Chemistry), Baba Mastnath University, Rohtak, Haryana, India

N. Kumar et al. (eds.), *Trends in Edible Packaging: A sustainable approaches for postharvest management of fruits and vegetables*,
https://doi.org/10.1007/978-3-032-13199-7_3

protein-based biopolymers offers significant potential for advancements in the medical and food industries, owing to their sustainability, functionality, and biocompatibility.

Keywords Protein · Polysaccharides · Polypeptides · Polynucleotides · Gelatin · Keratin

Introduction

Living organisms, such as bacteria, plants, algae, and animals, contain polymeric materials known as biopolymers (Vieira et al., 2011). They are often categorized as polysaccharides, polypeptides, and polynucleotides according to their monomeric unit (Choi et al., 2018). They are often extracted from waste products or the biological system using a range of chemical methods (Du et al., 2016). The processing conditions have a significant effect on their physiochemical properties, particularly the hierarchical assembly. High mechanical properties and superhydrophobicity are among the excellent attributes that the hierarchical structure of biopolymers imparts. Nevertheless, it is challenging to produce materials that mimic hierarchical assembly using the extracted biopolymer (Choi et al., 2018). Actually, a wide range of uses for extracted biopolymers exists, such as tissue engineering, food packaging, and the distribution of therapeutic compounds. Using the body's inherent capacity for healing, regenerative tissue engineering aims to restore damaged tissues or failing organs to better function. Scaffolds constructed from synthetic and biopolymers are utilized to provide a supportive environment that promotes tissue regeneration and cell proliferation. The low toxicity, biocompatibility, biodegradability, presence of biomimetic chemical entities, and release of innocuous fragments during disintegration make biopolymers an interesting class of materials. Therefore, the application of biopolymers in designs and systems for drug delivery and tissue engineering is common (Du et al., 2016).

The proteins utilized for the creation of films, Pickering emulsions, hydrogels, nano-polymers, microfibers, interconnecting porous scaffolds, and 3D-printed scaffolds include elastin, lactose, keratin, sericin, and fibroin. Electrospinning is an intriguing method for producing scaffolds designed for tissue engineering (Nagarajan et al., 2017). The regeneration of soft and hard tissues alike, such as neurons, bone, cartilage, and wound tissues, was assessed in relation to these mixes. This study clarifies that, in addition to animal-sourced proteins, there is a comprehensive examination of plant-based proteins—such as soy protein and zein—for medical applications. Biopolymers that have been extracted typically exhibit low mechanical properties (such as toughness and elasticity), insufficient film-forming ability, and increased brittleness (Nagarajan et al., 2018).

As stated by Chauhan et al. (2021), functional polymers possess atomic groups that are more polar or reactive than standard hydrocarbon chains. These materials exhibit enhanced properties, particularly in terms of stronger adhesion, separation capability, or chemical reactivity, when compared to their non-functional counterparts. The synthesis of such polymers frequently utilizes living anionic polymerization, a method that can be adapted. It allows for the creation of precisely defined polymers with functional groups positioned either along the chain or at their ends. These functional groups can be altered via reversible ionic mechanisms, chain extension, branching, or cross-linking reactions. Furthermore, functional macromolecular materials can be integrated into oligomeric or polymeric chains and joined with reactive copolymers (Chauhan et al., 2021). Polymerization or copolymerization can also be performed on monomers with functional groups that are sufficiently reactive.

The processing of bulk polymers often leads to chemical heterogeneity, which can provide various benefits, such as increased reactivity, phase separation, and better compatibility or molecular association. A major advantage of functionalized polymers is their capacity to create self-assemblies and supramolecular structures when mixed with different polymer architectures (Chauhan et al., 2021). It is possible to design these self-assembled materials so that they react to chemical or physical stimuli, paving the way for the creation of "smart" materials. Functional polymers are typically integrated into single linear backbones, such as block, graft, chain-end, and in-chain structures (Vana and Yagci et al., 2013).

These materials are essential for a variety of technical applications, including optics, electronics, and catalysis. They are also extensively employed in the synthesis of solid-phase oligonucleotides and membrane technologies, including applications in column chromatography. In the swiftly progressing domain of medicine, custom-designed functional polymers are essential—for example, in suture protection, dental fillings, wound dressings, bone cement, and hollow-fiber dialysis systems (Chauhan et al., 2021). Items like hydrogels and polymer-based inhibitors are crucial components in ophthalmic surgery as well.

There are three main approaches to polymer functionalization, which entail creating functional polymers: direct synthesis of functional polymers (in situ), post-functionalization (or post-polymer modification), and the transformation of functional groups.

Recent reviews focused on polymers based on proteins (Nitta et al., 2013; Elzoghby et al., 2012) identified several protein-based biopolymers that are appropriate for tissue engineering and drug delivery. (Ni et al., 2017) also underscored the importance of developing protein-based hydrogels for biomedicine applications. This review addresses recent advancements in polypeptides and their formulations, which are being utilized to develop scaffolds for tissue engineering. This assessment underscored both well-established and cutting-edge research areas in tissue engineering and drug delivery systems, encompassing subjects that had not been studied before, such as electrospinning, 3D printing, and the development of bio-inks from biopolymers. Based on the structure, proteins are classified into three categories (Fig. 3.1).

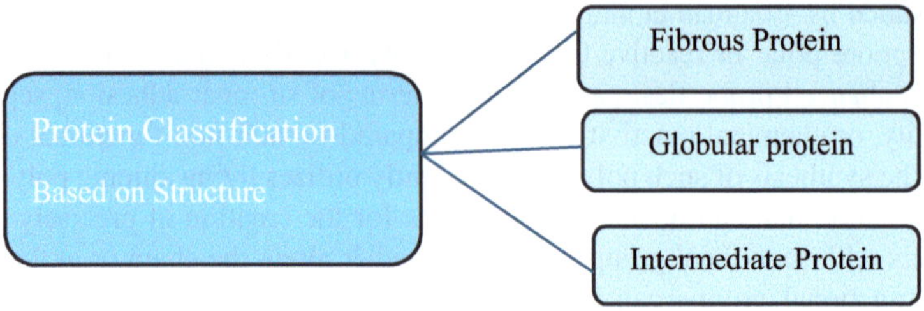

Fig. 3.1 Classifications of proteins based on their structure

Fibrous Protein

Fibrous proteins, also known as scleroproteins, form one of the three primary categories of protein structures, alongside globular and membrane proteins. They consist of elongated polypeptide chains arranged in either filamentous or sheet-like structures. Unlike globular proteins, which are generally spherical and soluble in water, fibrous proteins have low solubility in water (Shoulders & Raines, 2009).

These proteins primarily serve structural and protective roles in the body. They are key components of connective tissues, tendons, bone matrixes, and muscle fibers. In contrast, globular proteins perform more dynamic functions, such as enzymatic activity and molecular transport (Andreeva et al., 2014).

Globular Protein

Proteins can be classified into two primary types: globular and fibrous proteins. They are generally spherical or globular in shape and have a compact, folded structure. This structure makes globular proteins soluble in water, which is a key feature that distinguishes them from fibrous proteins (Prasad et al., 2023).

Intermediate Protein

Intermediate filaments (IFs) serve as structural elements of the cytoskeleton, present in the cells of both vertebrates and numerous invertebrates (Herrmann et al., 2007). Homologues of IF proteins have also been identified in the cephalochordate *Branchiostoma*, an invertebrate species (Karabinos et al., 1998). Composed of a family of related proteins that share common structural and sequence features, intermediate filaments were originally termed "intermediate" due to their average diameter of 10 nm, positioned between the narrower actin microfilaments (7 nm) and the wider microtubules (25 nm) found in cells (Herrmann et al., 2007). Animal intermediate filaments are categorized into six types due to the similarities in their amino

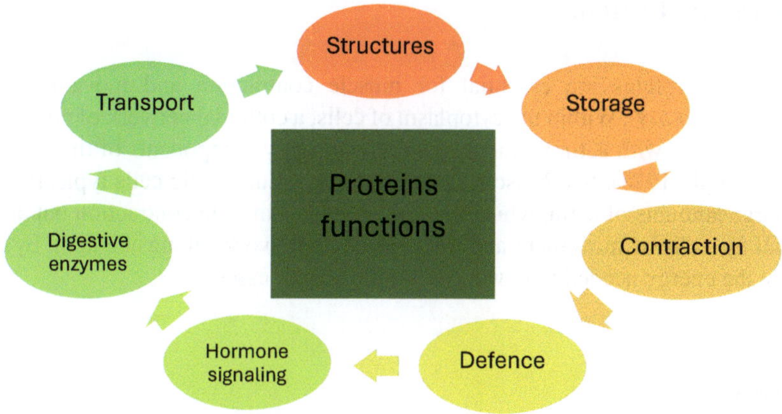

Fig. 3.2 Classification of proteins based on functions

acid sequences and protein structures. While the majority are cytoplasmic, a small number are located in the nucleus. Type V intermediate filaments are located in the nucleus and are known as nuclear lamins. Unlike microtubules, the distribution of intermediate filaments within cells does not closely correlate with the distribution of mitochondria or the endoplasmic reticulum (Soltys et al., 1992).

Classification of Proteins Based on Functions

Proteins are also categorized by the roles they perform in biological systems. All proteins can be broadly categorized into seven main types: antibodies, contractile proteins, enzymes, hormonal proteins, structural proteins, storage proteins, and transport proteins. The protein can be categorized into different types based on its functions (Fig. 3.2).

Antibodies

Because they shield the body against antigens like bacteria, viruses, and other external invaders, specialized proteins known as antibodies are vital to the immune system. Because they can travel via the bloodstream, antibodies help identify and get rid of threats throughout the body. They aid white blood cells in recognizing and getting rid of these intruders by binding to antigens and making them immobile.

Contractile Protein

Contractile proteins are essential for muscle contraction and movement (see *Molecular Motors*). Within the cytoplasm of cells, a colloidal network of contractile proteins exists, with actin filaments being the primary components of this network (Stossel et al., 1978). See Muscle Cells [Myocyte]. Eukaryotic cells typically contain large amounts of actin, which not only regulates muscle contraction but also is crucial for cellular movement and division. Myosin works alongside actin by providing the energy needed to power these cellular processes.

Enzymes

All known proteins, known as enzymes, act as catalysts for virtually all metabolic reactions in living organisms. They facilitate the production of essential biomolecules—including proteins, nucleic acids, carbohydrates, and lipids—and their conversion into other compounds and eventual breakdown when needed. Enzymes are indispensable to life; without them, the chemical reactions required to sustain life would occur too slowly to support biological processes (Koshland & Haurowitz, 2025).

Hormonal Protein

Hormonal proteins act as chemical messengers, regulating and coordinating various physiological processes across the body. For example, insulin is a hormonal protein that helps control blood sugar levels (Hernandez & Kravitz 2003). The types of protein and their functions are depicted in Table 3.1.

Growth Factors

Growth factors, a category of cytokines, are proteins characterized by their high specificity. They promote the division and differentiation of certain types of cells. In the domain of skeletal muscle hypertrophy, a key growth factor is the insulin-like growth factor (IGF), which is secreted by skeletal muscle. IGF plays a dual role by regulating insulin metabolism and stimulating protein synthesis, thereby supporting muscle growth and repair.

Table 3.1 Types and functions of proteins

Role	Examples	Function	Reference
Digestive enzyme	Amylase, lipase	Break down nutrients in food into small pieces that can be readily absorbed	Haurowitz (2025)
Transport	Hemoglobin	Carry substances throughout the body in blood or lymph	Koshland (2025)
Structures	Actin, tubulin	Build different structures, like the cytoskeleton	Koshland (2025)
Hormone signaling	Insulin, glucagon	Coordinate the activity of different body systems	Hernandez & Kravitz (2003)
Defense	Antibodies	Protect the body foreign pathogen	Haurowitz (2025)
Contraction	Myosin	Carry out muscle contraction	Koshland (2025)
Storage	Legume storage proteins, egg white (albumin)	Provide food for the early development of the embryo or the seedling	Hernandez & Kravitz (2003)

Testosterone

Testosterone, an androgen known primarily as the androgen, is chiefly responsible for the formation and upkeep of male reproductive organs and secondary sexual traits. Besides its reproductive functions, it affects multiple physiological systems, including the nervous system, skeletal muscles, bone marrow, skin, and hair.

Cortisol

Cortisol, a steroid hormone generated by the adrenal cortex of the kidneys, plays a crucial role in metabolic regulation and inflammation reduction during the body's stress response, helping control blood pressure. Because it has a steroid structure, cortisol can pass through cell membranes and interact directly with intracellular receptors.

Structural Protein

Structural proteins play a crucial role in maintaining and protecting the integrity of the animal body. One of the most well-known structural proteins is collagen, which is abundantly found in bones, skin, and connective tissues (Koshland & Haurowitz, 2025).

These proteins are also present within cells, where they contribute to the cytoskeleton—a framework that supports cell shape and organization. In addition to providing internal support, structural proteins can also be involved in cell movement. Their function becomes especially vital in larger or more complex cells, where maintaining shape and internal organization is critical.

Storage Proteins These proteins serve as reservoirs of amino acids and other essential nutrients, making them available for the body when needed. An example is ferritin, a storage protein that binds and stores iron for later use.

Transport Proteins These are carrier proteins that facilitate the movement of molecules throughout the body.

Hemoglobin It serves as a respiratory protein, transporting oxygen within the bloodstream, conveying oxygen from the lungs to tissues and organs (Koshland & Haurowitz, 2025).

Cytochromes are a distinct category of transport protein involved in the electron transport chain by carrying electrons, playing a key role in cellular energy production (see Adenosine Triphosphate, ATP) (Bailey et al., 2020).

Classification of Proteins Based on Structure

Proteins, which are heteropolymers, are formed by linking amino acids together in long chains. The sequences of these amino acids form essential components of proteins. As macromolecules, proteins are large and complex, and due to their size and nature, they fall under the acid-insoluble fraction when biochemical compounds are separated based on solubility. The structure of a protein can be described at four hierarchical levels, each providing insight into a different aspect of its shape and function (Fig. 3.3).

Primary Structure of Proteins

A protein's primary structure, which is essential for its function, is made up of its unique amino acid sequence. This sequence forms the foundation for all higher-order protein structures (Fig. 3.4), including secondary, tertiary, and quaternary configurations. An individual change in the amino acid sequence can result in significant effects. Mutations affecting the hemoglobin β-globin chain, like the substitution of valine for glutamic acid at position 6 (β6Glu → Val), can lead to sickle-cell anemia (Gambari et al., 2024).

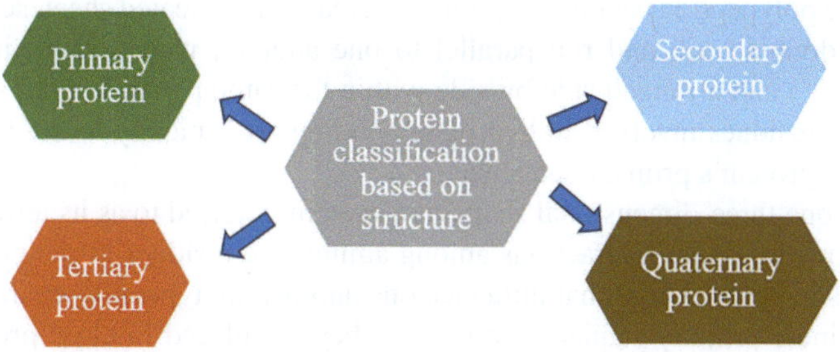

Fig. 3.3 Different types of protein structures

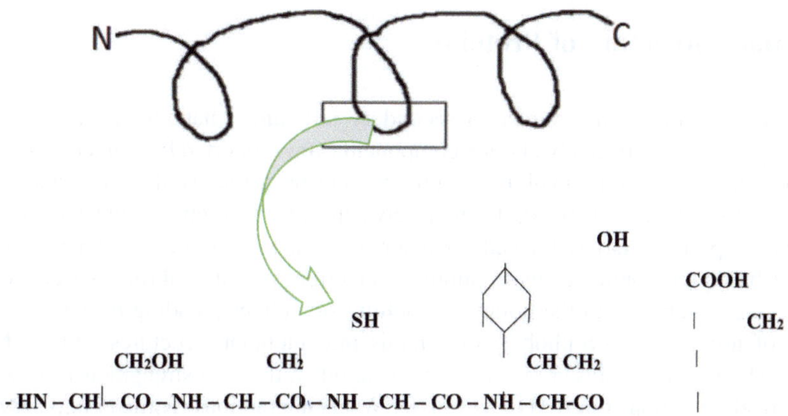

Fig. 3.4 Primary structure of protein

Secondary Structure of Proteins

Functional group interactions within the peptide-bonded backbone give rise to a protein's secondary structure. H2 bonds between the oxygen atom of a carbonyl group (C=O) on one amino acid and the H2 atom of an amine group (N-H) on another primarily maintain unique, regularly shaped structures such as alpha helices and beta sheets.

When the protein chain bends, polar groups in the backbone form hydrogen bonds, resulting in secondary structures, such as the α-helix and β-pleated sheet. In an α-helix, the backbone of the polypeptide spirals into a coil, while in a β-pleated sheet, segments of the protein chain are adjacent to each other, forming a sheet-like structure that is reinforced by hydrogen bonds. They turn sharply (approximately 180°) and arrange themselves next to each other in the same plane. Amino acid residues that participate in hydrogen bonding are generally situated close to one another in the protein's primary sequence in both structures.

As the chain bends, polar groups within the protein backbone form hydrogen bonds, leading to secondary structures such as the α-helix and β-pleated sheet. In an α-helix, the polypeptide backbone spirals, whereas in a β-pleated sheet, segments of the chain double back and run parallel to one another, with a sharp inclination (around 180°) and arranged side by side within the same plane. In both structures, amino acid residues involved in hydrogen bonding are typically located near each other in the protein's primary sequence.

The unique three-dimensional shape of a protein, referred to as its tertiary structure, is formed through interactions among amino acid residues as the polypeptide chain folds in space. Those that influence one another are typically separated in the protein's linear primary sequence; however, they are placed in close proximity to one another as the chain curves and collapses.

Tertiary Structure of Proteins

Tertiary structures, in contrast to secondary structures that rely exclusively on hydrogen bonds between backbone components, are supported by a diverse array of bonds and interactions involving R groups, and occasionally the backbone. This encompasses hydrogen bonds formed between polar side chains and their corresponding partial charges located on other R groups or the peptide backbone. In water-based environments, hydrophobic interactions play a vital role as well; water molecules interact with the polar, hydrophilic side chains, leading to the aggregation of nonpolar, hydrophobic side chains into globular structures. When these hydrophobic side chains come close to one another, they are strengthened by weak electrical attractions referred to as van der Waals interactions (Rittle et al., 2019).

Through a reaction that utilizes their sulfhydryl groups, covalent bonds can form disulfide bonds between the side chains of two cysteine residues. Known as disulfide bridges, these bonds establish robust links between various regions of a single polypeptide or between two distinct polypeptides. Ionic bonds, on the other hand, develop between R groups that possess full and opposite charges, such as those present in ionized acidic and basic side chains.

During the folding of a polypeptide into its functional 3D structure, amino acids whose side chains are nonpolar tend to group together in the protein's interior to evade contact with water. This hydrophobic core is stabilized by weak van der Waals forces. Moreover, the tertiary structure is bolstered by hydrogen bonds and ionic interactions involving charged or polar amino acids. Each of these interactions, considered separately, is relatively weak in the cellular environment, but together they play a vital role in establishing and maintaining the protein's unique shape (Scholten et al., 2024).

An established example is ribonuclease, which experiences denaturation with urea and the reducing agent mercaptoethanol. In these circumstances, the protein denatures and its disulfide bonds are broken. With the removal of urea and mercaptoethanol, ribonuclease spontaneously refolds into its native conformation and fully restores its function. This demonstrates that the primary structure alone contains all the details necessary to determine a protein's tertiary structure.

As proteins adopt their most stable conformations, they often pass through a series of intermediate structures that are not present in the final folded form. This process relies heavily on chaperonins, which are specialized protein molecules. They protect newly formed proteins from damaging chemical conditions in the cytoplasm and promote the correct and spontaneous folding of polypeptides. The molecular origins of various genetic disorders can stem from errors in protein folding.

Quaternary Structure of Proteins

Proteins' primary, secondary, and tertiary structures denote the organization of one polypeptide chain. However, there are proteins that consist of multiple polypeptide chains, called subunits, that come together to create a single functional entity. The quaternary structure, denoting the precise arrangement of these subunits, is reinforced by the same types of bonds and interactions present in tertiary structures.

Applications of Protein

Role of Proteins in the Meat Industry

Proteins are crucial to the formulation of many food items because they enhance texture, binding, stability, and water-holding capacity. In the meat industry, proteins are used as binders, extenders, or emulsifiers to improve the overall quality and yield of products like sausages, burgers, and meatballs. In order to enhance the texture, stability, and viscosity of items such as cheese, yogurt, ice cream, and milk-based drinks, the dairy industry commonly uses proteins like casein and whey. Similarly, in the baking industry, proteins like wheat gluten and soy protein improve the volume, structure, and elasticity of dough, enhancing the overall quality and texture of bread, cakes, and pastries (Bloom et al., 2005).

Growth of Plant-Based Protein Alternatives

In recent years, protein-based beverages have gained popularity due to their nutritional content and health benefits. Protein powders from sources like whey, soy, or pea are commonly added to meal replacements, nutritional smoothies, and sports drinks to provide essential amino acids and help with muscle repair. As the desire for plant-based diets increases, more and more proteins, such as soy, peas, lentils, and nuts, are being used to make meat and dairy alternatives. Because these plant-based proteins are produced and processed to mimic the taste and texture of animal-based products, vegetarians and vegans can choose from them. (Fernandes et al., 2010).

Versatility and Impact of Proteins in Food Products

Proteins have a significant impact on the taste, texture, nutritional value, and overall enjoyment of a wide variety of food products due to their versatility and use. Protein sources like casein and whey are commonly used in the production of infant formula to provide babies with a nutritionally balanced and easily digested diet. Protein powders and bars are also commonly used as dietary supplements by bodybuilders, athletes, and anybody else looking to increase their protein intake. These products often contain a variety of protein sources, including soy, whey, casein, and plant-based proteins (Ribeiro et al., 2010).

Proteins as Emulsifiers in Food Products

Proteins are also helpful in food preparation because of their functional characteristics. For instance, proteins like egg proteins or soy lecithin are used as emulsifiers in sauces, mayonnaise, and salad dressings. They improve the mouthfeel and texture of the product, stabilize the emulsion, and lessen separation. Proteins can also be supplemented with vitamins, minerals, essential amino acids, and other nutrients to enhance the nutritional profile of food products. This process is commonly used in beverages, bars, and fortified cereals (Bloom et al., 2005).

Essential Role of Proteins in Modern Food Development

Proteins have emulsifying and nutritious qualities, but they are also employed to change the texture of food items. Meat can be made more palatable by tenderizing it with hydrolyzed proteins or enzymes. Another protein called gelatin is frequently used to give desserts like mousses and puddings a smooth, creamy feel. These various uses demonstrate how essential proteins are to the creation and improvement of contemporary food products (Fernandes et al., 2010).

Impact of Enzymes on Food Quality and Functionality

Due to their capacity to function as catalysts and convert fundamental components into enhanced food items, enzymes have been of great significance in the field of food technology for a long time. Their main benefits are their efficiency at mild pH, temperature, and ion concentrations, as well as their substrate selectivity. When employed appropriately, enzymes can change and enhance the functional, nutritional, and sensory aspects of raw materials and final products. As a result, they are frequently utilized in the production and processing of various food products. Food scientists often select customized enzymes to improve different unit functions inside food manufacturing processes.

In a similar vein, proteins' nutritional value, practical properties, and ability to raise the general caliber of products make them indispensable to the food industry, such as the dairy industry, baking industry, juice industry, brewing industry, and meat processing. Proteins are essential components in many culinary applications due to these characteristics (Fernandes, 2010; Ribeiro et al., 2010).

Uses of Protein Biopolymers

Collagen in Cosmetics

Collagen is a crucial structural protein found in animals that is essential for maintaining tissue integrity and design, as well as for supplying the required mechanical support (Mitura et al., 2020). Instead of being a single, fundamental protein, collagen is a family of 29 genetically distinct proteins. The three most prevalent are various kinds of collagen, such as type I, which is located in bones, tendons, and skin; type II, located in cartilage; and type III, which is located in the skin and blood vessels. These types produce fibrillar structures that are crucial for maintaining the structural integrity of tissues.

The most common usage of type I collagen is in cosmetic applications due to its availability and properties. Type I collagen has an α-helical form and is linked together by a repeating tripeptide sequence. Each chain is composed of about 1000 amino acids. Collagen has a triple-helical configuration composed of three polypeptide chains. While covalent connections anchor the individual chains, these chains are linked via non-covalent interactions, which include hydrogen bonds, dipole-dipole interactions, ionic bonds, and van der Waals forces. Collagen's properties can be significantly influenced by its source. There are numerous creatures that can produce collagen. For instance, marine sponge collagen can be used as a substitute in cosmetic compositions, even though it usually has lower heat stability than collagen derived from chicken. Furthermore, bacterial fermentation methods using species like *Rhodopseudomonas palustris, Solibacter usitatus,* and *Methylobacterium* have been developed to produce collagen in a more controlled and sustainable manner.

The weak non-covalent links that keep the triple helix together dissolve when heated, whereas covalent bonds remain intact. The result is denaturation of collagen, which causes the helical chains to unravel and collapse into random coils to form gelatin. The water-soluble proteins that make up gelatin are essentially denatured collagen (Sionkowska et al., 2011). Since denaturation reveals more active groups, gelatin may form hydrogen bonds with water more effectively than natural collagen.

Collagen and gelatin are both extensively used in the pharmaceutical, cosmetic, and medical sectors due to their diverse range of functions and biocompatibility. They are frequently employed in the manufacturing of hydrogels, which are helpful in formulations for medicine administration, wound healing, and cosmetics. The degree of cross-linking, water content, and environmental pH are some of the factors that influence the thermal denaturation temperature, or melting point, of collagen. (Sionkowska et al., 2011).

Proteins Utilized in Edible Film and Coating Production

Protein-based packaging refers to the use of both plant- and animal-based proteins, including casein, zein, gluten, and gelatin, in food packaging. Because of their abundance, biodegradability, and ease of manufacturing, protein-based polymers have received a lot of attention from researchers seeking to develop sustainable food packaging films. These films can be made using a variety of methods, including solution casting, electrospinning, extrusion, and printing. The choice of production technique and certain process parameters has a significant impact on the final qualities of the films that are produced (Agarwal et al., 2022).

Gluten-Based Films

Gluten, a protein that occurs naturally in wheat, barley, and rye, is utilized for producing gluten-based films, which serve as a type of food packaging material (Guillaume et al., 2010). The two main types of gluten proteins, gliadin and glutenin, are classified based on their solubility in hydroalcoholic solutions. The usual method for producing these films involves isolating the gluten from wheat flour, then adding water and plasticizers to produce a uniform film-forming solution. Plasticizers enhance the flexibility and mechanical properties of the finished film (Gómez-Heincke et al., 2019).

Soy Protein as a Biopolymer for Film Formation

Films made from soy protein isolate (SPI) are known as soy protein-based films and provide a biodegradable and environmentally friendly option compared to conventional synthetic packaging materials (Koshy et al., 2015; González et al., 2013). These films' exceptional oxygen barrier properties help reduce oxidative damage, thereby prolonging the shelf life of packaged foods.

Zein Protein-Based Films

Zein, a prolamin protein that was extracted from maize, makes up between 60 and 70 percent of the protein in corn (Bayer et al., 2021). It is the perfect protein for making films and coatings for use in cooking because it is nontoxic and edible. Zein-based films are typically made by casting ethanol-based zein solutions, which are further strengthened by protein cross-linking and solvent evaporation (Moreno et al., 2019).

Casein Protein-Based Films

Casein, a phosphoprotein that precipitates at a pH of 4.6, is one of the primary proteins found in milk. The ability of its four primary subfractions—αS1-, αS2-, β-, and κ-caseins—to form films is influenced collectively. Published studies indicate that casein lacks a clear secondary structure and is an unstructured protein (Khan et al., 2021). Without much processing, casein-based films can be made directly from liquid solutions. Because of its ability to form micelles, chemical resistance, nontoxicity, and biodegradability, casein is thought to be an excellent choice for biodegradable food packaging films. Its utility in packaging applications is further enhanced by its capacity to bind small particles (Khan et al., 2021).

Casein's superior nutritional value, emulsifying capacity, and water solubility make it a promising biomaterial for filmmaking (Wagh et al., 2014). However, because native casein is somewhat insoluble, sodium caseinate, a sodium salt form of casein, is commonly used as an adjuvant in formulations to promote solubility and compatibility with other materials (Khan et al., 2021).

Furthermore, the amount of protein included in the formulation has a notable effect on casein- and whey-protein interactions that result in films, which in turn affects the performance and structural integrity of the finished films (del R. Moreira et al., 2011).

Gelatin Films for Sustainable Packaging

Gelatin-based films and coatings are widely used in a variety of industries, including food packaging, pharmaceuticals, and biomedical applications (Lu et al., 2022). These films' remarkable gas barrier properties, clarity, and flexibility make them perfect for packaging meat items, fruits, and vegetables, as well as confections (Luo et al., 2022).

Gelatin is created by hydrolyzing denatured collagen. The final product's molecular weight might vary from 65,000 to 300,000 g/mol, depending on the degree of hydrolysis. Gelatin is an essential raw material for the production of eco-friendly, biodegradable packaging films because of its long shelf life, biodegradability, biocompatibility, and uniform film-forming properties (Ramos et al., 2016).

Protein Films for Food Packaging

In the food industry, films—thin layers—are usually utilized as the main packaging material. A wide range of polymeric matrixes can be used to create them (Hassan et al., 2018). Films have been widely utilized to improve the sensory aspects of products, act as a barrier against moisture and gases, and add active substances to products to extend their shelf life (Chiralt et al., 2018). Films have been widely used to add active ingredients to items to increase their shelf life, enhance the sensory qualities of products, and serve as a barrier against gases and moisture (Flores-López et al., 2016). Because they exhibit 30 unique bio-based packaging qualities compared to other films, including transparency, ideal lipid and carbon-dioxide barrier qualities, and high water sensitivity, protein-based films are intriguing for usage in the food business (Chiralt et al., 2018). Additives, such as plasticizers (sorbitol, glycerol, and polyethylene glycol), are typically used in film-production processes to improve the properties of polymers in order to make the packaging more pliable and simpler to work with (Lacroix & Vu, 2014). The process of making films involves a number of steps, including changing the pH, adding salt, heating, enzyme modification, drying, using solvents, and applying mechanical forces. Covalent bonds, inter- and intramolecular forces, electrostatic interactions, and ionic forces are just a few of the numerous chemical reactions that take place during film development to create the three-dimensional polymer network that makes up the packaging (Han et al., 2005).

The characteristics of the native protein, such as the forces that comprise its structure and its initial amino acid sequence, as well as the methods used to extract it, are significant determinants of the final characteristics of the film that is created (Schmid et al., 2018). The structure of the protein and the method used to make the film will dictate the final characteristics of the packaging. Animal proteins, such as casein and whey protein from milk, have been the subject of much research due to their excellent nutritional value and outstanding functional characteristics. Generally speaking, milk protein films are cohesive, edible, flexible, transparent or translucent, and tasteless (Schmid et al., 2018). Two more protein types that are frequently employed are collagen and gelatin. Collagen can be found in the skin, tendons, and other connective tissues, which comprise up to 30% of the body. Collagen films are characterized by their high surface homogeneity, flexibility, low mechanical strength, and limited permeability to oxygen and humidity. Gelatin is used in films because it offers qualities including transparency, flexibility, resilience, and impermeability (Rocha et al., 2013). Additional examples are the myofibrillar proteins found in fish muscle, which create films with high water-vapor permeability (WVP), high tensile strength, and intermediate solubility (Lacroix et al., 2005).

Food packaging is made from a variety of vegetable-based proteins, including soy protein, maize zein, and wheat gluten. Gliadins and glutenins are two significant proteins that are separated from wheat gluten and are responsible for making gluten. Wheat gluten has a bright future in the business, primarily as adhesives, films, and coatings. Because the protein is only partially soluble in water, solvents like alcohol must be used in the process of making wheat gluten films. With their selective gas permeability, films made from this protein have a lot of potential for use as packaging, and are simple to employ to regulate respiratory exchange in the preservation of fruits and vegetables (Hassan et al., 2018). Because of the amino acid sequences that render it insoluble in water, the protein known as zein, which is present in corn, is hydrophobic. Due to the characteristics of the zein structure, the films produced from this protein have low WVP and water solubility. Soy protein is another protein that has been thoroughly studied for use in packaging. Soy protein may be dissolved in water by changing the pH and adding heat, unlike gluten and zein (Lacroix & Vu, 2014). Since soy protein is inexpensive and fully biodegradable, using it as a packaging substitute has several benefits. Films produced from separated soy protein show intermediate mechanical qualities and are somewhat oxygen-soluble at low humidity (Wang et al., 2017).

The physical properties of protein biopolymers are still inferior to those of various synthetic polymers used in the food industry, such as cellophane, high-density polyethylene (HDPE), and low-density polyethylene (LDPE), despite the fact that they have many advantages and great potential for utilization. In order to overcome these limitations and match the properties of natural and synthetic polymers, several strategies are used (Campos-Requena et al., 2017). Even though there is a very thorough search for biodegradable polymers with suitable properties for industrial applications, there are now very few commercially available packaging options. Cups, bags, straws, lids, gloves, cutlery, bottles, plates, and many other packaging materials are produced from biodegradable qualities. Nevertheless, the composition of most of these materials is usually patent-protected, and they are not solely made from biopolymers derived from natural and renewable sources. Most films either use natural polymers as an element in the packaging formulation process or contain additional biodegradable polymers from nonrenewable sources, such as poly(vinyl alcohol) (PVA) (Rydz et al., 2018). Carbohydrates are the most researched natural polymers in commercial applications due to their low cost and broad use. Although there are many potential applications for proteins, further studies is still required to enhance their packaging properties and production processes.

Conclusion

Proteins, being biopolymers with a high degree of adaptability, play an essential role across multiple fields such as biology, medicine, industry, and food. Biopolymers, such as polysaccharides, polypeptides, and polynucleotides, are derived from living organisms and often exhibit hierarchical structures that result in properties like high mechanical strength and superhydrophobicity. Even so, it remains challenging to produce synthetic versions of these natural assemblies. Protein-based biopolymers like collagen, gelatin, keratin, and fibroin are widely used in regenerative medicine due to their low toxicity, biodegradability, and compatibility with biological systems. Techniques like electrospinning and 3D printing have expanded their application to creating scaffolds for both soft and hard tissue engineering. In regenerative medicine, protein-based biopolymers such as collagen, gelatin, keratin, and fibroin are commonly employed because of their low toxicity, biodegradability, and compatibility with biological systems. Methods such as electrospinning and 3D printing have broadened their use for the development of scaffolds for both soft and hard tissue engineering. Conversely, synthetic functional polymers created through techniques such as anionic polymerization and post-functionalization provide precise control, reactivity, and responsiveness. This makes them useful for smart materials, drug delivery systems, and biosensors. On the other hand, synthetic functional polymers produced through methods like anionic polymerization and post-functionalization offer precise control, reactivity, and responsiveness. As a result, they can be utilized in smart materials, drug delivery systems, and biosensors. Proteins can be structurally categorized as fibrous, globular, or intermediate, and their functions and interactions are dictated by their primary to quaternary structures. It is crucial for proteins to fold correctly, as misfolding can impede functionality. In food manufacturing, proteins are involved in texture, taste, and preservation. Protein-based biopolymers like gluten, casein, soy protein, and zein are being acknowledged as eco-friendly packaging alternatives due to their film-forming ability and biodegradability. Because of their mechanical strength and biological compatibility, collagen and gelatin are used in pharmaceuticals, cosmetics, and packaging to create hydrogels and biodegradable films for drug delivery and wound care.

References

Agarwal, A., Shaida, B., Rastogi, M., & Singh, N. B. (2022). Food packaging materials with special reference to biopolymers-properties and applications. *Chemistry Africa, 6*, 0123456789. https://doi.org/10.1007/s42250-022-00446-w

Andreeva, A. (2014). SCOP2 prototype: A new approach to protein structure mining. *Nucleic Acids Research, 42*(Database issue), D310–D314.

Bailey, R., et al. (2020). *The role of proteins in food composition and nutrition*. Retrieved from https://www.example.com/role-of-proteins.

Bayer, I. S. (2021). Zein in food packaging. In *Sustainable food packaging technology* (pp. 199–224). Wiley. https://doi.org/10.1002/9783527820078.ch7

Bloom, M., Brown, R., & Smith, T. (2005). Functional roles of proteins in food processing. *Food Science and Technology Journal, 38*(4), 245–257.

Campos-Requena, V. H., Rivas, B. L., Pérez, M. A., Figueroa, C. R., Figueroa, N. E., & Sanfuentes, E. A. (2017). Thermoplastic starch/clay nanocomposites loaded with essential oil constituents as packaging for strawberries À in vivo antimicrobial synergy over Botrytis cinerea. *Postharvest Biology and Technology, 129*, 29–36.

Chauhan, N. P. S. (2021). Functionalized polymers.

Chiralt, A., González-Martínez, C., Vargas, M., & Atarés, L. (2018). Edible films and coatings from proteins. In R. Y. Yada (Ed.), *Protein-Based Films and Coatings* (pp. 477–500).

Choi, S. M., Chaudhry, P., Zo, S. M., & Han, S. S. (2018). Advances in protein-based materials: From origin to novel biomaterials. In H. J. Chun, C. H. Park, I. K. Kwon, & G. Khang (Eds.), *Cutting-edge enabling technologies for regenerative medicine* (p. 161). Springer Singapore.

Du, Y., Li, S., Zhang, Y., Rempel, C., & Liu, Q. (2016). *Journal of Applied Polymer Science, 133*, 29.

Elzoghby, O., Samy, W. M., & Elgindy, N. A. (2012). Protein-based nanocarriers as promising drug and gene delivery systems. *Journal of Controlled Release, 161*, 38–49.

Fernandes, A. C. (2010). Role of dietary proteins in food structure and functionality. *International Journal of Food Science and Nutrition, 61*(5), 389–397.

Flores-López, M. L., Cerqueira, M. A., de Rodríguez, D. J., & Vicente, A. A. (2016). Perspectives on utilization of edible coatings and nano-laminate coatings for extension of postharvest storage of fruits and vegetables. *Food Engineering Reviews, 8*, 292–305.

Gambari, R., Waziri, A. D., Goonasekera, H., & Peprah, E. (2024). Pharmacogenomics of drugs used in β-thalassemia and sickle-cell disease: From basic research to clinical applications. *International Journal of Molecular Sciences, 25*(8).

Gómez-Heincke, D., Martínez, I., Partal, P., Guerrero, A., & Gallegos, C. (2019). Development of antimicrobial active packaging materials based on gluten proteins. *Food Engineering Reviews*. https://doi.org/10.1007/s12393-019-09189-w

González, C. I., & Igarzabal, A. (2013). Soy protein—poly (lactic acid) bilayer films as biodegradable material for active food packaging. *Food Hydrocolloids, 33*(2), 289–296.

Guillaume, A., Pinte, J., Gontard, N., & Gastaldi, E. (2010). Wheat gluten-coated papers for bio-based food packaging: Structure, surface and transfer properties. *Food Research International, 43*(5), 1395–1401. https://doi.org/10.1016/j.foodres.2010.04.014

Han, J. H. (2005). Edible films and coatings: A review. In J. H. Han (Ed.), *Innovations in food packaging* (pp. 213–255).

Hassan, B., Ali, S., Chatha, S., Hussain, A. I., & Zia, K. M. (2018). Recent advances on polysaccharides, lipids and protein based edible films and coatings: A review. *International Journal of Biological Macromolecules, 109*, 1095–1107.

Haurowitz, F., Koshland, D.E. (2025, June 18). Protein. Encyclopedia Britannica. https://www.britannica.com/science/protein.

Hernandez, R. J., & Kravitz, L. (2003). The mystery of skeletal muscle hypertrophy. *ACSM's Health & Fitness Journal, 7*, 18–22.

Herrmann, E., Call, J., Hernàndez-Lloreda, M. V., Hare, B., & Tomasello, M. (2007). Humans have evolved specialized skills of social cognition: The cultural intelligence hypothesis. *Science, 317*(5843), 1360–1366. https://doi.org/10.1126/science.1146282. PMID: 17823346.

Karabinos, A., Riemer, D., Erber, A., & Weber, K. (1998). Homologues of vertebrate type I, II, and III intermediate filament (IF) proteins in an invertebrate:the IF multigene family of the cephalochordate Branchiostoma. *FEBS letters, 437*(1–2), 15–18.

Khan, M. R., Volpe, S., Valentino, M., Miele, N. A., Cavella, S., & Torrieri, E. (2021). Active casein coatings and films for perishable foods: structural properties and shelf-life extension. *Coatings, 11*(8). https://doi.org/10.3390/coatings11080899

Koshland, D.E., & Haurowitz, F. (2025, June 18). protein. *Encyclopedia Britannica*. https://www.britannica.com/science/protein.

Koshy, R. R., Mary, S. K., Thomas, S., & Pothan, L. A. (2015). Environment friendly green composites based on soy protein isolate—a review. *Food Hydrocolloids, 50*, 174–192. https://doi.org/10.1016/j.foodhyd.2015.04.023

Lacroix, M., & Cooksey, K. (2005). Edible films and coatings from animal-origin proteins. In H. H. Jung (Ed.), *Innovations in Food Packaging* (pp. 301–317).

Lacroix, M., & Vu, K. D. (2014). Edible coating and film materials. In H. Jung & Han (Eds.), *Innovations in Food Packaging* (pp. 325–350).

Lu, Y., Luo, Q., Chu, Y., Tao, N., Deng, S., Wang, L., & Li, L. (2022). Application of gelatin in food packaging: a review. *Polymers, 14*(3). https://doi.org/10.3390/polym14030436

Luo, Q., Hossen, M. A., Zeng, Y., Dai, J., Li, S., Qin, W., & Liu, Y. (2022). Gelatin-based composite films and their application in food packaging: a review. *Journal of Food Engineering, 313*(April 2021), 110762. https://doi.org/10.1016/j.jfoodeng.2021.110762

Mitura, S., Sionkowska, A., & Jaiswal, A. (2020). Biopolymers for hydrogels in cosmetics. *Journal of Materials Science: Materials in Medicine, 31*, 1–14.

Moreira, M. d. R., Pereda, M., Marcovich, N. E., & Roura, S. I. (2011). Antimicrobial effectiveness of bioactive packaging materials from edible chitosan and casein polymers: Assessment on carrot, cheese, and salami. *Journal of Food Science, 76*(1), 54–63. https://doi.org/10.1111/j.1750-3841.2010.01910.x

Moreno, M. A., Orqueda, M. E., Gómez-Mascaraque, L. G., Isla, M. I., & López-Rubi, A. (2019). Crosslinked electrospun zein-based food packaging coatings containing bioactive chilto fruit extracts. *Food Hydrocolloids, 95*, 496–505. https://doi.org/10.1016/j.foodhyd.2019.05.001

Nagarajan, S., Pochat-Bohatier, C., Balme, S., Miele, P., Kalkura, N. S., & Bechelany, M. (2017). *Pure and Applied Chemistry, 89*, 1799.

Nagarajan, S., Abessolo Ondo, D., Gassara, S., Bechelany, M., Balme, S., Miele, P., Kalkura, N., & Pochat-Bohatier, C. (2018). *Langmuir, 34*, 1542.

Ni, N., & Dumont, M.-J. (2017). *Waste and Biomass Valorization, 8*, 285.

Nitta, S. K., & Numata, K. (2013). *International Journal of Molecular Sciences, 14*, 1629.

Prasad, B., Shiv, A. K., Bharati, A. C., & Mallick, S. (2023). Proteins: Structure, properties, and importance. In *Handbook of biomolecules* (pp. 77–117).

Ramos, M., Valdés, A., Beltrán, A., & Garrigós, M. (2016). Gelatin-based films and coatings for food packaging applications. *Coatings, 6*(4), 41. https://doi.org/10.3390/coatings6040041

Ribeiro, A. C., Silva, L. M., Santos, R. J., & Oliveira, P. H. (2010). Application of proteins in food systems: Functional and nutritional aspects. *Journal of Food Science and Technology, 45*(3), 210–218.

Rittle, J., Field, M. J., Green, M. T., & Tezcan, F. A. (2019). An efficient, step-economical strategy for the design of functional metalloproteins. *Nature Chemistry, 11*(5), 434–441.

Rocha, M., Loiko, M. R., Tondo, E. C., & Prentice, C. (2013). Physical, mechanical and antimicrobial properties of Argentine anchovy (Engraulis anchoita) protein films incorporated with organic acids. *Food Hydrocolloids, 37*, 213–220.

Rydz, J., Musioł, M., Zawidlak-W, B., & Sikorska, W. (2018). Polymers for food packaging applications. In A. M. Grumezescu & A. M. Holban (Eds.), *Biopolymers for Food Design* (pp. 431–467).

Schmid, M., Merzbacher, S., & Müller, K. (2018). Time-dependent crosslinking of whey protein based films during storage. *Materials Letters, 215*, 8–10.

Scholten, N. R., Haandrikman, D., Tolhuis, J. O., Morandi, E., & Incarnato, D. (2024). SHAPEwarp-web: Sequence-agnostic search for structurally homologous RNA regions across databases of chemical probing data. *Nucleic Acids Research, 52*(W1), W362–W367.

Shoulders, M. D., & Raines, R. T. (2009). Collagen structure and stability. *Annual Review of Biochemistry, 78*, 929–958.

Sionkowska, A. (2011). Current research on the blends of natural and synthetic polymers as new biomaterials. *Progress in Polymer Science, 36*(9), 1254–1276.

Soltys, B. J., Gupta, R., et al. (1992). Functional and structural properties of food proteins: Applications in food processing. *Journal of Food Biochemistry, 16*(3), 205–218.

Stossel, T. P. (1978). Contractile proteins in cell structure and function. *Annual Review of Medicine, 29*(1), 427–457.

Vana, P., & Yagci, Y. (2013). *Living anion polymerization and functional polymer architectures*. In *Functionalized polymers: Synthesis, characterization and applications* (pp. specific pages). [Publisher information]. Functional polymers—Integrated through block, graft, chain-end, and in-chain structures—Can be tailored for self-assembly and responsiveness to physical or chemical stimuli onlinelibrary.wiley.com+8.

Vieira, M. G. A., Silva, M. A., Santos, L. O., & Beppu, M. M. (2011). Natural-based plasticizers and biopolymer films: A review. *European Polymer Journal, 47*, 254–263.

Wagh, Y. R., Pushpadass, H. A., Emerald, F. M. E., & Nath, B. S. (2014). Preparation and characterization of milk protein films and their application for packaging of Cheddar cheese. *Journal of Food Science and Technology, 51*(12), 3767–3775. https://doi.org/10.1007/s13197-012-0916-4

Wang, K., Wang, W., Ye, R., Liu, A., Xiao, J., Liu, Y., et al. (2017). Mechanical properties and solubility in water of corn starch-collagen composite films: Effect of starch type and concentrations. *Food Chemistry, 216*, 209–216.

Chapter 4
Lipid and Wax: Edible Coating and Films

Manoj Kumari, Sarita Kodan, and Ishu Khangwal

Abstract Lipid-based edible films and coatings are becoming a viable and efficient way to improve food quality and extend shelf life in the rapidly changing food packaging industry. The goal of this critical analysis is to provide a thorough overview of the lipids and waxes, and their applications and functions in food packaging. It investigates the special benefits of lipids, such as waxes, resins, and fatty acids, in offering efficient gas, water vapour, and microbiological barriers. Lipid-based composite films exhibit excellent mechanical, thermal, and barrier capabilities when combined with other biopolymers, including proteins and polysaccharides. The paper also discusses how these cutting-edge coatings are used to preserve a variety of fruits and vegetables, emphasising how they help retain firmness, regulate respiration rates, and lessen moisture loss. To meet consumer and regulatory concerns, the safety aspects of lipid-based coatings are also covered. Lipid-based films help minimise the loss of moisture and respiration, and reduce the growth of microbial activity by maintaining the quality of food. Lipid-based edible films maintain the shelf life of numerous food products, including vegetables and fruits. Lipids, particularly waxes, provide barriers against microbes and oxygen, and increase the preservation of various food products. By limiting access of oxygen, lipid-based edible films can slow down the rate of respiration of various food products, like fruits and vegetables—also extending the shelf life of fruits. Lipid-based films and coatings provide various benefits, certain challenges to overcome, like adhesion and flexibility of the film, mainly used in combination with polymers. Coatings integrate various lipids with several polymers, improving various mechanical, chemical, and thermal properties. Certain waxes, like beeswax and carnauba wax, improve water resistance in vegetables and fruits.

Keywords Biopolymers · Active packaging · Food preservation · Hydrophobicity · Composite films

---

M. Kumari · S. Kodan · I. Khangwal (✉)
Department of Life Sciences (Botany), Baba Mastnath University, Rohtak, Haryana, India

N. Kumar et al. (eds.), *Trends in Edible Packaging: A sustainable approaches for postharvest management of fruits and vegetables*,
https://doi.org/10.1007/978-3-032-13199-7_4

Introduction

Lipids are organic compounds derived from living organisms, such as insects, plants, and animals. These chemicals, however, require high temperatures for processing and are mostly insoluble or hardly soluble in water, but soluble in a few organic solvents such as methanol, hexane, diethyl ether, benzene, and chloroform. Simple, derived, and complex lipids are the three main types of lipids. Simple lipids can be further divided into oils, fats, and waxes. In contrast, there are two kinds of compound lipids: glycolipids and phospholipids. Derived lipids, the third class of lipids, are further divided into carotenoids, terpenes, and steroids. The most important among them in food and nutritional processes are simple lipids, especially fats and oils. Triglycerides make up the majority of oils and fats, which are common lipids found in a variety of food sources, including both plant and animal products (Zubair et al., 2023).

Natural polymer-based biodegradable packaging seems like a possible substitute for plastic packaging. To provide a protective barrier around the food surface against a variety of circumstances, materials are used in the manufacturing of edible films and coatings (Mohamed, 2023). Edible films and coatings are of tremendous interest in the food packaging sector because of their edible qualities, which can reduce packaging waste and show remarkable resistance to oxidation, moisture, solvent transfer, and scent transfer (Atta et al., 2023). They prolong the food's shelf life by preventing microbial development, decreasing particle clumping, and improving food safety and physical strength (Milani & Nemati, 2023). Prefabricated frameworks called edible films are used to envelop food items to prevent deterioration. They also act as carriers for active ingredients like antioxidants, flavours, and antimicrobials. On the other hand, edible coatings are usually applied directly to the food's surface by electro-spraying, which creates a thin and even coating, or spraying or dipping in liquid form, which forms micro-layers of film on the food's surface (Hassan et al., 2024).

Biopolymers like proteins, lipids, and polysaccharides are typically used to make films and coatings (Mohamed, 2023). Each set of materials has advantages and disadvantages. Protein and polysaccharide films generally have high mechanical properties and a robust barrier against oxygen at low and intermediate relative humidities. However, because they are hydrophilic, they have a weak barrier against water vapour (Ng et al., 2025; Rhim & Shellhammer, 2005). Lipid coatings offer superior moisture and oxygen barriers, guaranteeing food quality over time, whereas lipid films offer hydrophobicity and cohesiveness (Rhim and Shellhammer, 2005). Due to intrinsic restrictions, single-material packaging frequently falls short of optimal packaging standards. Thus, creating composites is a tactic to enhance food packaging's qualities. Various packaging materials can be combined to create composite films and coatings. Nonetheless, hydrocolloid and lipid-based films and coatings are more effective than composite films created by combining different materials (Yousuf et al., 2022). The structure, physical state (liquid or solid), chemical arrangement, hydrophobicity, and interactions with other components are some of the properties of lipid materials that affect their effectiveness in composite films and coatings (Pérez-Gago and Rhim, 2024; Rhim and Shellhammer, 2005).

The most widely utilised lipid molecules for creating edible films and coatings are waxes, fatty acids, resins, and neutral lipids (Pérez-Gago & Rhim, 2024). Compared to a solid lipid, a liquid lipid provides less protection against the passage of gases and vapours. Cereals, chocolates, dairy products, confections, frozen foods, fish, poultry, and dry goods can all be made into edible films using lipids (Milani & Nemati, 2023).

Waxes are hydrophobic substances composed of chains of carbon atoms that are long and medium in length. Depending on where it comes from, wax can be either natural or synthetic (Goslinska et al., 2019). They are better water barriers than other lipids because of their high hydrophobicity. Their high concentration of alcohols, long-chain fatty acids, and long-chain esters and alkanes causes their high hydrophobicity (Syahida, 2023). The application of wax coats to stop different fruits and vegetables from drying out has been the subject of numerous investigations.

Waxes

A broad definition of "waxes" dates back to 1975 when H. Bennett defined it as "an unctuous solid with varying degrees of gloss, slipperiness, and plasticity, which melts readily" (Bennett, 2024). There are various types of waxes described in Fig. 4.1. Waxes can be referred to as fatty components obtained from plants, insects, animal skins, or from a mineral origin. They can also be described as hydrophobic organic substances of medium- or long-chain length (Zhao et al., 2022). Generally, waxes contain very long-chain (>C18) fatty acid hydrocarbons, alcohols, aldehydes,

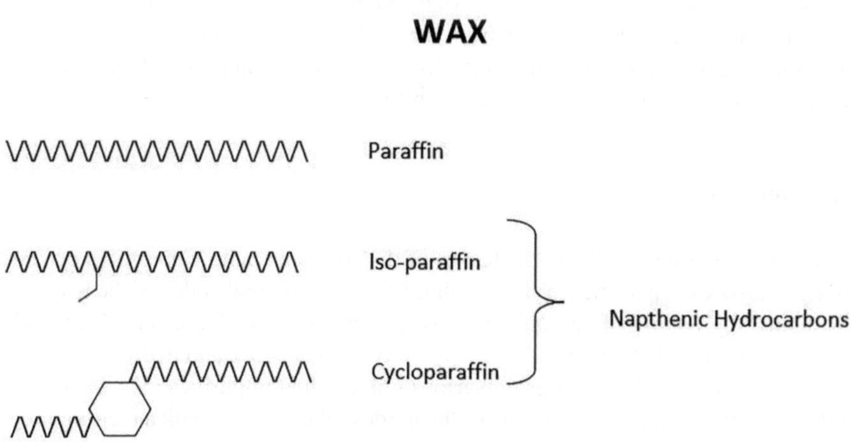

Fig. 4.1 Types of waxes and their schematic diagrams

ketones, esters, sterols, triterpenes, and flavonoids (Zhao et al., 2022). Waxes are insoluble in water and soluble in specific organic solvents. Since waxes are complex mixtures, their physical properties vary dramatically depending on their sources, and waxes are often used to describe specific products. Regardless of its broad definition and complex composition, there is no dispute that wax is broadly used in various industries. So far, there is no review focused on the aspect of food science. Therefore, in this chapter, classic waxes from various sources, as well as their food-related applications, are discussed.

Natural Waxes

Natural waxes originate from a variety of sources, including animals, plants, and minerals like petroleum, coal, and peat. The composition of waxes is largely determined by their source. Most natural waxes derived from animals and plants are composed of hydrocarbons, long-chain alcohols, long-chain carboxylic acids, and long-chain esters. Mineral waxes, on the other hand, consist almost entirely of long-chain hydrocarbons. Occasionally, the nature and classification of waxes can be confused with fats and gums. For instance, wool wax obtained from lanolin is often mistaken for a fat due to its properties, even though it is inherently a wax (Tinto et al., 2017).

Plant Waxes

Plant waxes typically serve as waterproofing agents, forming an amorphous layer on the outer surfaces of plants. They play a critical role in providing barrier protection against environmental stresses. Many plant waxes have received Generally Recognised as Safe (GRAS) status from the FDA (Marangoni et al., 2012), supporting their widespread use in food applications. Common examples include candelilla wax, carnauba wax, rice bran wax, and sunflower wax.

Carnauba Wax

Carnauba wax is derived from the leaves of the Brazilian palm *Copernicia cerifera* Martius, also known as the carnauba palm. It is found on both sides of the leaves. A distinguishing feature of carnauba wax is its high content of unusual compounds, such as unesterified alcohols (12%), x-hydroxy esters (14%), and hydroxylated cinnamic acid esters (30%), alongside various unidentified impurities (Vandeburg and Wilder, 1970). Carnauba wax is one of the hardest plant waxes, with a melting range of 82.5–83 °C, making it valuable as a hardening agent for raising the melting point of wax blends. It is the most commercially significant plant wax, used extensively

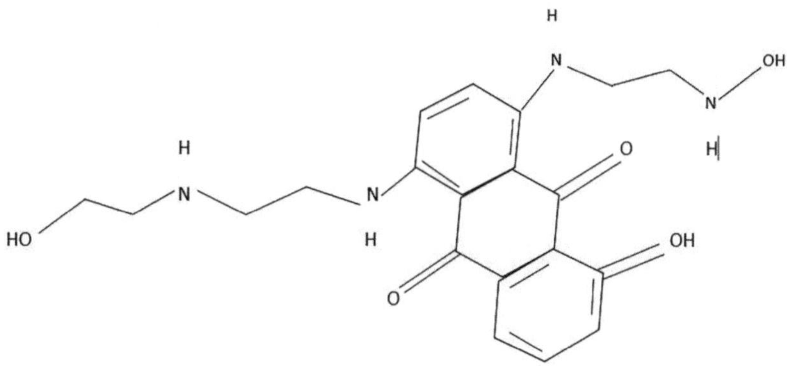

Fig. 4.2 Chemical structure of carnauba wax

in food products, confectionery coatings, cosmetics, automotive and furniture waxes (Tinto et al., 2017). Its use in food is regulated under FDA regulation 21 CFR 184.1978. This is the chemical structure of carnauba wax (Martinez-Hernandez et al., 2025) (Fig. 4.2).

Jojoba Wax

Jojoba wax is extracted from the seeds of the jojoba plant, *Simmondsia chinensis*. The terms "jojoba wax" and "jojoba oil" are often used interchangeably, as jojoba oil contains about 98% pure wax. Chemically, jojoba wax consists primarily of monoesters formed from long-chain fatty acids and alcohols, with a small portion of triglyceride esters. It is known for its long shelf life and strong resistance to high temperatures. Due to its composition and consistency, it is more accurately referred to as a liquid wax rather than an oil or fat (Fig. 4.3).

Candelilla Wax

Candelilla wax is primarily extracted from the leaves of *Euphorbia antisyphilitica* Zuccarini, a plant native to northern Mexico and southwestern Texas (Arato et al., 2014). Raw candelilla wax typically contains about 42% hydrocarbons, 39% wax, resin, and sitosteroyl esters, 8% free wax and resin acids, 6% lactones, and 5% free wax and resin alcohols (Fraps and Rather, 1910). Interestingly, most hydrocarbons in candelilla wax have an odd number of carbon atoms, with only trace amounts of even-numbered hydrocarbons. Approved by the FDA as a food additive, candelilla wax is used for glazing certain foods and for coating fruits via microemulsions (Hagenmaier and Baker, 1996) (Fig. 4.4).

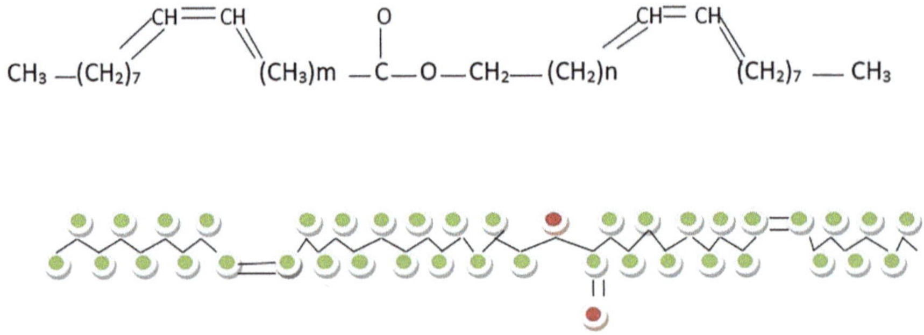

Fig. 4.3 Chemical structure of jojoba wax

a)

b)

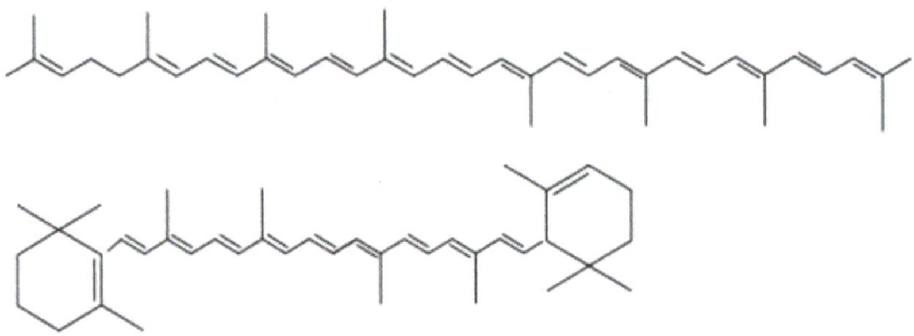

Fig. 4.4 Chemical structure of Candelilla wax (**a**) A straight long-chain hydrocarbon (**b**) A structure with a ring

Rice Bran Wax

High-melting vegetable wax is found in the husks of *Oryza sativa* (rice) (Cousins and Feuge, 1960). It is a by-product of rice bran oil dewaxing. Its composition is relatively uniform, mainly consisting of even-numbered aliphatic acids (wax acids) and higher alcohol esters, primarily C46–C62 esters formed from C20–C36 fatty alcohols and C20–C26 fatty acids. Additional components include free fatty acids (like palmitic acid), phospholipids, phytosterols, and squalene (Imam et al., 2014). The hydrocarbon content is generally low, around 2%. Rice bran wax has FDA approval as a safe food additive (21 CFR, 172.890), and it is widely used due to its low cost and abundance in Asia, particularly for coatings on fruits and vegetables, confectionery products, and chewing gum. Other uses include candles, textiles, lubricants, and cosmetics (Sabale et al., 2012) (Fig. 4.5).

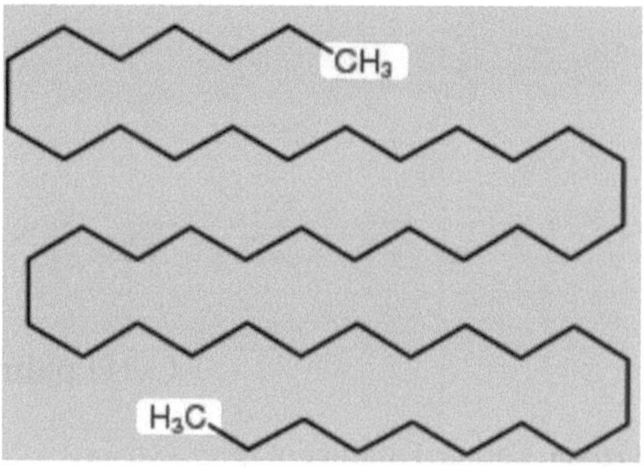

Fig. 4.5 Chemical structure of rice bran wax

Sunflower Wax

Sunflower wax is extracted from the seeds and hulls of *Helianthus annuus* (sunflower) (Matyushenskii, 1960), and is obtained through the winterisation of sunflower oil. Similar to rice bran wax, sunflower wax is a hard, high-melting wax mainly comprising long-chain saturated fatty esters (C40–C44). However, sunflower wax has not been granted GRAS status by the FDA, limiting its use primarily to cosmetics. It is commonly found in products like lipsticks, mascaras, lip balms, and emulsions, where it functions as a consistency enhancer and helps control hardness, texture, and mould release. It can also serve as a substitute for candelilla, carnauba, and rice bran waxes (Casas-Tinto et al., 2011).

Animal Waxes

Animal waxes are typically sourced from insects (such as bees, *Apis mellifera*), whales (*Physeter macrocephalus*), and sheep (*Ovis aries*), among others. *Beeswax* is a natural wax secreted primarily by honeybees (*A. mellifera*) for building honeycombs (Tulloch, 1957). Unhydrolysed beeswax is composed of about 71% esters, 15% hydrocarbons, 8% free fatty acids, and 6% other constituents (Tulloch, 1957). In its natural form, beeswax is brittle at lower temperatures, with a melting point between 62 °C and 65 °C. It is chemically inert and highly plastic, providing protection against corrosion, abrasion, and moisture loss. Beeswax is among the most widely used commercially available waxes (Öğütcü et al., 2015; Martins et al., 2017). It has diverse commercial applications, including metal casting, candle making, cosmetics, textiles, varnishes, and food processing, with food-related uses to be discussed later.Other examples of animal waxes are spermaceti (sourced from whales) and wool wax (from sheep).

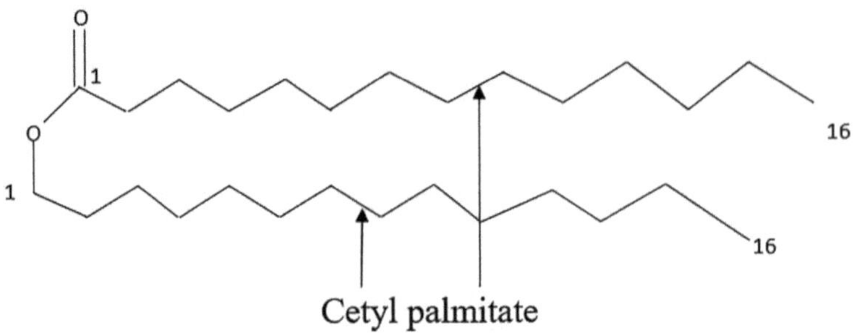

Fig. 4.6 Chemical structure of spermaceti wax

Spermaceti Wax

Spermaceti wax is obtained from the frontal organ in the head of the sperm whale (*P. macrocephalus*), an organ used for echolocation (Rudel and Morris, 1973). In the fifteenth century, spermaceti was utilised in medicine in England. However, due to the ban on whale hunting, it is no longer produced naturally and has been replaced with synthetic versions composed of pure acetal palmitate or jojoba-based mixtures (Crisp et al., 1984) (Fig. 4.6).

Wool Wax

Wool wax is secreted by the sebaceous glands in the skin of the sheep, serving to protect the skin and fleece (Horn, 1990). It is widely used in personal care products, including facial cosmetics and lip balms (Jover Martínez et al., 2025) (Fig. 4.7).

Petroleum Waxes

Petroleum waxes, also known as paraffin waxes, are derived from sources like petroleum, coal, or oil shale. These waxes consist of mixtures of hydrocarbons. They were first developed by Carl Reichenbach in Germany in 1830.

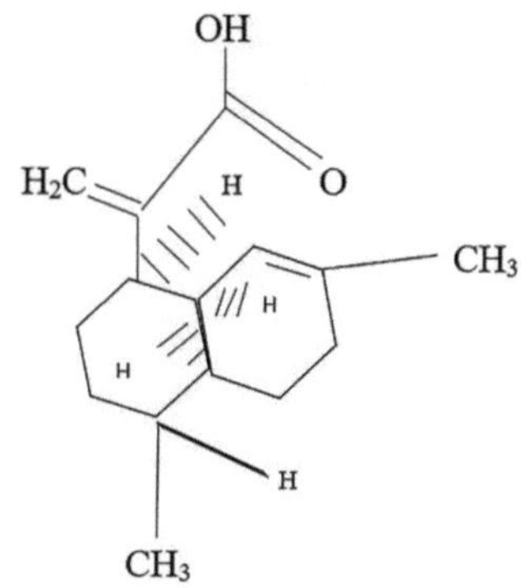

Fig. 4.7 Chemical structure of wool wax

Microbial Waxes

Microbial waxes are generated by microorganisms like bacteria and yeast. These waxes have attracted interest due to their potential in biotechnological fields such as biofuel production and bioremediation. Their composition can differ depending on the specific type of microorganism responsible for their synthesis.

Synthetic Waxes

Synthetic waxes are artificially produced waxes composed of a homologous series of long-chain aliphatic hydrocarbons that do not contain functional groups. These waxes are created through various synthetic processes using different raw materials. Common types of synthetic waxes include:

- Fatty acid amide waxes
- Fischer-Tropsch waxes
- Polyolefin waxes

Fatty Acid Amide Waxes

Fatty acid amide waxes are synthetic waxes produced by modifying palm and tallow waxes. These waxes act as solid lubricants and are commonly used to enhance resin surfaces due to the presence of an amide group. They are known for their chemical and thermal stability (Fig. 4.8).

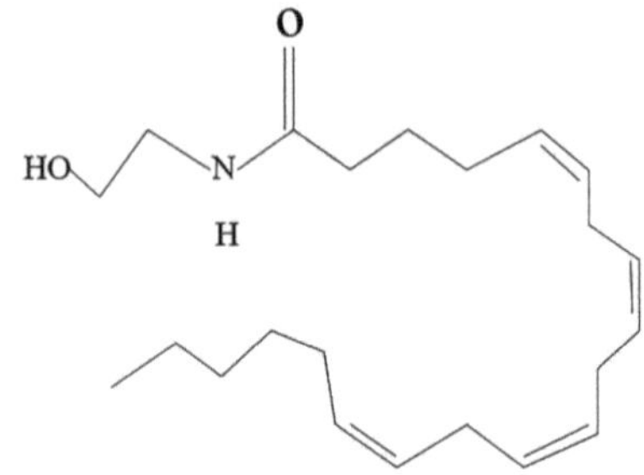

Fig. 4.8 Chemical structure of fatty acidamide wax

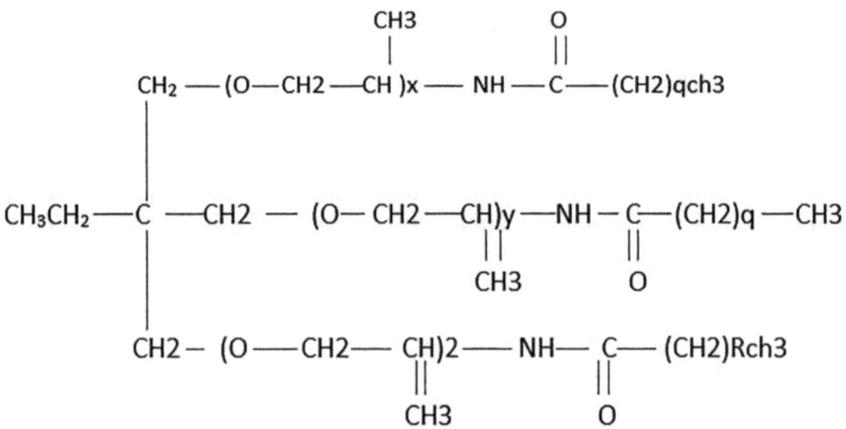

Fig. 4.9 Chemical structure of Fischer-Tropsch wax

Fischer-Tropsch Waxes

Fischer-Tropsch wax is a synthetic wax produced through the polymerisation of carbon monoxide under high pressure. It can be synthesised from different natural gases using high-temperature reactors and is considered the most widely used type of synthetic wax (Fig. 4.9).

Polyolefin Waxes (Fig. 4.10)

Polyolefin waxes are synthetic waxes produced through various manufacturing methods, including thermal degradation, polymerisation, and monomer synthesis. They are characterised by a low molecular weight and a high level of crystallinity. Common types of polyolefin waxes include:

- Polyethylene wax
- Polypropylene wax
- Polyamide

Fig. 4.10 Chemical structure of polyolefin wax

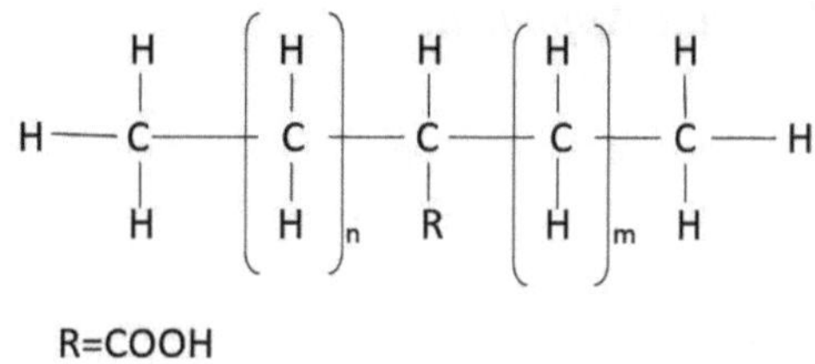

Applications of Lipids and Wax in Food Packaging

Lipids are deficient in a self-sustaining structure and are not polymers. Unlike polysaccharides and protein-based films, lipids have a superior water barrier since they are hydrophobic by nature. Wax films provide the best moisture barricade qualities of any lipid-based packaging. The glossy qualities of resins like wood and terpene resin are well-known. Since lipids are nonpolar, they can be added to composite films to increase the moisture barrier's overall strength. In addition to these, lipids have distinct flavours and greasy surfaces. Both plants and animals can yield edible waxes, which are lengthy chains of alcohols and esters. These coatings can lessen the permeability of water vapour since they are hydrophobic. Freshly cut fruits can be covered with gel that has been derived from the aloe vera plant. According to studies, the aloe vera coating helps preserve the firmness of the fruit, minimise weight loss, and form a moisture barrier. To increase the starch-based film's water barrier ability, wax is also applied. They have a semi-solid shape and range in colour from yellow to brown. The tomatoes and green chilies have been coated with a resin-based substance. In addition to providing gloss and transparency, this coating acts as an oxygen and water barrier (Cui et al., 2024). Plants and oil are used to extract fats and oils. Triglycerides are the main component of fat and oil. The elongation and water resistance of a film derived from palm fruit oil are good. Pork flesh has been coated with sunflower oil for its oxygen- and water vapor-blocking qualities. The essential oil has antibacterial qualities. They are hydrophobic and volatile. To enhance the packaging film's overall antibacterial qualities, essential oils from thyme, clove, and lemon peel are added to the biopolymer matrix (Sarraj et al., 2023).

Edible coatings are normally applied to the outer surface of products by dipping, brushing, etc. Polysaccharides, proteins, and lipids are the main sources of edible films and edible coatings. Multiple film-forming mechanisms of biopolymers, including intermolecular forces, e.g. cross-linking, disulfide bonds, covalent bonds, and ionic interactions. Polyanhydrides, poly-lactic acid, and polyvinyl alcohol are inedible biopolymers. Lipids, proteins, and polysaccharides are the three types of edible biopolymers. In polysaccharides, many modified starches, chitin, and pectin are included. In lipids, beeswax includes free fatty acids. In proteins, sunflower, casein, and pea are included..

Surface Appearance

Lipid coatings can improve the shine and overall appearance of food products, especially in confectionery and sugar products.

Flavour and Texture

Lipids play a crucial role in the development of food's flavour and texture, influencing both desirable and undesirable flavours and contributing to the overall sensory experience.

Shelf Life Extension

By reducing moisture loss and controlling respiration rates, lipid-based coatings can significantly extend the shelf life of fruits, vegetables, and other perishable foods.

Protection Against Abrasion

Lipids can help protect food from damage during handling and transport.

Heat Transfer Medium

Lipids can be used as a heat transfer medium during food processing, contributing to the desired texture and flavour.

Natural and Sustainable

Many waxes are derived from natural sources, making them a sustainable and environmentally friendly option for food packaging.

Protection During Transport

Wax coatings may protect foodstuffs during transport and handling.

Applications of Edible Coatings and Films

Fruits and Vegetables

The most widely used materials for a protective covering on fresh goods are waxes. Fruits and vegetables are the primary and most traditional applications for edible coatings and films. The impact of the various lipid-containing coatings and films on fruits and vegetables has been extensively studied. Lipids have been used for generations to increase the shelf life of fresh citrus fruits. Nowadays, the production and consumption of minimally processed fruits and vegetables have expanded due to the rising demand for convenience goods. Minimally processed foods, on the other hand, are very sensitive and susceptible to quality deterioration and the ensuing reduction in shelf life. One well-known method for increasing the shelf life of fresh-cut fruits and vegetables is the use of edible coatings and films. They can be used in conjunction with modified atmosphere packaging or on their own. Lipids such as beeswax, carnauba wax, and candelilla wax, as well as fatty acids, acetylated monoglycerides, and triglycerides, are used to provide edible coatings and films for fresh-cut agricultural goods (Miyazaki et al., 2012). Lipids prefer to form thicker, brittle coatings because of their hydrophobic properties, which also make it difficult for them to stick well to the wet surface of fresh-cut vegetables. Therefore, combining lipids with hydrophilic film-forming agents, such as hydrocolloid materials, is a superior way to use them in a coating or film for fresh-cut produce.

Animal-Based Products

Foods of animal origin are more susceptible to microbiological spoilage and chemical deterioration, which puts consumer health at risk and can result in losses for producers or processors. Fresh meat products have a short shelf life because of their high moisture, lipid, and protein content, which makes them prone to oxidation and rancidity (Zhou et al., 2010). These are the main reasons why meat quality deteriorates throughout preparation, transport, and storage, lowering its acceptability and shelf life. Therefore, the safety and quality of raw meat depend heavily on the right packaging technology and storage conditions. By reducing water loss, lowering lipid oxidation and discolouration, and improving product appeal in retail packaging, edible coatings and films can help preserve raw, frozen, and processed meat, poultry, and marine products. Lipid films and coatings have been applied to beef

Table 4.1 Lipid coatings and films applied to various food products

Food products	Lipid/wax	Key advantage	References
Plum	Rosehip oil	The application of rosehip oil delayed the ripening process and prolonged the plum's excellent characteristics. Day 14: Control fruit weight loss was 15.6%, whereas fruits that were treated displayed 10.04%.	Aringer et al. (2019)
Strawberry	Candelilla wax	It was discovered that biocontrol bacteria *(B. subtilis)* and candelilla wax films had a good ability to lessen the occurrence of strawberry deterioration. On day 6, the film + bacteria treatment reduced the decay by about 100% in comparison to the control.	Torres-Garcia et al. (2024)
Hygroscopic candy tablets	Sunflower oil, stearic acid	According to the study, the oil phase's composition was crucial for improved anti-browning. Coatings made of solid lipids showed higher effectiveness in lowering water vapour flow in contrast to coatings based on liquid lipids	Khan et al. (2014)
Persian limes	Candelilla wax, beeswax, oleic acid	Lower weight loss, good dark shade, maintenance of the green hue, and favourable physiochemical characteristics of Persian limes were all brought about by the coating	Carter et al. (2003)
Green bell paper	Safflower seed oil	The lipid coatings made from oilseeds significantly reduced moisture loss through the produce's surface. Up to 79% less weight was lost than with the control	Kuk et al. (2009)

products to improve qualities like softness. The majority of the time, natural waxes like beeswax, resins, acetoglycerides, fatty acids, and plant oils are used to make these films (Sun et al., 2021). Only a small number of publications addressing the efficacy of lipids as preventive coatings for meat, poultry, and seafood were found, despite the potential of lipid materials to improve the qualities of packaging films and coatings. In some regions of the world, meat processors used microcrystalline wax derived from petroleum to make frozen meat products during the past century (Table 4.1).

There are numerous other lipid or non-lipid coatings, but there are only a few studies on the use of waxes, fats, and oils as preventive coatings on meat products (Wu et al., 2021). Typically, lipids are combined with proteins and polysaccharides to create blend coatings and films that preserve the qualities of meat, poultry, and fish products. Composite coatings and films involving various hydrocolloids and lipid materials have been tested for their ability to preserve meat products, including precooked meat products. According to Malik et al. (2017), the addition of lipids to edible coatings and films can improve their hydrophobic, cohesive, and flexible qualities while also giving them a good water barrier. This prolongs the freshness, colour, aroma, tenderness, and microbiological safety of both fresh and processed meat and seafood (Malik et al., 2017).

Cereal Products

By decreasing water absorption and inhibiting microbiological development, edible coatings and films also make it possible to prolong the shelf life of baked goods. The crisp texture of all bakery products, including biscuits, snacks, and breakfast cereals, contributes to their allure to consumers. Water activity and moisture content have a significant impact on the texture of some meals, such as cereals and baked goods. Dry bread goods lose their desired crispness when their water activity rises. Complete quality loss may arise from the loss of crunchy texture and subsequent softening brought on by increasing relative humidity conditions during storage.

Retarding moisture transfer is frequently required to extend the shelf life of many foods. By reducing the water vapor transfer between dry bakery goods and their surroundings, edible coatings can help products last longer on the shelf. Cereal-based goods can also be coated for a number of additional reasons. For example, adding food additives to make products more consumer-friendly or fortifying cereal items with active substances are significant developments that can be accomplished with coatings and films (Rhim and Wang, 2014). Once more, there is a dearth of research on the application of lipid or lipid-containing coatings and films and how they affect cereal products. There might only be a small number of studies available. Bravin et al. conducted one such study on an edible composite layer made of polysaccharides and lipids to extend the shelf life of dry baked goods. They came to the conclusion that coated crackers had a longer shelf life based on the results.

Dairy Products

A daily diet must include dairy products, including milk, cheese, and yogurt. Cheese is primarily high in fats, proteins, and vitamins. The dairy product's shelf life is extended, mass transfer is avoided, and the ripening process is regulated by edible films and coatings surrounding it (Leandro et al., 2019). Film-forming techniques include casting, whereas edible coating techniques include dipping, brushing, spraying, and electrostatic brushing. The cheese should be completely covered with film or coating to extend its shelf life. Since the dipping process necessitates an efficient forming solution in the tank, brushing is recommended for small-sized cheese packaging. The dipping procedure is recommended if the cheese has an uneven form. It will coat the whole surface of the cheese easily. The spraying method is used when a uniform and thin coating is required on the cheese. Electrostatic spraying provides higher efficiency and reduces coating solution wastage. Casting is used to form edible films, and after that, they are wrapped around the cheese so that the films completely cover the cheese. Films and coatings create a barrier between cheese and the surrounding environment, hence improving the cheese's shelf life (Cruz-Diaz et al. 2019). Edible films made from whey protein show better gas barrier properties compared to polysaccharides and other protein-based films. Heat treatment can improve the film-forming ability of whey protein-based films. Heat treatment causes the whey protein's structure to open up and more intra- and inter-bonding to take place. Whey protein is combined with plasticiser (glycerol), pH-adjusting agents, and other chemicals. A film-forming solution is then created using the solvent. Following that, the films are dried for eighteen hours at about 35 °C in a hot oven. Cheese slices can then be sandwiched between the films. Because whey protein-based films are transparent, customers can see how good the cheese is (Costa et al., 2018). The whey protein matrix is supplemented with antimicrobial agents. These substances will stop yeast and bacteria from growing. Compared to agents introduced in the whey protein matrix, antimicrobial agents applied directly to the cheese will have a reduced activation tendency. A certain kind of microbe is the target of each antimicrobial agent. Lactic acid works well against moulds and yeast, lysozyme works well against Escherichia coli, oregano essential oil works well against Penicillium, and rosemary essential oil works well against Mucor. Cheese's propensity to react with oxygen is lessened by antioxidant compounds. Cheese's colour and flavour will alter if it comes into contact with oxygen (Cruz-Diaz et al. 2019). Ricotta cheese is kept at 4 °C after being wrapped in an edible packaging film made of chitosan and whey protein. Even after 21 days, there were substantially fewer mesophilic and psychotropic bacteria, and the cheese was still in acceptable condition (Di Pierro et al., 2011). Staphylococcus species, Pseudomonas species, Enterobacteriaceae, yeast, and mould did not grow appreciably for 60 days when whey protein isolates containing antimicrobials like natamycin and lactic acid were used to package semi-hard bovine cheese (Ramos et al., 2012). Due to its high fat content, butter experiences lipid oxidation, which shortens its shelf life. When kept between 2 °C and 5 °C, an edible coating made of corn starch with ginger oil as an active component lowers lipid oxidation (Arshad et al., 2020). The development of microbes on the food surface gives paneer a short shelf life. The addition of cinnamon essential oil as an antimicrobial agent to the sodium alginate edible coating matrix preserves the paneer's quality and extends its shelf life (Raju and Sasikala, 2016).

Functions of Lipid and Wax in Food Packaging

Improved Thermal and Mechanical Packaging Systems

Mechanical strength and thermal properties are crucial for food safety and commercialisation in the food packaging industry. The ability of materials to display particular characteristics under the influence of heat and force is referred to as their mechanical properties. The impact of lipids and their derivatives on the mechanical and thermal characteristics of food packaging materials is examined. Lipids generally strengthen the barrier, but their type and concentration can have an impact on how well films or coatings work mechanically. Since excessive addition and incompatibility with the matrix may result in stiff, weak, and brittle films, which reduce elongation and elastic modulus as well as tensile strength, lipid ingredients must be tuned based on the application (Zubair et al., 2021). The mechanical characteristics of pistachio globulin protein-based films were impacted by the addition of fatty acids, as reported by Zahedi et al. (2010). The elongation at break was considerably decreased (35–70%), and the ultimate tensile strength was also shown to be decreasing. However, according to Wang et al. (2016), adding epoxy castor oil to soy protein-based films enhanced their mechanical qualities by almost a factor of two and raised their elongation at break by almost 23%. This enhancement in the material's structural analysis was ascribed to the hydrogen bonding between long chains of epoxidised castor oil and soy protein isolates (Wang et al., 2016). Nevertheless, a thorough analysis of the literature revealed that little is known about the effects of adding lipids to different biopolymers to enhance their mechanical qualities for food packaging. Nonetheless, it was found that adding hydrophilic substances to lipid materials, such as proteins, starch, chitosan, and cellulose, can improve their mechanical and thermal characteristics (Zubair et al., 2021). The usage of ultrasonic technology in filmmaking has grown recently to enhance the qualities of polymer films, examining the impact of ultrasonics on films based on lauric acid and starch. The mechanical performance of the films was enhanced by this investigation. Lauric acid and starch complexed more when ultrasonic amplitudes reached 40%. This is explained by better dispersion of lauric acid and the release of amylose because of starch grain disintegration. Although the tensile strength increased from 6.2 to 8.31 MPa, complex formation was disrupted, and the tensile strength decreased with a further increase in ultrasonic amplitude (70%) (Liu et al., 2018). Additionally, a study by Kang et al. (2020) demonstrated that ultrasonication enhances the mechanical characteristics of starch-lipid composite films.

Improved Barrier Properties; Water Vapour Permeability

When evaluating or forecasting the shelf life of food goods, the barrier qualities of food packaging film are crucial. In order to maintain food quality, one of the main purposes of food packaging film is to slow down the movement of molecules from food to the environment. The permeability and transfer of molecules like organic vapour, water vapour, or gases through a film can be ascertained by measuring these characteristics. Due to its substantial influence on food freshness, crispiness, or dehydration prevention, depending on the type of food, WVP is one of these that is thoroughly researched in biodegradable films (Devi et al., 2024). Lipid-based materials' hydrophobic and crystalline characteristics directly affect their barrier qualities, which change depending on the type of chemical functionality, aliphatic chain length, and degree of unsaturation. Depending on the method of film production, barrier characteristics are also influenced by film structure. The characteristics of the penetrant molecule also influence the film's permeability and interaction with its constituent parts (Zubair et al., 2021). Due to their nonpolar properties and incredibly low WVP values, lipid-based materials are efficient moisture barriers (Tyagi et al., 2021). Commonly used lipids include fatty acids, oils, and waxes (Khodaei et al., 2021). Composite films have been created using a range of hydrocolloids, such as cellulose, chitosan, pectin, starch, alginates, and carrageenan. It was claimed that, because of their enhanced hydrophobicity, the WVP of pistachio globulin protein films containing fatty acids dropped dramatically as the fatty acid content rose. When compared to films without fatty acids, the WVP dropped by roughly 3% (Wang et al., 2014). When lipids like palmitic or caproic acid are added to fish proteins film (Pereira et al., 2020), oleic acid is added to wheat bran cellulose/wheat gluten composite films (Shen et al., 2022), and stearic acid is added to rice starch-ι-carrageenan edible films (Thakur et al., 2016), the results are generally consistent with decreased WVP and solubility of the food packaging films. Conversely, when the hydrophobicity of the components increased and the WVP decreased, waxes were shown to be among the most effective. A good food packaging material should have excellent moisture barriers for sodium caseinate-based films, starch-based gluten films, and chitosan-based films. Biopolymer films with added waxes like candelilla wax (Kowalczyk et al., 2018), beeswax (Wardhono et al., 2019), and carnauba wax (de Oliveira Filho et al., 2021) have been found to have these qualities. Similarly, Nurul Syahida et al. (2020) investigated the use of fish gelatin-based films by adding palm wax and how it affected the food packaging's water-resistance characteristics. The WVP of the films was found to be considerably decreased ($p < 0.05$) by the addition of 15% palm wax. Water adsorption is decreased as a result of the hydrophobic wax's inhibition of hydrogen bonds between water molecules as it adheres to the gelatin network (Nurul Syahida et al., 2020). Depending on the kind of lipid and the relative polarity of the support polymer, the emulsified films have the biggest impact on the water vapour barrier (Wang et al., 2014). It is crucial to remember that pinholes or cracks in the film potentially affect the barrier property of biopolymer films by facilitating the transport of water vapour (Kowalczyk et al., 2018).

Antioxidant Activities

Although active chemicals are being added to the polymer matrix to create packaging materials with antioxidant qualities, the carrier itself may also have antioxidative qualities (Kowalczyk et al., 2018). Because of their antibacterial and antioxidant properties, films and coatings are useful for maintaining the quality of fruits and vegetables after harvest (Devi et al., 2025). Natural antioxidants are preferred to be integrated into biopolymers in active food packaging to avoid off-flavours and unwanted textures, extending shelf life. Antioxidants based on lipids, such as lipophenols, phospholipids, phenolipids, and essential oils, are utilised in food packaging. These components' antioxidant qualities are attributed to a flavonoid and a phenolic compound (Zubair et al., 2021).

Antimicrobial Properties

Antimicrobial systems, also referred to as agents, are active packaging systems in which the product, the ambient, and the packaging material all work together to prevent the growth of germs in the food. The best food packaging should preserve the product's structural integrity in addition to shielding it from impurities and food-borne illnesses (Duda-Chodak and Tarko, 2023). Covalently bonding the active ingredient to a macromolecule used in the production of packaging materials and physically combining microbial agents into a polymer matrix during processing are two methods employed in the development of antimicrobial packaging systems (Dini, 2016).

A range of antimicrobial agents is used to treat edible films and coatings. Three methods are used to activate these antimicrobial compounds in food: release, immobilisation, and the absorption system during storage, all of which improve food safety (Abdollahzadeh et al., 2021). The effects of several plant and animal fats, such as corn oil, sunflower oil, olive oil, and butter, on the environmental qualities of chitosan packaging materials were examined (Akyuz, 2018). Another study looked at the influence of fat and oil unsaturation on the characteristics of films. It was found that olive oil/chitosan films had the highest antimicrobial properties of all films, comparable to the commercial antibiotic gentamicin (Akyuz, 2018). Waxes such as carnauba wax, beeswax, or candelilla wax have been used in the development of bioactive coatings.

Hydrophobicity

Made up of waxes, resins, fats, sterols, fat-soluble vitamins, and other natural components, lipids are safe, edible, and hydrophobic. They are one of the main energy sources and are essential to the membrane's and cell's operations. The degree of unsaturation, the length of the aliphatic chain, and the type of chemical functionality all affect the barrier, which is directly influenced by the lipid hydrophobic nature and structure (Yousuf et al., 2022). They can be used to store and transport energy-rich substances like triglycerides or cholesterol because of their ability to form stable structures like lipid droplets or lipoproteins due to their lipid hydrophobicity. Hydrophobic materials that can be successfully applied to lipid-based edible films and coatings include vegetable oils, acetoglycerides and fatty acids, natural waxes including rice bran wax, candelilla wax, carnauba wax, and beeswax, and resins like wood rosin and shellac (Rhim and Shellhammer, 2005). Lipids are said to be very good at preventing moisture transfer between food and packaging because of their low polarity. As the concentration of the hydrophobic (lipid) phase rises, the material's WVP falls. Waxes exhibited superior barrier qualities among lipids because they had superior hydrophobic components (Zubair et al., 2021). Beeswax and coffee were combined to create an edible super-hydrophobic covering, which was then tested for hydrophilicity using contact angle measurements. The apparent contact angle of this coating may remain above 150° (hydrophobic) during further heating and flushing. With numerous potential uses in the field of functional food packaging, this thermo-resistant edible super-hydrophobic coating offers a viable remedy for the issue of standard edible super-hydrophobic coatings' inability to withstand high temperatures (Zhang et al., 2019). The WVP of bio-based food packaging materials was also lowered by the addition of oils. Oils' hydrophobic qualities increase the material's hydrophobic phase, which reduces water diffusion through coatings or films (Zubair et al., 2021).

Conclusion and Future Perspectives

Because edible films and coatings are naturally occurring, inexpensive, and renewable, they have been demonstrated to improve food quality and shelf life by regulating moisture, oxygen, lipids, fragrance, and flavour chemicals in food systems while lowering the use of synthetic plastic packaging. Lipid-based edible coatings preserve food quality by acting as efficient barriers against moisture and, to a lesser extent, oxygen. Food, agriculture, and medicinal applications are just a few of the many uses for lipid-based coatings. However, lipid coatings quickly oxidise and impart a lipid flavour that diminishes the food's attractiveness and sensory attributes. Applying certain lipids can be difficult if solvents or high temperatures are needed, since they are solid at room temperature. Therefore, extensive research and development efforts are needed to address these coating-related problems in order to make them appealing and useful and to optimise their use for humankind. Furthermore, there is still very little use of lipid coatings and films for new food items or unusual substrates, like dairy products, meat substitutes, or ready-to-eat food. Nonetheless, improvements in barrier technology and research could have a big impact on prolonging food shelf life. Researchers must conduct additional research and characterise lipid coatings and films as possible edible packaging. Composite coatings and films may also be produced to combine the advantages of lipids and hydrocolloid components. The efficiency of edible coatings and films relies on their components' characteristics. To improve food quality and shelf life, it is therefore essential to utilise the proper composition and quantities. It is therefore essential to use innovative technologies to improve control over the properties and performance of coatings or films. Additionally, little is known about how lipids affect the characteristics of edible films. Future studies should therefore concentrate on analysing the important characteristics of biodegradable edible films that contain different lipid components. To evaluate their potential for enhancing food quality and extending food shelf life, more interaction mechanisms should be investigated and applied to food systems.

References

Abdollahzadeh, E., Nematollahi, A., & Hosseini, H. (2021). Composition of antimicrobial edible films and methods for assessing their antimicrobial activity: A review. *Trends in Food Science & Technology, 110*, 291–303.

Akyuz, D. (2018). Measuring technological pedagogical content knowledge (TPACK) through performance assessment. *Computers & Education, 125*, 212–225.

Arato, A., Wehrer, M., Biró, B., & Godio, A. (2014). Integration of geophysical, geochemical and microbiological data for a comprehensive small-scale characterization of an aged LNAPL-contaminated site. *Environmental Science and Pollution Research, 21*(15), 8948-8963.

Aringer, M., Costenbader, K., Daikh, D., Brinks, R., Mosca, M., Ramsey-Goldman, R., … & Johnson, S. R. (2019). European League Against Rheumatism/American College of Rheumatology classification criteria for systemic lupus erythematosus. *Arthritis & Rheumatology, 71*(9), 1400–1412.

Arshad, Z., Robaina, M., & Botelho, A. (2020). The role of ICT in energy consumption and environment: an empirical investigation of Asian economies with cluster analysis. *Environmental Science and Pollution Research, 27*(26), 32913–32932.

Atta, O. M., Manan, S., Shahzad, A., Ul-Islam, M., Ullah, M. W., & Yang, G. (2023). Biobased materials for active food packaging: A review. *Food Hydrocolloids, 125*, 107419.

Bennett, R. E. (2024). Personalizing assessment: dream or nightmare? *Educational Measurement: Issues and Practice, 43*(4), 119–125. https://doi.org/10.1111/emip.12652

Carter, C., Chaires-Martinez, L., Ponce-De-Leon, L., Pelayo, C., & Bosquez-Molina, E. (2003). Edible coatings from candelilla wax microemulsions. *International Journal of Food Science & Technology, 38*, 1–15.

Casas-Tinto, S., Zhang, Y., Sanchez-Garcia, J., Gomez-Velazquez, M., Rincon-Limas, D. E., & Fernandez-Funez, P. (2011). The ER stress factor XBP1s prevents amyloid-β neurotoxicity. *Human Molecular Genetics, 20*(11), 2144-2160.

Costa, M. J., Maciel, L. C., Teixeira, J. A., Vicente, A. A., Cerqueira, M. A. (2018). Use of edible films and coatings in cheese preservation: Opportunities and challenges. *Food Research International, 107*, 84–92.

Cousins, E. R., & Feuge, R. O. (1960). Hydrogenation of methyl oleate in solvents. *Journal of the American Oil Chemists' Society, 37*(9), 435-438.

Cui, Y., Cheng, Y., Xu, Z., Li, B., Tian, W., & Zhang, J. (2024). Cellulose-Based Transparent Edible Antibacterial Oxygen-Barrier Coating for Long-Term Fruit Preservation. *Advanced Science, 11*(48), 2409560.

de Oliveira Filho, J. G., Miranda, M., Ferreira, M. D., & Plotto, A. (2021). Nanoemulsions as edible coatings: a potential strategy for fresh fruits and vegetables preservation. *Foods, 10*(10), 2438.

Di Pierro, P., Sorrentino, A., Mariniello, L., Giosafatto, C. V. L., & Porta, R. (2011). Chitosan/whey protein film as active coating to extend Ricotta cheese shelf-life. *LWT-Food Science and Technology, 44*(10), 2324–2327.

Devi, L. S., Jaiswal, A. K., & Jaiswal, S. (2024). Lipid incorporated biopolymer based edible films and coatings in food packaging: A review. *Current Research in Food Science, 8*, 100720.

Devi, L. S., Mukherjee, A., Das, M. R., & Kumar, S. (2025). Antimicrobial efficacy of eugenol and neem oil nanoemulsions against foodborne pathogens and food spoilage fungi. *Food Bioengineering, 4*(1), 101–112.

Dini, I. (2016). Use of essential oils in food packaging. In Essential oils in food preservation, flavor and safety (pp.139–147). Academic Press.

Duda-Chodak, A., & Tarko, T. (2023). Possible side effects of polyphenols and their interactions with medicines. *Molecules, 28*(6), 2536.

Fraps, G. S., & Rather, J. B. (1910). Constitution of Candelilla Wax. *Industrial & Engineering Chemistry, 2*(11), 454–455.

Goslinska, M., Selmer, I., Kleemann, C., Kulozik, U., Smirnova, I., & Heinrich, S. (2019). Novel technique for measurement of coating layer thickness of fine and porous particles using focused ion beam. *Particuology, 42*, 190–198.

Hagenmaier, R. D., & Baker, R. A. (1996). Edible coatings from candelilla wax microemulsions. *Journal of Food Science, 61*(3), 562–565.

Hassan, Q., Viktor, P., Al-Musawi, T. J., Ali, B. M., Algburi, S., Alzoubi, H. M., … & Jaszczur, M. (2024). The renewable energy role in the global energy Transformations. *Renewable Energy Focus, 48*, 100545.

Horn, B. K. (1990). Relative orientation. *International Journal of Computer Vision, 4*(1), 59–78.

Imam, M. U., Ishaka, A., Ooi, D. J., Zamri, N. D. M., Sarega, N., Ismail, M., & Esa, N. M. (2014). Germinated brown rice regulates hepatic cholesterol metabolism and cardiovascular disease risk in hypercholesterolaemic rats. *Journal of Functional Foods, 8*, 193–203.

Jover Martínez, A., Lemmens, L. H., Fried, E. I., Guðmundsdóttir, G. R., & Roefs, A. (2025). Validation of a transdiagnostic psychopathology ecological momentary assessment protocol in a university student sample. *Psychological Assessment, 37*(1-2), 46.

Kang, D., Choi, H., Kim, J. H., & Choi, J. (2020). Spatial epidemic dynamics of the COVID-19 outbreak in China. *International Journal of Infectious Diseases, 94*, 96–102.

Khalil, A. A., Enezei, H. H., Aldelaimi, T. N., & Mohammed, K. A. (2024). Advances in diagnosis and treatment ofbasal cell carcinoma. *Journal of Craniofacial Surgery, 35*(2), e204-e208

Khan, M. I. R., Asgher, M., & Khan, N. A. (2014). Alleviation of salt-induced photosynthesis and growth inhibition by salicylic acid involves glycinebetaine and ethylene in mungbean (Vigna radiata L.). *Plant Physiology and Biochemistry, 80*, 67–74.

Khodaei, D., Javanmardi, F., & Khaneghah, A. M. (2021). The global overview of the occurrence of mycotoxins in cereals: A three-year survey. *Current Opinion in Food Science, 39*, 36–42.

Kowalczyk, D., Kazimierczak, W., Zięba, E., Mężyńska, M., Basiura-Cembala, M., Lisiecki, S., Karaś, M., & Baraniak, B. (2018). Ascorbic acid- and sodium ascorbate-loaded oxidized potato starch films: Comparative evaluation of physicochemical and antioxidant properties. *Carbohydrate Polymers, 181*, 317–326. https://doi.org/10.1016/j.carbpol.2017.10.063

Kowalczyk, J. B., Royer, D. L., Miller, I. M., Anderson, C. W., Beerling, D. J., Franks, P. J., ... & Ramezani, J. (2018). Multiple proxy estimates of atmospheric CO_2 from an early Paleocene rainforest. *Paleoceanography and Paleoclimatology, 33*(12), 1427–1438.

Kuk, J. L., Saunders, T. J., Davidson, L. E., & Ross, R. (2009). Age-related changes in total and regional fat distribution. *Ageing Research Reviews, 8*(4), 339–348.

Leandro, C. G., Barboza, E. G., Caron, F., & de Jesus, F. A. (2019). GPR trace analysis for coastal depositional environments of southern Brazil. *Journal of Applied Geophysics,162*, 1–12.

Liu, R., Lehman, J., Molino, P., Petroski Such, F., Frank, E., Sergeev, A., & Yosinski, J. (2018). An intriguing failing of convolutional neural networks and the coordconv solution. *Advances in Neural Information Processing Systems, 31.*

Malik, W. U., Javed, M., & Hassan, S. T. (2017). Influence of transformational leadership components on job satisfaction and organizational commitment. *Pakistan Journal of Commerce and Social Sciences (PJCSS), 11*(1), 147–166.

Marangoni, A. G., Acevedo, N., Maleky, F., Peyronel, F., Mazzanti, G., Quinn, B., & Pink, D. (2012). Structure and functionality of edible fats. *Soft Matter, 8*(5), 1275–1300.

Martins, G. D. O., Gomes, C. A. S., Brocardo, J., Pedroso, J. V., Camilo, J. L. A., Silva, L. M. U., … & Rodrigues, S. M. C. V. (2017). Perfil dos alunos à saída da escolaridade obrigatória.

Martinez-Hernandez, G. B., Navarro-Martinez, A., & Lopez-Gomez, A. (2025). Carnauba wax enriched with encapsulated plant essential oils into native β-cyclodextrins as an innovative wax coating on citrus fruits. *Industrial Crops and Products, 225*, 120461.

Matyushenskii, B. V. (1960). Oil-wax components of the sunflower-seed hulls and corncobs. *Uchenye Zapiski-Kishinevskii Gosudarstvennyi Universitet, 56*, 81–83.

Milani, J. M., & Nemati, A. (2023/2024/2025). Lipid-based edible films and coatings: A review of recent advances and applications. *Journal of Packaging Technology and Research, 6*(1), 11–22.

Miyazaki, T., Plotto, A., Baldwin, E. A., Reyes-De-Corcuera, J. I., & Gmitter Jr, F. G. (2012). Aroma characterization of tangerine hybrids by gas-chromatography–olfactometry and sensory evaluation. *Journal of the Science of Food and Agriculture, 92*(4), 727–735.

Mohamed, N. (2023). Current trends in AI and ML for cybersecurity: A state-of-the-art survey. *Cogent Engineering, 10*(2), 2272358.

Ng, M., Gakidou, E., Lo, J., Abate, Y. H., Abbafati, C., Abbas, N., … & Azargoonjahromi, A. (2025). Global, regional, and national prevalence of adult overweight and obesity, 1990–2021, with forecasts to 2050: a forecasting study for the Global Burden of Disease Study 2021. *The Lancet, 405*(10481), 813–838.

Öğütcü, M., Arifoğlu, N., & Yılmaz, E. (2015). Preparation and characterization of virgin olive oil-beeswax oleogel emulsion products. *Journal of the American Oil Chemists' Society, 92*(4), 459–471.

Pereira, L. J., Pereira, C. V., Murata, R. M., Pardi, V., & Pereira-Dourado, S. M. (2020). Biological and social aspects of Coronavirus Disease 2019 (COVID-19) related to oral health. *Brazilian Oral Research, 34*, e041.

Pérez-Gago, M. B., & Rhim, J. W. (2024). Edible coating and film materials: Lipid bilayers and lipid emulsions. In *Innovations in food packaging* (pp. 325–350). Academic Press.

Raju, A., & Sasikala, M. S. (2016). Natural antimicrobial edible film for preservation of paneer. *Biosciences Biotechnology Research Asia, 13*(2), 1083–1088.

Ramos, O. L., Fernandes, J. C., Silva, S. I., Pintado, M. E., & Malcata, F. X. (2012). Edible films and coatings from whey proteins: a review on formulation, and on mechanical and bioactive properties. *Critical Reviews in Food Science and Nutrition, 52*(6), 533–552.

Rhim, J. W., & Wang, L. F. (2014). Preparation and characterization of carrageenan-based nanocomposite films reinforced with clay mineral and silver nanoparticles. *Applied Clay Science, 97*, 174–181.

Rhim, J. W., & Shellhammer, T. H. (2005). Lipid-based edible films and coatings. In *Innovations in food packaging* (pp. 362–383). Academic Press.

Rudel, L. L., & Morris, M. D. (1973). Determination of cholesterol using o-phthalaldehyde. *Journal of Lipid Research, 14*(3), 364–366.

Sabale, P., Bhimani, B., Prajapati, C., & Sabalea, V. (2012). An overview of medicinal plants as wound healers. *Journal of Applied Pharmaceutical Science, 2*(11), 143–150.

Sarraj, A., Hassan, A. E., Abraham, M. G., Ortega-Gutierrez, S., Kasner, S. E., Hussain, M. S., … & Albers, G. W. (2023). Trial of endovascular thrombectomy for large ischemic strokes. *New England Journal of Medicine, 388*(14), 1259–1271.

Shen, G., Yu, G., Wu, H., Li, S., Hou, X., Li, M., et al. (2022). Incorporation of lipids into wheat bran cellulose/wheat gluten composite film improves its water resistance properties. *Membranes, 12*(1), 18.

Sun, H., Lei, C., Xu, J., & Li, R. (2021). Foliar uptake and leaf-to-root translocation of nanoplastics with different coating charge in maize plants. *Journal of Hazardous Materials, 416*, 125854.

Syahida, S. N., Ismail-Fitry, M. R., Ainun, Z. M. A. A., & Hanani, Z. A. N. (2020). Effects of palm wax on the physical, mechanical and water barrier properties of fish gelatin films for food packaging application. *Food Packaging and Shelf Life, 23*, 100437.

Syahida, Z. S. (2023). Pengaruh Total Assets Turnover (TATO) dan Return On Assets (ROA) Terhadap Profit Growth pada saham yang terdaftar di Jakarta Islamic Index (JII): Studi di PT. Erajaya Swasembada Tbk. *Periode 2012-2021 (Doctoral dissertation, UIN Sunan Gunung Djati Bandung).*

Thakur, S., Singh, L., Wahid, Z. A., Siddiqui, M. F., Atnaw, S. M., & Din, M. F. M. (2016). Plant-driven removal of heavy metals from soil: uptake, translocation, tolerance mechanism, challenges, and future perspectives. *Environmental Monitoring and Assessment, 188*(4), 206.

Tinto, V. (2017). Through the eyes of students. *Journal of College Student Retention: Research, Theory & Practice, 19*(3), 254–269.

Torres-García, J. R., Leonardo-Elias, A., Angoa-Pérez, M. V., Villar-Luna, E., Arias-Martínez, S., Oyoque-Salcedo, G., & Oregel-Zamudio, E. (2024). Bacillus subtilis edible films for strawberry preservation: antifungal efficacy and quality at varied temperatures. *Foods, 13*(7), 980

Tulloch, D. M. (1957). The logic of positive terms and the transcendental notion of being. *Mind, 66*(263), 351–362.

Tyagi, P., Salem, K. S., Hubbe, M. A., & Pal, L. (2021). Advances in barrier coatings and film technologies for achieving sustainable packaging of food products—a review. *Trends in Food Science & Technology, 115*, 461–485.

Vandeburg, L. E., & Wilder, E. A. (1970). The structural characteristics of carnauba wax. *Journal of the American Oil Chemists' Society, 47*(12), 514–518.

Wang, Y., Jodoin, P. M., Porikli, F., Konrad, J., Benezeth, Y., & Ishwar, P. (2014). CDnet 2014: An expanded change detection benchmark dataset. *In Proceedings of the IEEE conference on computer vision and pattern recognition workshops* (pp. 387–394).

Wang, Q., Wang, Z., Awasthi, M. K., Jiang, Y., Li, R., Ren, X., & Zhang, Z. (2016). Evaluation of medical stone amendment for the reduction of nitrogen loss and bioavailability of heavy metals during pig manure composting. *Bioresource Technology, 220*, 297–304.

Wardhono, E. Y., Pinem, M. P., Kustiningsih, I., Agustina, S., Oudet, F., Lefebvre, C., … & Guénin, E. (2019). Cellulose nanocrystals to improve stability and functional properties of emulsified film based on chitosan nanoparticles and beeswax. *Nanomaterials, 9*(12), 1707.

Wu, J., Liang, B., Chen, C., Wang, H., Fang, Y., Shen, S., … & Zheng, X. (2021). SARS-CoV-2 infection induces sustained humoral immune responses in convalescent patients following symptomatic COVID-19. *Nature Communications, 12*(1), 1813.

Yousuf, B., Sun, Y., & Wu, S. (2022). Lipid and lipid-containing composite edible coatings and films. *Food Reviews International, 38*(sup1), 574–597.

Zahedi, P., Rezaeian, I., Ranaei-Siadat, S. O., Jafari, S. H., & Supaphol, P. (2010). A review on wound dressings with an emphasis on electrospun nanofibrous polymeric bandages. *Polymers for Advanced Technologies, 21*(2), 77–95.

Zhang, S., Yao, L., Sun, A., & Tay, Y. (2019). Deep learning based recommender system: A survey and new perspectives. *ACM Computing Surveys (CSUR), 52*(1), 1–38.

Zhang, W., Zhang, Y., Cao, J., & Jiang, W. (2023a). Improving the performance of edible food packaging films by using nanocellulose as an additive. *International Journal of Biological Macromolecules, 166*, 23188–23296.

Zhang, Y., Bi, J., Wang, S., Cao, Q., Li, Y., Zhou, J., & Zhu, B. W. (2023b). Functional food packaging for reducing residual liquid food: Thermo-resistant edible super-hydrophobic coating from coffee and beeswax. *Journal of Colloid and Interface Science, 533*, 742–749.

Zhao, Z., Ju, Y., Kou, M., Tian, M., Christensen, M. J., Zhang, X., & Nan, Z. (2022). Cuticular wax modification by epichloë endophyte in Achnatherum inebrians under different soil moisture availability. *Journal of Fungi, 8*(7), 725.

Zhou, T., Lu, Y., & Wang, B. (2010). Integrating TTF and UTAUT to explain mobile banking user adoption. *Computers in Human Behavior, 26*(4), 760–767.

Zubair, M., Wang, S., & Ali, N. (2021). Advanced approaches to breast cancer classification and diagnosis. *Frontiers in Pharmacology, 11*, 632079.

Zubair, M., Pradhan, R. A., Arshad, M., & Ullah, A. (2023). Recent advances in lipid derived bio-based materials for food packaging applications. *Macromolecular Materials and Engineering, 306*(7), 2000799.

Chapter 5
Composite Packaging

Himanshu, Nishant Kumar, Pratibha, and Himani

Abstract Composite packaging is a practical substitute for synthetic petroleum-based polymers, including polyethylene terephthalate, polyethylene, polypropylene, and polyethylene, due to their superior performance and environmental friendliness. Covalent or noncovalent connections, such as electrostatic, hydrophobic, hydrogen bonding, or van der Waals interactions, bring together active and matrix molecules to form composite packaging. The ability to integrate the unique properties of several materials is the main advantage of composite packaging. This results in improved mechanical strength, barrier qualities, biodegradability, and usefulness. The properties of composite packaging are affected by the kind and proportion of the materials used. Composite packaging may be enhanced with active ingredients such as antioxidants, minerals, and antimicrobials to provide additional benefits, including reducing lipid oxidation, inhibiting microbial development, and increasing nutritional value. Using these materials in food packaging allows for active and intelligent packaging, which has benefits including longer shelf life and real-time freshness indications. In addition, their biodegradable nature greatly aids in reducing plastic waste and achieving environmental goals. Common techniques for preparing composites, such as solvent casting and extrusion, are covered in this chapter. Research into composite packaging for perishable foods, including fruits and vegetables, focuses on its biodegradability, mechanical strength, and barrier properties.

Keywords Composite · Casting methods · Extrusion · Biopolymers · Antimicrobial activity · Molecular interactions

Himanshu · Himani (✉)
Department of Botany, Baba Mast Nath University, Rohtak, Haryana, India

N. Kumar
Amity Food and Agriculture Foundation (AFAF), Amity University, Noida, Uttar Pradesh, India

Pratibha
Centre of Management and Humanities (CMH), Punjab Engineering College (Deemed to be University), Chandigarh, India

N. Kumar et al. (eds.), *Trends in Edible Packaging: A sustainable approaches for postharvest management of fruits and vegetables*,
https://doi.org/10.1007/978-3-032-13199-7_5

Introduction

A protective barrier is formed around food products by food packaging, which assists in food preservation (Wani et al., 2024). In spite of protecting food against contamination and other forms of physical, biological, chemical, and environmental damage, proper packaging also keeps perishable food products fresh for longer periods of time during transit and storage (Haghighi et al., 2021). The loss of food, its shortened shelf life, and the increased risk of foodborne illnesses are all consequences of microbial contamination (Haghighi et al., 2019). Food deterioration resulted in economic losses exceeding 25% of the consumed food (Chawla et al., 2021). Polyethylene, polypropylene, polystyrene, and polyethylene terephthalate are among the most popular synthetic petroleum-derived polymers for packaging because they are long-lasting, inexpensive, resistant to degradation, and have excellent mechanical properties (Kumar, 2024). There are major environmental problems due to the widespread use of these polymers, which are not biodegradable and do not regenerate themselves (Cazón & Vázquez, 2021). Renewable, biological resource–based biodegradable food packaging has emerged as an eco-friendly alternative to traditional plastic packaging in the past few years (Amin et al., 2021). Bacteria, fungi, and algae are just a few examples of the microorganisms that may decompose the biodegradable packaging into carbon dioxide, water, and biomass (Weligama Thuppahige & Karim, 2022). These biodegradable materials improve the quality, safety, and longevity of packaged food while reducing the negative impact of plastics on the environment (Jahangiri et al., 2024). Proteins, polysaccharides, and other food biopolymers, together with a variety of functional components such as lipids, phospholipids, and inorganic particles, make up biodegradable packaging (Alibekov et al., 2024). As a whole, the new packaging is safer, more biodegradable, and better for the environment than traditional plastics. Composite packaging can be enriched with active components such as antioxidants, nutrients, and antimicrobials to offer additional benefits like reducing lipid oxidation, increasing nutritional value, and inhibiting microbial proliferation (Khalid & Arif, 2022; Ribeiro et al., 2021).

Composting under controlled conditions can speed up the biodegradation of these materials (Cheng et al., 2024). Covalent or noncovalent connections, such as electrostatic, hydrophobic, hydrogen bonding, or van der Waals interactions, join active and matrix molecules in composite packaging (Wang et al., 2022). The sharing of electron pairs between two atoms is what constitutes a covalent bond. This means that covalent bonds are often stronger and more stable than other types of bonding (Li et al., 2021). It is common practice to use specialised chemical or enzymatic activities to facilitate the development of new covalent connections among the molecules used in packaging. Some functional groups on active and matrix molecules undergo commonly used reactions such as esterification, etherification, amidation, and glycosylation (Deng et al., 2020). In order for molecules to form noncovalent connections, they must include certain functional groups. These groups may be either hydrophobic (involving nonpolar groups), polar (using hydrogen bonding), or electrostatic (involving charged groups). Retention and release of active components may be controlled by interacting with matrix molecules, which restrict their mobility inside the packaging (Wang et al., 2022; Benbettaïeb et al., 2019). Composite films containing thyme extract polyphenols bound to chitosan/pea starch were shown to reduce polyphenol release rate (Talón et al., 2017). Also, by cross-linking the chitosan molecules within the biopolymer matrix, tannic acid incorporation into the films might reduce polyphenol release. By adjusting the film matrix characteristics, such as swelling, biodegradation, and porosity, the rate of active ingredient release from this packaging material may be controlled (Almasi et al., 2021). Two active substances, oregano oil and tea polyphenol, and two matrix molecules, zein and gelatin, have been described in prior research as forming composite packaging with the goal of controlling the release of these active components (Chen et al., 2020). These active chemicals and matrix molecules formed hydrogen and hydrophobic bonds. After the active components were added, the films' microstructure changed, showing small spherical droplets dispersed throughout a biopolymer matrix (Wang et al., 2022; Chen et al., 2020).

The cationic polysaccharide chitosan (CS) is the second most common biopolymer in nature, according to several studies. It is used in packaging because it is nontoxic, biodegradable, biocompatible, and antibacterial, and has excellent film-forming ability. It is also soluble in acidic aqueous environments (Lee et al., 2023). However, there are a lot of drawbacks to using only chitosan in packaging materials. It is brittle, sensitive to humidity, has moderate bio-functionality, and poor mechanical and thermal properties. As a result, it has a shorter shelf life than other packaging materials and cannot be used for sustainable food packaging (Bhowmik et al., 2022, 2024). Using polyphenol chemicals, which are recognised for their antioxidant, antibacterial, and improved UV protection, has been shown in recent studies to enhance the qualities of CS (Chen et al., 2022a; Prabhakar et al., 2022). It is easy to separate tannic acid (TA) from plant tissues and green tea, among other polyphenols. In addition, tannic acid (TA) may facilitate many interactions with various materials, including hydrogen bonding, π-π interactions, coordination bonding, dynamic covalent bonding, and more, due to its five gallol and five catechol groups (Pan et al., 2022)."Bananas suffered less moisture loss and enzymatic browning when packaged in a film made by combining pure chitosan (CS) with multifunctional tannic acid (TA). Because of their exceptional mechanical, antioxidant, and antibacterial properties, the CS-TA composite films were shown to be appropriate active packaging materials for extended food preservation (Lee et al., 2023). To further improve its antibacterial adhesion and antibacterial properties, the sulfobetaine-modified chitosan (CS-SBMA) is synthesised by grafting copolymerised betaine methacrylate sulfonate onto the chitosan chain. Based on the available evidence, CS-SBMA/polyvinyl alcohol (PVA) films made using the casting process demonstrated exceptional antibacterial qualities, including the ability to effectively prevent bacterial adherence and control biofilms (Yin et al., 2024).

Methods of Preparation of Composite Packaging

There are primarily two distinct methods for preparing edible films: the wet approach (solvent casting) and the dry procedure (extrusion) (Chen et al., 2022b).

Extrusion

The technique of extrusion is widely used in the packaging industry, and it makes use of the thermoplastic properties of polymer materials. An adequately constructed nozzle is used to extrude a melted solid from a substance that has low moisture content. The material is heated above the glass transition temperature until it is completely combined. Film formation occurs when the resulting material is cooled (Wang et al., 2022). Biodegradable film production by extrusion often involves many steps, including feeding, mixing, hydration, and extrusion.

Polymers and plasticisers are introduced into the feed zone during the extrusion process by injection, and then they are compressed using air to reduce their moisture content. After that, the ingredients are sent to the kneading zone to be kneaded under high pressure and heat. The material is then heated to a point where it becomes molten and the forces generated by the rotating extrusion screw are what push it out of the nozzle. Blending, rather than stacking, the polymer components used to make films is different from coextrusion. The capacity to process a wide variety of materials, cheap production costs, high throughput, and continuous production are only a few of the possible benefits of extrusion over other film-forming processes (Wang et al., 2022). Extrusion facilitates simple monitoring and optimisation through the adjustment of parameters like temperature, feed rate, and screw speed. Moreover, extruded PLA films often have greater ductility and reduced tensile strength compared to solvent-cast films. Elongation at break levels of 500% is achieved in flexible packaging that is created via blown extrusion. The improved alignment of polymer chains compared to solvent casting is thought to be caused by the extensional flow that occurs during extrusion. In comparison with solvent-cast films, extruded films usually have a more constant thickness and shape (Foong et al., 2024). Different types of extrusion, i.e., cast film, blown film, and reactive extrusion, are illustrated in Fig. 5.1 (Foong et al., 2024).

Solvent Casting

The solvent casting process and the extrusion procedure are the two main ways that edible films may be made. The three main steps of the solvent casting method are as follows: first, the biopolymer is dissolved in an appropriate solvent with any additives added; second, the mixture is poured into a mould, Petri dish, or glass plate

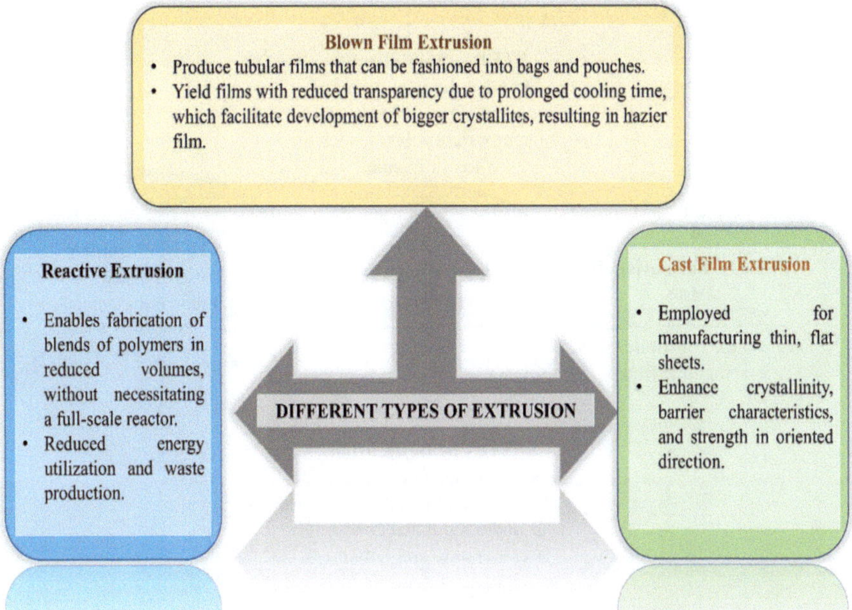

Fig. 5.1 Different types of extrusion processes for preparing composite packaging

that has been previously coated with Teflon; and lastly, the mixture is dried using an air dryer, vacuum, hot-air oven or microwave dryer, and finally, the film is removed and various steps followed during the solvent casting method are shown in Fig. 5.2 (Basumatary et al., 2023).

The film created by the low-temperature process of solvent casting is isotropically aligned, flat, clear, uniform, and defect-free. The main drawbacks are as follows: limited moulding configurations, long drying times, scaling up to commercial scale, denaturation of thermolabile substances, film characteristic variability caused by changes in processing parameters like temperature and evaporation rate, and the inclusion of potentially harmful solvents in the biopolymer matrix (Karki et al., 2016; Suhag et al., 2020). Film production mostly makes use of solvent casting because of its simplicity and low cost (Suhag et al., 2020).

Extending the shelf life of fruits and vegetables is possible through the use of these coatings or film-forming procedures, which involve creating homogeneous layers on the surfaces of food products. These layers can inhibit biological deterioration in addition to mechanical, physical, chemical, thermal, optical, gas, or moisture migration. The characteristics of the target product, the coating process, and the formulation parameters are all controlled to produce a film on the surface of fruits and vegetables that is uniform in thickness and has good adherence (Sharma et al., 2024). Table 5.1 shows many instances of composite films formed via solvent casting.

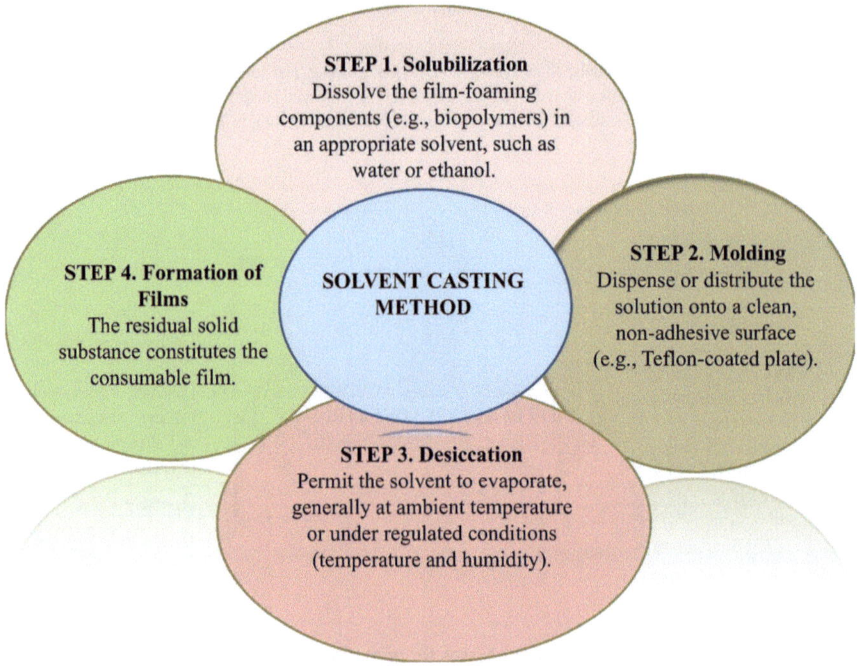

Fig. 5.2 Process of the solvent casting method for the preparation of composite packaging

Spinning Method

The spinning processes utilised in materials science encompass electrospinning, rotary jet spinning, and solution blow spinning. The electrospinning technique is probably the most prevalent among them. Electrospinning, a sophisticated nonmechanical fibre fabrication process, employs high voltage to generate nanofiber films. Rotating-jet spinning technology is a production method that utilises centrifugal force to facilitate the deposition of dissolved polymer solutions as fibres on a designated surface while regulating the fibre arrangement by a directed airstream. The edible coating incorporating natural antibacterial agents, produced via rotary jet spinning technology, has already been demonstrated to efficiently inhibit microbial development on avocado surfaces and preserve the fruit's weight. Solution blow spinning (SBS) is an innovative process that efficiently generates nanofibrous films characterised by elevated specific surface area and porosity from polymers. Compared to electrospinning, SBS does not require a high-voltage power supply and demonstrates considerably greater efficiency, resulting in its widespread application in the food packaging sector. The fibres generated using spinning technology can attain the nanoscale, and by integrating the characteristics of several polymers, can yield films with superior physical and chemical properties. Furthermore, it can mitigate the problem of material functional degradation induced by thermal exposure (Pan et al., 2025).

Table 5.1 Formation of composite films using the solvent casting method with their applications

Film	Film formation method	Application	References
Cellulose/chitosan blend composite films with tannic acid and gallic acid	Solvent casting method	Composite films made of cellulose and chitosan that included tannic acid (TA) and gallic acid (GA) greatly increased the cherry tomatoes' shelf life by improving their mechanical strength, the water contact angle, barrier properties, and antibacterial effectiveness.	Hou et al. (2025)
CeO_2/chitosan/PVA composite films	Solvent casting method	Composite films made of CeO2/chitosan/PVA showed antibacterial and unique physicochemical characteristics, allowing tomatoes to stay good for an extended period.	Ali et al. (2024a)
Gum acacia/pectin/pullulan (GPPL) polysaccharide composite film	Solvent casting method	The ivy gourd's shelf life was extended by up to 23 days when a polysaccharide composite film was made.	Suresh et al. (2022)
Modified corn starch/gelatin composite film/coating	Solution casting method	The cucumbers' shelf life was extended by up to 16 days by the composite film.	Kumar et al. (2021)
Chitosan/Enoki mushroom foot polysaccharide composite film	Solution casting method	The film that was made had superior antioxidant and antibacterial properties, which meant that blueberries could be kept for a longer time.	Zhao et al. (2023)
N-phosphorylated chitosan (NPCS-CA)/ polyvinyl alcohol (PVA) cinnamon essential oil (CEO) composite films, Gallic acid (GA), Peach gum (PG), polyethylene (PE)	Solvent casting method	Composite films made of NPCS-CA and PVA-CEO showed remarkable antibacterial activity, UV barrier properties in the 200–300 nm region, and significantly decreased oxygen, carbon dioxide, and water vapour permeability. At 25 °C, it extended the mangoes' shelf life.	Pan et al. (2023)

Properties of Composite Packaging Material

Barrier Properties

The water vapour permeability, gas permeability, and light transmittance are important features of food packaging materials (Trinh et al., 2023). Intricately related to the polymer matrix's composition and structure, diffusivity, solubility, and permeability are the principal parameters impacting the mass transfer of substances via packaging materials (Zhang et al., 2022). The gas or vapour permeation through a polymeric substrate begins with the surface of the polymer being absorbed by penetrating species, continues with their solubility and diffusion inside the matrix of the polymer, and finally ends with their desorption from the opposite surface. The permeability of a material can be used to compare how well different materials act as barriers under similar testing circumstances, like temperature, relative humidity, and pressure, but only for films that are thicker than 10 μm, since it is calculated by multiplying the transmission rate by the thickness of the testing membrane. Factors such as micro-morphology (crystallisation, phase separation, etc.), testing conditions (temperature, humidity, etc.), and sample preparation methods (casting, biaxial orientation, annealing, etc.) may influence the barrier efficiency of materials (Wu et al., 2021).

Managing additional permeants, such as CO_2 and scent, is critical in food packing, but has received less attention than the required strong barrier against water vapour and oxygen vapour. Important properties for food packaging include gas barrier, mechanical performance, heat-sealing capacity, printability, resistance to ultraviolet light, and antibacterial properties (Zhong et al., 2020). Research on composite active packaging materials has also been carried out. Integrating eugenol (active component) into starch and chitosan composites (matrix materials) by solvent casting and solvent-evaporation processes has resulted in the synthesis of biodegradable films (Zheng et al., 2019). The optimisation of the matrix composition resulted in a reduction of the water vapour permeability (WVP) of the films from about 2.00×10^{-10} g m^{-1} $Pascal^{-1}$ s^{-1} to 1.29×10^{-10} g m^{-1} $Pascal^{-1}$ s^{-1}. The addition of eugenol to the films also improved their hydrophobicity, antioxidant capabilities, antibacterial efficiency, and flexibility.

Mechanical Properties

A variety of applications rely on the mechanical qualities of packing materials, which include things like Young's modulus, elongation at break, yield stress, yield strain, and tensile strength (Anukiruthika et al., 2020). Certain active substances may improve the mechanical characteristics of biodegradable films, which are influenced by molecular interactions (Wu et al., 2019). Citric acid, according to the research, significantly improves the mechanical characteristics of potato starch-chitosan (PS-CS) composite films by acting as a cross-linker. From a starting point of 9.7 MPa, the films' tensile strength increased from 0 to a 15% citric acid concentration; however, further increases resulted in a decrease in tensile strength. The decrease in attractive interactions between the molecules of the matrix may be due to the plasticising effect of citric acid at high concentrations, which would explain the phenomenon. Biodegradable films may have less mechanical strength after adding active compounds, according to some research. Adding apple polyphenols to chitosan films reduced the films' tensile strength and break elongation, and improved their antioxidant and antibacterial properties (Sun et al., 2017).

Degradability

Most studies on active packaging materials focus on how bioactive they are, ignoring how biodegradable they are (Chen et al., 2022b). Photographically degradable, oxidatively degradable, biodegradable, and hydrolytically degradable are the four main ways in which packing materials fail. The effects of environmental variables, such as pH, temperature, and humidity, on different packaging materials must be carefully considered (Wang et al., 2022).

Application of Composite Packaging for Enhancing the Shelf Life of Fruits and Vegetables

Composite packaging materials also have several uses in the food industry. The shelf life of bananas, avocados, papayas, and strawberries may be extended by using composite films made of egg protein and cellulose nanomaterials (Jung et al., 2020). Composites containing organic porous chitosan microspheres embedded in a poly (L-lactic acid) matrix have been shown to increase the shelf life of food packaging by controlling gas penetration (Zhou et al., 2021). There is a huge opportunity to create value-added and ecologically friendly structural composites from waste by using orange peels as a bio filler in polymer composites. Orange peels are biocompatible, biodegradable, cheap, and widely available. Environmentally friendly chitosan/PVA composite films enhanced with orange peel are solvent-cast. Orange peel enhanced the chitosan/PVA films' properties, making them more flexible, thermally stable, and water vapour permeable. It enhanced the antioxidant and UV-visible light barrier properties of neat films by decreasing their transparency. Bioactive food packaging that is kind to the environment might be made using chitosan, polyvinyl alcohol, and orange peel composite sheets (Terzioğlu et al., 2021).

Products that are functional, biosafe, and environmentally beneficial may benefit from peach gum (PG) as a polymeric feedstock. The mechanical and antibacterial properties of PG make it an unappealing choice for food packaging. To get over these restrictions, a dual cross-linked network was created by adding glycerol and chitosan (CS) to the PG matrix. When CS was added to the PG matrix, the Young's modulus rose from 277.62 to 925.89 MPa, and the tensile strength increased from 5.96 to 39.94 MPa, according to the results. Glycerol also greatly improved extension. Ionic and hydrogen bonding between the biopolymers were responsible for these enhancements. Additionally, colonies of *Staphylococcus aureus* were decreased by 99.9% and those of *Escherichia coli* by 99.2% after applying the composite films. The use of CS resulted in films that were more amorphous, which enhanced their flexibility, stiffness, and barrier properties. Strawberries preserved in a composite polyelectrolyte film were fresher than those wrapped in plain old plastic (Chen et al., 2024).

Functional composite films infused with curcumin and made from chitosan (CS) and Tenebrio molitor larvae protein (TMP) are another example. Curcumin was found to be physically distributed throughout the polymer film, according to the results. When compared to the control film (CS/TMP), the curcumin composite film with 0.3% addition showed a 27.39% increase in elongation at break (EBA), a 37.04% improvement in water vapour barrier qualities, and strong antioxidant and UV-blocking capabilities. The created composite films for packaging treatment prolonged the ripening and quality preservation of blueberries, reduced the rate of deterioration and weight loss during storage, and outperformed the control film and polyethylene (PE) films (Liu et al., 2024). In addition, composite films (CNSP) made of NPCS and PVA have shown UV-blocking and antibacterial properties, with citric acid serving as the cross-linking agent and functional component. Results showed that the composite film could withstand loads up to 20.4 MPa, had an elongation at break of 6.77 percent, and 70% of UV light barrier transmittance. The utilisation of these composite membranes in strawberry preservation proved beneficial in sustaining freshness by minimising water loss, suppressing microbial proliferation, and prolonging shelf life. Furthermore, the composite film exhibited biosafety (Lai et al., 2024).

Prior studies have shown the development of an innovative bioactive binary composite film made of gelatin and dextran, which has different amounts of citronellal and α-tocopherol added to it. The amounts of these ingredients range from 0% to 1.2% by weight. Research showed that the film with Cit and Toc significantly reduced the growth of spoilage and harmful bacteria, as well as free radicals (2, 2-diphenyl-1-picrylhydrazyl and 2, 2′-azino-bis(3-ethylbenzthiazoline-6-sulfonic acid)) (Yue et al., 2024). In recent years, the interest of researchers has increased in the utilisation of fruit and vegetable waste to develop edible coatings and films for food packaging applications (Table 5.2).

Conclusion

Composite packaging has rapidly evolved into a critical segment of modern packaging solutions, offering a blend of materials that combine the best properties of different substances to meet the increasingly complex demands of industries and consumers. Its versatility, strength, and efficiency make it an indispensable choice in various sectors such as food and pharmaceuticals. By integrating biopolymers, nanoparticles, and bioactive compounds, composite materials exhibit improved mechanical strength, barrier properties, biodegradability, and active functionalities, such as antimicrobial effects and extended shelf life. Methods of preparation, including casting and extrusion, provide various advantages depending on the desired properties and scale of production. Casting is favoured for its simplicity, while extrusion is more suited for industrial-scale applications. Furthermore, the

Table 5.2 Innovative food packaging films from fruit and vegetable waste materials

S. No.	Material used	Significant outcomes	References
1.	Apple pomace and polyvinyl alcohol (PVA)	Produced biocomposite films with antioxidant qualities, showing potential for application in active food packaging.	Gaikwad et al. (2016)
2.	Pomegranate rind, starch, and polyvinyl alcohol (PVA)	Produced bioactive packaging films that improve the shelf life of packed food by having antibacterial and antioxidant qualities.	Dhaundiyal et al. (2024)
3.	Tomato by-product, pomegranate peel, and banana peel extract	Highlighted the mechanical, antibacterial, and antioxidant qualities of fruit and vegetable wastes, as well as their potential for creating biodegradable packaging materials.	Ali et al. (2024b)
4.	Sweet lime peel, starch, and polyvinyl alcohol (PVA)	Produced biodegradable films with enhanced mechanical qualities and potential for use in food packaging.	Singha et al. (2023)
5.	Banana peel, carboxymethyl cellulose, and starch	Developed biodegradable films with enhanced mechanical properties	Verma et al. (2024)
6.	Banana starch, chitosan tapioca starch, xanthan gum, and yam starch	Biodegradable starch-based packaging, both native and modified, was developed to increase the shelf life and safety of fruits and vegetables.	Bangar et al. (2023)

potential of composite packaging for providing superior protection, extending shelf life, and maintaining product integrity, while simultaneously addressing cost-effectiveness, renders it indispensable across sectors such as food and pharmaceuticals.

References

Ali, M. H., Dutta, S. K., Sultana, M. S., et al. (2024a). Green synthesized CeO2 nanoparticles-based chitosan/PVA composite films: Enhanced antimicrobial activities and mechanical properties for edible berry tomato preservation. *International Journal of Biological Macromolecules, 280*, 135976.

Ali, M. Q., Ahmad, N., Azhar, M. A., Munaim, M. S. A., Hussain, A., & Mahdi, A. A. (2024b). An overview: Exploring the potential of fruit and vegetable waste and by-products in food biodegradable packaging. *Discover Food, 4*(1), 130.

Alibekov, R. S., Urazbayeva, K. U., Azimov, A. M., et al. (2024). Advances in biodegradable food packaging using wheat-based materials: Fabrications and innovations, applications, potentials, and challenges. *Foods, 13*, 2964.

Almasi, H., Jahanbakhsh Oskouie, M., & Saleh, A. (2021). A review on techniques utilized for design of controlled release food active packaging. *Critical Reviews in Food Science and Nutrition, 61*, 2601–2621.

Amin, U., Khan, M. U., Majeed, Y., et al. (2021). Potentials of polysaccharides, lipids and proteins in biodegradable food packaging applications. *International Journal of Biological Macromolecules, 183*, 2184–2198.

Anukiruthika, T., Sethupathy, P., Wilson, A., et al. (2020). Multilayer packaging: Advances in preparation techniques and emerging food applications. *Comprehensive Reviews in Food Science and Food Safety, 19*, 1156–1186.

Bangar, S. P., Scott Whiteside, W., Suri, S., Barua, S., & Phimolsiripol, Y. (2023). Native and modified biodegradable starch-based packaging for shelf-life extension and safety of fruits/vegetables. *International Journal of Food Science and Technology, 58*(2), 862–870.

Basumatary, I. B., Mukherjee, A., & Kumar, S. (2023). Chitosan-based composite films containing eugenol nanoemulsion, ZnO nanoparticles and Aloe vera gel for active food packaging. *International Journal of Biological Macromolecules, 242*, 124826.

Benbettaïeb, N., Karbowiak, T., & Debeaufort, F. (2019). Bioactive edible films for food applications: Influence of the bioactive compounds on film structure and properties. *Critical Reviews in Food Science and Nutrition, 59*, 1137–1153.

Bhowmik, S., Agyei, D., & Ali, A. (2022). Bioactive chitosan and essential oils in sustainable active food packaging: Recent trends, mechanisms, and applications. *Food Packaging and Shelf Life, 34*, 100962.

Bhowmik, S., Agyei, D., & Ali, A. (2024). Enhancement of mechanical, barrier, and functional properties of chitosan film reinforced with glycerol, COS, and gallic acid for active food packaging. *Sustainable Materials and Technologies, 41*, e01092.

Cazón, P., & Vázquez, M. (2021). Bacterial cellulose as a biodegradable food packaging material: A review. *Food Hydrocolloids, 113*, 106530.

Chawla, R., Sivakumar, S., & Kaur, H. (2021). Antimicrobial edible films in food packaging: Current scenario and recent nanotechnological advancements-a review. *Carbohydrate Polymer Technologies and Applications, 2*, 100024.

Chen, X., Xiao, J., Cai, J., & Liu, H. (2020). Phase separation behavior in zein-gelatin composite film and its modulation effects on retention and release of multiple bioactive compounds. *Food Hydrocolloids, 109*, 106105.

Chen, T., Shen, Y., Wu, D., et al. (2022a). Biodegradable films of chitosan and tea polyphenols catalyzed by laccase and their physical and antioxidant activities. *Food Bioscience, 46*, 101513.

Chen, W., Ma, S., Wang, Q., et al. (2022b). Fortification of edible films with bioactive agents: A review of their formation, properties, and application in food preservation. *Critical Reviews in Food Science and Nutrition, 62*, 5029–5055.

Chen, Y., Ye, Y., Zhu, Z., et al. (2024). Preparation and characterization of peach gum/chitosan polyelectrolyte composite films with dual cross-linking networks for antibacterial packaging. *International Journal of Biological Macromolecules, 261*, 129754.

Cheng, J., Gao, R., Zhu, Y., & Lin, Q. (2024). Applications of biodegradable materials in food packaging: A review. *Alexandria Engineering Journal, 91*, 70–83.

Deng, L., Li, X., Miao, K., et al. (2020). Development of disulfide bond crosslinked gelatin/ε--polylysine active edible film with antibacterial and antioxidant activities. *Food and Bioprocess Technology, 13*, 577–588.

Dhaundiyal, A., Jamwal, V., Mittal, A., Pal, K., Tripathi, R., Paul, N., Rameshpathy, M., & Guimarães Junior, M. (2024). Harnessing pomegranate rind waste to develop novel bioactive packaging films using citric acid crosslinked polyvinyl alcohol (PVA)/starch-based polymer matrix. *Biomass Conversion and Biorefinery*, 1–20.

Foong, H. L., Ismail-Fitry, M. R., Noranizan, M. A., et al. (2024). Recent advances in extruded polylactic acid-based composites for food packaging: A review. *International Journal of Biological Macromolecules*, 131340.

Gaikwad, K. K., Lee, J. Y., & Lee, Y. S. (2016). Development of polyvinyl alcohol and apple pomace bio-composite film with antioxidant properties for active food packaging application. *Journal of Food Science and Technology, 53*, 1608–1619.

Haghighi, H., De Leo, R., Bedin, E., et al. (2019). Comparative analysis of blend and bilayer films based on chitosan and gelatin enriched with LAE (lauroyl arginate ethyl) with antimicrobial activity for food packaging applications. *Food Packaging and Shelf Life, 19*, 31–39.

Haghighi, H., Gullo, M., La China, S., et al. (2021). Characterization of bio-nanocomposite films based on gelatin/polyvinyl alcohol blend reinforced with bacterial cellulose nanowhiskers for food packaging applications. *Food Hydrocolloids, 113*, 106454.

Hou, T., Venkatesan, R., Dhilipkumar, T., et al. (2025). Enhancing cherry tomato packaging: Evaluation of physicochemical, mechanical, and antibacterial properties in tannic acid and gallic acid crosslinked cellulose/chitosan blend films. *International Journal of Biological Macromolecules, 285*, 138276.

Jahangiri, F., Mohanty, A. K., & Misra, M. (2024). Sustainable biodegradable coatings for food packaging: Challenges and opportunities. *Green Chemistry, 26*, 4934.

Jung, S., Cui, Y., Barnes, M., et al. (2020). Multifunctional bio-nanocomposite coatings for perishable fruits. *Advanced Materials, 32*, 1908291.

Karki, S., Kim, H., Na, S.-J., et al. (2016). Thin films as an emerging platform for drug delivery. *Asian Journal of Pharmaceutical Sciences, 11*, 559–574.

Khalid, M. Y., & Arif, Z. U. (2022). Novel biopolymer-based sustainable composites for food packaging applications: A narrative review. *Food Packaging and Shelf Life, 33*, 100892.

Kumar, S. (2024). Prospects and challenges of nanomaterials in sustainable food preservation and packaging: A review. *Discover Nano, 19*, 1–18.

Kumar, R., Ghoshal, G., & Goyal, M. (2021). Biodegradable composite films/coatings of modified corn starch/gelatin for shelf life improvement of cucumber. *Journal of Food Science and Technology, 58*, 1227–1237.

Lai, W., Wang, L., Pang, Y., et al. (2024). Preparation of citric acid cross-linked chitosan quaternary phosphonium/polyvinyl alcohol composite film and its application in strawberry preservation. *Food Chemistry, 455*, 139908.

Lee, S. J., Gwak, M. A., Chathuranga, K., et al. (2023). Multifunctional chitosan/tannic acid composite films with improved anti-UV, antioxidant, and antimicrobial properties for active food packaging. *Food Hydrocolloids, 136*, 108249.

Li, S., Sun, J., Yan, J., et al. (2021). Development of antibacterial nanoemulsions incorporating thyme oil: Layer-by-layer self-assembly of whey protein isolate and chitosan hydrochloride. *Food Chemistry, 339*, 128016.

Liu, M., Zou, X., Wu, X., et al. (2024). Preparation of chitosan/Tenebrio molitor larvae protein/curcumin active packaging film and its application in blueberry preservation. *International Journal of Biological Macromolecules, 275*, 133675.

Pan, W., Qi, X., Xiang, Y., et al. (2022). Facile formation of injectable quaternized chitosan/tannic acid hydrogels with antibacterial and ROS scavenging capabilities for diabetic wound healing. *International Journal of Biological Macromolecules, 195*, 190–197.

Pan, Q., Zhou, C., Yang, Z., et al. (2023). Preparation and characterization of functionalized chitosan/polyvinyl alcohol composite films incorporated with cinnamon essential oil as an active packaging material. *International Journal of Biological Macromolecules, 235*, 123914.

Pan, J.-N., Sun, J., Shen, Q.-J., et al. (2025). Fabrication, properties, and improvement strategies of edible films for fruits and vegetables preservation: A comprehensive review. *Food Innovation and Advances, 4*, 43–52.

Prabhakar, P., Sen, R. K., Mayandi, V., et al. (2022). Mussel-inspired chemistry to design biodegradable food packaging films with antimicrobial properties. *Process Safety and Environment Protection, 162*, 17–29.

Ribeiro, A. M., Estevinho, B. N., & Rocha, F. (2021). Preparation and incorporation of functional ingredients in edible films and coatings. *Food and Bioprocess Technology, 14*, 209–231.

Sharma, S., Nakano, K., Kumar, S., & Katiyar, V. (2024). Edible packaging to prolong postharvest shelf-life of fruits and vegetables: A review. *Food Chemistry Advances, 4*, 100711.

Singha, P., Rani, R., & Badwaik, L. S. (2023). Sweet lime peel-, polyvinyl alcohol-and starch-based biodegradable film: Preparation and characterization. *Polymer Bulletin, 80*(1), 589–605.

Suhag, R., Kumar, N., Petkoska, A. T., & Upadhyay, A. (2020). Film formation and deposition methods of edible coating on food products: A review. *Food Research International, 136*, 109582.

Sun, L., Sun, J., Chen, L., et al. (2017). Preparation and characterization of chitosan film incorporated with thinned young apple polyphenols as an active packaging material. *Carbohydrate Polymers, 163*, 81–91.

Suresh, S. N., Puspharaj, C., Natarajan, A., & Subramani, R. (2022). Gum acacia/pectin/pullulan-based edible film for food packaging application to improve the shelf-life of ivy gourd. *International Journal of Food Science and Technology, 57*, 5878–5886.

Talón, E., Trifkovic, K. T., Vargas, M., et al. (2017). Release of polyphenols from starch-chitosan based films containing thyme extract. *Carbohydrate Polymers, 175*, 122–130.

Terzioğlu, P., Güney, F., Parın, F. N., et al. (2021). Biowaste orange peel incorporated chitosan/polyvinyl alcohol composite films for food packaging applications. *Food Packaging and Shelf Life, 30*, 100742.

Trinh, B. M., Chang, B. P., & Mekonnen, T. H. (2023). The barrier properties of sustainable multiphase and multicomponent packaging materials: A review. *Progress in Materials Science, 133*, 101071.

Verma, P., Rani, R., Das, D., Rai, K. K., Gogoi, P., & Badwaik, L. S. (2024). Transformation of banana peel into biodegradable film added with starch and carboxymethyl cellulose and its characterization. *Sustainable Chemistry and Pharmacy, 37*, 101356.

Wang, Q., Chen, W., Zhu, W., et al. (2022). A review of multilayer and composite films and coatings for active biodegradable packaging. *Npj Science of Food, 6*, 18.

Wani, A. K., Anjli, A., & Tripathi, I. (2024). Interaction of food and packaging. In *Intelligent packaging* (pp. 289–318). Elsevier.

Weligama Thuppahige, V. T., & Karim, M. A. (2022). A comprehensive review on the properties and functionalities of biodegradable and semibiodegradable food packaging materials. *Comprehensive Reviews in Food Science and Food Safety, 21*, 689–718.

Wu, H., Lei, Y., Lu, J., et al. (2019). Effect of citric acid induced crosslinking on the structure and properties of potato starch/chitosan composite films. *Food Hydrocolloids, 97*, 105208.

Wu, F., Misra, M., & Mohanty, A. K. (2021). Challenges and new opportunities on barrier performance of biodegradable polymers for sustainable packaging. *Progress in Polymer Science, 117*, 101395.

Yin, M., Li, J., Wang, H., et al. (2024). Development of anti-bacterial adhesion and antibacterial sulfobetaines modified chitosan/polyvinyl alcohol composite films as packaging materials. *International Journal of Biological Macromolecules, 260*, 129465.

Yue, C., Fan, F., Chen, Y., et al. (2024). Preparation and characterization of bioactive binary composite film based on gelatin–dextran incorporated with citronellal and α-tocopherol. *Food Frontiers, 5*, 130–141.

Zhang, M., Biesold, G. M., Choi, W., et al. (2022). Recent advances in polymers and polymer composites for food packaging. *Materials Today, 53*, 134–161.

Zhao, J., Wang, Y., Li, J., et al. (2023). Preparation of chitosan/Enoki mushroom foot polysaccharide composite cling film and its application in blueberry preservation. *International Journal of Biological Macromolecules, 246*, 125567.

Zheng, K., Xiao, S., Li, W., et al. (2019). Chitosan-acorn starch-eugenol edible film: Physico-chemical, barrier, antimicrobial, antioxidant and structural properties. *International Journal of Biological Macromolecules, 135*, 344–352.

Zhong, Y., Godwin, P., Jin, Y., & Xiao, H. (2020). Biodegradable polymers and green-based antimicrobial packaging materials: A mini-review. *Advanced Industrial and Engineering Polymer Research, 3*, 27–35.

Zhou, Z., Ma, J., Li, K., et al. (2021). A plant leaf-mimetic membrane with controllable gas permeation for efficient preservation of perishable products. *ACS Nano, 15*, 8742–8752.

Chapter 6
Properties of Edible Coatings and Films

B. G. U. Janith, A. R. S. A. Athauda, T. M. R. Rusarani, M. D. A. M. Perera, J. M. C. M. Jayasekara, and H. M. P. C. Kumarihami

Abstract This book chapter provides a comprehensive overview of the physico-chemical, mechanical, optical, thermal, and functional properties of edible coatings and films, with a focus on their role in postharvest management of fruits and vegetables. Derived from natural biopolymers such as polysaccharides, proteins, and lipids, edible coatings act as biodegradable barriers that regulate moisture, gas, solute, and aroma transfer, while enhancing visual appeal, microbial safety, and shelf life. The chapter explores the structural attributes, including thickness, tensile strength, elasticity, transparency, surface morphology, and adhesion, and how these influence overall coating performance. It examines barrier properties against water vapor, oxygen, ethylene, aroma compounds, and solutes, emphasizing the impact of formulation strategies, plasticizers, nanomaterials, and composite systems. Thermal behavior, solubility, pH stability, and biodegradability are also addressed in the context of food safety, environmental sustainability, and regulatory compliance. The chapter highlights emerging trends such as nano-reinforced and multilayer coatings and identifies key challenges in scalability, sensory compatibility, and processing technologies. By integrating material science with practical postharvest applications, this chapter aims to guide the development of sustainable and consumer-acceptable edible coatings tailored for diverse fresh produce commodities.

Keywords Barrier properties · Biodegradable packaging · Edible coatings · Mechanical properties · Nanocomposites · Postharvest preservation

B. G. U. Janith · A. R. S. A. Athauda · T. M. R. Rusarani · M. D. A. M. Perera
H. M. P. C. Kumarihami (✉)
Department of Crop Science, Faculty of Agriculture, University of Peradeniya, Peradeniya, Sri Lanka
e-mail: prathibhani@agri.pdn.ac.lk

J. M. C. M. Jayasekara
Department of Food Science and Technology, Faculty of Agriculture, University of Peradeniya, Peradeniya, Sri Lanka

N. Kumar et al. (eds.), *Trends in Edible Packaging: A sustainable approaches for postharvest management of fruits and vegetables*,
https://doi.org/10.1007/978-3-032-13199-7_6

Introduction

Over recent decades, increasing concerns about environmental pollution, food safety, and postharvest losses have accelerated research in sustainable packaging, with particular attention to reducing environmental impact and extending the shelf life of perishables (Lin & Zhao, 2007; Dhall, 2013; Ma et al., 2024). Among these innovations, edible coatings and films have emerged as promising alternatives. These consumable barriers enhance physicochemical stability, microbial safety, and the visual appeal of fresh produce (Pérez-Gago & Krochta, 2005; Tapia et al., 2008; Valencia-Chamorro et al., 2011; Liyanapathiranage et al., 2023).

Edible coatings/films are developed from natural biopolymers such as polysaccharides (e.g., starch, alginate, cellulose, chitosan), proteins (e.g., whey, casein, soy, gelatin), lipids (e.g., waxes, fatty acids), or their composites (Ghanbarzadeh et al., 2010; Amin et al., 2021; Devi et al., 2024). They are commonly applied via dipping, spraying, or brushing, depending on the product surface and the desired functionality (Dhumal & Sarkar, 2018; Kumarihami et al., 2022; Liyanapathiranage et al., 2023). These coatings regulate moisture loss, gas exchange, and solute diffusion, the key factors in ripening, spoilage, and sensory deterioration of fresh produce (Hong & Krochta, 2006; Valencia-Chamorro et al., 2011; Liu et al., 2022).

Many edible coatings are fortified with bioactive compounds, such as antimicrobials, antioxidants, or nutraceuticals, to further support food safety, nutritional value, and shelf life extension (Kong et al., 2010; Alvarez et al., 2013; Ranjith et al., 2022). Functionally, these coatings offer a biodegradable substitute for plastic packaging and play a critical role in mitigating spoilage, dehydration, and microbial decay during postharvest handling (Pérez-Gago & Krochta, 2005; Kumarihami et al., 2022; Ma et al., 2024). The coating's effectiveness is largely determined by its material composition. Lipid-based coatings offer superior water vapor barriers but are often brittle and less effective against gas exchange. Conversely, protein- and polysaccharide-based coatings offer excellent gas barriers but are prone to water vapor permeability due to their hydrophilic nature (Bourtoom, 2008; Elsabee & Abdou, 2013; Liu et al., 2022; Devi et al., 2024). To overcome these limitations, composite formulations incorporating essential oils, nanofillers, or emulsifiers have been developed to enhance flexibility, antimicrobial activity, and barrier functionality (Arnon-Rips & Poverenov, 2018; Chawla et al., 2021). The functional properties of an edible coating or film, including mechanical properties (e.g., tensile strength, elongation), barrier properties (e.g., water vapor and gas permeability), optical clarity, and adhesion to food surfaces, must be optimized to specific produce types and meet regulatory and sensory standards (Raghav et al., 2016; Kumari et al., 2017; Wu et al., 2024). The efficacy of edible coatings is influenced by environmental factors such as temperature, relative humidity, and pH, requiring careful formulation and application (Bourtoom, 2008; Lorevice et al., 2016; Kumarihami et al., 2022). Despite significant advancements, commercial-scale implementation remains challenging, including formulation optimization, cost-efficiency, sensory compatibility, and regulatory approval (Mkandawire & Aryee, 2018; Rossi-Márquez et al., 2023; Ma et al., 2024).

This chapter reviews the physicochemical, mechanical, optical, and functional properties of edible coatings and films, emphasizing their formulation strategies, characterization techniques, and interactions with fruits and vegetables. It also examines processing methods, scalability challenges, regulatory considerations, and the future role of nanotechnology and bioactive delivery in edible coating systems. The chapter aims to bridge the gap between fundamental materials science and practical postharvest management to advance the development of sustainable, safe, and consumer-acceptable packaging solutions for fresh produce.

Properties of Edible Coatings/Films

Edible coatings and films possess interrelated material properties that define their suitability and effectiveness for fresh fruits and vegetables in postharvest systems (Bourtoom, 2008; Vargas et al., 2008; Falguera et al., 2011; Dhall, 2013; Krishnan et al., 2025). These properties determine how coatings interact with produce surfaces, respond to environmental conditions, and influence quality retention during storage and distribution (Lin & Zhao, 2007; Valencia-Chamorro et al., 2011; Ma et al., 2024). Edible coatings are primarily formulated from biopolymers such as polysaccharides, proteins, and lipids. These coatings exhibit diverse structural and functional characteristics vital for the effective preservation of fresh produce. Their performance is assessed through key property domains (Amin et al., 2021; Rossi-Márquez et al., 2023; Devi et al., 2024; Wu et al., 2024).

Mechanical Properties (e.g., Tensile Strength, Flexibility, Adhesion)

Affect durability and film integrity during the handling and storage of fresh fruits and vegetables (Ghanbarzadeh et al., 2010; Dhumal & Sarkar, 2018; Calva-Estrada et al., 2019).

Barrier Properties (e.g., Water Vapor and Gas Permeability)

Regulate moisture loss, respiration rate, and aroma retention in produce, directly impacting shelf life and ripening (Galus & Lenart, 2013; Lorevice et al., 2016; Salazar et al., 2019; Devi et al., 2022).

Optical Properties (e.g., Transparency, Gloss, Color Stability)

Influence consumer perception and visual appeal of produce (García et al., 2000; Rhim et al., 2000; Kumari et al., 2017; Xu & Li, 2023).

Chemical and Functional Properties (e.g., Solubility, pH Stability, Antioxidant, Antimicrobial Activity)

Determine how coatings respond to processing and contribute to food safety and the nutritional enhancement of fruits and vegetables (Jiang et al., 2012; Alvarez et al., 2014; Kritchenkov et al., 2020).

Biodegradability and Edibility

Reflect environmental sustainability and regulatory compliance, particularly as alternatives to plastic packaging for produce (Tharanathan, 2003; Siracusa et al., 2008; Rhim et al., 2013; Hassan et al., 2018; Krishnan et al., 2025).

These properties are dynamic and depend on the choice of biopolymers, incorporation of functional additives (e.g., essential oils, probiotics), and external factors such as temperature, humidity, and the specific characteristics of produce surface (Vargas et al., 2008; Otoni et al., 2017; Chawla et al., 2021; Devi et al., 2022; Sharma et al., 2024). For example, effective coatings must allow adequate respiration in climacteric fruits (e.g., mango, papaya) to prevent anaerobic conditions, while maintaining firmness and appearance in delicate items like strawberries (Ali et al., 2011; Velickova et al., 2013; Mendy et al., 2019; Ranjith et al., 2022).

A comprehensive understanding of these fundamental properties is essential for designing edible coatings that are technically effective, commercially viable, regulatorily acceptable, and consumer-friendly for fresh fruits and vegetables (Pavlath & Orts, 2009; Espitia et al., 2014; Matloob et al., 2023; Mikus & Galus, 2025). The following sections provide detailed insights into these properties, their measurement techniques, influencing factors and specific relevance to postharvest preservation of fruits and vegetables.

Physical and Mechanical Properties

Thickness

The thickness of edible coatings and films is a fundamental parameter affecting their physical, mechanical, and barrier properties, ultimately influencing their effectiveness in extending the shelf life of fresh produce (García et al., 2000; Bourtoom, 2008; Ghanbarzadeh et al., 2010; Dhumal & Sarkar, 2018; Devi et al., 2022). An optimal thickness ensures sufficient protection while maintaining flexibility and avoiding negative impacts on texture or sensory qualities. Research studies reported that films between 20–50 μm provide an ideal balance between barrier efficacy and mechanical adaptability (Jiménez et al., 2012; Galus & Kadzińska, 2015; Basiak et al., 2017; Devi et al., 2022). For example, lipid-based coatings with thicknesses below 30 μm were reported to reduce water vapor transmission by up to 40% compared to uncoated controls, significantly improving moisture retention without affecting flexibility (Galus & Kadzińska, 2015; Hassan et al., 2018; Devi et al., 2024; Wu et al., 2024; Krishnan et al., 2025). This moisture barrier improvement is particularly advantageous for extending the shelf life of highly perishable fruits and vegetables. Conversely, excessive film thickness can lead to undesirable structural issues. Polysaccharide-based coatings exceeding 80 μm thickness are often associated with increased brittleness and cracking tendencies (Vargas et al., 2008; Lorevice et al., 2016; Kumari et al., 2017; Devi et al., 2022; Sharma et al., 2024). Vargas et al. (2008) reported a 35% reduction in elongation at break for thick alginate coatings applied to strawberries, indicating that over-thickening compromises film elasticity and adherence. Advanced application techniques, such as electrospraying and electrohydrodynamic atomization, allow for precision-controlled coating deposition, achieving uniform thickness with variations of less than ±5 μm (Fabra et al., 2016; Alizadeh-Sani et al., 2021; Jamuna et al., 2024; Lopez-Polo et al., 2024). This precision is critical for maintaining mechanical consistency; for example, an increase in Young's modulus of approximately 12 MPa per 10 μm thickness has been observed up to 60 μm, signifying enhanced film rigidity and load-bearing capacity (Colussi et al., 2017; Podshivalov et al., 2017; Rossi-Márquez et al., 2023). Additionally, multilayered film systems (e.g., Layer-by-Layer assembly) allow nanoscale thickness adjustments to modulate properties such as gas permeability and mechanical strength (Fabra et al., 2016; Arnon-Rips & Poverenov, 2018; Wu et al., 2024). Such innovations enable customization of the coatings architecture to meet specific functional requirements across different commodity types. In conclusion, while thin coatings (20–50 μm) generally offer the best combination of moisture retention and flexibility, careful consideration of application techniques and material composition is essential to avoid the drawbacks of excessive thickness.

Tensile Strength (TS)

Tensile strength (TS) is a fundamental mechanical property defining the maximum stress that an edible coating material can withstand while being stretched or pulled before breaking (Tharanathan, 2003; Bourtoom, 2008; Falguera et al., 2011). In edible coatings and films, TS is an important mechanical parameter that indicates the structural integrity of the material, its resistance to deformation, and its durability during handling, storage, and application (Tharanathan, 2003; Siracusa et al., 2008; Falguera et al., 2011; Dhumal & Sarkar, 2018; Devi et al., 2022). High TS values are desirable for films intended to serve as protective barriers during the transportation, storage, and distribution of fresh produce, particularly in conditions that involve mechanical stress or deformation (Siracusa et al., 2008; Galus & Lenart, 2013; Lorevice et al., 2016; Basiak et al., 2017; Devi et al., 2022).

TS is influenced by several factors, including the type and concentration of the biopolymer used (e.g., polysaccharides, proteins, lipids), the presence and type of plasticizers, reinforcements (e.g., nanoparticles, fibers), the degree of cross-linking, and the processing conditions, such as drying temperature and film-forming technique (Abdorreza et al., 2011; Muscat et al., 2012; Souza et al., 2013; Basiak et al., 2017; Calva-Estrada et al., 2019; Gupta et al., 2024; Wu et al., 2024). For example, protein-based films such as those made from whey protein isolate (WPI), soy protein, or gelatin, typically exhibit higher TS due to their extensive intermolecular bonding, whereas polysaccharide-based films may require cross-linkers or blending to enhance their mechanical strength (Ghanbarzadeh et al., 2010; Galus & Lenart, 2013; Rhim et al., 2013; Lorevice et al., 2016; Calva-Estrada et al., 2019; Gupta et al., 2024). Higher TS indicates superior mechanical robustness, but may compromise flexibility (Sothornvit & Krochta, 2005; Kumari et al., 2017; Devi et al., 2022; Rossi-Márquez et al., 2023; Xu & Li, 2023).

Optimizing TS is essential for balancing mechanical performance with functional requirements in food packaging applications. The wheat gluten films plasticized with glycerol had a TS of ~4–6 MPa, comparable to some synthetic polymers. However, excessive plasticizer content can weaken TS by disrupting polymer cohesion (Abdorreza et al., 2011; Muscat et al., 2012; Fabra et al., 2014; Cabello et al., 2015; Devi et al., 2024). Polysaccharide-based films (e.g., starch, chitosan) generally show lower TS unless reinforced with nanoparticles (e.g., cellulose nanocrystals) or cross-linked with agents like citric acid (Galus & Lenart, 2013; Souza et al., 2013; Lorevice et al., 2016; Goudarzi & Shahabi-Ghahfarrokhi, 2018; Alizadeh-Sani et al., 2021; Wu et al., 2024).

Accurate measurement of TS is typically conducted using a texture analyzer or universal testing machine, following standardized protocols such as ASTM D882 or ASTM D638, where force at break is normalized by cross-sectional area. Understanding and optimizing TS is essential for designing edible films that are not only functionally effective, but also resilient enough to maintain integrity throughout the supply chain (Matloob et al., 2023; Rossi-Márquez et al., 2023; Wu et al., 2024).

Tensile Elasticity

Tensile elasticity is a critical mechanical property that determines the flexibility, stretchability, and durability of edible coatings during postharvest handling, storage, and transportation. This property influences the coating's ability to deform under mechanical stress without cracking or delaminating from the product surface. Among biopolymer-based coatings, protein-based coatings, particularly those derived from whey protein isolates, consistently exhibit superior tensile elasticity, with elongation at break values exceeding 60%. In contrast, polysaccharide-based films (e.g., starch, chitosan) typically show lower elongation (below 40%) due to their rigid, highly ordered molecular structures (Galus & Lenart, 2013; Schmid, 2013; Lorevice et al., 2016; Devi et al., 2022; Xu & Li, 2023). This makes protein-based films more suitable for coating applications requiring high flexibility. Plasticizers, such as glycerol, play a pivotal role in tensile elasticity by disrupting hydrogen bonds between polymer chains and increasing their mobility. For example, adding 15–20% (w/w) glycerol to gelatin films has been shown to increase elongation at break by approximately 35%, as the polymer matrix becomes more flexible and extensible (Abdorreza et al., 2011; Muscat et al., 2012; Galus & Kadzińska, 2016; Devi et al., 2024). However, excessive plasticizer content (>25%) may cause phase separation or excessive softening, reducing tensile strength and mechanical stability (Sothornvit & Krochta, 2005; Vu & Lumdubwong, 2016; Podshivalov et al., 2017; Goudarzi & Shahabi-Ghahfarrokhi, 2018; Rossi-Márquez et al., 2023). Composite coatings incorporating nano-reinforcements such as cellulose nanocrystals (CNCs), chitin nanowhiskers, or zinc oxide nanoparticles have demonstrated improved elasticity while maintaining strength. For instance, adding 1–3% CNCs to biopolymer matrixes can produce films with ~50% elongation at break and tensile strengths of 12–15 MPa, offering an optimal balance between flexibility and durability (Lorevice et al., 2016; Etxabide et al., 2022; Wu et al., 2024). To further enhance the tensile elasticity of edible coatings while maintaining structural integrity, advanced fabrication techniques such as electrospinning and multilayered film designs have emerged as innovative solutions (Fabra et al., 2016; Arnon-Rips & Poverenov, 2018; Alizadeh-Sani et al., 2021; Lopez-Polo et al., 2024). These methods allow precise control over film architecture, enabling customization of mechanical properties for specific fruit and vegetable packaging requirements. In conclusion, achieving optimal tensile elasticity in edible films and coatings requires a multifaceted formulation strategy that carefully balances:

Biopolymer Selection

Prioritizing flexible protein-based materials (e.g., whey protein) over rigid polysaccharides to enhance inherent stretchability (Gounga et al., 2007; Galus & Lenart, 2013; Schmid, 2013; Lorevice et al., 2016; Devi et al., 2022; Xu & Li, 2023).

Plasticizer Optimization

Incorporating moderate concentrations (15–20%) of plasticizers like glycerol to improve chain mobility while avoiding excessive softening (>25%) that compromises strength (Gounga et al., 2007; Abdorreza et al., 2011; Muscat et al., 2012; Galus & Kadzińska, 2016; Vu & Lumdubwong, 2016; Devi et al., 2024).

Nanofiller Reinforcement

Integrating nano-additives (e.g., cellulose nanocrystals, ZnO nanoparticles) to enhance elasticity–strength synergy without sacrificing flexibility (Lorevice et al., 2016; Yang et al., 2018; Etxabide et al., 2022; Wu et al., 2024).

Advanced Fabrication Techniques

Leveraging electrospinning or multilayered designs to fine-tune mechanical properties for specific product requirements (Fabra et al., 2016; Arnon-Rips & Poverenov, 2018; Alizadeh-Sani et al., 2021; Lopez-Polo et al., 2024; Mikus & Galus, 2025).

By strategically combining these elements, edible coatings can achieve superior mechanical resilience, maintaining flexibility for stress adaptation while ensuring durability throughout postharvest handling. This balance is critical for extending shelf life and minimizing damage to perishable fruits and vegetables during storage and transport. Future innovations, such as 3D-printed films and bioactive nanocomposites, could further improve performance and sustainability, reducing food waste in the supply chain (Rhim et al., 2013; Gupta et al., 2024; Sharma et al., 2024; Wu et al., 2024).

Elongation at Break (EAB) and Flexibility

Elongation at break (EAB) and flexibility are critical parameters in evaluating the mechanical performance of edible coatings. EAB measures the extent to which a film can be stretched before rupture, reflecting its ductility and ability to endure tension without failure (Sothornvit & Krochta, 2005; Falguera et al., 2011; Galus & Lenart, 2013; Lorevice et al., 2016; Devi et al., 2022). Flexibility, closely related to EAB, indicates a film's capacity to bend or deform without cracking, which is essential for conforming to irregular food surfaces and resisting mechanical stress during handling, packaging, storage, and transport (Falguera et al., 2011; Galus & Kadzińska, 2015; Basiak et al., 2017; Mikus & Galus, 2025). A higher EAB indicates greater stretchability, meaning the film is less likely to crack under tension. Films with high EAB values, such as polysaccharides or protein-based coatings plasticized with agents like glycerol, tend to exhibit superior flexibility, making them ideal for applications requiring resilience and adaptability (Abdorreza et al., 2011; Muscat et al., 2012; Galus & Kadzińska, 2016; Xu & Li, 2023). For instance, protein or polysaccharide-based films (e.g., pectin) plasticized with 30% glycerol have demonstrated EAB values around 80%, indicating excellent stretchability and flexibility, making them more suitable for flexible packaging (Sothornvit & Krochta, 2005; Espitia et al., 2014; Lorevice et al., 2016; Devi et al., 2022; Wu et al., 2024). In contrast, lipid-based films (e.g., beeswax) are generally more brittle, often exhibiting EAB values below 10% (Ghanbarzadeh et al., 2010; Valencia-Chamorro et al., 2011; Galus & Kadzińska, 2015; Basiak et al., 2017; Hassan et al., 2018; Sharma et al., 2024). Optimizing flexibility often involves balancing EAB with tensile strength (TS), as increasing plasticizer content enhances flexibility but can compromise film strength (Abdorreza et al., 2011; Muscat et al., 2012; Galus & Kadzińska, 2016; Rossi-Márquez et al., 2023; Devi et al., 2024). To address this, dual plasticizer systems such as glycerol and sorbitol are employed to improve both EAB and TS simultaneously, making the coating both robust and pliable (Falguera et al., 2011; Sanyang et al., 2016; Vu & Lumdubwong, 2016; Podshivalov et al., 2017).

Young's Modulus and Stiffness

Young's modulus and stiffness are fundamental mechanical properties used to evaluate the performance of edible coatings, alongside tensile strength and elongation at break (Bourtoom, 2008; Jiménez et al., 2012; Galus & Lenart, 2013; Basiak et al., 2017; Devi et al., 2022). Young's modulus specifically quantifies a material's stiffness, representing its resistance to deformation under tensile stress. A higher Young's modulus indicates a more rigid and less flexible material, which is desirable in applications requiring structural integrity (Lacroix & Le Tien, 2005; Rhim et al., 2013; Lorevice et al., 2016; Xu & Li, 2023; Mikus & Galus, 2025). Stiffness, though often used interchangeably with Young's modulus, generally refers to a material's resistance to bending or stretching, and is directly influenced by its modulus value (Sothornvit et al., 2010; Calva-Estrada et al., 2019; Rossi-Márquez et al., 2023; Devi et al., 2024).

It is closely related to Young's modulus, as a higher Young's modulus implies a stiffer material. For example, zein-based films demonstrate high stiffness, with Young's modulus values in the range of 1–2 GPa, due to dense protein packing (Bourtoom, 2008; Fabra et al., 2014; Galus & Kadzińska, 2016; Xu & Li, 2023; Wu et al., 2024). In contrast, starch-based films are softer, with values ranging from 0.1 to 0.5 GPa, reflecting their greater deformability (Ghanbarzadeh et al., 2010; Jiménez et al., 2012; Muscat et al., 2012; Devi et al., 2022; Sharma et al., 2024). Modifying film composition can further influence stiffness; for instance, cross-linking casein films with transglutaminase increases Young's modulus by 40%, enhancing their suitability for rigid packaging applications (Lacroix & Le Tien, 2005; Arnon-Rips & Poverenov, 2018; Calva-Estrada et al., 2019; Rossi-Márquez et al., 2023; Mikus & Galus, 2025). Additionally, the incorporation of nanofillers such as montmorillonite in chitosan matrixes restricts polymer chain mobility, resulting in improved stiffness and mechanical reinforcement (Sothornvit et al., 2010; Rhim et al., 2013; Alizadeh-Sani et al., 2021; Devi et al., 2022; Wu et al., 2024).

Transparency and Optical Clarity

Transparency and optical clarity are fundamental quality attributes of edible films and coatings, particularly for fresh produce, minimally processed foods, and ready-to-eat products, where visual appeal influences consumer acceptance. These characteristics determine whether the coating maintains the natural, appealing look of the food product while providing functional benefits. Polysaccharide-based films, such as pullulan and hydroxypropyl methylcellulose (HPMC), typically achieve high transparency (>90% transmittance at 600 nm), preserving the natural appearance of foods like berries and cut fruits (Tongdeesoontorn et al., 2011; Galus & Lenart, 2013; Eroglu et al., 2014; Lorevice et al., 2016; Devi et al., 2022; Xu & Li, 2023). However, incorporating lipid components to improve moisture barrier properties often compromises clarity due to light scattering effects. For instance, adding >10% beeswax to chitosan films increases haze and decreases transparency by up to a 40% (Valencia-Chamorro et al., 2011; Galus & Kadzińska, 2016; Lorevice et al., 2016; Devi et al., 2022; Sharma et al., 2024). Advanced approaches like nanoemulsion technology help resolve this trade-off by using ultra-fine lipid droplets (<100 nm) that maintain high transparency (>85%) even at significant lipid concentrations (15%) (Acevedo-Fani et al., 2015; Hasan et al., 2020; Alizadeh-Sani et al., 2021; Devi et al., 2024; Wu et al., 2024; Mikus & Galus, 2025). This trade-off between moisture resistance and clarity necessitates careful compositional balancing. Moreover, film thickness plays a pivotal role in determining optical clarity. Research on pectin-based films has revealed a strong positive correlation ($r = 0.8$) between film thickness and haze index, indicating that thicker coatings tend to exhibit reduced transparency due to increased light diffraction (Cabello et al., 2015; Lorevice et al., 2016; Podshivalov et al., 2017; Saberi et al., 2017; Devi et al., 2022; Wu et al., 2024). Maintaining coating thickness below 50 μm is therefore critical to preserving visual quality. Therefore, optimizing material selection, lipid incorporation methods, and coating thickness is crucial to achieving both functional performance and the visual appeal of food.

Color and Appearance

The visual properties of edible coatings, particularly color and surface characteristics, play a pivotal role in consumer perception and product marketability. As the first attribute assessed by consumers, appearance directly influences purchasing decisions. Protein-based coatings, such as those formulated from whey protein isolates or zein, often develop a slight yellowish hue due to Maillard reaction byproducts and residual chromophores, resulting in total color difference (ΔE^*) values between 3.5 and 5.0 (Pérez-Gago and Krochta, 2005; Han et al., 2008; Galus & Kadzińska, 2016; Xu & Li, 2023; Wu et al., 2024). In contrast, polysaccharide-based films (e.g., alginate, pullulan, pectin) exhibit minimal color interference ($\Delta E^* < 1.5$), making them ideal for applications where preserving the natural appearance of fresh produce is critical (Pérez-Gago and Krochta, 2005; Rojas-Graü et al., 2007; Tapia et al., 2008; Falguera et al., 2011; Galus & Lenart, 2013; Espitia et al., 2014; Devi et al., 2022). To enhance visual appeal while adding functional benefits, natural pigments such as anthocyanins (from saffron petals or berries) or chlorophyll extracts (from spinach) can be incorporated into coatings. Anthocyanin-fortified chitosan films, for instance, demonstrate exceptional color stability, retaining ~85% of their original hue after 14 days of refrigerated storage, with negligible shifts in chromaticity under cold conditions (Alizadeh-Sani et al., 2021; Devi et al., 2022; Rossi-Márquez et al., 2023; Wu et al., 2024). These bioactive pigments not only improve visual appeal but also serve functional roles as radical (antioxidants) scavengers and pH-sensitive indicators, offering real-time quality monitoring (Alizadeh-Sani et al., 2021; Pirsa et al., 2021; Armghan Khalid et al., 2022; Li et al., 2022). Surface morphology also significantly impacts perceived quality. Coatings with low surface roughness ($Ra < 0.8$ μm) exhibit superior gloss, a key determinant of freshness. Emulsion-based systems, particularly those stabilized with nanostructured lipids or emulsifiers, enhance light reflectance by ~30% compared to conventional solvent-cast films, further improving visual appeal (Galus & Kadzińska, 2015; Lorevice et al., 2016; Devi et al., 2024; Wu et al., 2024; Mikus & Galus, 2025). However, excessive lipid incorporation (>15% w/w) can induce undesirable opacity or whitening due to phase separation and crystalline domain formation, emphasizing the need for precise formulation control (Morillon et al., 2002; Ghanbarzadeh et al., 2010; Galus & Lenart, 2013; Basiak et al., 2017; Sharma et al., 2024). For light-sensitive produce (e.g., apples, pears), photoinduced discoloration remains a challenge. Recent advances in nanotechnology address this issue through UV-blocking coatings. The inclusion of titanium dioxide (0.1% w/w) or zinc oxide nanoparticles reduces discoloration by up to 60% during ambient storage while concurrently improving antimicrobial and barrier properties (Yang et al., 2018; Kritchenkov et al., 2020; Alizadeh-Sani et al., 2021; Wu et al., 2024). These nanomaterials also contribute to improved antimicrobial and barrier properties, making them multifunctional components in edible packaging systems. In summary, color and appearance are multifactorial attributes influenced by polymer type (proteins vs. polysaccharides), pigment integration (natural or synthetic), surface properties (roughness, gloss), and functional additives (UV blockers). Balancing these parameters is essential for designing coatings that maintain or enhance the natural visual quality of fruits and vegetables while providing protective functionality.

Surface Morphology (Smoothness, Porosity)

The surface morphology of edible coatings, including smoothness, roughness, and porosity, plays a critical role in their barrier efficacy, optical properties, and sensory appeal. A well-structured surface minimizes moisture and gas transfer while enhancing product quality. Atomic force microscopy (AFM) and scanning electron microscopy (SEM) analyses reveal that solvent-cast protein-based films, particularly those made from whey protein isolate, exhibit ultra-smooth surfaces (Ra < 50 nm) due to their homogeneous polymer networks (Gounga et al., 2007; Janjarasskul et al., 2011; Galus & Lenart, 2013; Schmid, 2013; Schmid et al., 2017; Xu & Li, 2023; Wu et al., 2024). In contrast, polysaccharide-based coatings (e.g., alginate or pectin) tend to be rougher (Ra = 50–200 nm), reflecting their fibrous microstructure and variable intermolecular interactions (García et al., 2000; Ghanbarzadeh et al., 2010; Lorevice et al., 2016; Devi et al., 2022; Sharma et al., 2024). Controlling porosity is essential for tailoring the permeability properties of edible films. For moderately respiring fruits and vegetables, films with pore diameters of 5–15 μm facilitate adequate gas exchange without compromising barrier function. Zein-based coatings, with natural pore sizes around 8 μm, exhibit up to 40% higher oxygen permeability than nonporous films, making them ideal for high-respiration commodities (Fabra et al., 2014, 2016; Galus & Kadzińska, 2016; Devi et al., 2022; Gupta et al., 2024). Incorporating nanomaterials, such as cellulose nanocrystals or chitin nanowhiskers, enhances surface morphology by reducing pore size (up to 60%) and surface roughness (~30%), thereby improving water vapor resistance and mechanical strength (Lorevice et al., 2016; Otoni et al., 2016; Otoni et al., 2017; Alizadeh-Sani et al., 2021; Etxabide et al., 2022; Wu et al., 2024; Mikus & Galus, 2025). SEM studies confirm that nanocomposite films exhibit denser, smoother microstructures due to improved matrix packing.

Emulsion-based coatings containing lipid or essential oil droplets often develop bimodal pore distributions. Larger pores (20–50 μm) can disrupt matrix continuity, increasing water vapor transmission rates by up to 35% (Sánchez-González et al., 2011; Galus & Kadzińska, 2015; Lorevice et al., 2016; Hassan et al., 2018; Rossi-Márquez et al., 2023; Devi et al., 2024). To achieve precise porosity and surface control, advanced fabrication methods such as electrospinning have been explored. Electrospinning enables precise control over porosity (5–80 μm) and surface roughness (<100 nm), making it possible to engineer coatings for specific gas and moisture permeability requirements while maintaining optical transparency and structural integrity (Fabra et al., 2016; Dhumal & Sarkar, 2018; Alizadeh-Sani et al., 2021; Lopez-Polo et al., 2024; Wu et al., 2024).

Adhesion and Cohesion

Adhesion and cohesion are fundamental properties that determine the mechanical performance, structural integrity, and functional longevity of edible coatings. These factors influence film formation, substrate compatibility, crack resistance, and overall durability during application, storage, and consumption. Adhesion determines a coating's ability to adhere uniformly to food surfaces. Protein-based coatings, particularly those derived from whey protein isolate or gelatin, exhibit superior peel strength (0.8–1.2 N/cm) compared to polysaccharide-based films (0.3–0.6 N/cm). This is attributed to their amphiphilic nature and capacity to form hydrogen bonds with hydrophilic substrates (Gounga et al., 2007; Han et al., 2008; Elsabee & Abdou, 2013; Galus & Lenart, 2013; Galus & Kadzińska, 2016; Devi et al., 2022). In contrast, polysaccharide coatings (e.g., alginate, pectin) often require surface activation techniques (e.g., plasma treatment, chemical etching) to improve adhesion, especially on hydrophobic or waxy surfaces like citrus or apples (Chauhan et al., 2015; Lorevice et al., 2016; Gupta et al., 2024; Sharma et al., 2024; Mikus & Galus, 2025).

Cohesion reflects the internal bonding strength of the film matrix, which resists cracking and delamination during handling. Plasticizers like glycerol enhance flexibility but can compromise cohesion at higher concentrations (>20%). For instance, gelatin films with excessive glycerol exhibit up to 40% reduction in cohesive strength due to disrupted intermolecular interactions (Muscat et al., 2012; Fakhouri et al., 2013; Sánchez-Ortega et al., 2014; Galus & Kadzińska, 2016; Devi et al., 2024; Wu et al., 2024).

Surface pretreatment methods, such as cold plasma, ultrasound, or enzymatic hydrolysis, can improve adhesion by 25–50% through increasing substrate roughness, surface energy, and functional group availability on the substrate (Dhall, 2013; Chauhan et al., 2015; Alizadeh-Sani et al., 2021; Lopez-Polo et al., 2024; Mikus & Galus, 2025). These methods are particularly effective for hydrophobic or smooth-skinned fruits like tomatoes, apples, and citrus, ensuring uniform coating anchorage.

Composite coatings combining hydrophilic polymers (e.g., alginate or pectin) with hydrophobic agents (e.g., beeswax or shellac) optimize adhesion-cohesion balance, achieving >90% surface coverage and < 5% delamination on apples after 14 days (Chauhan et al., 2015; Lorevice et al., 2016; Devi et al., 2022; Gupta et al., 2024; Sharma et al., 2024). Recent innovations in nanocomposite coatings, particularly those containing 1–3% silica or ZnO nanoparticles, enhance adhesion (via surface roughness) and cohesion (through polymer-nanofiller interactions), improving mechanical resilience (Rhim et al., 2013; Yang et al., 2018; Alizadeh-Sani et al., 2021; Wu et al., 2024). Optimizing adhesion and cohesion requires a multifaceted approach, including formulation tuning (e.g., biopolymer selection, plasticizer content), substrate modification (e.g., plasma treatment), and functional enhancements (e.g., composite or nanocomposite design). These strategies collectively ensure coating durability, efficacy, and consumer acceptability in edible film applications.

Light Barrier Properties

Light barrier properties of edible coatings are critical for preserving the quality of photosensitive fruits and vegetables, which are vulnerable to UV-induced oxidative degradation, pigment bleaching, and nutrient loss. Edible films/coatings that block harmful UV radiation while maintaining sufficient visible light transmission can extend the shelf life while preserving product appearance and consumer appeal.

Protein-derived coatings, particularly those made from whey protein isolates and zein, exhibit strong UV attenuation (85–90%) due to their aromatic amino acids (tyrosine, tryptophan, phenylalanine), which absorb UV light efficiently. These films/coatings also maintain relatively high visible light transmittance (70–80%), ensuring minimal impact on product visibility (Hong & Krochta, 2006; Schmid et al., 2012; Schmid, 2013; Galus & Kadzińska, 2016; Wang et al., 2019; Devi et al., 2022; Xu & Li, 2023). To further enhance light barrier performance, nanocomposite coatings have been developed using UV-scattering and UV-absorbing nanomaterials such as titanium dioxide (TiO_2) and zinc oxide (ZnO). TiO_2 nanoparticles can achieve >99% UV attenuation while maintaining transparency in the visible spectrum (Youssef et al., 2016; Yang et al., 2018; Kritchenkov et al., 2020; Alizadeh-Sani et al., 2021; Wu et al., 2024). These nanocomposites are ideal for fresh produce packaging, where UV protection must not compromise visual appeal.

The thickness of the coating plays a pivotal role in optimizing the light barrier and mechanical properties. A 40–60 μm thickness film provides the best balance between: UV shielding effectiveness, mechanical integrity, and consumer-acceptable appearance (Schmid, 2013; Fabra et al., 2016; Lorevice et al., 2016; Wu et al., 2024; Mikus & Galus, 2025). Excessive thickness reduces flexibility and transparency, while insufficient thickness fails to provide adequate protection. In addition to synthetic nanoparticles, plant-derived compounds like polyphenols and anthocyanins offer dual functionality: UV absorption (similar to synthetic nanoparticles) and antioxidant activity, which help mitigate oxidative and microbial spoilage (Hasan et al., 2020; Alizadeh-Sani et al., 2021; Celik et al., 2021; Rossi-Márquez et al., 2023). These are particularly promising for sustainable, multifunctional edible packaging.

The light barrier properties of edible coatings can be significantly enhanced by material selection (e.g., protein-based films for inherent UV absorption), nanocomposite integration (e.g., TiO_2 for near-total UV blocking), and structural optimization (e.g., controlled thickness). These strategies are essential for preserving nutritional value, color, and quality while extending the shelf life of UV-sensitive fresh commodities.

Barrier Properties

Barrier Properties Against Moisture

Edible coatings play a crucial role in controlling moisture migration between food products and their surrounding environment, thereby preserving key quality attributes such as texture, weight, and overall quality of fresh and minimally processed fruits and vegetables. Moisture loss or gain can lead to undesirable changes in food products, such as wilting, hardening, or microbial growth, making moisture barrier efficiency a key factor in edible coatings/films development (Morillon et al., 2002; Ghanbarzadeh et al., 2010; Galus & Lenart, 2013; Hassan et al., 2018; Devi et al., 2022).

Among various biopolymer classes, lipid-based coatings, such as those made from beeswax, carnauba wax, or shellac, are particularly effective due to their inherent hydrophobicity. This property renders them highly resistant to water vapor transmission, allowing these materials to form semipermeable barriers that reduce transpiration water loss and delay dehydration-related deterioration. For instance, Morillon et al. (2002) reported that wax-based coatings significantly reduced water loss in apple fruits by up to 50% compared to uncoated fruits, showcasing their efficacy in moisture retention. According to Devi et al. (2022), carnauba wax-based composite films and coatings significantly reduce water vapor permeability (WVP) in various fruits and vegetables by forming a continuous, hydrophobic barrier. Specifically, these coatings have been shown to reduce weight loss in produce such as cucumbers, tomatoes, and mangoes by up to 60% during storage, depending on storage conditions and formulation composition. Carnauba wax forms a dense, crystalline structure upon drying, which effectively blocks moisture transfer while preserving fruit/vegetable quality and prolonging shelf life by minimizing dehydration (Devi et al., 2022; Sharma et al., 2024; Wu et al., 2024).

Although lipid-based materials offer superior moisture resistance, they often lack mechanical strength. To address this, composite films combining lipids with polysaccharides (e.g., starch, chitosan) or proteins (e.g., whey protein, gelatin) have been developed (Jiménez et al., 2012; Galus & Kadzińska, 2015; Basiak et al., 2017; Devi et al., 2022; Gupta et al., 2024). For example, starch-beeswax blends have been shown to improve moisture resistance while maintaining structural integrity (Jiménez et al., 2012; Lorevice et al., 2016; Sharma et al., 2024; Wu et al., 2024; Mikus & Galus, 2025). Moreover, the inclusion of biopolymers (e.g., starch, chitosan) in carnauba wax composites enhances adhesion and mechanical integrity while maintaining excellent moisture barrier performance (Galus & Lenart, 2013; Hassan et al., 2018; Devi et al., 2022; Rossi-Márquez et al., 2023; Gupta et al., 2024). Devi et al. (2022) emphasize that carnauba wax–chitosan blends exhibit synergistic effects, as chitosan provides structural integrity, while carnauba wax enhances hydrophobicity. Such composites have been successfully applied to extend the shelf life of perishables like mangoes and strawberries by reducing moisture loss by 30–50%.

Despite the effectiveness of lipid-based and composite coatings/films, a key challenge in their formulation is balancing water barrier efficiency with other functional properties, such as oxygen permeability and mechanical strength (Galus & Kadzińska, 2015; Hassan et al., 2018; Sharma et al., 2024; Wu et al., 2024; Mikus & Galus, 2025). Although higher lipid content improves hydrophobicity, it can also reduce film flexibility and increase brittleness.

Recent advances in nanotechnology, such as incorporating nanocellulose or clay nanoparticles, have shown promise in improving moisture barrier properties without compromising other functionalities (Otoni et al., 2017; Haghighi et al., 2020; Alizadeh-Sani et al., 2021; Devi et al., 2024; Wu et al., 2024). Chitosan–fatty acid nanocomposites (e.g., chitosan–stearic acid) have demonstrated improved moisture and gas barrier properties due to their hydrophobic interactions and nanostructured matrix (Otoni et al., 2017; Haghighi et al., 2020; Alizadeh-Sani et al., 2021; Devi et al., 2024; Mikus & Galus, 2025). For example, carnauba wax–nanocellulose composites have shown 40% reduction in WVP compared to pure wax films, making them promising for commercial applications (Devi et al., 2022; Gupta et al., 2024; Wu et al., 2024).

Barrier Properties Against Oxygen and Gases

One of the most vital functional attributes of edible coatings and films is their ability to block gases, particularly oxygen (O_2), carbon dioxide (CO_2), and ethylene (C_2H_4). Edible coatings and films play a crucial role in food preservation by inhibiting oxygen diffusion, thereby slowing oxidative reactions that lead to lipid rancidity, enzymatic browning, color degradation, and nutrient loss, while also modulating respiration rates and ethylene-mediated ripening in fresh produce, extending the shelf life of perishable products (Vargas et al., 2008; Pavlath & Orts, 2009; Rhim et al., 2009; Falguera et al., 2011; Galus & Kadzińska, 2015).

Protein-based films, especially those formulated from whey protein isolate (WPI), soy protein, and gelatin, demonstrate superior oxygen barrier properties due to their dense, cohesive polymeric networks and the hydrophilic nature of proteins, which reduces the diffusion of nonpolar molecules like oxygen (Galus & Kadzińska, 2015; Schmid et al., 2017; Xu & Li, 2023; Devi et al., 2024; Wu et al., 2024).

Notably, whey protein films have exhibited oxygen permeability values significantly lower than those of synthetic polymers such as low-density polyethylene, making them highly effective in preventing oxidative spoilage. Incorporating montmorillonite (MMT) nanoplatelets into WPI matrixes results in up to a tenfold reduction in oxygen permeability by creating a complex diffusion path (Rhim et al., 2013; Schmid et al., 2017; Alizadeh-Sani et al., 2021; Li et al., 2022). Plasticizer type also impacts gas barrier properties; sorbitol-plasticized WPI films exhibit better oxygen resistance than those plasticized with glycerol due to lower hygroscopicity (Schmid et al., 2017).

Gelatin-based films, particularly in multilayer structures, enhance oxygen resistance. Laminated edible films composed of sodium alginate, gelatin, and WPI achieve significantly lower oxygen permeability than monolayer films (Xu & Li, 2023; Wu et al., 2024). These improvements result from extensive hydrogen bonding and polymer interactions, although their performance declines under high-humidity conditions (Falguera et al., 2011; Hassan et al., 2018; Devi et al., 2022).

Polysaccharide-based films (e.g., chitosan, starch, alginate, cellulose derivatives) also demonstrate commendable oxygen barrier properties due to compact, hydrogen-bonded structures (Vargas et al., 2008; Elsabee & Abdou, 2013; Otoni et al., 2017; Sharma et al., 2024; Mikus & Galus, 2025). However, high relative humidity compromises their performance by increasing water absorption, which plasticizes the matrix and facilitates gas diffusion (Vargas et al., 2008; Falguera et al., 2011; Otoni et al., 2017; Devi et al., 2024; Wu et al., 2024). To overcome humidity sensitivity, nanocomposite films using fillers like montmorillonite, cellulose nanocrystals, or metal oxides significantly enhance barrier properties (Rhim et al., 2013; Vaezi et al., 2019; Alizadeh-Sani et al., 2021; Li et al., 2022; Wu et al., 2024; Mikus & Galus, 2025). In addition, integrating essential oils or plant extracts improves both gas barrier functionality and antimicrobial effects (Pranoto et al., 2005; Vargas et al., 2008; Alvarez et al., 2014; Rossi-Márquez et al., 2023; Devi et al., 2024).

Edible films also modulate the permeability of CO_2 and ethylene, critical for climacteric fruits' postharvest physiology (Rhim et al., 2009; Falguera et al., 2011; Velickova et al., 2013; Hassan et al., 2018; Yousuf et al., 2018). While hydrophilic matrixes (e.g., polysaccharides) allow moderate CO_2 transmission, lipid-based coatings (e.g., waxes, fatty acids) provide better gas resistance (Galus & Kadzińska, 2015; Devi et al., 2022; Gupta et al., 2024; Sharma et al., 2024; Wu et al., 2024). Protein–lipid composite films combine structural strength with enhanced gas resistance (Yousuf et al., 2018; Devi et al., 2024; Mikus & Galus, 2025).

According to Dhall (2013), factors such as coating thickness, uniformity, and environmental conditions (temperature and humidity) influence barrier efficacy. Edible coatings create semipermeable layers that reduce respiration and delay ripening. Chitosan-based coatings, in particular, can carry antimicrobial agents and also act as gas barriers (Ali et al., 2011; Dhall, 2013; Velickova et al., 2013; Hassan et al., 2018; Devi et al., 2024).

Ethylene permeability is particularly important for shelf life extension. Chitosan and lipid-based coatings reduce ethylene transmission, slowing ripening and senescence (Vargas et al., 2008; Ali et al., 2011; Velickova et al., 2013; Galus & Kadzińska, 2015; Hassan et al., 2018). Multilayer films combining hydrophilic and hydrophobic layers further enhance this barrier effect (Hassan et al., 2018; Mikus & Galus, 2025). Ethylene scavenging is also possible with functional composites, for example, starch or chitosan films with potassium permanganate-loaded halloysite nanotubes that delay ripening in bananas and tomatoes (Joung et al., 2021; Devi et al., 2024). Similarly, pectin-essential oil coatings offer combined ethylene suppression and antimicrobial protection (Alvarez et al., 2014; Celik et al., 2021; Wu et al., 2024).

In conclusion, edible coatings and films serve as multifunctional barriers by inhibiting O_2, modulating CO_2, and limiting ethylene diffusion. Their performance depends on polymer selection, nanofiller use, lipid integration, and fabrication strategy. These properties are vital for extending shelf life, maintaining quality, and reducing food waste in postharvest supply chains.

Barrier Properties Against Aroma and Flavor

Edible films and coatings play a crucial role in preserving the sensory quality and freshness of food products by acting as barriers against volatile compounds, especially aroma and flavor molecules. Research demonstrates that these biodegradable packaging materials can significantly reduce the permeability of volatile compounds while maintaining food sensory properties (Falguera et al., 2011; Espitia et al., 2014; Galus & Kadzińska, 2015; Devi et al., 2022; Wu et al., 2024). The effectiveness of these barriers depends on the film's composition, matrix structure, and incorporation of lipophilic or hydrophobic elements. Protein-based films, such as those made from whey protein isolate, exhibit excellent aroma barrier properties due to their dense molecular structure and low free volume (Calva-Estrada et al., 2019; Xu & Li, 2023; Wu et al., 2024; Mikus & Galus, 2025). Chitosan-based coatings have proven especially effective for fresh fruits, not only providing good barrier properties but also antimicrobial effects that help maintain flavor quality during storage (Ali et al., 2011; Elsabee & Abdou, 2013; Hassan et al., 2018; Devi et al., 2022; Gupta et al., 2024). Polysaccharide and protein-based films, due to their hydrophilic nature, generally exhibit poor resistance to volatile aroma compounds. However, this limitation can be significantly addressed by adding lipids or waxes. For example, incorporating beeswax into chitosan coatings reduces permeability and improves volatile compound retention in strawberries (Velickova et al., 2013; Galus & Kadzińska, 2015; Devi et al., 2022; Sharma et al., 2024; Wu et al., 2024). Similarly, carnauba wax, when integrated into biopolymer coatings, improves aroma retention by increasing overall film hydrophobicity (Devi et al., 2022; Gupta et al., 2024; Mikus & Galus, 2025). The lamination of biopolymers such as whey protein isolate, gelatin, and sodium alginate improves structural integrity and reduces the diffusion of volatile compounds by creating tortuous diffusion paths (Xu & Li, 2023; Wu et al., 2024). These multilayered systems are increasingly adopted for their ability to enhance aroma barrier performance.

Additionally, films based on pectin and other polysaccharides can offer partial aroma barrier functions, especially when plasticized and cross-linked appropriately, although their barrier capacity remains lower than synthetic plastics (Espitia et al., 2014; Devi et al., 2024; Wu et al., 2024). Active packaging systems incorporating essential oils (e.g., cinnamon, oregano) not only inhibit microbial spoilage but also help mask off-odors, thereby indirectly preserving aroma (Souza et al., 2013; Alvarez et al., 2014; Celik et al., 2021; Rossi-Márquez et al., 2023; Wu et al., 2024). Nevertheless, precise formulation is required to prevent these additives from altering the food's original sensory profile. Lipid-incorporated biopolymer coatings have demonstrated promising aroma barrier improvements, with lipid type and concentration critically influencing volatile compound diffusion rates (Devi et al., 2024; Gupta et al., 2024; Sharma et al., 2024; Wu et al., 2024; Mikus & Galus, 2025).

The incorporation of nanoparticles such as nano clays, montmorillonite, and zinc oxide enhances molecular packing and creates complex diffusion paths, thereby reducing film porosity and retarding volatile migration (Rhim et al., 2009; Rhim et al., 2013; Schmid et al., 2017; Vaezi et al., 2019). These technological innovations in edible films and coatings support aroma retention while fulfilling growing sustainability and food quality demands in postharvest systems.

Barrier Properties Against Solute Transport

Edible coatings and films serve as effective barriers not only to gases and moisture but also to solute transport, including salts, food additives, and pigments, which can influence food quality, stability, and sensory attributes. Controlling the diffusion of these solutes is essential for preserving food quality, stability, and sensory characteristics. The ability of these materials to limit solute migration depends on their composition, microstructure, and interactions with diffusing molecules (Falguera et al., 2011; Galus & Kadzińska, 2015; Hassan et al., 2018; Devi et al., 2022; Wu et al., 2024).

Polysaccharide-based films, such as those derived from chitosan, starch, or alginate, exhibit selective permeability to solutes due to their hydrophilic nature and porous matrix (Pavlath & Orts, 2009; Falguera et al., 2011; Elsabee & Abdou, 2013; Otoni et al., 2017; Sharma et al., 2024). For instance, chitosan films have demonstrated efficacy in reducing the migration of salts and acidic compounds in coated fruits, thereby maintaining flavor balance and preventing tissue degradation (Ali et al., 2011; Elsabee & Abdou, 2013; Devi et al., 2022; Gupta et al., 2024; Wu et al., 2024). Similarly, alginate-based coatings can act as barriers to pigment diffusion, preserving the visual appeal of processed foods (Pavlath & Orts, 2009; Velickova et al., 2013; Devi et al., 2022; Sharma et al., 2024; Mikus & Galus, 2025). However, under high-humidity conditions, the swelling of hydrophilic matrixes can increase solute permeability, necessitating the use of composite or cross-linked formulations (Otoni et al., 2017; Devi et al., 2024; Wu et al., 2024).

Protein-based films, particularly those made from whey protein isolate (WPI) or gelatin, form denser polymer networks that more effectively hinder solute diffusion than polysaccharide-based counterparts (Galus & Kadzińska, 2015; Schmid et al., 2017; Xu & Li, 2023; Mikus & Galus, 2025; Wu et al., 2024). WPI films reduce the leaching of antioxidants or colorants from food surfaces, while gelatin films cross-linked with tannins or polyphenols show further enhancement in solute barrier properties through tighter molecular packing (Schmid et al., 2017; Rossi-Márquez et al., 2023; Xu & Li, 2023; Wu et al., 2024).

Lipid-based coatings, such as carnauba wax or beeswax, are particularly effective in blocking nonpolar solutes (e.g., oil-soluble pigments or hydrophobic additives) due to their inherent hydrophobicity (Jiménez et al., 2012; Galus & Kadzińska,

2015; Devi et al., 2022; Gupta et al., 2024; Sharma et al., 2024). Devi et al. (2022) reported that carnauba wax composites reduced pigment migration in coated fruits by up to 40% compared to uncoated controls, which is critical for maintaining color in strawberries and mangoes. However, lipid films alone often lack mechanical strength and adhesion, hence the preference for emulsified or multilayer systems (Galus & Kadzińska, 2015; Devi et al., 2024; Wu et al., 2024).

Composite and nanocomposite coatings offer highly tailored solute transport regulation for fresh produce. Chitosan–lipid composites, for example, have proven effective on fresh-cut fruits like apples and pears by limiting salt uptake from antimicrobial dips and preserving tissue firmness (Hassan et al., 2018; Yousuf et al., 2018; Devi et al., 2022; Sharma et al., 2024; Wu et al., 2024). In tropical fruits such as mangoes and pineapples, starch-beeswax blends have shown strong resistance to pigment leaching and flavor loss due to the semicrystalline structure of beeswax (Jiménez et al., 2012; Devi et al., 2022; Mikus & Galus, 2025).

Nanocomposite formulations further elevate barrier performance, especially for delicate commodities such as berries and leafy greens. For example, cellulose nanocrystal-reinforced WPI films applied to strawberries have been shown to reduce anthocyanin migration by 40–60% compared to conventional coatings (Rhim et al., 2013; Schmid et al., 2017; Li et al., 2022; Devi et al., 2024; Wu et al., 2024). Similarly, montmorillonite-doped alginate films have demonstrated the ability to minimize electrolyte leakage in spinach by forming a complex nanostructured diffusion network (Rhim et al., 2013; Otoni et al., 2017; Vaezi et al., 2019; Wu et al., 2024; Mikus & Galus, 2025). These systems are especially beneficial for precut vegetables treated with calcium-based firming agents, as they help retain functional minerals and mitigate undesirable texture changes (Otoni et al., 2017; Schmid et al., 2017; Devi et al., 2024; Mikus & Galus, 2025).

Overall, the development of composite and nanocomposite edible films and coatings enables precise control over solute transport in fresh produce. These systems are essential for maintaining color stability, nutrient retention, and sensory quality throughout postharvest storage and distribution.

Thermal Properties

Thermal properties of edible films and coatings are crucial for maintaining their structural integrity during food processing operations such as drying, pasteurization, and sterilization (García et al., 2000; Rhim et al., 2009; Otoni et al., 2017). These properties determine the films' performance under thermal stress and their suitability for different food applications. Key thermal characteristics include the melting point, glass transition temperature, thermal degradation, and thermal transitions.

The melting point determines the temperature at which films soften or melt, directly affecting their applicability in high-temperature food processing (García et al., 2000; Rhim et al., 2009; Calva-Estrada et al., 2019; Wu et al., 2024). The melting behavior varies significantly among different biopolymers, with protein-based films typically exhibiting higher melting points than polysaccharide-based ones (Calva-Estrada et al., 2019; Xu & Li, 2023; Mikus & Galus, 2025). The glass transition temperature (Tg) indicates the transition from a glassy, brittle state to a rubbery, flexible state and is influenced by the use of plasticizers such as glycerol and xylitol (Sothornvit & Krochta, 2005; Abdorreza et al., 2011; Muscat et al., 2012; Galus & Lenart, 2013; Devi et al., 2022). While these additives enhance flexibility, they often reduce mechanical strength and thermal resistance.

Thermal degradation temperature, typically analyzed through thermogravimetric analysis (TGA), provides insights into the decomposition behavior and thermal stability of edible films (Rhim et al., 2013; Sanyang et al., 2016; Schmid et al., 2017; Vaezi et al., 2019; Gupta et al., 2024). Most biopolymer films show multiphase degradation, with initial weight loss from water evaporation followed by breakdown of the polymer matrix.

Differential scanning calorimetry (DSC) is widely used to evaluate thermal transitions such as endothermic (melting) and exothermic (crystallization) events, essential for optimizing processing conditions (García et al., 2000; Rhim et al., 2009; Schmid et al., 2017; Mikus & Galus, 2025). For example, gelatin films exhibit distinct endothermic peaks corresponding to helix-coil transitions, while starch films show gelatinization endotherms (García et al., 2000).

Protein-based and nanocomposite films generally demonstrate superior thermal stability compared to polysaccharide-based ones, making them more suitable for heat-processed foods (Rhim et al., 2013; Schmid et al., 2017; Vaezi et al., 2019; Gupta et al., 2024). In contrast, starch and chitosan films often require modifications, such as cross-linking or nanoparticle incorporation, to enhance their heat resistance (Rhim et al., 2013; Vaezi et al., 2019; Devi et al., 2024; Wu et al., 2024).

Nanocomposite films, especially those with montmorillonite nanoplatelets or ZnO nanoparticles, offer improved thermal stability due to the barrier effect and increased structural density provided by the dispersed nanoparticles (Rhim et al., 2013; Vaezi et al., 2019; Alizadeh-Sani et al., 2021; Wu et al., 2024).

Understanding these thermal properties is vital for developing edible coatings that can endure food processing and storage conditions while maintaining their protective functions (Otoni et al., 2017; Devi et al., 2024; Mikus & Galus, 2025). Future research should focus on the development of novel plasticizers that can improve thermal resistance without compromising flexibility, as well as optimizing nanocomposite designs for heat sensitive and high-temperature applications (Gupta et al., 2024; Wu et al., 2024).

Chemical and Biochemical Properties

Solubility and Biodegradability

The solubility of edible coatings is a critical functional property that determines their application methods, ease of removal, and interaction with food surfaces. Water-soluble polysaccharides such as pullulan, alginate, and carboxymethyl cellulose (CMC) are particularly suitable for minimally processed or ready-to-eat products, as they can be easily removed through rinsing or cooking (García et al., 2000; Ghanbarzadeh et al., 2010; Galus & Lenart, 2013; Necas & Bartosikova, 2013; Devi et al., 2022). In contrast, protein-based coatings like zein, which are soluble in ethanol and other organic solvents, are better suited for alcohol-based application methods and offer superior resistance in moist environments (Han et al., 2008; Fabra et al., 2014; Galus & Kadzińska, 2015; Xu & Li, 2023; Wu et al., 2024). The solubility profile of edible coatings/films can be modified through various approaches. Incorporating lipid components or essential oils can reduce water solubility and enhance moisture resistance, but may reduce film transparency and flexibility (Ghanbarzadeh et al., 2010; Galus & Kadzińska, 2016; Devi et al., 2022; Sharma et al., 2024; Wu et al., 2024). Conversely, increasing plasticizer content (e.g., glycerol or sorbitol) improves solubility and flexibility, but may weaken mechanical integrity and barrier properties (Ghanbarzadeh et al., 2010; Abdorreza et al., 2011; Muscat et al., 2012; Fakhouri et al., 2013; Devi et al., 2024; Mikus & Galus, 2025).

From an environmental perspective, biodegradability represents a key advantage of edible coatings over conventional packaging. Most biopolymers used in edible films, including chitosan, alginate, gelatin, starch, and cellulose derivatives, undergo complete microbial or enzymatic degradation within 28–45 days under controlled aerobic composting conditions, depending on polymer structure, environmental parameters, and the presence of additives (Tharanathan, 2003; Siracusa et al., 2008; Elsabee & Abdou, 2013; Ruggero et al., 2019; Kalita et al., 2021). The degradation process is polymer-specific: chitosan-based films degrade efficiently via lysozyme and microbial chitinase activity, while starch and gelatin coatings are broken down by amylases and proteases, respectively (Elsabee & Abdou, 2013; Mujtaba et al., 2019; Kalita et al., 2021; Devi et al., 2022; Wu et al., 2024). Even when enhanced with biodegradable nanofillers like chitin nanowhiskers or lignin nanoparticles, these materials maintain their compostability while gaining improved functional properties (Yang et al., 2018; Alizadeh-Sani et al., 2021; Etxabide et al., 2022; Wu et al., 2024). Several factors influence degradation rates, including crystallinity, degree of cross-linking, and surface area. More complex structures, such as highly cross-linked or multilayer films, tend to degrade more slowly due to limited enzymatic accessibility (Lacroix & Le Tien, 2005; Siracusa et al., 2008; Rhim et al., 2013; Mikus & Galus, 2025). The solubility and biodegradability characteristics of edible coatings are essential for achieving both functional performance and environmental sustainability. Through strategic selection of biopolymers and formulation techniques, it is possible to develop coatings that satisfy specific application requirements while preserving a strong biodegradability profile. This balance between food quality preservation and ecological responsibility makes edible coatings a promising alternative to plastic packaging (Lacroix & Le Tien, 2005; Siracusa et al., 2008; Ruggero et al., 2019; Kalita et al., 2021; Gupta et al., 2024).

pH Stability

pH stability is an essential functional attribute of edible coatings, as these materials must maintain functionality across the broad pH spectrum (2.0–9.0) encountered in food systems—from highly acidic fruits and pickled products to neutral dairy items and alkaline-cooked foods. Effective coatings must preserve their structural integrity and barrier properties without undergoing significant dissolution or degradation under varying pH conditions.

Polysaccharide-based films, particularly those made from alginate, pectin, and their composites, exhibit excellent pH-responsive characteristics. These biopolymers are stable at neutral pH and show limited, controlled swelling in acidic environments. Optimized alginate-pectin composite films have demonstrated minimal dissolution (<10%) even at pH 3.0, making them particularly suitable for acidic food applications, including berry coatings and fermented vegetable protection (García et al., 2000; Rojas-Graü et al., 2007; Tapia et al., 2008; Wu et al., 2024).

Cross-linking strategies significantly enhance pH resistance. The incorporation of multivalent cations like calcium (Ca^{2+}) forms stable ionic bridges between polysaccharide chains, resulting in a robust three-dimensional network. Studies have shown that this approach can improve pH stability by 30–40%, reducing film degradation in both acidic and alkaline environments (García et al., 2000; Ghanbarzadeh et al., 2010; Galus & Lenart, 2013; Gupta et al., 2024; Sharma et al., 2024). The density of cross-linking can be adjusted to control dissolution rates and adapt the coatings to specific food system needs.

Protein-based coatings, such as those derived from whey, casein, or gelatin, show unique pH-dependent behavior. Near their isoelectric points (pI), these films become more brittle and lose mechanical strength. For example, whey protein films experience reduced integrity at pH 5.2, the pI of β-lactoglobulin (Hong & Krochta, 2006; Schmid, 2013; Xu & Li, 2023; Devi et al., 2024; Wu et al., 2024). To mitigate this issue, stabilizers such as polyols or compatible polysaccharides can be added to maintain structure across a wider pH range.

Nanocomposite technology has further improved the pH stability of edible coatings. Incorporating functional nanoparticles like zinc oxide or cellulose nanocrystals into film matrixes enhances not only the mechanical and barrier properties but also resistance to pH-induced degradation. This is attributed to matrix densification and decreased water absorption capacity (Youssef et al., 2016; Alizadeh-Sani et al., 2021; Etxabide et al., 2022; Wu et al., 2024).

In conclusion, developing pH-stable edible coatings requires careful consideration of both base material selection and modification strategies. Cross-linked polysaccharide systems currently offer the most versatile solution, demonstrating reliable performance across acidic, neutral, and moderately alkaline food matrixes. Innovations in nanocomposite formulations continue to expand the pH stability range while retaining other essential functional properties such as flexibility, strength, and biodegradability.

Compatibility with Additives

Edible coatings must effectively incorporate and retain functional additives like antimicrobials and antioxidants without phase separation. Recent studies show that emulsion-based coatings (e.g., whey protein–carboxymethylcellulose) can encapsulate both hydrophilic (nisin) and hydrophobic (thymol) compounds with >85% encapsulation efficiency due to their amphiphilic nature (Sánchez-González et al., 2011; Galus & Kadzińska, 2015; Otoni et al., 2016; Al-Tayyar et al., 2020). The addition of 1–2% Tween 80 as an emulsifier further improves additive dispersion uniformity by reducing interfacial tension between incompatible compounds (Hasan et al., 2020; Wu et al., 2024).

The successful integration of functional additives, including antimicrobials, antioxidants, nutraceuticals, and pH indicators into edible coatings relies on achieving optimal compatibility between the bioactive compounds and the coating matrix. This ensures uniform dispersion, chemical stability, and controlled release kinetics for sustained functionality throughout storage (Rojas-Graü et al., 2007; Galus & Kadzińska, 2015; Arnon-Rips & Poverenov, 2018; Devi et al., 2024; Mikus & Galus, 2025).

Amphiphilic matrixes such as whey protein–carboxymethylcellulose blends or chitosan–essential oil emulsions have demonstrated the ability to encapsulate both hydrophilic and lipophilic compounds efficiently, achieving encapsulation efficiencies >85% (Sánchez-González et al., 2011; Galus & Kadzińska, 2015; Otoni et al., 2016; Devi et al., 2022; Wu et al., 2024). This capability is key for retaining additive functionality during film formation and storage.

The use of surfactants, such as Tween 80 (1–2%), plays a crucial role in stabilizing emulsions, minimizing particle aggregation, and enhancing homogeneous integration of additives into the matrix (Al-Tayyar et al., 2020; Hasan et al., 2020; Mikus & Galus, 2025). These formulations are especially useful for achieving sustained release kinetics, which directly contributes to shelf life extension of coated produce.

Incorporation of nanoemulsified essential oils and plant extracts, such as oregano oil, rosemary extract, or green tea catechins into biopolymer matrixes has shown efficacy in enhancing film functionality without compromising mechanical or barrier properties. Chitosan coatings enriched with *Origanum vulgare* essential oil or ZnO nanoparticles maintained long-lasting antifungal activity against *Aspergillus* and *Rhizopus* spp., common postharvest pathogens in fruits (Dos Santos et al., 2012; Souza et al., 2013; Alizadeh-Sani et al., 2021; Wu et al., 2024).

The interaction between additives and coating matrix composition determines release kinetics. Hydrophilic matrixes favor rapid release of water-soluble compounds, whereas hydrophobic or cross-linked structures enable slower, controlled diffusion (Rojas-Graü et al., 2007; Quezada-Gallo, 2009; Schmid, 2013; Devi et al., 2024; Wu et al., 2024). More advanced approaches, such as multilayer or layer-by-layer (LbL) assemblies, further refine additive integration by enabling the sequential integration of complementary functional agents (Arnon-Rips & Poverenov, 2018).

In conclusion, the effective incorporation of additives into edible coatings requires careful selection of biopolymer matrixes, emulsifiers, and processing conditions to ensure compatibility, stability, and functional performance. Advances in emulsion systems, nanoencapsulation, and multifunctional nanocomposites are expanding the potential of edible coatings as active packaging systems that not only protect food but also enhance its safety, quality, and shelf life.

Interaction with Food Matrix

The interfacial interactions between edible coatings and food surfaces critically determine their functional performance, influencing adhesion quality, moisture regulation, and sensory attributes. The compatibility between the coating material and the food surface determines the uniformity of application, barrier effectiveness, and consumer acceptability.

Protein-derived films, such as those based on whey protein isolate and gelatin, generally exhibit excellent adhesion to hydrophilic food surfaces. This behavior is attributed to their high wettability. It is commonly evidenced by low contact angle values (<30–40°) on substrates such as cut fruits and vegetables, depending on film composition and surface characteristics. This strong wettability enables the formation of continuous, defect-free films that effectively minimize moisture loss. Research shows such coatings reduce dehydration by 20–30% compared to uncoated samples, significantly preserving texture and visual quality in high-respiration produce like strawberries, apples, and melons (Hong & Krochta, 2006; Lin & Zhao, 2007; Ghidelli et al., 2014; Xu & Li, 2023; Wu et al., 2024).

Polysaccharide-based coatings (alginate, pectin, CMC-carboxymethylcellulose) also interact well with hydrophilic matrixes, but may require surface pretreatment or plasticizer optimization to enhance adhesion and minimize delamination. Cross-linking with divalent cations (e.g., calcium, Ca^{2+}) can improve adhesion by enhancing polymer-substrate compatibility and mechanical strength (Ghanbarzadeh et al., 2010; Galus & Lenart, 2013; Sharma et al., 2024; Mikus & Galus, 2025). The integration of functional additives (e.g., essential oils, phenolic compounds) into edible coatings can influence food surface interactions. While these additives enhance antimicrobial and antioxidant properties, they may also increase film hydrophobicity, potentially compromising adhesion and overall film integrity (Otoni et al., 2016;

Al-Tayyar et al., 2020; Alizadeh-Sani et al., 2021; Wu et al., 2024). However, nanoemulsion technologies and amphiphilic carrier systems mitigate this issue by maintaining interfacial stability while preserving additive functionality.

The organoleptic impact of coatings at the food interface is also a critical consideration. Poor adhesion can result in peeling or flaking, negatively affecting consumer perception of freshness and quality. Conversely, coatings that effectively integrate and preserve key sensory attributes, such as color, gloss, and texture, significantly enhance sensory appeal and consumer satisfaction (Robles-Sánchez et al., 2013; Galus & Kadzińska, 2016; Devi et al., 2022; Wu et al., 2024).

Optimizing the interactions between edible coatings and food matrixes is crucial for effective moisture control, targeted additive delivery, and ultimately, consumer satisfaction. While protein-based coatings typically offer superior performance on cut produce surfaces, novel composite and nanostructured formulations are now allowing for customized adhesion and functionality across diverse food types, advancing the versatility and commercial viability of edible packaging.

Optical Properties

Edible coatings play a vital role in maintaining the visual appeal and nutritional quality of fresh produce. Two key optical properties, transparency and gloss, directly influence consumer perception, while UV absorbance provides crucial protection against degradation. These optical properties are influenced by several formulation and processing factors.

Transparency and Gloss

The optical properties of edible coatings, particularly transparency and gloss, are critical for ensuring product marketability and consumer acceptance. Consumers strongly associate glossy, clear coatings with freshness and high quality.

Among biopolymers, pullulan-based coatings demonstrate exceptional transparency, with light transmittance exceeding 90% at 600 nm, and gloss values above 80 gloss units (GU), making them comparable to commercial synthetic films (Eroglu et al., 2014; Galus & Kadzińska, 2015; Wu et al., 2024). These characteristics are attributed to low crystallinity, excellent film formation, and smooth surface morphology.

Surface roughness plays a significant role in gloss development. Films with an average surface roughness (R_a) below 0.2 μm, achieved through careful drying and plasticizer optimization, exhibit enhanced specular reflectance. This results in a visually glossy finish on coated fruits like apples and strawberries (Galus & Kadzińska, 2015; Devi et al., 2022; Sharma et al., 2024; Mikus & Galus, 2025).

Even emulsion-based coatings with well-dispersed lipid droplets can also improve gloss if surface uniformity is maintained.

Protein-based films from whey or casein generally exhibit moderate transparency and gloss, but sometimes may develop a slight yellow hue depending on drying temperature and protein denaturation (Hong & Krochta, 2006; Han et al., 2008; Galus & Lenart, 2013). The formulation challenge is to balance optical clarity with mechanical strength and barrier properties, particularly when incorporating functional additives or processing at elevated temperatures.

UV Absorbance

Protecting fresh produce from ultraviolet (UV) radiation is crucial for minimizing the photo-degradation of sensitive nutrients such as vitamin C, anthocyanins, and carotenoids. Edible coatings with UV-absorbing capabilities act as protective barriers, thereby reducing oxidative deterioration and prolonging shelf life.

Chitosan-based coatings, especially those enhanced with zinc oxide (ZnO) and titanium dioxide (TiO2) nanoparticles, show excellent UV-blocking properties. For example, chitosan coatings containing only 0.1% ZnO nanoparticles can block up to 95% of UV-B radiation while allowing about 75% of visible light transmission, thus maintaining both nutritional integrity and product visibility (Youssef et al., 2016; Wu et al., 2024).

These coatings are particularly beneficial for light-sensitive produce, such as strawberries and kiwifruit. Studies indicate that such coatings reduce vitamin C degradation by up to 60% over 14 days of storage compared to uncoated controls (Lin et al., 2020; Wu et al., 2024). In addition to synthetic nanoparticles, natural UV-blocking alternatives such as plant-based polyphenols, anthocyanins, and lignin nanoparticles are gaining attention. These compounds provide dual benefits: UV protection and antioxidant activity, enhancing the coating's biofunctionality while aligning with clean-label consumer trends (Yang et al., 2018; Alizadeh-Sani et al., 2021).

The latest advances in edible coating design aim to synergize optical and functional properties, maintaining high transparency and gloss for consumer appeal, while integrating selective UV filtration for nutrient preservation. Achieving this dual performance requires strategic biopolymer selection, nanoparticle incorporation, and processing optimization to fine-tune surface morphology and light interaction. Edible coatings that combine high gloss, clarity, and UV protection offer a promising solution to shelf life extension, nutrient retention, and consumer satisfaction in the fresh produce sector.

Sensory and Organoleptic Properties

Edible coatings play a crucial role in maintaining and enhancing the sensory attributes of fresh produce. Their design must carefully consider factors like taste, odor, color, and consumer perception to ensure optimal product quality and acceptance.

Taste and Odor Retention

Maintaining the natural flavor and aroma of coated fruits and vegetables is a key quality determinant. Ideal edible coatings act as barriers to oxidative changes and aroma compound loss, preventing the development of off-flavors while preserving the product's sensory integrity. Whey protein isolate (WPI) coatings significantly reduce volatile loss and oxidative deterioration in fresh-cut apples and strawberries, maintaining sensory scores >7/9 on the hedonic scale (Han et al., 2008; Ghidelli et al., 2014; Wu et al., 2024). Soy protein coatings on fresh-cut eggplant effectively delay aroma degradation during cold storage, preserving sensory quality (Ghidelli et al., 2014). Chitosan coatings, particularly when enriched with essential oils, inhibit microbial spoilage and enzymatic reactions that cause off-odors. For example, chitosan combined with *Origanum vulgare* oil retained the fresh aroma of grapes and suppressed sour or fermented notes (Dos Santos et al., 2012; Wu et al., 2024). Starch-based coatings have demonstrated excellent performance in preserving aroma volatiles in various fruits, including sweet cherries and papaya (Souza et al., 2013; Dong & Wang, 2018; Devi et al., 2022; Wu et al., 2024). Coated sweet cherries with starch–ginseng extract retained ~80% of limonene and linalool, compared to only 40–50% in uncoated controls (Dong & Wang, 2018; Devi et al., 2022). In papaya, coatings made from Aloe vera and CMC delayed ripening and off-odor formation, enhancing freshness perception (Zillo et al., 2018; Mendy et al., 2019; Wu et al., 2024).

Colorants and Pigments

Incorporating natural pigments into edible coatings enhances both visual appeal and functionality, such as antioxidant activity and pH-indicating properties. *Spirulina*-infused alginate coatings (0.3%) improved the perceived freshness of apples and increased consumer preference by 40%, while boosting antioxidant capacity (ORAC = 2.8 μmol TE/g) and maintaining color stability ($\Delta E < 2$) over 14 days (Alizadeh-Sani et al., 2021; Etxabide et al., 2022; Wu et al., 2024). Anthocyanin-enriched coatings derived from red cabbage, saffron petals, or berries serve as natural pH indicators, shifting color from red at pH < 4 to blue at pH > 7, providing both freshness cues and functional appeal in line with clean-label trends (Quirós-Sauceda et al., 2014; Yang et al., 2018; Alizadeh-Sani et al., 2021; Wu et al., 2024).

Consumer Acceptance

Texture is a primary sensory attribute influencing consumer acceptance. Studies show that 90% of consumers reject coatings that reduce natural crispness by more than 15% (Vargas et al., 2008; Wu et al., 2024). Coatings must therefore maintain the inherent firmness of produce to ensure acceptability.

Gloss is another key factor. Coatings with gloss levels ≥60 gloss units (GU) significantly enhance perceived freshness, particularly in fruits like apples and strawberries (Galus & Kadzińska, 2015; Devi et al., 2022; Mikus & Galus, 2025). Similarly, optimal porosity (10–15 μm) facilitates controlled aroma release, aligning with olfactory thresholds and improving flavor perception (Sharaf Eddin et al., 2019; Wu et al., 2024).

For lipid-based coatings, flavor masking is essential. Waxy or greasy mouthfeel often leads to consumer rejection, with hedonic scores dropping below the 6.5/9 threshold (Han et al., 2008; Sharaf Eddin et al., 2019; Wu et al., 2024).

Coating thickness also influences consumer preference. Studies indicate that coatings thinner than 40 μm achieve a high acceptance rate of 78%, whereas thicker coatings (>80 μm) are accepted by only 32% of participants, primarily due to noticeable alterations in texture and visibility (Dhumal & Sarkar, 2018; Wu et al., 2024).

Achieving high consumer acceptance requires a balanced integration of attributes, including visual clarity, flavor preservation, textural neutrality, and functional additives, all carefully tailored to the specific sensory expectations of each food product.

Functional and Bioactive Properties

Edible coatings provide more than just a protective barrier; they actively contribute to preserving fresh produce by exhibiting various functional and bioactive properties. These properties, either inherent to the biopolymers or introduced through incorporated agents, enhance food safety, extend shelf life, and even enhance nutritional value.

Antimicrobial Properties

Edible coatings help combat microbial spoilage through two main mechanisms: the intrinsic antimicrobial activity of certain biopolymers and the incorporation of potent antimicrobial agents.

Intrinsic Antimicrobial Properties of Biopolymers

Several biopolymers used in edible coatings, such as chitosan, whey protein, and alginate, exhibit inherent antimicrobial properties, making them suitable for preserving the microbial quality of fruits and vegetables. Chitosan, due to its cationic nature, interacts with the negatively charged microbial cell membranes, leading to cell disruption and death. Applications on grapes, guava, and papaya have demonstrated strong antimicrobial activity, delaying spoilage and extending shelf life (Kong et al., 2010; Dos Santos et al., 2012; Batista Silva et al., 2018; Mendy et al., 2019). Whey protein isolate (WPI) has also been shown to limit microbial growth on fresh-cut fruits by incorporating lactoferrin or activating lactoperoxidase systems, with reported improvements in microbial safety of apples and melons (Min et al., 2007; Han et al., 2008; Wu et al., 2024).

Incorporated Antimicrobial Agents

Edible coatings can be enhanced with natural antimicrobials, such as essential oils (e.g., thyme, oregano), plant extracts, and nanoparticles. The chitosan coatings with oregano oil reduced fungal spoilage in grapes by more than 4 log CFU/g (Dos Santos et al., 2012). Similarly, papaya and melons coated with chitosan-ZnO nanocomposites exhibited reduced microbial load and longer shelf life (Wu et al., 2024). Essential oil-loaded starch or alginate films have also reduced yeast and mold growth in fruits such as strawberries, mangoes, and cherries, while maintaining postharvest quality (Rodríguez-García et al., 2016; Dong & Wang, 2018; Wu et al., 2024).

Antioxidant Properties

Edible coatings enriched with antioxidants reduce oxidative stress by scavenging free radicals and limiting reactive oxygen species (ROS). Chitosan–green tea extract coatings have maintained phenolic content and color in apples, reducing browning and lipid oxidation (Kumarihami et al., 2022; Wu et al., 2024).

Starch- and alginate-based coatings with added tocopherols or polyphenols preserved firmness and color in papaya and strawberries during refrigerated storage (Alvarez et al., 2018; Zillo et al., 2018; Devi et al., 2022; Wu et al., 2024). Curcumin-loaded gelatin coatings also significantly slowed lipid oxidation in mangoes, extending shelf life (Maqbool et al., 2011).

Anti-Browning Properties

Enzymatic browning caused by polyphenol oxidase (PPO) can be controlled by edible coatings through oxygen barriers and enzyme inhibition. Whey protein coatings with citric acid inhibit PPO via copper ion chelation, reducing browning in apples and potatoes (Le Tien et al., 2001; Han et al., 2008). Chitosan coatings enriched with ascorbic acid or thyme oil have effectively delayed browning in mushrooms and bananas (Pérez-Gago et al., 2005; Jiang et al., 2012; Robles-Sánchez et al., 2013). Alginate and gellan-based coatings, when used with anti-browning agents, maintained visual quality in papaya and mangoes (Rojas-Graü et al., 2007; Tapia et al., 2008).

Bioactive Compounds/ Properties

Edible coatings also act as carriers for bioactive compounds, adding value beyond preservation. These include anthocyanins, carotenoids, and polyphenols, offering antioxidant, anti-inflammatory, and nutritional enhancement benefits. Pectin coatings with blueberry extract improved anthocyanin retention in raspberries, boosting antioxidant levels and visual quality (Quirós-Sauceda et al., 2014). *Spirulina*-loaded alginate coatings increased the antioxidant values (ORAC increase) of apples while maintaining freshness (Etxabide et al., 2022).

Coatings incorporating curcumin or saffron anthocyanins, such as gelatin–curcumin films for mangoes or chitosan–saffron films for melons, offer dual benefits of color enhancement and oxidative protection (Maqbool et al., 2011; Wang et al., 2019; Alizadeh-Sani et al., 2021).

Edibility and Biodegradability

Edible coatings play a crucial role in extending the shelf life of fruits and vegetables while addressing key considerations related to safety, environmental impact, allergenicity, and dietary restrictions.

Safety and Regulatory Aspects

Edible coatings for fruits and vegetables must comply with stringent safety standards set by leading regulatory bodies such as the U.S. Food and Drug Administration (FDA), the European Food Safety Authority (EFSA), and Codex Alimentarius. All components used in edible coating formulations must be classified as Generally Recognized as Safe (GRAS) or equivalent under relevant food safety legislation. In the United States, the Code of Federal Regulations mandates that substances used in edible coatings for direct food contact be GRAS (Necas & Bartosikova, 2013; Maringgal et al., 2020). This includes well-established biopolymers such as chitosan, alginate, pectin, cellulose derivatives, and carboxymethyl cellulose. These have been extensively used in coatings for a variety of fresh produce, including apples, papayas, strawberries, and guavas.

For example, chitosan coatings applied to papaya, grapes, and guava are not only effective in prolonging shelf life but are also non-toxic, biodegradable, and suitable for postharvest application (Kean & Thanou, 2010; Dos Santos et al., 2012; Batista Silva et al., 2018). Carrageenan-based coatings have similarly demonstrated safety and effectiveness when applied to apples, mangoes, and citrus (Necas & Bartosikova, 2013; Maringgal et al., 2020).

The inclusion of emulsifiers and stabilizers in coatings, such as polydextrose, sorbitan monostearate, sucrose fatty acid esters, cocoa butter, and castor oil, is permitted for fruit coatings within regulatory limits, especially when intended to enhance coating flexibility, gloss, and moisture resistance (Maringgal et al., 2020).

Approved film-forming agents in Europe include gum arabic, pectin, shellac, beeswax, candelilla wax, and carnauba wax, recognized for their non-toxicity, edibility, and efficacy in shelf life extension without sensory alteration (Maringgal et al., 2020).

Environmental Impact

Edible coatings made from biopolymers like starch, alginate, pectin, gelatin, and chitosan offer a sustainable alternative to synthetic plastic packaging. Starch-based films have been shown to degrade fully within 28–40 days in compost, with no microplastic residue (Ghanbarzadeh et al., 2010). Whey protein coatings also degrade efficiently under similar conditions (Schmid et al., 2012). Chitosan coatings completely degrade in seawater within 8–12 weeks, without generating persistent residues (Mujtaba et al., 2019). Alginate coatings have demonstrated positive effects on soil microbiota, enhancing rhizosphere diversity rather than disrupting it (Ruggero et al., 2019). Biodegradation proceeds through enzymatic depolymerization, bioassimilation, and mineralization, yielding CO_2, H_2O, and biomass (Siracusa et al., 2008). The degradation rate depends on polymer type: starch–gelatin blends degrade in ~40 days, pure starch films in ~55 days, and gelatin films in ~65 days (Ghanbarzadeh et al., 2010).

Allergenicity

Protein-based coatings made from whey, soy, and wheat (gluten) require clear labeling due to allergenic potential. Whey protein coatings applied to fruits like apples or melons could trigger allergic responses in lactose-intolerant or dairy-allergic individuals. Gluten-based coatings must be avoided for celiac or gluten-free labeling (Hischenhuber et al., 2006).

Dietary Restrictions

Gelatin, while offering excellent mechanical and barrier properties, is not suitable for vegan, vegetarian, or certain religious diets unless derived from approved sources. Plant-based film-formers like pectin, agar, and cellulose derivatives are widely accepted across dietary categories (Tapia et al., 2008; Karim & Bhat, 2009; Saha & Bhattacharya, 2010). These plant-based alternatives have been applied successfully to apples, citrus, papaya, and mangoes, providing barrier protection without conflicting with dietary or ethical standards (Tapia et al., 2008; Saha & Bhattacharya, 2010).

Cost-Effectiveness and Ease of Application

Edible coatings are a practical and economically viable solution for preserving fruits and vegetables. Their low production cost, straightforward processing, and compatibility with circular economy models make them particularly attractive for large-scale postharvest applications.

Production Feasibility

The production of edible coatings is scalable, cost-efficient, and compatible with standard food processing equipment such as homogenizers, spray dryers, and film applicators (Han et al., 2008; Tapia et al., 2008; Schmid et al., 2012). Chitosan-based coatings, which have been applied successfully to papaya, grapes, and guava, can be industrially processed using spray-drying and casting techniques. These methods are well-suited for existing infrastructure and can reduce processing costs relative to synthetic packaging (Mujtaba et al., 2019; Haghighi et al., 2020).

Pectin coatings, often extracted from citrus peel waste, represent a sustainable and cost-effective option, achieving high reproducibility and compatibility with commercial coating systems. Their successful application to apples, mangoes, and citrus fruits supports their relevance in postharvest treatment chains (Rojas-Graü et al., 2007; Tapia et al., 2008). Additionally, alginate, whey protein, and starch-based coatings are widely compatible with existing food industry workflows. Their biodegradability, availability, and low processing requirements make them appealing for fruit packaging operations involving apples, mangoes, and strawberries (Han et al., 2008; Tapia et al., 2008; Schmid et al., 2012).

Application Methods

Several application techniques are available for applying edible coatings on fresh produce, each with distinct operational benefits. Spray coating is the most widely adopted method in commercial packing houses for fruits like apples, citrus, and papaya (Fig. 6.1). It allows for precise control over coating thickness (20–40 μm) and achieves over 95% uniform coverage, ensuring consistent film properties (Han et al., 2008; Tapia et al., 2008; Schmid et al., 2012; Dhumal & Sarkar, 2018; Wu et al., 2024). It is highly compatible with high-throughput conveyor systems.

Dipping, while more labor-intensive and prone to material loss, is effective for ensuring complete surface coverage, especially for irregularly shaped or soft fruits such as strawberries and mango slices (Rojas-Graü et al., 2007; Han et al., 2008; Tapia et al., 2008; Galus & Kadzińska, 2015; Devi et al., 2022).

Electrospinning offers advanced functional benefits, producing nanofiber coatings with enhanced barrier properties, such as a 60% reduction in oxygen permeability compared to dipped coatings. While currently more suited for research or

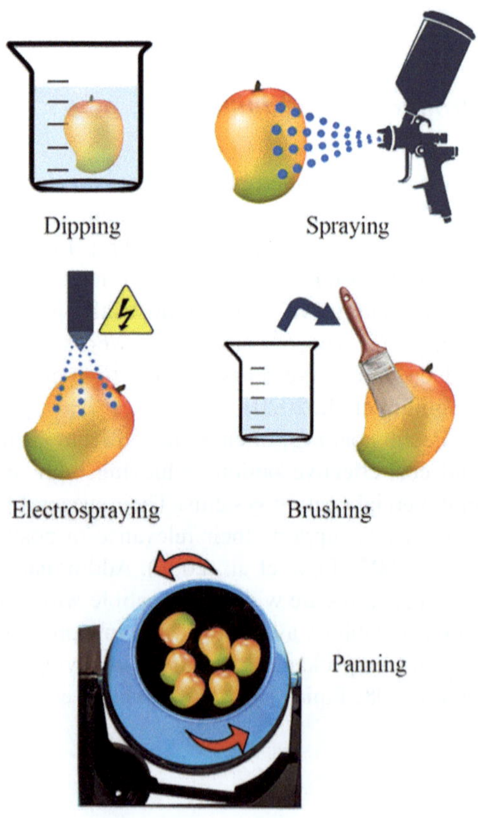

Fig. 6.1 Application methods of edible coatings

high-value specialty applications, it has been explored for preserving fresh-cut fruits (Fabra et al., 2016; Dhumal & Sarkar, 2018; Yang et al., 2018; Wu et al., 2024).

The selection of the application method depends on the specific fruit characteristics, desired shelf life, and processing scale, with methods like spray coating preferred for firm, smooth fruits (e.g., apples, citrus), and dipping or electrospinning more suitable for irregular or fragile items like berries or cut melons (Rojas-Graü et al., 2007; Tapia et al., 2008; Fabra et al., 2016; Dhumal & Sarkar, 2018; Wu et al., 2024).

Sustainability in the Circular Economy

Edible coatings strongly support circular economy principles by converting agricultural and food processing byproducts into value-added materials, while reducing reliance on fossil-based plastics. Whey protein coatings, derived from cheese production waste, reduce dairy industry waste and provide excellent moisture and oxygen barrier properties (Han et al., 2008; Schmid et al., 2012). These approaches demonstrate how edible coatings can valorize waste streams into functional packaging solutions. Similarly, pectin extracted from citrus peels, starch from potato waste, and cellulose from fruit pomace have been effectively used to coat papaya, strawberries, tomatoes, and citrus fruits, enabling zero-waste processing systems (Tapia et al., 2008; Ruggero et al., 2019; Kalita et al., 2021; Wu et al., 2024).

These strategies significantly reduce landfill use, lower greenhouse gas emissions, and promote resource efficiency. They align with sustainability goals by decreasing synthetic packaging reliance, enhancing biodegradability, and increasing the value of fruit and vegetable supply chains through environmentally sound innovation (Tharanathan, 2003; Ruggero et al., 2019; Kalita et al., 2021; Wu et al., 2024).

Factors Influencing Coatings/Films Properties

Formulation

The growing interest in bilayer and composite coatings arises from their ability to deliver enhanced functional properties compared to single-component systems. These coatings often combine proteins, polysaccharides, and lipids to create a synergistic effect, improving mechanical strength, gas permeability, and moisture barrier capacity (Galus & Kadzińska, 2015; Arnon-Rips & Poverenov, 2018; Yousuf et al., 2018; Etxabide et al., 2022). For example, hydrocolloids like pectin and gelatin provide robust matrixes with excellent mechanical integrity, while the addition of hydrophobic lipids such as beeswax or carnauba wax significantly enhances water vapor resistance due to their hydrophobic nature (Tapia et al., 2008; Yousuf et al., 2018; Devi et al., 2022).

In fruits like guava, mango, and papaya, these composites preserve color, firmness, and gas barrier functionality. For example, a chitosan–lemongrass essential oil coating preserved guava quality by minimizing texture loss and off-flavor formation through controlled volatile release (Dos Santos et al., 2012).

Additives

Additives play a critical role in improving the functionality and stability of edible coatings. These include plasticizers, emulsifiers, antimicrobials, anti-softening agents, cross-linkers, and nutraceuticals, each offering tailored functional improvements (Massilia et al., 2010; Galus & Kadzińska, 2015; Otoni et al., 2016; Kocira et al., 2021; Devi et al., 2022).

Antimicrobial Agents

Microbial spoilage significantly limits the shelf life of fruits and vegetables. Chitosan, sorbic acid, nisin, and essential oils have been widely used to inhibit microbial growth (Dos Santos et al., 2012; Rhim et al., 2013; Alvarez et al., 2014; Sharma et al., 2019). For example, chitosan coatings enriched with oregano oil reduced fungal spoilage in grapes, while lactic acid and sodium benzoate were effective for papaya and strawberries (Dos Santos et al., 2012; Alvarez et al., 2013). Coatings also provide a sustained release matrix for these agents (Quezada-Gallo, 2009).

Plasticizers

Plasticizers like glycerol, sorbitol, and polyethylene glycol are used to reduce brittleness and improve flexibility. These have been successfully used in essential oils for apples, tomatoes, and citrus fruits (Abdorreza et al., 2011; Muscat et al., 2012; Schmid, 2013; Galus & Kadzińska, 2015).

Emulsifiers/Surfactants

Emulsifiers improve the wettability and distribution of coatings on fruit surfaces, promoting film uniformity. Tween 20 has been widely used in chitosan and starch-based coatings for apples and strawberries, helping to reduce ripening rates (Lin &

Zhao, 2007; Chauhan et al., 2015; Galus & Kadzińska, 2015; Otoni et al., 2016). Emulsifiers are also vital for enhancing the stability of multilayer or composite films, especially those containing oils or other bioactives.

Anti-softening Agents

Loss of firmness is a common issue in fresh produce like strawberries, guava, and papaya. Coatings with calcium salts, particularly chitosan–calcium chloride, have preserved firmness in bananas and apples by reinforcing cell wall structure (Martín et al., 2006; Yousuf et al., 2018; Zhang et al., 2019; Devi et al., 2022; Kumarihami et al., 2022).

Nutraceuticals

Edible coatings can also serve as effective delivery systems for functional nutrients such as vitamins, antioxidants, and minerals. A xanthan gum-based coating containing calcium and vitamin E successfully enhanced nutritional retention in baby carrots (Mei et al., 2002). Similarly, alginate coatings with incorporated ascorbic acid preserved vitamin C content and delayed ripening in papaya (Tapia et al., 2008; Zillo et al., 2018).

Environmental Factors

Environmental conditions during coating preparation and application, such as pH, temperature, and relative humidity (RH), significantly impact the final film properties. Both pH and heating temperature influence a film's tensile strength and solubility. Mung bean protein films showed improved strength at pH 9.5 and 75 °C, though solubility and vapor permeability decreased under these same conditions. Conversely, very high pH (>10) can reduce film integrity due to electrostatic repulsion among protein chains (Bourtoom, 2008; Rhim et al., 2000). High relative humidity during storage promotes plasticization and film softening, especially in starch and alginate-based films, due to water molecule interactions (Hong & Krochta, 2006; Abdorreza et al., 2011; Galus & Lenart, 2013).

Processing Methods

Various processing methods are used to prepare edible films and coatings, each influencing the final structure and properties of the film.

Solvent Casting

Solvent casting is a widely used method for preparing protein and polysaccharide coatings, both in lab-scale and commercial settings. This technique yields homogeneous, transparent films with low water vapor permeability and has been applied in developing coatings for apples, bananas, and papayas (Hernandez-Izquierdo & Krochta, 2008; Dangaran et al., 2009; Fakhouri et al., 2013; Galus & Kadzińska, 2015).

Extrusion

Extrusion is energy-efficient, scalable, and ideal for solvent-free film production, increasingly used in commercial produce trays and wraps (Dangaran et al., 2009; Rhim et al., 2013; Vaezi et al., 2019).

Spinning

Spinning and electrospinning produce nanofiber films with enhanced barrier and mechanical properties. These have been used on berries and cut fruits (Fabra et al., 2016; Dhumal & Sarkar, 2018; Yang et al., 2018;).

Characterization Techniques of Coatings / Films Properties

Edible coatings are crucial for preserving perishable fruits and vegetables, extending their shelf life, and maintaining quality. A coating's effectiveness depends on its physical, chemical, mechanical, and functional properties, which are all influenced by its molecular structure, size, and composition (Han et al., 2008; Raghav et al., 2016; Bratovcic, 2021; Etxabide et al., 2022; Kumarihami et al., 2022). A combination of microscopy, spectroscopy, and barrier testing methods is commonly used to characterize these properties, particularly when applied to fresh fruits, including apples, strawberries, mangoes, papayas, and grapes.

Microscopy Surface and Structural Evaluation

Microscopy techniques help assess surface morphology, film integrity, and nanostructure dispersion in edible coatings. Scanning electron microscopy (SEM) is commonly used to observe surface features such as roughness and porosity in coatings on fruits like papaya and apple (García et al., 2000; Sharma et al., 2019). Transmission electron microscopy (TEM) allows nanoscale visualization of embedded ZnO particles or essential oil carriers in coatings applied to strawberries or melons (Sharma et al., 2019). Atomic force microscopy (AFM) evaluates nanoscale roughness and adhesion, contributing to optical and barrier quality, especially for coatings on smooth produce like apples (Sharma et al., 2019). Scanning tunneling microscopy (STM), while rare in food research, can assess atomic-scale surface uniformity and has potential for future edible coating analysis (Sharma et al., 2019).

Mechanical Testing: Strength, Elongation, and Puncture Resistance

Mechanical tests assess the durability and functional performance of edible coatings under physical stress, particularly during handling, packaging, and storage. These tests are typically performed according to standardized methods such as American Society for Testing and Materials (ASTM) D882. Tensile strength and elongation at break are key indicators of a coating's resistance to stretching and its flexibility. For example, studies on chitosan–starch films applied on guava and papaya have shown that adding glycerol as a plasticizer significantly enhances both tensile strength and elasticity, improving coating performance (Sothornvit & Krochta, 2005; Galus & Lenart, 2013; Fakhouri et al., 2013). Puncture resistance testing evaluates the force required to rupture the film using a mechanical probe and is particularly useful for soft or high-moisture produce like tomatoes. This test simulates real-world mechanical impacts and helps assess the coating's ability to protect fruits from physical damage during postharvest operations (Dangaran et al., 2009; Rhim et al., 2013; Galus & Kadzińska, 2015). These tests ensure that coatings maintain their structural integrity throughout postharvest operations.

Barrier Property Testing: Water Vapor and Gas Permeability

Gravimetric and manometric tests following ASTM E96 and ASTM D1653 are used to assess water vapor permeability (WVP), especially in high-moisture fruits like strawberries or melons, which are susceptible to dehydration (Tapia et al., 2008; Carneiro-da-Cunha et al., 2009; Sánchez-Tamayo et al., 2021). Gas permeability measurements help control respiration in climacteric fruits like mangoes, adjusting O_2 and CO_2 exchange rates to delay ripening (Galus & Kadzińska, 2015; Sánchez-Tamayo et al., 2021). The swelling index testing evaluates water absorption resistance. A lower index improves coating performance in produce like papaya (Tapia et al., 2008; Carneiro-da-Cunha et al., 2009). Barrier properties depend on film thickness, polymer polarity, and continuity (Hong & Krochta, 2006; Sánchez-Tamayo et al., 2021).

Spectroscopic Analysis: Fourier transform infrared (FTIR), UV–Vis, and Fluorescence Spectroscopy

Spectroscopy offers molecular-level insight into film composition. For example, Fourier transform infrared (FTIR) spectroscopy detects chemical interactions in coatings. It confirmed the green tea extract incorporation in chitosan films through characteristic peaks (Bratovcic, 2021; Kumarihami et al., 2022). UV–Vis spectroscopy is used to assess the transparency and opacity of edible coatings at 600 nm. Transparent coatings are ideal for visually appealing fruits like apples and strawberries. However, incorporating natural pigments like anthocyanins into coatings can intentionally increase opacity, provide better light protection while still maintaining acceptable visual quality (Hong & Krochta, 2006; Apriliyani et al., 2020; Etxabide et al., 2022). Fluorescence spectroscopy monitors the stability of bioactives like antioxidants in films, particularly for antioxidant-enriched coatings (Alizadeh-Sani et al., 2021).

Challenges and Future Prospects

Despite the significant potential of edible films and coatings, their widespread adoption in the fruit and vegetable industry faces several obstacles, ranging from economic and regulatory concerns to the complexities of emerging technologies and growing consumer expectations.

Scale-Up and Commercialization

A major barrier to commercial-scale adoption is the high production cost, especially when incorporating functional additives like essential oils or nanomaterials. Advanced application techniques such as precision spraying and electrospinning require significant capital investment in equipment and trained labor (Dhall, 2013; Maringgal et al., 2020). Although edible coatings have successfully extended the shelf life and preserved the quality of fruits like papaya, strawberries, guava, and mangoes (Maqbool et al., 2011; Mendy et al., 2019), their industrial uptake remains constrained due to cost, scalability, and regulatory concerns. To address this, researchers emphasize the need for low-cost, biodegradable, and food-grade materials such as pectin and starch (Tapia et al., 2008; Quirós-Sauceda et al., 2014). Recent innovations such as composite coatings with essential oils, UV-blocking nanoparticles, and multilayer systems have enhanced functionality (Arnon-Rips & Poverenov, 2018; Alizadeh-Sani et al., 2021).

Regulatory Concerns

Edible coatings fall under regulatory oversight as food additives or packaging materials, depending on their composition and function. According to U.S. and European frameworks, all ingredients, including base polymers, plasticizers, and functional additives, must be food-grade, safe, and hygienically produced (Maringgal et al., 2020). However, coatings made from allergenic proteins (milk, soy, gluten, or egg) raise consumer safety concerns and require proper labeling. For example, whey-based coatings used on apple slices or cut melons must be clearly labeled to avoid consumer health risks (Franssen & Krochta, 2003). Due to this regulatory complexity, international commercialization becomes challenging, demanding rigorous allergen disclosure and risk assessment at every stage of coating development and marketing.

Nanotechnology and Smart Coatings

Nanotechnology offers powerful tools for enhancing the functional properties of edible coatings, particularly through micro- and nanoencapsulation. This technique improves stability, facilitates targeted release, and enhances the delivery of bioactive compounds. In produce applications, encapsulated essential oils and antioxidants (e.g., curcumin, anthocyanins) have demonstrated extended efficacy in grapes, papaya, and strawberries (Alizadeh-Sani et al., 2021; Etxabide et al., 2022). Alginate, derived from brown seaweed, is one of the most common encapsulation matrixes used in fruit coatings due to its biocompatibility and gel-forming ability (Gharsallaoui et al., 2007). Other systems include layer-by-layer (LbL) nanostructured coatings, which alternate charged layers of chitosan, pectin, and gelatin, providing enhanced gas barrier properties and mechanical stability (Weiss et al., 2006; Arnon-Rips & Poverenov, 2018). Nanoemulsions offer superior coating uniformity, stability, and effective delivery of lipophilic bioactives. They are effective in improving antioxidant retention and the appearance of fresh-cut fruits. However, their long-term safety, metabolism, and labeling requirements still need further evaluation (Quirós-Sauceda et al., 2014; Hasan et al., 2020). Moreover, advanced application techniques such as ultrasound-assisted application, plasma treatment, and electrospinning are emerging as scalable methods. These techniques improve coating adherence and homogeneity while reducing material waste and environmental impact (Ashtari et al., 2019).

Sustainability and Consumer Trends

Sustainability is a key driving force behind the development of edible coatings for fruits and vegetables. Utilizing agricultural byproducts such as whey, pectin, and potato starch aligns with circular economy goals and reduces reliance on traditional plastic packaging (Tapia et al., 2008; Schmid et al., 2012; Ruggero et al., 2019). There is also growing consumer interest in functional foods. Integrating probiotics into edible coatings is emerging as a promising strategy to enhance both preservation and health benefits, particularly in ready-to-eat fruit products (Tapia et al., 2008; Quirós-Sauceda et al., 2014). However, incorporating bioactives like glutathione, N-acetylcysteine, or vitamin E can sometimes lead to undesirable sensory issues such as off-flavors or bitterness (Rojas-Graü et al., 2007; Dhall, 2013). Furthermore, coatings that over-restrict gas exchange in climacteric fruits like mangoes, papaya, or guava can induce anaerobic respiration. This process triggers the formation of ethanol and acetaldehyde, leading to unpleasant off-flavors, softening, and textural degradation. Therefore, an ongoing challenge is finding the right balance between gas permeability, sensory quality, and shelf life extension. This requires a deeper understanding of mass transfer, coating–food matrix interactions, and long-term performance across diverse produce types (Dhall, 2013).

Conclusion

Edible coatings and films have emerged as a promising, sustainable technology for the postharvest preservation of fruits and vegetables. They offer a range of benefits that align with modern demands for food safety, environmental responsibility, and consumer health. These coatings are primarily composed of polysaccharides, proteins, and lipids, allowing them to function as semipermeable barriers. This functionality allows them to regulate gas exchange, minimize moisture loss, reduce microbial spoilage, and effectively retain both the nutritional and sensory quality of fresh produce. The functional properties of edible coatings, including barrier performance, optical clarity, antimicrobial efficacy, and mechanical strength, are directly influenced by their formulation components and processing methods. Recent advancements in composite coatings, which combine the strengths of multiple biopolymers, have shown significant improvements in maintaining the quality of perishables such as papaya, mango, guava, strawberries, and apples. The incorporation of functional additives like antioxidants, antimicrobials, probiotics, and nutraceuticals further enhances product stability and health appeal. However, careful formulation is essential to avoid negative impacts on flavor, texture, or appearance.

Sophisticated characterization techniques, including SEM, TEM, AFM, FTIR, UV–Vis spectroscopy, and permeability testing, are crucial for gaining vital insights into the structure-function relationships of these coatings. This understanding then allows for their precise optimization, tailored to the specific needs of different commodities. Nonetheless, complex factors like environmental conditions, coating–food matrix interactions, and processing variables must be meticulously addressed to ensure product integrity and consistent performance under commercial conditions. Despite their considerable potential, edible coatings face practical challenges related to scalability, regulatory compliance, and cost-effectiveness. The broad commercial application demands not only the economical production of coating materials but also the deployment of efficient application methods and clear regulatory frameworks, particularly concerning allergenicity and ingredient labeling. The emergence of nanotechnology, smart delivery systems, and eco-design principles offers exciting avenues for future innovation, provided that safety and consumer acceptance are rigorously evaluated. Moreover, the success of edible coatings will depend on interdisciplinary collaboration, spanning food chemistry, material science, microbiology, and engineering. This collaborative effort is essential to develop coatings that are affordable, scalable, safe, and precisely tailored to the diverse physiological needs of fresh produce. By striking the right balance of functionality, sustainability, and consumer appeal, edible coatings can play a transformative role in cultivating a more resilient and environmentally responsible global food system.

References

Abdorreza, M. N., Cheng, L. H., & Karim, A. A. (2011). Effects of plasticizers on thermal properties and heat sealability of sago starch films. *Food Hydrocolloids, 25*(1), 56–60. https://doi.org/10.1016/j.foodhyd.2010.05.005

Acevedo-Fani, A., Salvia-Trujillo, L., Rojas-Graü, M. A., & Martín-Belloso, O. (2015). Edible films from essential-oil-loaded nanoemulsions: Physicochemical characterization and antimicrobial properties. *Food Hydrocolloids, 47*, 168–177. https://doi.org/10.1016/j.foodhyd.2015.01.032

Ali, A., Muhammad, M. T. M., Sijam, K., & Siddiqui, Y. (2011). Effect of chitosan coatings on the physicochemical characteristics of Eksotika II papaya (*Carica papaya* L.) fruit during cold storage. *Food Chemistry, 124*(2), 620–626. https://doi.org/10.1016/j.foodchem.2010.06.085

Alizadeh-Sani, M., Tavassoli, M., McClements, D. J., & Hamishehkar, H. (2021). Multifunctional halochromic packaging materials: Saffron petal anthocyanin loaded-chitosan nanofiber/methyl cellulose matrices. *Food Hydrocolloids, 111*, 106237. https://doi.org/10.1016/j.foodhyd.2020.106237

Al-Tayyar, N. A., Youssef, A. M., & Al-Hindi, R. R. (2020). Edible coatings and antimicrobial nanoemulsions for enhancing shelf life and reducing foodborne pathogens of fruits and vegetables: A review. *Sustainable Materials and Technologies, 26*, e00215. https://doi.org/10.1016/j.susmat.2020.e00215

Alvarez, M. V., Ponce, A. G., & Moreira, M. D. R. (2013). Antimicrobial efficiency of chitosan coating enriched with bioactive compounds to improve the safety of fresh cut broccoli. *LWT - Food Science and Technology, 50*(1), 78–87. https://doi.org/10.1016/j.lwt.2012.06.021

Alvarez, M. V., Ortega-Ramirez, L. A., Gutierrez-Pacheco, M. M., Bernal-Mercado, A. T., Rodriguez-Garcia, I., Gonzalez-Aguilar, G. A., Ponce, A., Moreira, M. D. R., Roura, S. I., & Ayala-Zavala, J. F. (2014). Oregano essential oil-pectin edible films as anti-quorum sensing and food antimicrobial agents. *Frontiers in Microbiology, 5*, 699. https://doi.org/10.3389/fmicb.2014.00699

Alvarez, M. V., Ponce, A. G., & Moreira, M. R. (2018). Influence of polysaccharide-based edible coatings as carriers of prebiotic fibers on quality attributes of ready-to-eat fresh blueberries. *Journal of the Science of Food and Agriculture, 98*(7), 2587–2597. https://doi.org/10.1002/jsfa.8751

Amin, U., Khan, M. U., Majeed, Y., Rebezov, M., Khayrullin, M., Bobkova, E., Shariati, M. A., Chung, I. M., & Thiruvengadam, M. (2021). Potentials of polysaccharides, lipids and proteins in biodegradable food packaging applications. *International Journal of Biological Macromolecules, 183*(2021), 2184–2198. https://doi.org/10.1016/j.ijbiomac.2021.05.182

Apriliyani, M. W., Purwadi, A. M., & Ikhwan, A. D. (2020). Characteristics of moisture content, swelling, opacity and transparency with addition chitosan as edible films/coating base on casein. *Advance Journal of Food Science and Technology, 1*(18), 9–14.

Armghan Khalid, M., Niaz, B., Saeed, F., Afzaal, M., Islam, F., Hussain, M., Mahwish, Muhammad Salman Khalid, H., Siddeeg, A., & Al-Farga, A. (2022). Edible coatings for enhancing safety and quality attributes of fresh produce: A comprehensive review. *International Journal of Food Properties, 25*(1), 1817–1847. https://doi.org/10.1080/10942912.2022.2107005

Arnon-Rips, H., & Poverenov, E. (2018). Improving food products' quality and storability by using layer by layer edible coatings. *Trends in Food Science & Technology, 75*, 81–92. https://doi.org/10.1016/j.tifs.2018.03.003

Ashtari, M., Khademi, O., Soufbaf, M., Afsharmanesh, H., & Askari Sarcheshmeh, M. A. (2019). Effect of gamma irradiation on antioxidants, microbiological properties and shelf life of pomegranate arils cv. 'Malas Saveh'. *Scientia Horticulturae, 244*, 365–371. https://doi.org/10.1016/j.scienta.2018.09.067

ASTM. (2012). *Standard test method for tensile properties of thin plastic sheeting* (ASTM D882-12).

ASTM. (2017). *Standard test method for tensile properties of plastics* (ASTM D638-14).

Basiak, E., Lenart, A., & Debeaufort, F. (2017). Effect of starch type and lipids on the properties of edible films. *International Journal of Biological Macromolecules, 98*, 348–356. https://doi.org/10.1016/j.ijbiomac.2017.01.122

Batista Silva, W., Cosme Silva, G. M., Santana, D. B., Salvador, A. R., Medeiros, D. B., Belghith, I., Da Silva, N. M., Cordeiro, M. H. M., & Misobutsi, G. P. (2018). Chitosan delays ripening and ROS production in guava (Psidium guajava L.) fruit. *Food Chemistry, 242*, 232–238. https://doi.org/10.1016/j.foodchem.2017.09.052

Bourtoom, T. (2008). Edible films and coatings: Characteristics and properties. *International Food Research Journal, 15*(3), 237–248. https://sid.ir/paper/681164/en

Bratovcic, A. (2021). Physical–chemical, mechanical and antimicrobial properties of bionanocomposite films and edible coatings. *International Journal for Research in Applied Sciences and Biotechnology, 8*(5), 151–156. https://doi.org/10.31033/ijrasb.8.5.22

Cabello, S. P., Takara, E. A., Marchese, J., & Ochoa, N. A. (2015). Influence of plasticizers in pectin films: Microstructural changes. *Materials Chemistry and Physics, 162*, 491–497. https://doi.org/10.1016/j.matchemphys.2015.06.019

Calva-Estrada, S. J., Jiménez-Fernández, M., & Lugo-Cervantes, E. (2019). Protein-based films: Advances in the development of biomaterials applicable to food packaging. *Food Engineering Reviews, 11*, 78–92. https://doi.org/10.1007/s12393-019-09189-w

Carneiro-da-Cunha, M. G., Cerqueira, M. A., Souza, B. W., Souza, M. P., Teixeira, J. A., & Vicente, A. A. (2009). Physical properties of edible coatings and films made with a polysaccharide from Anacardium occidentale L. *Journal of Food Engineering, 95*(3), 379–385. https://doi.org/10.1016/j.jfoodeng.2009.05.020

Celik, N., Kiremitler, N. B., Ruzi, M., & Onses, M. S. (2021). Waxing the soot: Practical fabrication of all-organic superhydrophobic coatings from candle soot and carnauba wax. *Progress in Organic Coatings, 153*, 106169. https://doi.org/10.1016/j.porgcoat.2021.106169

Chauhan, O. P., Nanjappa, C., Ashok, N., Ravi, N., Roopa, N., & Raju, P. S. (2015). Shellac and Aloe vera gel-based surface coating for shelf-life extension of tomatoes. *Journal of Food Science and Technology, 52*(2), 1200–1205. https://doi.org/10.1007/s13197-013-1035-6

Chawla, R., Sivakumar, S., & Kaur, H. (2021). Antimicrobial edible films in food packaging: Current scenario and recent nanotechnological advancements-a review. *Carbohydrate Polymer Technologies and Applications, 2*, 100024. https://doi.org/10.1016/j.carpta.2020.100024

Colussi, R., Pinto, V. Z., El Halal, S. L. M., Biduski, B., Prietto, L., Castilhos, D. D., da Rosa Zavareze, E., & Dias, A. R. G. (2017). Acetylated rice starches films with different levels of amylose: Mechanical, water vapor barrier, thermal, and biodegradability properties. *Food Chemistry, 221*, 1614–1620. https://doi.org/10.1016/j.foodchem.2016.10.129

Dangaran, K., Tomasula, P. M., & Qi, P. (2009). Structure and function of protein-based edible films and coatings. In K. C. Huber & M. E. Embuscado (Eds.), *Edible films and coatings for food applications* (pp. 25–56). Springer. https://doi.org/10.1007/978-0-387-92824-1_2

Devi, L. S., Kalita, S., Mukherjee, A., & Kumar, S. (2022). Carnauba wax-based composite films and coatings: Recent advancement in prolonging postharvest shelf-life of fruits and vegetables. *Trends in Food Science & Technology, 129*, 296–305. https://doi.org/10.1016/j.tifs.2022.09.019

Devi, L. S., Jaiswal, A. K., & Jaiswal, S. (2024). Lipid incorporated biopolymer based edible films and coatings in food packaging: A review. *Current Research in Food Science, 8*, 100720. https://doi.org/10.1016/j.crfs.2024.100720

Dhall, R. K. (2013). Advances in edible coatings for fresh fruits and vegetables: A review. *Critical Reviews in Food Science and Nutrition, 53*(5), 435–450. https://doi.org/10.1080/10408398.2010.541568

Dhumal, C. V., & Sarkar, P. (2018). Composite edible films and coatings from food-grade biopolymers. *Journal of Food Science and Technology, 55*(11), 4369–4383. https://doi.org/10.1007/s13197-018-3402-9

Dong, F., & Wang, X. (2018). Guar gum and ginseng extract coatings maintain the quality of sweet cherry. *LWT, 89*, 117–122. https://doi.org/10.1016/j.lwt.2017.10.035

Dos Santos, N. S. T., Athayde Aguiar, A. J. A., De Oliveira, C. E. V., Veríssimo De Sales, C., De Melo, E., Silva, S., Sousa Da Silva, R., Stamford, T. C. M., & De Souza, E. L. (2012). Efficacy of the application of a coating composed of chitosan and Origanum vulgare L. essential oil to control Rhizopus stolonifer and Aspergillus Niger in grapes (Vitis labrusca L.). *Food Microbiology, 32*(2), 345–353. https://doi.org/10.1016/j.fm.2012.07.014

Elsabee, M. Z., & Abdou, E. S. (2013). Chitosan based edible films and coatings: A review. *Materials Science and Engineering: C, 33*(4), 1819–1841. https://doi.org/10.1016/j.msec.2013.01.010

Eroglu, E., Torun, M., Dincer, C., & Topuz, A. (2014). Influence of pullulan-based edible coating on some quality properties of strawberry during cold storage. *Packaging Technology and Science, 27*(10), 831–838. https://doi.org/10.1002/pts.2077

Espitia, P. J. P., Du, W.-X., Avena-Bustillos, R. d. J., Soares, N. d. F. F., & McHugh, T. H. (2014). Edible films from pectin: Physical-mechanical and antimicrobial properties—A review. *Food Hydrocolloids, 35*, 287–296. https://doi.org/10.1016/j.foodhyd.2013.06.005

Etxabide, A., Kilmartin, P. A., Maté, J. I., & Gómez-Estaca, J. (2022). Characterization of glucose-crosslinked gelatin films reinforced with chitin nanowhiskers for active packaging development. *Lwt, 154*, 112833. https://doi.org/10.1016/j.lwt.2021.112833

Fabra, M. J., Lopez-Rubio, A., & Lagaron, J. M. (2014). Nanostructured interlayers of zein to improve the barrier properties of high barrier polyhydroxyalkanoates and other polyesters. *Journal of Food Engineering, 127*, 1–9. https://doi.org/10.1016/j.jfoodeng.2013.11.022

Fabra, M. J., López-Rubio, A., & Lagaron, J. M. (2016). Use of the electrohydrodynamic process to develop active/bioactive bilayer films for food packaging applications. *Food Hydrocolloids, 55*, 11–18. https://doi.org/10.1016/j.foodhyd.2015.10.026

Fakhouri, F. M., Costa, D., Yamashita, F., Martelli, S. M., Jesus, R. C., Alganer, K., Collares-Queiroz, F. P., & Innocentini-Mei, L. H. (2013). Comparative study of processing methods for starch/gelatin films. *Carbohydrate Polymers, 95*(2), 681–689. https://doi.org/10.1016/j.carbpol.2013.03.027

Falguera, V., Quintero, J. P., Jiménez, A., Muñoz, J. A., & Ibarz, A. (2011). Edible films and coatings: Structures, active functions and trends in their use. *Trends in Food Science & Technology, 22*(6), 292–303. https://doi.org/10.1016/j.tifs.2011.02.004

Franssen, L. R., & Krochta, J. M. (2003). 12-Edible coatings containing natural antimicrobials for processed foods. In S. Roller (Ed.), *Natural antimicrobials for the minimal processing of foods, A volume in Woodhead Publishing Series in Food Science, technology and nutrition* (pp. 250–262). Woodhead Publishing, ISBN 9781855736696.. https://doi.org/10.1533/9781855737037.250

Galus, S., & Kadzińska, J. (2015). Food applications of emulsion-based edible films and coatings. *Trends in Food Science & Technology, 45*(2), 273–283. https://doi.org/10.1016/j.tifs.2015.07.011

Galus, S., & Kadzińska, J. (2016). Whey protein edible films modified with almond and walnut oils. *Food Hydrocolloids, 52*, 78–86. https://doi.org/10.1016/j.foodhyd.2015.06.013

Galus, S., & Lenart, A. (2013). Development and characterization of composite edible films based on sodium alginate and pectin. *Journal of Food Engineering, 115*(4), 459–465. https://doi.org/10.1016/j.jfoodeng.2012.03.006

García, M. A., Martino, M. N., & Zaritzky, N. E. (2000). Microstructural characterization of plasticized starch-based films. *Starch-Stärke, 52*(4), 118–124. https://doi.org/10.1002/1521-379X(200006)52:4<118::AID-STAR118>3.0.CO;2-0

Ghanbarzadeh, B., Almasi, H., & Entezami, A. A. (2010). Physical properties of edible modified starch/carboxymethyl cellulose films. *Innovative Food Science & Emerging Technologies, 11*(4), 697–702. https://doi.org/10.1016/j.ifset.2010.06.001

Gharsallaoui, A., Roudaut, G., Chambin, O., Voilley, A., & Saurel, R. (2007). Applications of spray-drying in microencapsulation of food ingredients: An overview. *Food Research International, 40*(9), 1107–1121. https://doi.org/10.1016/j.foodres.2007.07.004

Ghidelli, C., Mateos, M., Rojas-Argudo, C., & Pérez-Gago, M. B. (2014). Extending the shelf life of fresh-cut eggplant with a soy protein–cysteine based edible coating and modified atmosphere packaging. *Postharvest Biology and Technology, 95*, 81–87. https://doi.org/10.1016/j.postharvbio.2014.04.007

Goudarzi, V., & Shahabi-Ghahfarrokhi, I. (2018). Photo-producible and photo-degradable starch/TiO2 bionanocomposite as a food packaging material: Development and characterization. *International Journal of Biological Macromolecules, 106*, 661–669. https://doi.org/10.1016/j.ijbiomac.2017.08.058

Gounga, M. E., Xu, S. Y., & Wang, Z. (2007). Whey protein isolate-based edible films as affected by protein concentration, glycerol ratio and pullulan addition in film formation. *Journal of Food Engineering, 83*(4), 521–530. https://doi.org/10.1016/j.jfoodeng.2007.04.008

Gupta, D., Lall, A., Kumar, S., Patil, T. D., & Gaikwad, K. K. (2024). Plant-based edible films and coatings for food-packaging applications: Recent advances, applications, and trends. *Sustainable Food Technology, 2*, 1428–1455. https://doi.org/10.1039/D4FB00110A

Haghighi, H., Leugoue, S. K., Pfeifer, F., Siesler, H. W., Licciardello, F., Fava, P., & Pulvirenti, A. (2020). Development of antimicrobial films based on chitosan-polyvinyl alcohol blend enriched with ethyl lauroyl arginate (LAE) for food packaging applications. *Food Hydrocolloids, 100*, 105419. https://doi.org/10.1016/j.foodhyd.2019.105419

Han, J. H., Hwang, H. M., Min, S., & Krochta, J. M. (2008). Coating of peanuts with edible whey protein film containing α-tocopherol and ascorbyl palmitate. *Journal of Food Science, 73*(8), E349–E355. https://doi.org/10.1111/j.1750-3841.2008.00910.x

Hasan, S. M. K., Ferrentino, G., & Scampicchio, M. (2020). Nanoemulsion as advanced edible coatings to preserve the quality of fresh-cut fruits and vegetables: A review. *International Journal of Food Science & Technology, 55*(1), 1–10. https://doi.org/10.1111/ijfs.14273

Hassan, B., Chatha, S. A. S., Hussain, A. I., Zia, K. M., & Akhtar, N. (2018). Recent advances on polysaccharides, lipids and protein based edible films and coatings: A review. *International Journal of Biological Macromolecules, 109*, 1095–1107. https://doi.org/10.1016/j.ijbiomac.2017.11.097

Hernandez-Izquierdo, V. M., & Krochta, J. M. (2008). Thermoplastic processing of proteins for film formation—A review. *Journal of Food Science, 73*(2). https://doi.org/10.1111/j.1750-3841.2007.00636.x

Hischenhuber, C., Crevel, R., Jarry, B., Mäki, M., Moneret-Vautrin, D. A., Romano, A., Troncone, R., & Ward, R. (2006). Safe amounts of gluten for patients with wheat allergy or coeliac disease. *Alimentary Pharmacology & Therapeutics, 23*(5), 559–575. https://doi.org/10.1111/j.1365-2036.2006.02768.x

Hong, S.-I., & Krochta, J. M. (2006). Oxygen barrier performance of whey-protein-coated plastic films as affected by temperature, relative humidity, base film and protein type. *Journal of Food Engineering, 77*(3), 739–745. https://doi.org/10.1016/j.jfoodeng.2005.07.034

Jamuna, P., Vijayalatha, K. R., Devi, A. N., Jegadeeswari, V., Kumar, R. A., Geetha, K., Lakshmi, I. G., Maheswari, T. U., & Anitha, T. (2024). Edible coatings: Classification, applications and innovations in food preservation. *Plant Science, 11*(sp4), 01–13. https://doi.org/10.14719/pst.5263

Janjarasskul, T., Tananuwong, K., & Krochta, J. M. (2011). Whey protein film with oxygen scavenging function by incorporation of ascorbic acid. *Journal of Food Science, 76*(9), E561–E568. https://doi.org/10.1111/j.1750-3841.2011.02409.x

Jiang, T., Feng, L., & Zheng, X. (2012). Effect of chitosan coating enriched with thyme oil on postharvest quality and shelf life of shiitake mushroom (Lentinus edodes). *Journal of Agricultural and Food Chemistry, 60*(1), 188–196. https://doi.org/10.1021/jf202638u

Jiménez, A., Fabra, M. J., Talens, P., & Chiralt, A. (2012). Edible and biodegradable starch films: A review. *Food and Bioprocess Technology, 5*(6), 2058–2076. https://doi.org/10.1007/s11947-012-0835-4

Joung, J., Boonsiriwit, A., Kim, M., & Lee, Y. S. (2021). Application of ethylene scavenging nanocomposite film prepared by loading potassium permanganate-impregnated halloysite nanotubes into low-density polyethylene as active packaging material for fresh produce. *LWT - Food Science and Technology, 145*, 111309. https://doi.org/10.1016/j.lwt.2021.111309

Kalita, N. K., Sarmah, A., Bhasney, S. M., Kalamdhad, A., & Katiyar, V. (2021). Demonstrating an ideal compostable plastic using biodegradability kinetics of poly (lactic acid)(PLA) based green biocomposite films under aerobic composting conditions. *Environmental Challenges, 3*, 100030. https://doi.org/10.1016/j.envc.2021.100030

Karim, A. A., & Bhat, R. (2009). Fish gelatin: Properties, challenges, and prospects as an alternative to mammalian gelatins. *Food Hydrocolloids, 23*(3), 563–576. https://doi.org/10.1016/j.foodhyd.2008.07.002

Kean, T., & Thanou, M. (2010). Biodegradation, biodistribution and toxicity of chitosan. *Advanced Drug Delivery Reviews, 62*(1), 3–11. https://doi.org/10.1016/j.addr.2009.09.004

Kocira, A., Kozłowicz, K., Panasiewicz, K., Staniak, M., Szpunar-Krok, E., & Hortyńska, P. (2021). Polysaccharides as Edible films and coatings: Characteristics and influence on fruit and vegetable quality—A review. *Agronomy, 11*(5), 813. https://doi.org/10.3390/agronomy11050813

Kong, M., Chen, X. G., Xing, K., & Park, H. J. (2010). Antimicrobial properties of chitosan and mode of action: A state of the art review. *International Journal of Food Microbiology, 144*(1), 51–63. https://doi.org/10.1016/j.ijfoodmicro.2010.09.012

Krishnan, R., Misra, M., Subramanian, J., & Mohanty, A. (2025). Emerging trends and application of edible coating as a sustainable solution for postharvest management in stone fruits: A comprehensive review. *Comprehensive Reviews in Food Science and Food Safety, 24*(3), e70179. https://doi.org/10.1111/1541-4337.70179

Kritchenkov, A. S., Egorov, A. R., Volkova, O. V., Zabodalova, L. A., Suchkova, E. P., Yagafarov, N. Z., Kurasova, M. N., Dysin, A. P., Kurliuk, A. V., Shakola, T. V., & Khrustalev, V. N. (2020). Active antibacterial food coatings based on blends of succinyl chitosan and triazole betaine chitosan derivatives. *Food Packaging and Shelf Life, 25*, 100534. https://doi.org/10.1016/j.fpsl.2020.100534

Kumari, M., Mahajan, H., Joshi, R., & Gupta, M. (2017). Development and structural characterization of edible films for improving fruit quality. *Food Packaging and Shelf Life, 12*, 42–50. https://doi.org/10.1016/j.fpsl.2017.02.003

Kumarihami, H. P. C., Kim, Y. H., Kwack, Y. B., Kim, J., & Kim, J. G. (2022). Application of chitosan as edible coating to enhance storability and fruit quality of Kiwifruit: A review. *Scientia Horticulturae, 292*, 110647. https://doi.org/10.1016/j.scienta.2021.110647

Lacroix, M., & Le Tien, C. (2005). Edible films and coatings from nonstarch polysaccharides. In *Innovations in food packaging* (pp. 338–361). Academic Press. https://doi.org/10.1016/B978-012311632-1/50052-8

Le Tien, C., Vachon, C., Mateescu, M.-A., & Lacroix, M. (2001). Milk protein coatings prevent oxidative browning of apples and potatoes. *Journal of Food Science, 66*(4), 512–516. https://doi.org/10.1111/j.1365-2621.2001.tb04594.x

Li, N., Zhou, Z., Wu, F., Lu, Y., Jiang, D., Zhong, L., & Xie, F. (2022). Development of pH-indicative and antimicrobial films based on polyvinyl alcohol/starch incorporated with ethyl lauroyl arginate and mulberry anthocyanin for active packaging. *Coatings, 12*(10), 1392. https://doi.org/10.3390/coatings12101392

Lin, D., & Zhao, Y. (2007). Innovations in the development and application of edible coatings for fresh and minimally processed fruits and vegetables. *Comprehensive Reviews in Food Science and Food Safety, 6*(3), 60–75. https://doi.org/10.1111/j.1541-4337.2007.00018.x

Lin, Y., Li, N., Lin, H., Lin, M., Chen, Y., Wang, H., Ritenour, M. A., & Lin, Y. (2020). Effects of chitosan treatment on the storability and quality properties of longan fruit during storage. *Food Chemistry, 306*, 125627. https://doi.org/10.1016/j.foodchem.2019.125627

Liu, S., Li, L., Li, B., Zhu, J., & Li, X. (2022). Size effect of carnauba wax nanoparticles on water vapor and oxygen barrier properties of starch-based film. *Carbohydrate Polymers, 296*, 119935. https://doi.org/10.1016/j.carbpol.2022.119935

Liyanapathiranage, A., Dassanayake, R. S., Gamage, A., Karri, R. R., Manamperi, A., Evon, P., Jayakodi, Y., Madhujith, T., & Merah, O. (2023). Recent developments in Edible films and coatings for fruits and vegetables. *Coatings, 13*(7), 1177. https://doi.org/10.3390/coatings13071177

Lopez-Polo, J., Muñoz-Shuguli, C., Vidal, M. P., & Vidal, C. P. (2024). Electrospun edible films and coatings: Development, functionality and food applications. *Trends in Food Science & Technology, 143*, 104253. https://doi.org/10.1016/j.tifs.2023.104253

Lorevice, M. V., Otoni, C. G., de Moura, M. R., & Mattoso, L. H. C. (2016). Chitosan nanoparticles on the improvement of thermal, barrier, and mechanical properties of high-and low-methyl pectin films. *Food Hydrocolloids, 52*, 732–740. https://doi.org/10.1016/j.foodhyd.2015.08.003

Ma, M., Liu, Y., Zhang, S., & Yuan, Y. (2024). Edible coating for fresh-cut fruit and vegetable preservation: Biomaterials, functional ingredients, and joint non-thermal technology. *Food, 13*(23), 3937. https://doi.org/10.3390/foods13233937

Maqbool, M., Ali, A., Alderson, P. G., Zahid, N., & Siddiqui, Y. (2011). Effect of a novel edible composite coating based on gum Arabic and Chitosan on biochemical and physiological responses of banana fruits during cold storage. *Journal of Agricultural and Food Chemistry, 59*(10), 5474–5482. https://doi.org/10.1021/jf200623m

Maringgal, B., Hashim, N., Mohamed Amin Tawakkal, I. S., & Muda Mohamed, M. T. (2020). Recent advance in edible coating and its effect on fresh/fresh-cut fruits quality. *Trends in Food Science & Technology, 96*, 253–267. https://doi.org/10.1016/j.tifs.2019.12.024

Martín, B. O., Soliva, F. R., & Oms, O. G. (2006). *Handbook of fruits and fruit processing*. John Wiley & Sons.

Massilia, R. M. R., Melgar, J. M., & Tapia, M. S. (2010). Edible films and coatings for post-harvest products with reduced humidity. *Stewart Postharvest Review, 6*(3), 1–10. https://doi.org/10.2212/spr.2010.3.3

Matloob, A., Ayub, H., Mohsin, M., Ambreen, S., Khan, F. A., Oranab, S., Rahim, M. A., Khalid, W., Nayik, G. A., Ramniwas, S., & Ercisli, S. (2023). A review on edible coatings and films: Advances, composition, production methods, and safety concerns. *ACS Omega, 8*(32), 28932–28944. https://doi.org/10.1021/acsomega.3c03459

Mei, Y., Zhao, Y., Yang, J., & Furr, H. C. (2002). Using Edible coating to enhance nutritional and sensory qualities of baby carrots. *Journal of Food Science, 67*(5), 1964–1968. https://doi.org/10.1111/j.1365-2621.2002.tb08753.x

Mendy, T. K., Misran, A., Mahmud, T. M. M., & Ismail, S. I. (2019). Application of Aloe vera coating delays ripening and extend the shelf life of papaya fruit. *Scientia Horticulturae, 246*, 769–776. https://doi.org/10.1016/j.scienta.2018.11.054

Mikus, M., & Galus, S. (2025). Extending the shelf life of apples after harvest using edible coatings as active packaging—A review. *Applied Sciences, 15*(2), 767. https://doi.org/10.3390/app15020767

Min, S., Krochta, J. M., & Rumsey, T. R. (2007). Diffusion of thiocyanate and hypothiocyanite in whey protein films incorporating the lactoperoxidase system. *Journal of Food Engineering, 80*(4), 1116–1124. https://doi.org/10.1016/j.jfoodeng.2006.09.003

Mkandawire, M., & Aryee, A. N. (2018). Resurfacing and modernization of edible packaging material technology. *Current Opinion in Food Science, 19*, 104–112. https://doi.org/10.1016/j.cofs.2018.03.010

Morillon, V., Debeaufort, F., Blond, G., Capelle, M., & Voilley, A. (2002). Factors affecting the moisture permeability of lipid-based Edible films: A Review. *Critical Reviews in Food Science and Nutrition, 42*(1), 67–89. https://doi.org/10.1080/10408690290825466

Mujtaba, M., Morsi, R. E., Kerch, G., Elsabee, M. Z., Kaya, M., Labidi, J., & Khawar, K. M. (2019). Current advancements in chitosan-based film production for food technology; A review. *International Journal of Biological Macromolecules, 121*, 889–904. https://doi.org/10.1016/j.ijbiomac.2018.10.109

Muscat, D., Adhikari, B., Adhikari, R., & Chaudhary, D. S. (2012). Comparative study of film forming behaviour of low and high amylose starches using glycerol and xylitol as plasticizers. *Journal of Food Engineering, 109*(2), 189–201. https://doi.org/10.1016/j.jfoodeng.2011.10.019

Necas, J., & Bartosikova, L. (2013). Carrageenan: A review. *Veterinární Medicína, 58*(4), 187–205. https://doi.org/10.17221/6758-VETMED

Otoni, C. G., Avena-Bustillos, R. J., Olsen, C. W., Bilbao-Sáinz, C., & McHugh, T. H. (2016). Mechanical and water barrier properties of isolated soy protein composite edible films as affected by carvacrol and cinnamaldehyde micro and nanoemulsions. *Food Hydrocolloids, 57*, 72–79. https://doi.org/10.1016/j.foodhyd.2016.01.012

Otoni, C. G., Avena-Bustillos, R. J., Azeredo, H. M. C., Lorevice, M. V., Moura, M. R., Mattoso, L. H. C., & McHugh, T. H. (2017). Recent advances on edible films based on fruits and vegetables—A review. *Comprehensive Reviews in Food Science and Food Safety, 16*(5), 1151–1169. https://doi.org/10.1111/1541-4337.12281

Pavlath, A. E., & Orts, W. (2009). Edible films and coatings: Why, what, and how? In *Edible films and coatings for food applications* (pp. 1–23). Springer. https://doi.org/10.1007/978-0-387-92824-1_1

Pérez-Gago, M. B., & Krochta, J. M. (2005). Emulsion and bi-layer edible films. In J. H. Han (Ed.), *Innovations in food packaging, A volume in food science and technology* (Vol. 22, pp. 384–402. ISBN 9780123116321.). Academic Press. https://doi.org/10.1016/B978-012311632-1/50054-1

Pirsa, S., Sani, I. K., & Mirtalebi, S. S. (2021). Nano-biocomposite based color sensors: Investigation of structure, function, and applications in intelligent food packaging. *Food Packaging and Shelf Life, 31*, 100789. https://doi.org/10.1016/j.fpsl.2021.100789

Podshivalov, A., Zakharova, M., Glazacheva, E., & Uspenskaya, M. (2017). Gelatin/potato starch edible biocomposite films: Correlation between morphology and physical properties. *Carbohydrate Polymers, 157*, 1162–1172. https://doi.org/10.1016/j.carbpol.2016.10.079

Pranoto, Y., Rakshit, S. K., & Salokhe, V. M. (2005). Enhancing antimicrobial activity of chitosan films by incorporating garlic oil, potassium sorbate and nisin. *LWT-Food Science and Technology, 38*(8), 859–865. https://doi.org/10.1016/j.lwt.2004.09.014

Quezada-Gallo, J.-A. (2009). Delivery of food additives and antimicrobials using Edible films and coatings. In K. C. Huber & M. E. Embuscado (Eds.), *Edible films and coatings for food applications* (pp. 315–333). Springer New York. https://doi.org/10.1007/978-0-387-92824-1_11

Quirós-Sauceda, A. E., Ayala-Zavala, J. F., Olivas, G. I., & González-Aguilar, G. A. (2014). Edible coatings as encapsulating matrices for bioactive compounds: A review. *Journal of Food Science and Technology, 51*(9), 1674–1685. https://doi.org/10.1007/s13197-013-1246-x

Raghav, P. K., Agarwal, N., & Saini, M. (2016). Edible coating of fruits and vegetables: A review. In *International Journal of Scientific Research and Modern Education (IJSRME)*. ISSN: Vol. I. www.rdmodernresearch.com

Ranjith, F. H., Adhikari, B., Muhialdin, B. J., Yusof, N. L., Mohammed, N. K., Ariffin, S. H., & Hussin, A. S. M. (2022). Peptide-based edible coatings to control postharvest fungal spoilage of mango (Mangifera indica L.) fruit. *Food Control, 135*, 108789. https://doi.org/10.1016/j.foodcont.2021.108789

Rhim, J. W., Gennadios, A., Handa, A., Weller, C. L., & Hanna, M. A. (2000). Solubility, tensile, and color properties of modified soy protein isolate films. *Journal of Agricultural and Food Chemistry, 48*(10), 4937–4941.

Rhim, J.-W., Hong, S.-I., & Ha, C.-S. (2009). Tensile, water vapor barrier and antimicrobial properties of PLA/nanoclay composite films. *LWT - Food Science and Technology, 42*(2), 612–617. https://doi.org/10.1016/j.lwt.2008.02.015

Rhim, J.-W., Park, H.-M., & Ha, C.-S. (2013). Bio-nanocomposites for food packaging applications. *Progress in Polymer Science, 38*(10–11), 1629–1652. https://doi.org/10.1016/j.progpolymsci.2013.05.008

Robles-Sánchez, R. M., Rojas-Graü, M. A., Odriozola-Serrano, I., González-Aguilar, G., & Martin-Belloso, O. (2013). Influence of alginate-based edible coating as carrier of antibrowning agents on bioactive compounds and antioxidant activity in fresh-cut Kent mangoes. *LWT - Food Science and Technology, 50*(1), 240–246. https://doi.org/10.1016/j.lwt.2012.05.021

Rodrigues, D. C., Cunha, A. P., Brito, E. S., Azeredo, H. M., & Gallao, M. I. (2016). Mesquite seed gum and palm fruit oil emulsion edible films: Influence of oil content and sonication. *Food Hydrocolloids, 56*, 227–235. https://doi.org/10.1016/j.foodhyd.2015.12.018

Rojas-Graü, M. A., Tapia, M. S., Rodríguez, F. J., Carmona, A. J., & Martin-Belloso, O. (2007). Alginate and gellan-based edible coatings as carriers of antibrowning agents applied on fresh-cut Fuji apples. *Food Hydrocolloids, 21*(1), 118–127. https://doi.org/10.1016/j.foodhyd.2006.03.001

Rossi-Márquez, G., Dávalos-Saucedo, C. A., & Di Pierro, P. (2023). Edible films and coatings applied in the food industry. *Coatings (2079-6412), 13*(4). https://doi.org/10.3390/coatings13040670

Ruggero, F., Gori, R., & Lubello, C. (2019). Methodologies to assess biodegradation of bioplastics during aerobic composting and anaerobic digestion: A review. *Waste Management & Research: The Journal for a Sustainable Circular Economy, 37*(10), 959–975. https://doi.org/10.1177/0734242X19854127

Saberi, B., Vuong, Q. V., Chockchaisawasdee, S., Golding, J. B., Scarlett, C. J., & Stathopoulos, C. E. (2017). Physical, barrier, and antioxidant properties of pea starch-guar gum biocomposite edible films by incorporation of natural plant extracts. *Food and Bioprocess Technology, 10*(12), 2240–2250. https://doi.org/10.1007/s11947-017-1995-z

Saha, D., & Bhattacharya, S. (2010). Hydrocolloids as thickening and gelling agents in food: A critical review. *Journal of Food Science and Technology, 47*(6), 587–597. https://doi.org/10.1007/s13197-010-0162-6

Salazar, A. S. S., Cavazos, P. A. S., Paz, H. M., & Fragoso, A. V. (2019). External factors and nanoparticles effect on water vapor permeability of pectin-based films. *Journal of Food Engineering, 245*, 73–79. https://doi.org/10.1016/j.jfoodeng.2018.09.002

Sánchez-González, L., Vargas, M., González-Martínez, C., Chiralt, A., & Cháfer, M. (2011). Use of essential oils in bioactive edible coatings: A review. *Food Engineering Reviews, 3*(1), 1–16. https://doi.org/10.1007/s12393-010-9031-3

Sánchez-Tamayo, M. I., Vélez Pasos, C., & Ochoa-Martínez, C. I. (2021). Methods for gas permeability measurement in edible films for fruits and vegetables: A review. In Food science and technology (Brazil) (Vol. 41(4), pp. 807–815). Sociedade Brasileira de Ciencia e Tecnologia de Alimentos, SBCTA. https://doi.org/10.1590/fst.07520.

Sanyang, M. L., Sapuan, S. M., Jawaid, M., Ishak, M. R., & Sahari, J. (2016). Recent developments in sugar palm (Arenga pinnata) based biocomposites and their potential industrial applications: A review. *Renewable and Sustainable Energy Reviews, 54*, 533–549. https://doi.org/10.1016/j.rser.2015.10.037

Schmid, M. (2013). Properties of cast films made from different ratios of whey protein isolate, hydrolysed whey protein isolate and glycerol. *Materials, 6*(8), 3254–3269. https://doi.org/10.3390/ma6083254

Schmid, M., Dallmann, K., Bugnicourt, E., Cordoni, D., Wild, F., Lazzeri, A., & Noller, K. (2012). Properties of whey-protein-coated films and laminates as novel recyclable food packaging materials with excellent barrier properties. *International Journal of Polymer Science, 2012*, 1–7. https://doi.org/10.1155/2012/562381

Schmid, M., Noller, K., & Müller, K. (2017). Improvement of food packaging-related properties of whey protein isolate-based nanocomposite films and coatings by addition of montmorillonite nanoplatelets. *Frontiers in Materials, 4*, 35. https://doi.org/10.3389/fmats.2017.00035

Sharaf Eddin, A., Ibrahim, S. A., & Tahergorabi, R. (2019). Egg quality and safety with an overview of edible coating application for egg preservation. *Food Chemistry, 296*, 29–39. https://doi.org/10.1016/j.foodchem.2019.05.182

Sharma, P., Shehin, V. P., Kaur, N., & Vyas, P. (2019). Application of edible coatings on fresh and minimally processed vegetables: A review. *International Journal of Vegetable Science, 25*(3), 295–314. https://doi.org/10.1080/19315260.2018.1510863

Sharma, S., Nakano, K., Kumar, S., & Katiyar, V. (2024). Edible packaging to prolong postharvest shelf-life of fruits and vegetables: A review. *Food Chemistry Advances*, 100711. https://doi.org/10.1016/j.focha.2024.100711

Siracusa, V., Rocculi, P., Romani, S., & Rosa, M. D. (2008). Biodegradable polymers for food packaging: A review. *Trends in Food Science & Technology, 19*(12), 634–643. https://doi.org/10.1016/j.tifs.2008.07.003

Sothornvit, R., & Krochta, J. M. (2005). Plasticizers in edible films and coatings. In J. H. Han (Ed.), *Innovations in food packaging, A volume in food science and technology* (Vol. 23, pp. 403–433. ISBN 9780123116321.). Academic Press. https://doi.org/10.1016/B978-012311632-1/50055-3

Sothornvit, R., Hong, S.-I., An, D. J., & Rhim, J.-W. (2010). Effect of clay content on the physical and antimicrobial properties of whey protein isolate/organo-clay composite films. *LWT - Food Science and Technology, 43*(2), 279–284. https://doi.org/10.1016/j.lwt.2009.08.010

Souza, A. C. D., Goto, G. E. O., Mainardi, J. A., Coelho, A. C. V., & Tadini, C. C. (2013). Cassava starch composite films incorporated with cinnamon essential oil: Antimicrobial activity, microstructure, mechanical and barrier properties. *LWT-Food Science and Technology, 54*(2), 346–352. https://doi.org/10.1016/j.lwt.2013.06.017

Tapia, M. S., Rojas-Graü, M. A., Carmona, A., Rodríguez, F. J., Soliva-Fortuny, R., & Martin-Belloso, O. (2008). Use of alginate- and gellan-based coatings for improving barrier, texture and nutritional properties of fresh-cut papaya. *Food Hydrocolloids, 22*(8), 1493–1503. https://doi.org/10.1016/j.foodhyd.2007.10.004

Tharanathan, R. N. (2003). Biodegradable films and composite coatings: Past, present and future. *Trends in Food Science & Technology, 14*(3), 71–78. https://doi.org/10.1016/S0924-2244(02)00280-7

Tongdeesoontorn, W., Mauer, L. J., Wongruong, S., Sriburi, P., & Rachtanapun, P. (2011). Effect of carboxymethyl cellulose concentration on physical properties of biodegradable cassava starch-based films. *Chemistry Central Journal, 5*(1), 6. https://doi.org/10.1186/1752-153X-5-6

Vaezi, K., Asadpour, G., & Sharifi, H. (2019). Effect of ZnO nanoparticles on the mechanical, barrier, and optical properties of thermoplastic cationic starch/montmorillonite biodegradable films. *International Journal of Biological Macromolecules, 124*, 519–529. https://doi.org/10.1016/j.ijbiomac.2018.11.142

Valencia-Chamorro, S. A., Palou, L., del Río, M. A., & Pérez-Gago, M. B. (2011). Antimicrobial edible films and coatings for fresh and minimally processed fruits and vegetables: A review. *Critical Reviews in Food Science and Nutrition, 51*(9), 872–900. https://doi.org/10.1080/10408398.2010.485705

Vargas, M., Pastor, C., Chiralt, A., McClements, D. J., & González-Martínez, C. (2008). Recent advances in edible coatings for fresh and minimally processed fruits. *Critical Reviews in Food Science and Nutrition, 48*(6), 496–511. https://doi.org/10.1080/10408390701537344

Velickova, E., Winkelhausen, E., Kuzmanova, S., Moldão-Martins, M., & Alves, V. D. (2013). Impact of chitosan-beeswax edible coatings on the quality of fresh strawberries (Fragaria ananassa cv Camarosa) under commercial storage conditions. *LWT - Food Science and Technology, 52*(2), 80–92. https://doi.org/10.1016/j.lwt.2013.02.004

Vu, H. P. N., & Lumdubwong, N. (2016). Starch behaviors and mechanical properties of starch blend films with different plasticizers. *Carbohydrate Polymers, 154*, 112–120. https://doi.org/10.1016/j.carbpol.2016.08.034

Wang, K., Lim, P. N., Tong, S. Y., & Thian, E. S. (2019). Development of grapefruit seed extract-loaded poly(ε-caprolactone)/chitosan films for antimicrobial food packaging. *Food Packaging and Shelf Life, 22*, 100396. https://doi.org/10.1016/j.fpsl.2019.100396

Weiss, J., Takhistov, P., & McClements, D. J. (2006). Functional materials in food nanotechnology. *Journal of Food Science, 71*(9). https://doi.org/10.1111/j.1750-3841.2006.00195.x

Wu, Y., Wu, H., & Hu, L. (2024). Recent advances of proteins, polysaccharides and lipids-based Edible films/coatings for food packaging applications: A review. *Food Biophysics, 19*, 29–45. https://doi.org/10.1007/s11483-023-09794-7

Xu, J., & Li, Y. (2023). Wheat gluten–based coatings and films: Preparation, properties, and applications. *Journal of Food Science, 88*, 582–594. https://doi.org/10.1111/1750-3841.16454

Yang, H., Li, X. I. A., & Lu, G. (2018). Effect of carnauba wax–based coating containing glycerol monolaurate on decay and quality of sweet potato roots during storage. *Journal of Food Protection, 81*(10), 1643–1650. https://doi.org/10.4315/0362-028X.JFP-18-017

Youssef, A. M., EL-Sayed, S. M., EL-Sayed, H. S., Salama, H. H., & Dufresne, A. (2016). Enhancement of Egyptian soft white cheese shelf life using a novel chitosan/carboxymethyl cellulose/zinc oxide bionanocomposite film. *Carbohydrate Polymers, 151*, 9–19. https://doi.org/10.1016/j.carbpol.2016.05.023

Yousuf, B., Qadri, O. S., & Srivastava, A. K. (2018). Recent developments in shelf-life extension of fresh-cut fruits and vegetables by application of different edible coatings: A review. *LWT - Food Science and Technology, 89*, 198–209. https://doi.org/10.1016/j.lwt.2017.10.051

Zhang, C., Long, Y.-H., Wang, Q.-P., Li, J.-H., An, H.-M., Wu, X.-M., & Li, M. (2019). The effect of preharvest 28.6% chitosan composite film sprays for controlling the soft rot on kiwifruit. *Horticultural Science, 46*(4), 180–194. https://doi.org/10.17221/84/2018-HORTSCI

Zillo, R. R., Da Silva, P. P. M., De Oliveira, J., Da Glória, E. M., & Spoto, M. H. F. (2018). Carboxymethylcellulose coating associated with essential oil can increase papaya shelf life. *Scientia Horticulturae, 239*, 70–77. https://doi.org/10.1016/j.scienta.2018.05.025

Chapter 7
Mechanisms of Edible Packaging in Food Protection

Khaoula Elhadef, Bouthaina Ben Hadj Hmida, Sarra Akermi, Moufida Chaari, and Slim Smaoui

Abstract Edible packaging can be used to decrease food deterioration. Removing oxygen from the packaging headspace and/or adding antioxidants to films and coatings to improve the antioxidant qualities of food surfaces are the two main ways to prevent oxidation. This chapter examines the mechanisms of edible packaging in food protection and highlights its potential as a sustainable alternative to conventional packaging. It begins by exploring the composition and structure of edible packaging, focusing on polysaccharide-based (chitosan, pectin, starch, alginate, gum, cellulose), protein-based (soy protein isolate, whey protein isolate, gelatin), and lipid-based (waxes, resins, fats, oils) materials. The chapter then addresses the protective mechanisms of edible packaging, including barrier properties (moisture, oxygen), antimicrobial activity, antioxidant functions, and physical protection (tensile strength, coating functions). Applications in diverse food categories—fruits, vegetables, meat, poultry, and seafood are discussed. Furthermore, innovative approaches, such as smart packaging, nano-encapsulation, and bio-nanocomposites, are presented. The chapter concludes with a focus on biodegradability and sustainability, emphasizing edible packaging's potential to enhance food protection while reducing environmental impact.

Keywords Edible packaging · Biodegradable films · Food protection · Sustainable packaging · Natural biopolymers

---

K. Elhadef · S. Akermi · M. Chaari · S. Smaoui (✉)
Laboratory of Microbial and Enzymatic Biotechnologies and Biomolecules (LBMEB), Center of Biotechnology of Sfax (CBS), University of Sfax, Sfax, Tunisia
e-mail: slim.smaoui@cbs.rnrt.tn

B. B. H. Hmida
Laboratory of Biochemistry and Enzymatic Engineering of Lipases, National Engineering School of Sfax, University of Sfax, Sfax, Tunisia

N. Kumar et al. (eds.), *Trends in Edible Packaging: A sustainable approaches for postharvest management of fruits and vegetables*,
https://doi.org/10.1007/978-3-032-13199-7_7

Introduction

During production, processing, distribution, and storage, food is susceptible to deterioration due to chemical and microbiological processes. One of the promising strategies to address this challenge is the use of edible packaging, which offers multiple benefits, including food preservation, quality maintenance, safety enhancement, and consumer information transfer (Jurić et al., 2024). Edible packaging is defined as a thin, edible layer that can be coated directly on food or serve as a barrier between the food and its surrounding environment (Kumar et al., 2022).

Oxidation remains one of the primary causes of food deterioration, negatively impacting sensory attributes such as flavor and color (Zhang et al., 2024). While synthetic antioxidants like butylated hydroxyanisole (BHA) and butylated hydroxytoluene (BHT) are commonly used in edible films due to their cost-effectiveness and stability, concerns over potential health risks have led to stricter regulations and a growing interest in natural antioxidant alternatives (Eranda et al., 2024). Consequently, the incorporation of natural antioxidants derived from vegetable extracts is being explored to enhance the stability of lipids and lipid-containing foods while preserving sensory and nutritional qualities (Kasaai, 2025). Additionally, the use of encapsulation techniques can help mitigate the strong flavors of certain natural antioxidants, such as essential oils, when incorporated into edible film formulations. Ambient temperature and water activity significantly impact food shelf life (Hedegaard et al., 2024). Although controlled atmosphere and hypobaric storage can be beneficial for preserving fruits and vegetables, these methods are often costly. In contrast, edible coatings offer a cost-effective alternative and can complement storage techniques like a low temperature and controlled atmosphere systems. The oxygen barrier properties of edible films depend on factors such as temperature, polymer chain interactions, hydrophilic and lipophilic properties, and atmospheric humidity (Shah et al., 2024). For instance, hydrocolloid-based films are often superior oxygen barriers compared to synthetic polymers due to their dense network of covalent, hydrogen, and van der Waals bonds (Bao et al., 2025).

The increasing global demand for sustainable and environmentally friendly food packaging has also driven innovation within the food industry. Edible packaging, composed primarily of natural and biodegradable materials like polysaccharides, proteins, and lipids, addresses the need for both environmental sustainability and food protection (Devi et al., 2024). These materials, known for their film-forming capabilities, can effectively preserve food quality, prolong shelf life, and, in some cases, provide nutritional or sensory benefits. The unique advantage of edible packaging lies in its dual functionality: it serves as both a protective layer and a consumable component, reducing waste and promoting a circular economy.

This chapter aims to comprehensively review the mechanisms by which edible packaging protects food, focusing on its composition and structure, including polysaccharide-based (e.g., chitosan, pectin, starch), protein-based (e.g., soy protein isolate, whey protein isolate, gelatin), and lipid-based (e.g., waxes, resins, fats, oils) materials. It also examines the fundamental protective mechanisms—such as barrier properties, antimicrobial activity, antioxidant functions, and physical protection—that make edible coatings effective in extending the shelf life of various food products. Furthermore, the chapter explores the practical applications of edible packaging across diverse food categories, including fruits, vegetables, meat, poultry, and seafood. It also highlights innovative approaches like smart and active packaging, nano-encapsulation, and bio-nanocomposites. By bridging scientific insights with practical applications, this chapter seeks to underscore the transformative potential of edible packaging in the food industry, fostering the development of sustainable and efficient food protection strategies.

Composition and Structure of Edible Packaging

Research on innovative food packaging methods is crucial to advancing food safety and quality. Edible packaging is becoming more prevalent. Sustainable, biodegradable materials are commonly used in edible packaging to cover or wrap food in a consumable manner. At least two elements are required for edible packaging: a solvent, typically water, and a biopolymer-based matrix that will form a cohesive structure (Hamed et al., 2022). Based on the fundamental biopolymer used in their formulation, edible packaging compositions may be divided into four major categories, as seen in Fig. 7.1.

Polysaccharide-Based Edible Packaging

High-molecular-mass polysaccharides are the primary component of polysaccharide-based edible films and coatings. These materials interact with intramolecular and intermolecular hydrogen bonds to generate a dense networked film with favorable mechanical properties. They are hydrophilic, which makes them a weak barrier toward moisture loss in fruits, but they have strong barrier qualities against O_2 and CO_2. To reduce the amount of plastic packaging that is commonly used, biopolymers are employed to create edible films and coatings (Dinika et al., 2020).

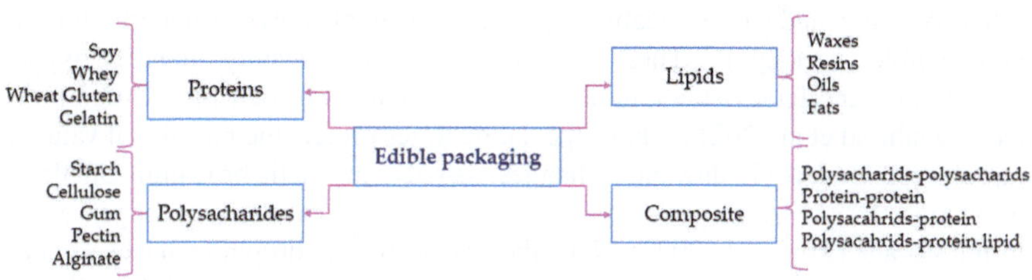

Fig. 7.1 The different edible packaging compositions

Chitosan

Chitosan, a sustainable biopolymer, is extracted from chitin, which is naturally found in yeast, fungi, and crustaceans. Its linear structure is made up of N-acetyl-D-glucosamine (acetylated units) and β-(1–4)-linked D-glucosamine (deacetylated groups) (Adiletta et al., 2021). According to Chaudhary et al. (2020), chitosan is a harmless, biodegradable, biocompatible, and film-forming substance that has antifungal, antibacterial, and anticancer features. Eco-friendliness, good strength, and selective gas permeability (CO_2 and O_2) are characteristics of chitosan films (Kocira et al., 2021). They limit dehydration damage, prevent microbiological growth, postpone fruit enzymatic browning, and lower oxygen levels in packaging. Furthermore, chitosan improves emulsification, which stabilizes color, texture, and flavor (Wu et al., 2024). Chitosan films and coatings have been applied to several food products to assess their practical efficacy in terms of physical and biological protection. Pavinatto et al. (2020) successfully created glycerol-containing chitosan-based films to preserve strawberries against fungal infections. The films' high hydrophobicity allowed the protection of the strawberries' flavor, texture, and appearance. Also, Duran and Kahve (2020) examined how beef storage was affected by vacuum packaging and a chitosan-based film. The findings demonstrated that the composite film successfully decreased total volatile basic nitrogen values, the growth of lactic acid, and thermophilic aerobic bacteria. It also totally inhibited *S. aureus*. During cold storage, chitosan containing *Mentha piperita L.* essential oil (EO) preserved the catechins and procyanidins B1 and B2 in the mango cultivar "*Tommy Atkins*" and postponed color changes in the peel and pulp (de Oliveira et al., 2020).

Pectin

Several research studies have demonstrated that pectin has great potential for use in food packaging as an edible film. Pectin is a complex gel-forming biopolymer made up of high-molecular-mass glycanogalacturonans with a few rhamnose and arabinose residues in the core chain connected to 1,4-linked α-D-galacturonic acid units, and galactose and xylose in the side chains. It is frequently used as a component in fruit coatings, either alone or in combination with other polysaccharides, essential oils, or plant materials (Shiekh et al., 2021). Given its biodegradability, biocompatibility, swelling, and gas permeability, pectin is a suitable polymer material for creating edible packaging. These qualities may significantly reduce oxygen consumption and food oxidation, ultimately extending the shelf life of food products (Panahirad et al., 2021). These special qualities protect the nutritional value of unpreserved food, avoid unwanted changes, namely enzymatic browning, off-flavor production, loss of scent, slow lipid migration, and quickly stop pathogen attack during storage (Roy et al., 2023). Nevertheless, pectin's hydrophilia makes its coatings and films more brittle and increases their water vapor permeability (WVP). Pectin edible packaging is therefore commonly used with other materials to reduce hydrophilicity in high-moisture situations (Nastasi et al., 2022). Chitosan-pectin edible films and coatings for dry food were made by (Merino et al., 2023). Chitosan and pectin interact through electrostatic forces, which enhance the films by reducing their thermal stability, lowering their water vapor and oxygen permeability, and improving their elastic modulus. This composite packaging also demonstrated excellent transparency and strong antioxidant capacity.

Starch

The primary components of starch, which account for 98–99% of the dry weight of plant crops, are branching amylopectin and linear amylose. D-glucopyranose units are joined by α-1,4-glycosidic bonds to form amylose, while amylopectin is made up of both a straight-chain fraction (α-1,4-glycosidic bonds) and a branching part (α-1,6-glycosidic bonds) (Thakur et al., 2019). Amylose's linear structure tends to align itself parallel to the hydroxyl groups, which promotes hydrogen bonding and makes coating films more hydrophobic. Hence, starches that include more amylose have improved rigidity, extension, gas barrier, and film-forming qualities (Shiekh et al., 2021). In addition to being non-toxic, the starch films have no flavor, color, or odor. Nevertheless, a significant drawback of starch is its high WVP, which also degrades the coating's mechanical qualities. Consequently, hydrophobic components are frequently included during the film-making process (Kocira et al., 2021). To manage and extend the postharvest quality of *Cavendish* bananas during ripening at 20 °C, an edible covering made of rice starch and sucrose esters was created. The findings demonstrated that the coating was effective in delaying the manufacture of ethylene and reducing respiration rate. The coated fruit's shelf life was increased by 12 days in comparison to the control, which ripened in 7 days and showed no change after Day 6 (Thakur et al., 2019). To cover apples throughout a 15-day storage period, starch-based films, including citric pectin and Feijoa Peel Flour (FPF), were developed. The film successfully inhibited *Salmonella Typhimurium*, *P. aeruginosa*, and *Escherichia coli*. After five days, steady weight retention was noted, indicating that this packaging method helped to preserve the quality of the apples during storage (Sganzerla et al., 2020).

Alginate

Alginate is a naturally occurring polymer that is frequently extracted from algae. It is made up of linear, unbranched polymer chains of β-D-mannuronic acid and α-L-guluronic acid residues connected by one to four glycosidic linkages. When *Shirazi* thyme EO was added to alginate, fresh pistachios (*Pistacia vera* L.) had higher phenolic levels and antioxidant activity, as well as less mold and yeast growth (Hashemi et al., 2021a). Likewise, the combination of alginate and rhubarb extract suppressed *Penicillium expansum* while maintaining the physiological and sensory characteristics of coated peaches (*Prunus persica*) (Li et al., 2019).

Meanwhile, according to Kapetanakou et al., (2019) the growth of *A. carbonarius* on coated apple and pear slices was suppressed by sodium alginate combined with cinnamon essential oil (0.9%, v/v).

Gum

Gums are polysaccharides that can form a gel when they absorb water. They are inexpensive, safe, renewable, and biodegradable, and are used to create edible packaging. Gum Arabic (GA) edible coatings with Shirazi thyme were evaluated as possible postharvest methods to preserve the fresh pistachios' quality over 36 days at 2 °C. The findings demonstrated that the edible coating increased the PAL activity, total phenolics, free fatty acid content, and peroxide values while decreasing color alteration and PPO activity of the samples (Hashemi et al., 2021b). In another study, guava fruits were treated with GA (10%), oleic acid (1%), and cinnamon essential oil (1%). They were then kept in a cool environment for 28 days. By avoiding chilling damage, the combination treatment prevented weight loss, retained firmness, decreased browning, and retained quality. In addition to better total soluble solids (TSS) and TSS/tartaric acid (TA) ratios, coated fruits displayed increased amounts of bioactive substances (phenolics and flavonoids) and antioxidant activity (Etemadipoor et al., 2020). Similar findings were reported by Yang ZhiKun et al. (2019), who discovered that GA-based coatings combined with roselle extracts successfully safeguarded the quality of blueberries during freezing by increasing phenolic content and decreasing anthocyanin degradation and enzyme activity (POD and PPO).

Cellulose

The macromolecular polysaccharide cellulose is constructed of β-1,4 glycosidic bonds that join D-glucopyranose. Its excellent mechanical, thermal, and barrier qualities are combined with its non-hazardous, moisture-proof, and biodegradable nature. Modified cellulose, such as carboxymethyl cellulose (CMC), hydroxypropyl cellulose (HPC), and bacterial cellulose (BC), is typically used to create coatings since native cellulose is poorly soluble (Panahirad et al., 2021). For fresh-cut guava, (Forato et al., 2015) made edible coatings using cashew gum (CG) and CMC, employing glycerol as a plasticizer. The CMC/CG/glycerol coating was able to prolong the guava's shelf life, postpone the change in color and tissue deterioration, and decrease mass loss compared to the uncoated guava. The goal of the research by Nongnual et al. (2024) was to formulate crosslinked carboxymethyl cellulose (CMC) films with citric acid. These clear films served incredibly well to keep fresh bananas. They improved stiffness during storage, decreased weight loss, slowed ripening by preventing amyloplast breakdown, and preserved color.

Protein-Based Edible Packaging

Amino acids are joined by peptide bonds to form proteins. The core carbon of amino acids is joined to the amino, carboxyl, hydrogen, and radical groups. Naturally occurring substances, proteins can take the shape of globular or fibrous proteins (Akhila et al., 2022).

Soy Protein Isolate

Soy protein isolate (SPI) is a cheap globular protein, is biodegradable, and has strong film-forming and oil and oxygen barrier qualities. Its films are smooth-textured and bendable. Because of its high-water permeability, it can be deployed to coat fruits and vegetables and to stop the breakdown of lipids and transpiration of moisture in meat (Wu et al., 2024). Zareie et al. (2020) discovered that adding 2.5% chitosan to SPI-based films greatly enhanced their tensile strength (TS) and elongation at break (EB). They also found that the edible film surface was more fluid, denser, and had fewer pores. Liu et al. (2024) investigated how to improve the tensile strength and barrier qualities of soybean protein isolate (SPI) films for use in food packaging. The researchers greatly enhanced the films' mechanical characteristics, light-blocking capacity, water resistance, and barrier efficacy by combining SPI with camellia oil body-based oleogel (COBO). The SPI-COBO films showed excellent outcomes in avoiding a rise in total soluble solids, postponing weight loss, and maintaining the texture of strawberries throughout storage. The red dragon fruit peel (RDF) extract was added to the soy protein isolate (SPI) film matrix to create an active environmental packaging. When compared to the control film, the use of betalain-rich RDF extract greatly enhanced the film's antioxidant and UV-blocking properties (Prajapati et al., 2024).

Whey Protein Isolate (WPI)

It is commonly known that the casein coagulation procedure used by the dairy industry produces a significant amount of liquid waste as a by-product. This liquid, identified as dairy whey (DW), is a water-soluble protein that is yellow-green in color and is obtained during the cheese-making process after casein is extracted. There are several creative ways to turn the DW fraction into useful items. Using its protein content as a source of biopolymers to make edible films or coatings for food is one of them (Kandasamy et al., 2021). WGP is a hydrophobic protein made of a mixture of polypeptide units in the shape of a globular protein, which can generate hydrophobic associations with disulfide bonds to form sticky films and coatings as it contains -OH and -SH (Wang et al., 2021). Because of its extensibility, viscoelasticity, and anti-oxidation qualities, WGP films can be applied as a gas barrier over food surfaces to decrease oxygen and moisture permeability and increase food shelf life. However, its use in food packaging is restricted by its shortcomings, which include poor mechanical qualities, fragility, and low water resistance. WGP is frequently mixed with other substances to create edible films to get around these drawbacks.

Galus et al. (2021) evaluated how edible coatings made with WPI and essential oils (lemon and lemongrass) influenced the quality and sensory characteristics of fresh-cut pears. Pears coated with EO + WPI exhibited less variation in texture and color. Seiwert et al. (2021) prepared an edible film by adding xylan and transglutaminase to WPI. The results indicated that the films had r of fungal strain growth, a substantial rise in tensile stress and film strength, and an important reduction in WVP, oxygen permeability (OP), and elongation. Dong et al. (2022) altered WGP films by adding tartaric acid (TA), egg white protein (EWP), (CMC), and apple pectin (AP). The WGP/AP film had superior mechanical qualities with increased homogeneity. The film based on WGP/CMC had notable improvements in thermal stability. Furthermore, the EWP-based film had outstanding physical strength and notable improvements in transparency and light-shielding qualities. Finally, films based on WGP/TA had outstanding UV and water resistance. Additionally, WGP and birch powder were incorporated into films and coatings aimed at preserving fruits (Luo et al., 2022). These coatings demonstrated outstanding resistance to oxygen and water vapor, along with remarkable water-repellent, UV-blocking, antimicrobial, and degradable properties.

Gelatin

Gelatin, derived from collagen, is a transparent, scentless, water-soluble protein that is frequently utilized in edible films because of its affordability, excellent biocompatibility, and high safety. One of the drawbacks of natural gelatin is its limited antioxidant and antibacterial capabilities. Incorporating certain active components can enhance their antioxidant and antibacterial functions (Wu et al., 2024). Roy et Rhim (2020) developed an edible film made from gelatin and curcumin. The findings indicated that, in comparison to the standalone gelatin film, the composite film exhibited greater UV resistance, enhanced antioxidant properties, and a marked increase in antibacterial effectiveness against Escherichia coli and foodborne pathogens. The study of Kingwascharapong et al. (2020) examined the impact of a gelatin-based coating containing frog skin oil (FSO) on persimmon quality. The coated fruit exhibited a notable decrease in weight loss, pH, total soluble solids, and firmness.

Lipid-Based Edible Packaging

Waxes

Waxes consist of hydrophobic substances composed of long or medium chains of carbon atoms. They can be either natural or manufactured. Waxes are better water barriers than other lipids thanks to their excellent hydrophobicity, due to their high concentration of long-chain fatty acids, alcohols, esters, and alkanes (Syahida et al., 2020). Numerous studies have investigated applying wax coatings to different fruits and vegetables. When compared to uncoated fruits, sapodilla fruits with an edible coating made of chitosan and beeswax had 17 days longer shelf life (Pashova, 2023). The goal of the research by Formiga et al. (2019 was to increase the shelf life of red guavas known as "Pedro Sato" by 8-days, applying edible coatings of beeswax (BW) and HPMC. The HPMC+BW film delayed the ripening process, preserved the fruit's green hue, and improved firmness. Similar outcomes were found by Sousa et al. (2021) using the same edible coatings to prolong the freshness of "Palmer" mangoes. The coatings effectively preserved key qualities (color, firmness, nutritional content), oxidative stress, and anthracnose incidence..

Resins

Except for shellac, which is produced by female *Kerria lacca* on trees, resins are nonvolatile plant products that either naturally drip onto a plant's surface (surface resins) or can be obtained through cutting or infection (interior resins). In addition to being fusible and lacking sharp melting points, they are amorphous, inert, and get sticky when warmed (Devi et al., 2024). Shellac resin is composed of esters that contain both hydroxyl and carboxylic functional groups. It is a common coating substance because of its strong water resistance and barrier-forming qualities. Shellac and tannic acid were used to create composite coatings. According to Ma et al., the antifungal activity of mangoes was enhanced, lipid oxidation was decreased, and the postharvest storage period of mangoes was prolonged. Nevertheless, the food packaging industry has taken notice of gum rosin, a waste product of pine resin, as a natural additive (compatibilizers, stabilizers, plasticizers) with interesting qualities like an improvement in the matrix of viscoelastic and tensile properties for food packaging applications.

Fats and Oils

Despite their shared chemical structure, fats and oils exhibit distinct physical characteristics: fats remain solid at room temperature, whereas oils are liquid. Both are primarily made up of mixed triglycerides, which are created when one glycerol molecule bonds with three fatty acid molecules. Essential fatty acids commonly used in edible films and coatings include linolenic acid, oleic acid, and stearic acid. These fatty acids are employed as dispersants and emulsifiers to enhance the mechanical qualities of edible films and coatings, increase the stability of emulsions, and demonstrate a high resistance to water vapor (Nehchiri et al., 2021). A composite covering made of olive and ginger oils can prolong the preservation time of oranges and tomatoes and protect them from deterioration (Atta et al., 2022). Fangfang et al. created virgin coconut oil films made with potato starch that enhanced their water-repellent, mechanical, and antibacterial qualities.

On the whole, composite packaging is advantageous since it combines edible packaging materials to improve biodegradability, mechanical strength, and barrier qualities. These coatings or films fall into various categories, including protein-protein, polysaccharide-protein, protein-polysaccharide-lipid, and polysaccharide-polysaccharide combinations. Each kind makes use of the distinct qualities of its constituent parts to enhance packaging performance, increasing its efficacy for food preservation and sustainability.

Mechanisms of Food Protection by Edible Packaging

Barrier Properties

Moisture Barrier

Edible packaging acts as a protective barrier between food products and their surrounding environment, which often contains gases (such as oxygen) and water vapor (relative humidity), thus promoting the growth of food spoilage microorganisms. By forming a moisture barrier, edible films and coatings help mitigate this issue. Polysaccharide- and protein-based films and coatings are inherently hydrophilic, resulting in high water-vapor permeability and making them suitable for use in low relative humidity conditions (Cui et al., 2023). In contrast, films made from lipids and waxes are hydrophobic, effectively providing a water vapor barrier even in high-humidity environments. Additionally, these lipid-based films serve as an oxygen barrier (Yousuf et al., 2022). Polysaccharide- and protein-based films exhibit excellent gas barrier properties in low humidity conditions; however, as the humidity level increases, these materials tend to absorb moisture, consequently diminishing their oxygen barrier efficiency (Chen et al., 2024).

Oxygen Barrier

Enzymatic browning of fresh-cut produce, oxidation of lipids and food ingredients, and discoloration of myoglobin pigment in freshly cut meat pieces are the main causes of food degradation or spoiling (Hasan., 2022). Edible packaging with low oxygen permeability (OP) can be used for oxygen-sensitive foods, helping to maintain their quality and prolong their shelf life. Additionally, it aids in lowering the use of numerous pricey, high-end, non-recyclable, oxygen-blocking polymers. To have good barrier qualities, hydrophilic films and hydrocolloid-based films are utilized (Shah et al., 2024). Furthermore, advancements in the formulation of edible films have made it possible to create modified atmospheres, thereby reducing the respiration rate of horticultural products and the ethylene production of physiologically active climacteric produce during storage and distribution (Alonso-Salinas et al., 2024). Hydrocolloid-based films generally demonstrate exceptional gas barrier properties under low relative humidity conditions. In contrast, under high RH conditions, plasticized hydrocolloid-based films and hydrophilic ethylene vinyl alcohol films are preferred due to their enhanced barrier performance (Kumar et al., 2022).

Antimicrobial and Antioxidant Mechanisms

Among the most popular bioadditives for the production of functional films are secondary compounds with antimicrobial and antioxidant properties. By preventing bacterial development, antimicrobial drugs can be released gradually, extending food shelf life and maintaining food quality (Gutierrez et al., 2009). The active ingredients in active food packaging are processed to create biopolymers. The types of active agents vary depending on the intended use and how they interact chemically and/or physically with polymer matrixes (Guo et al., 2019). Furthermore, the primary goal of active food packaging is to prolong the shelf life of food by shielding it from microbial contamination. Generally speaking, the migration of active chemicals from the package to the food is the sole basis for active packaging inhibitory mechanisms (Tian et al., 2013). Depending on the food types and storage conditions, migration assays can identify the active agent movement at a specific temperature and time (Ribeiro-Santos et al., 2017). A variety of factors, including food contents (Zinoviadou et al., 2009), ambient factors such as temperature and humidity, packaging physicochemical characteristics, and thickness (Huang et al., 2013), can affect the migration rate and mechanisms. Essential oils (EOs), as secondary compounds, consist of various active natural metabolites with low molecular weights, including monoterpenes and sesquiterpenes. Due to their diverse chemical composition and functionality, EOs may exhibit varying behaviors when used as active agents in food packaging applications. Among the most extensively studied EOs for food packaging are rosemary, cinnamon (cinnamaldehyde), tea tree, lavender, thyme, lemon, and citrus oils. Additionally, extracts from blueberry (*Vaccinium corymbosum L.*), grape seed, and green tea have demonstrated strong inhibitory effects against key foodborne pathogens, including *Listeria monocytogenes*, *Staphylococcus* spp., *Salmonella enteritidis*, *Salmonella Typhimurium*, *Escherichia coli*, and *Campylobacter jejuni* (Perumalla et al., 2011; Shen et al., 2014). Cinnamon essential oil has demonstrated antimicrobial activity against *E. coli*, S. aureus, and S. fluorescence when incorporated into chitosan nanofiber and whey protein films for active packaging (Mohammadi et al., 2020). Another widely used EO in active food packaging is rosemary, derived from *Salvia rosmarinus or Rosmarinus officinalis*, a Mediterranean perennial plant. Encapsulating rosemary EO into carboxylmethylcellulose has shown antimicrobial and antioxidant effects, particularly against Pseudomonas spp. and lactic acid bacteria, as observed in rosemary-coated smoked eel packaging (Choulitoudi et al., 2017). The bioactivity of functional compounds is closely linked to their release from the polymer matrix, which largely depends on the physical interactions between the food and the packaging containing these compounds, as shown by Leelaphiwat et al. (2022). Furthermore, water diffusion from the food surface into the biopolymer network can increase the swelling of hydrophilic macromolecular chains, thereby enhance macromolecular mobility and facilitating the release of bioactive compounds (Serino et al., 2021).

Similar to antimicrobial effects, the antioxidant action mechanism involves the release and dispersion of antioxidants within the packaging medium. These antioxidants act by absorbing undesirable compounds, including oxygen, food-derived chemicals, and oxidative radicals, from both the food surface and the packaging environment (Oun et al., 2017). Antioxidant agents are primarily used in fresh foods, such as meat, fish, and fruits, as well as processed and raw food products. Regarding the antioxidant activity, fruits and vegetables are frequently subjected to microbial attack, primarily from phytopathogenic fungi, which affect their antioxidant capacity. By creating a natural barrier against bacterial flora, antifungal substances, including organic acids, different plant extracts, and essential oils, may increase the shelf life of fruits and vegetables after harvest (Sapper et al., 2018). Essential oils may enhance the hygroscopic properties of the polysaccharide-based film in addition to its antibacterial and antioxidant properties (Marín et al., 2017). Table 7.1 lists more of the research on the antioxidant and/or antibacterial qualities of other secondary chemicals.

Physical Protection and Mechanical Properties

Edible packaging, including films and coatings, serves as a protective layer for food products, shielding them from physical damage caused by mechanical factors such as compression and impact. To assess the mechanical strength of edible films, standard tests are employed, including tensile strength, tearing strength, elongation at break, and abrasion resistance. Compared to conventional plastic-based packaging, edible films generally exhibit lower tensile strength but may demonstrate better elongation properties. However, the permeation of unwanted particles and aroma compounds can compromise food quality and consumer acceptance (Ahmed et al., 2022). Edible films function as moisture barriers, minimizing nutrient migration between the food and the external environment. By limiting oxygen transfer, these films help slow down oxidation, thereby extending the food product's shelf life. However, high atmospheric moisture can negatively affect the film's physical properties, as moisture interacts with the edible polymer matrix, compromising tensile

Table 7.1 Natural products as antioxidant and antimicrobial agents in active biopackaging

Natural products	Packaging system	Role	Bioassay	References
Origanum vulgare L., *Rosmarinus officinalis* L.	Chitosan	Antimicrobial, antioxidant	*Bacillus subtilis, Escherichia coli*	Delgado Ospina et al. (2019)
Citrus sinensis	Gelatin	Antimicrobial, antioxidant	*S. aureus, Pseudomonas aeruginosa, Salmonella enterica, Listeria monocytogenes*	Jridi et al. (2019)
Licorice	Carboxymethyl xylan	Antimicrobial, antioxidant	*Enterococcus faecalis, L. monocytogenes*	Luís et al. (2019)
Carvacrol	Sodium alginate	Antimicrobial	*Trichoderma* sp.	Cheng et al. (2019)
Licorice essential oil	Zein-based films	Antimicrobial, antioxidant	*L. monocytogenes*	Luís et al. (2019)
Cinnamomum cassia, rosemary officinalis, Ocimum basilicum	Whey protein	Antimicrobial, antifungal, antioxidant	*E. coli, S. aureus, Penicillium*	Ribeiro-Santos et al. (2017)

strength, structural integrity, and barrier performance. Polysaccharide- and protein-based edible films are inherently hydrophilic, which makes them prone to reduced structural integrity under high-humidity conditions, thus diminishing their protective capacity. To counteract this, hydrophobic waxes, such as carnauba and candelilla, are often incorporated into the biopolymer matrix. These waxes effectively create a water vapor barrier, preserving the film's structural integrity even in humid environments (Cortés-Rodríguez et al., 2020).

Edible Coatings for Different Food Categories

Edible films and coatings are eco-friendly and biodegradable, aligning with the growing consumer demand for sustainable packaging solutions. As environmental awareness increases, the use of edible packaging has gained significant traction. These films and coatings are widely applied to various food products, including cheese, meat, fish, poultry, and fresh produce. By acting as selective barriers to water and oxygen, edible packaging effectively enhances the shelf life of food items compared to conventional plastic packaging. Table 7.2 highlights the diverse functions of edible films and coatings in food packaging applications (Kumar et al., 2022).

Table 7.2 Functional roles and properties of edible films and coatings for food protection

Edible film, coating for food	Function	Properties of edible film
	Sustainable	Edible and bio-based
		Reduces plastic waste
	Shelf life	More demand in the market
		Reduces food product waste
	Antimicrobial	Prevention of microbial growth
		Enhances food's shelf life
	Antioxidant	Prevents enzymatic browning
		More consumer acceptance
	Protection	Water and gas barrier
		Protection from light and ultraviolet light

Meat Products

Edible films and coatings offer a protective barrier for meat, seafood, and poultry products, which are highly perishable due to their high water content. These coatings help prevent shrinkage loss, control microbial growth, and reduce discoloration and oxidative off-flavors in meat products. Various edible biopolymers are utilized to coat meat products, providing both preservation and quality enhancement (Gaikwad et al., 2020). Sodium alginate, a natural polysaccharide biopolymer extracted from brown algae, is hydrophilic and exhibits excellent film-forming properties. It also offers good oxygen barrier capabilities. Initially, alginate films were used to control the respiration rate of fruits and vegetables, and more recently, they have been applied to meat products. When combined with active and antimicrobial agents, these edible films help enhance the shelf life of meat products (Kontominas et al., 2021). The U.S. Food and Drug Administration (FDA) has classified alginate as a generally recognized as safe (GRAS) substance, and the European Food Safety Authority (EFSA) has approved it for use on chicken breast, fish fillets, and beef muscle. However, due to its hydrophilic nature, alginate films have low moisture barrier properties. When exposed to water, the films absorb moisture and swell, which can compromise the integrity and effectiveness of the packaging. Nonetheless, this issue can be addressed through nanotechnology by incorporating montmorillonite clay and cellulose nanoparticles into the alginate matrix, which significantly enhances its water barrier properties. To improve the flexibility of the edible film, plasticizers such as glycerol or sorbitol are added. Additionally, essential oils like oregano, thyme, cinnamon, and rosemary are incorporated as antimicrobial agents to prevent the growth of microorganisms such as *Escherichia coli*, *Enterobacteriaceae*, and *Salmonella Typhimurium*, thereby extending the shelf life of meat products. This multifunctional edible packaging film serves as a sustainable alternative to conventional nondegradable polymer films and represents the future of the food packaging industry (Gheorghita et al., 2020). Seafood is a highly perishable product with a short shelf life. During transportation and storage, seafood is susceptible to contamination from the surrounding environment, which can lead to foodborne illnesses and a decline in both quality and nutritional value, ultimately reducing consumer acceptance (Dehghani et al., 2018). The postharvest processing of shrimp is often done manually, increasing the risk of contamination by *Listeria monocytogenes.* The addition of antimicrobial agents in catfish-based gelatin coatings has been shown to inhibit microbial growth and extend the shelf life of fresh white shrimp by 10 days. Similarly, gelatin-based coatings enriched with cinnamon as an active antimicrobial agent help slow microbial growth and prolong the shelf life of fresh rainbow trout (Andevari & Rezaei, 2011). Furthermore, whey protein isolate-based edible coatings containing clove essential oil as an active agent have been found to enhance quality and extend the shelf life of chicken breast fillets (Fernández-Pan et al., 2014).

Fruits and Vegetables

Fruits and vegetables are key components of a healthy diet, and consumers today have higher expectations for their food—demanding not only safety and nutritional value but also variety and extended shelf life. For the food industry, preserving the quality of minimally processed or cut fruits and vegetables is crucial. Factors such as moisture loss, microbial growth, respiration, and ethylene production significantly contribute to the deterioration of these products (Kumar et al., 2021). Aloe vera gel (AVG), a natural biopolymer primarily composed of polysaccharides, has been explored as a solution to help maintain product quality.

Aloe vera gel (AVG) acts as a semipermeable membrane, regulating water and oxygen transfer. It reduces the respiration rate of fruits and vegetables, helping to maintain their weight and overall quality. With its potent antioxidant and antimicrobial properties, this edible coating enhances the shelf life of produce. When applied, the coating reduces the oxygen available for the metabolism of the fruit or vegetable, thereby delaying ripening and aging, which extends their shelf life. Moreover, even if the coating is consumed, it poses no harm to human health. Aloe vera gel's natural antimicrobial properties help inhibit the growth of microorganisms, further preventing spoilage and improving the shelf life of fruits and vegetables (Nicolau-Lapeña et al., 2021).

Applying aloe vera gel (AVG) to freshly cut papaya stored at 28 °C effectively minimizes weight loss while also preserving the fruit's firmness (Mendy et al., 2019). Similarly, a coating composed of aloe vera gel, carnauba wax, and 1% glycerol applied to fresh-cut mango stored at 40 °C helps maintain firmness and reduces weight loss (Dussán-Sarria et al., 2023). Additionally, the use of chitosan nanoparticle-based edible coatings significantly enhances the appearance and extends the shelf life of strawberries (Abu and Gedanken, 2021). In the case of cucumbers, coating with wax (a combination of shellac and carnauba wax) effectively limits moisture loss (Li et al., 2018). Furthermore, when aloe vera gel is applied to tomatoes and stored at 11 °C, the fruit retains its firmness within

Table 7.3 Impact of various edible/active packaging on the preservation of fruits and vegetables

Edible packaging	Fruits and vegetable	Functions	References
Arabic gum	Tomato	Reduction in respiration rate and ethylene emission while preserving ascorbic acid and other quality attributes	Kurubas et al. (2022)
Aloe vera gel, chitosan, calcium chloride	Mango	Preservation of ascorbic acid and suppression of polyphenol oxidase enzymatic activity	Hajebi Seyed et al. (2021)
Sodium alginate, calcium chloride	Apple	Mitigation of chill injury, decreased respiration rate and physiological weight loss, and enhanced antioxidant activity during storage	Duong et al. (2022)
Chitosan, *Prosopis juliflora* water-soluble leaf ethanolic	Strawberries	Lowering of surface microbial count	Saleh et al. (2022)
Chitosan, tea tree oil, calcium chloride	Red bell pepper	The coating demonstrated efficacy against *salmonella enterica* and *listeria monocytogenes*, while preserving flavonoid content and maintaining antimicrobial activity	Sathiyaseelan et al. (2021)

acceptable limits even after 14 days of storage (Chrysargyris et al., 2016). Table 7.3 lists more research on the impact of edible/active packaging on the preservation of fruits and vegetables.

Innovative Approaches and Emerging Technologies

Active and Smart Edible Packaging

Nowadays, the formulation of edible packaging with natural antibacterial and antioxidant compounds can significantly increase food items' shelf life by reducing oxidation while maintaining better quality characteristics (Fig. 7.2). To improve food preservation, EU Commission Regulation No. 450/2009 allows the purposeful use of active substances in food packaging to facilitate the release or absorption of compounds. Active-FilmTM, BIOPAC, ATCO, ATOX, ZERO2, Biomaster, CeloxTM, Food-touch, SANDRY, FreshPax, Pure Temp, PEAKfresh, and Green Box are examples of readily accessible active packaging (Drago et al., 2020). These active packaging methods contribute to food safety by reducing moisture loss, decreasing water vapor transmission rates, inhibiting the creation of ethylene, and strengthening antibacterial and antioxidant activities. Naturally occurring active substances, such as plant extracts and essential oils, can be added to edible coatings and films to enhance their useful properties. They reduce oxidative stress and the

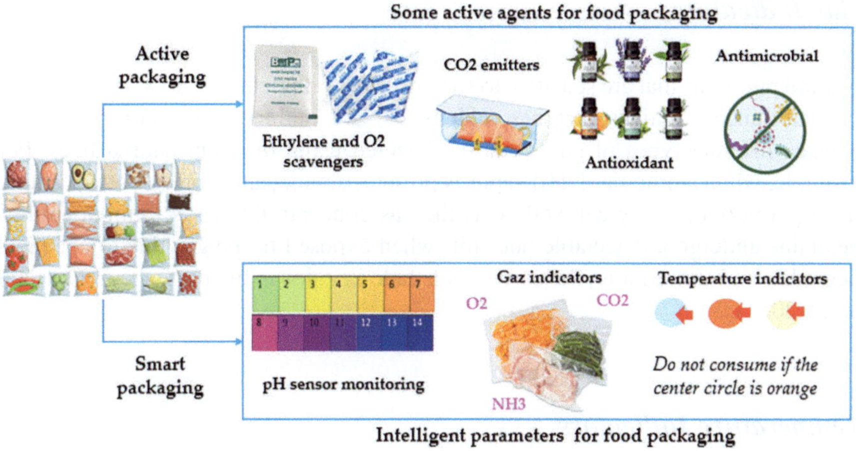

Fig. 7.2 Classification of active packaging systems and common indicators in smart packaging

spread of bacteria by acting as antioxidants and/or antibacterials. The membranes inside and outside of microbial cells are damaged and disrupted when active chemicals are added (Kumar et al., 2023). There are two forms of active packaging: sachets or pads placed inside packages, and ingredients mixed into packaging materials. It works by changing the atmosphere in the package, making it an advanced packaging system..

In the case of a systemic change in the packaged product (alteration in the quality, safety, or maturity), smart packaging components are used to indicate such modifications. By providing novel techniques for guaranteeing safety and real-time product quality monitoring for perishable commodities, the incorporation of indicators into smart packaging has completely transformed the food sector. These cutting-edge technologies give customers and distributors information on food conditions, including vitality, gas presence, and environmental variables, allowing them to efficiently track product integrity across the supply chain (D'Almeida et al., 2024).

pH Indicators

When the pH of a packaged product significantly changes, this type of indicator reveals the change in a measured way. Enzymatic activities, chemical interactions, microbial development, or the presence of volatile chemicals in the packaged food cause these pH shifts. A proton (H+) exchange across the pigments and their surroundings is the principle of this colorimetric pH sensor. Numerous botanical sources, such as anthocyanins, carotenoids, betalains, and chlorophylls, have been used to create pH-sensitive indicators. The molecular composition of these natural pigments and extrinsic variables like temperature, oxygen concentration, and light contact affect their reaction to changes in pH (Sani et al., 2021).

Gas Indicators

Natural pigments that are sensitive to various gases (ethylene, O_2, CO_2, H_2S) or volatile substances (amines, ketones, aldehydes) are included in smart packaging. Hence, different types of gases can be detected using different natural hues. For example, when ammonia (NH_3) vapor is present, anthocyanins break down, turning from purple/violet to green/yellow as the gas concentration rises. Similar to this, betalains undergo a detectable hue shift when exposed to oxygen due to chemical degradation. Furthermore, the carotenoids fade in color in the presence of O_2 (Sani et al., 2021).

Temperature Indicators

By determining whether packaged goods have been exposed to high temperatures for extended periods of time, time-temperature indicators (TTIs) are typically used to monitor the safety and quality of those foods throughout storage and transportation. They can be categorized into different groups based on their functions: Critical temperature/time integrators (CTTI) provide information on whether a food product has been heated above or cooled below a specific temperature for an extended period of time; temperature-time indicators (TTIs) provide the full temperature versus time profile of a food product over its lifetime; and critical temperature indicators (CTI) provide information on whether a food product has been heated above, or cooled below a specified temperature over its lifetime.

Nanomaterials and Bio-Nanocomposites

Nanotechnology has greatly influenced food packaging. Soon, it is anticipated that 25% of all food packaging will be made of nanopackaging. Because nanoparticles range in size from 1 to 100 nm, they have improved biological efficacy, ability to adsorb, and a surface-to-volume ratio (Bhavya et al., 2024). Bio-nanocomposites are generated from biobased polymers serving as a matrix and nanoparticles or nanofibers as reinforcement agents to enhance thermal and mechanical features, flexibility, gas barrier performance, biocompatibility, biodegradability, environmental friendliness, and cost efficiency. Thus, nano-packaging improves UV barriers, fixes packaging tears, and lessens food waste and shortages to benefit consumer health (Siciliano et al., 2024). Various types of nanomaterials (organic and inorganic) have been employed in nanopackaging to extend the main properties of the packaged food.

Organic Nanomaterials

Nanocellulose

Nanocellulose is generated by cell-free systems, different microbial strains, and via the breakdown of vegetal and animal cellulose. As a tunable substance, it is frequently used in food packaging, either alone or in combination with other materials. 3D reticular network fiber structure, hydrophilicity, extremely high crystallinity, transmission of light, a large specific surface area, multifunctionality, and the capacity to be molded into intricate 3D shapes are the main potential properties of nanocellulose. It is also amenable to a variety of chemical and physical modifications (Gulati et al., 2024).

Nanochitosan

Nanochitosan enhances the physicochemical attributes of chitosan and boosts the coating's barrier properties and functionality due to the greater surface area provided by the nanoparticles. These nanoparticles' unique characteristics that promote emulsification and crosslinking allow them to form an ideal smart delivery system. Chitosan nanoparticles are known for their strong antimicrobial properties, biosafety, accessibility, biological activity, polycationicity, high adsorptive ability, anticancer activity, immunostimulatory properties, and sustainability. Food packaging applications for foodborne pathogens are facilitated by the functionalization of the nanoparticles by the positive amino and the negative hydroxyl bonds of chitosan (Karthik et al., 2022).

Nanostarch

The components of starch nanoparticles are either amorphous or alternate crystalline-and-amorphous regions. They are new materials with distinct biological and physicochemical characteristics from starch biopolymers. They also have a high rate of biological access, dissolution, reaction surface, and absorption capacity. Chemical, physical, or enzymatic techniques, including enzyme debranching, thermosonication, nanoprecipitation, milling, and ultrasonication, can be used to develop starch nanoparticles (Perera et al., 2022). The growing use of nano-starch-based packaging is attributed to its edible, sustainable, and affordable nature, demonstrating high antimicrobial and antioxidant activities, which helped to increase the nutritive value, quality, and storage period of packaged products (Bhavya et al., 2024).

Inorganic Nanomaterials

Metal Oxide–Based Nanomaterials

Metal oxide nanoparticles are structures at the nanoscale that consist of metal atoms linked to oxygen molecules. The extraordinary properties of metal oxide nanoparticles are due to their smaller size compared to those of microscopic particles. Metal oxide nanocomposites have an enormous potential to provide superior performance, renewability, sustainability, and other parameters in food packaging applications. Thus, it would contribute to the retention of the quality and reliability of packed food products (Nandhini et al., 2024).

Nanoclays

Nanoclays, a form of refined clay, consist of layered mineral silicates piled up together, and they are now regarded as the most commercially applicable nanomaterials all around the world. They have been recognized as economical nanofillers to enhance the mechanical, thermal, and barrier characteristics of polymer nanocomposites for food packaging. At present, multiple clay minerals, which are grouped into four primary classes based on their structure, chemical features, and sources, include chlorite, MMT/smectite, illite, and kaolinite (Ahari et al., 2021).

Nanohybrids

Materials that combine inorganic nanoparticles (metal, metal oxides) and organic nanomaterials are known as nanohybrid nanocomposite films. The disadvantages of individual nanoparticles can be solved by nanohybrids. The most common combinations of nanoparticles are inorganic–inorganic, organic–organic, or organic–inorganic. Numerous techniques, including microwave irradiation, laser ablation, electrochemical synthesis, electrospinning, mechanical synthesis, chemical synthesis, and biological synthesis, are used to create hybrid nanoparticles (Sant'anna Andrade et al., 2024). Recent advancements in bio-nanocomposite films incorporating nanohybrids are highlighted in Table 7.4.

Conclusion

Edible packaging has emerged as a promising and sustainable alternative to conventional synthetic polymer-based packaging, offering a practical solution to both environmental challenges and food protection needs. By utilizing natural, biodegradable materials such as polysaccharides, lipids, and proteins, edible films and coatings provide functional benefits that extend the shelf life of food products. These benefits

Table 7.4 Application of different nanomaterial types in food packaging

Nanomaterials	Packaging metrics	Propreties of bio-nanocomposite film	References
Organic nanomaterials			
Nanocellulose	Purple sweet potato anthocyanin/oregano EO	Inhibited *E. coli* and *L. monocytogenes*	Chen et al. (2020)
		Improved protection against ultraviolet and visible light	
	Chitosan/Curcumin	Improved antibacterial activity, oxidation resistance, crystallinity, and UV blocking	Zhang et al. (2021)
Nanochitosan	Methylcellulose/barberryanthocyanins	Improved water barrier properties	Alizadeh-Sani et al. (2021)
		Reduced UV–vis light transmittance	
		pH-monitoring color indicator for real-time quality tracking	
	ε-polylysine	Boosted moisture content and water solubility	Lin et al. (2018)
		Prevented S. typhimurium and S. enteritidis	
Nanostarch	Amaranth protein isolate	Enhanced Young's modulus and tensile strength	Condes et al. (2018)
		Decreased elongation at break reduced	
		Reduced water solubility from 80% to 40%	
	Mango kernelstarch	Increased the tensile strength by and Young's modulus by 90% and 120% respectively	Oliveira et al. (2020)
Inorganic nanomaterials			
Metal based nanomaterials			
Silver nanoparticles (AgNPs)	Oligomeric lactic acid/poly lactic acid	Enhanced antimicrobial activity, mechanical, and thermal features	Sonseca et al. (2019)
		Biodegradability after 28 days	
Sulfur NanoParticules (SNPs)	Alginate	Upgraded the tensile and water vapor barrier characteristics by 12% and 41%, respectively,	Priyadarshi et al. (2021)
		Complete bactericidal effect was observed for *L. monocytogenes* in 12 hours	

(continued)

Table 7.4 (continued)

Nanomaterials	Packaging metrics	Propreties of bio-nanocomposite film	References
Metal oxide–based nanomaterials			
Zinc oxide NPs (ZnONPs)	Chitosan-gelatin polymer	Showed antimicrobial activity against both Gram-negative and -positive bacteria	Kumar et al. (2020)
		Decreased slightly while boosted elongation at break	
Silicon dioxide NPs (SiO2NPs)	Chitosan and D-α-tocopheryl polyethylene glycol 1000 succinate	Improved free radical scavenging activity and enahanced antimicrobial activity	Bi et al. (2019)
		Increased the oxidative stability	
Nanoclays			
Montmorillonite nanoclay	Pectin-carum capsicum essential oils-β-carotene	Demonstrated high flexibility and firmness	Asdagh and Pirsa (2020)
		Increased the oxidative stability of butter during storage	
Graphene oxide (GO)	Poly (D- glucosamine)	Improved tensile strength, thermal stability, and electrical conductivity	Shekhar et al. (2020)
Nanohybrids			
Zn NPs-MgO-NPS	Alginate	Inhibited L. monocytogenes in packed cold salmon (4 days)	Vizzini et al. (2020)
Nanocrystalline cellulose- chitin-whiskers	Poly lactic acid	Increase tensile strength, mechanical, and barriercharacteristics	Xu et al. (2020)
		Lowered oxygen transmission	
Ag NPs-cellulosenanocrystals	CMC	Extend the shelf life of strawberries up to 7 days	He et al. (2021)
		Showed antibacterial activities against *E. coli* and *S. aureus*	

include antimicrobial and antioxidant properties, moisture and gas barrier functions, and enhanced mechanical strength. The efficacy of edible packaging materials largely depends on the inherent characteristics of the film-forming biopolymers. Composite edible packaging, including binary and ternary systems, shows even greater potential by maintaining freshness, color, vitamins, minerals, firmness, and other sensory and nutritional attributes. This is achieved through barrier properties that reduce water transpiration and gas transmission. Additionally, incorporating bioactive compounds like essential oils and plant extracts imparts natural antimicrobial properties, inhibiting the growth of harmful microorganisms. To build a more sustainable and resilient food system, replacing traditional synthetic polymer-based coatings with biodegradable edible films is essential. Continued interdisciplinary collaboration and integration of scientific innovations with practical applications will be crucial in advancing the development and adoption of edible packaging, aligning with global sustainability goals and consumer preferences.

References

Abu Salha, B., & Gedanken, A. (2021). Extending the shelf life of strawberries by the sonochemical coating of their surface with nanoparticles of an edible anti-bacterial compound. *Applied Nano, 2*(1), 14–24.

Adiletta, G., Di Matteo, M., & Petriccione, M. (2021). Multifunctional role of chitosan edible coatings on antioxidant systems in fruit crops: A review. *International Journal of Molecular Sciences, 22*(5), 2633.

Ahari, H., Anvar, A. A., Ataee, M., & Naeimabadi, M. (2021). Employing nanosilver, nanocopper, and nanoclays in food packaging production: A systematic review. *Coatings, 11*(5), 509.

Ahmed, M. W., Haque, M. A., Mohibbullah, M., Khan, M. S. I., Islam, M. A., Mondal, M. H. T., & Ahmmed, R. (2022). A review on active packaging for quality and safety of foods: Current trends, applications, prospects and challenges. *Food Packaging and Shelf Life, 33*, 100913.

Akhila, V., & Badwaik, L. S. (2022). Recent advancement in improvement of properties of polysaccharides and proteins based packaging film with added nanoparticles: A review. *International Journal of Biological Macromolecules, 203*, 515–525.

Alizadeh-Sani, M., Tavassoli, M., Mohammadian, E., Ehsani, A., Khaniki, G. J., Priyadarshi, R., & Rhim, J. W. (2021). pH-responsive color indicator films based on methylcellulose/chitosan nanofiber and barberry anthocyanins for real-time monitoring of meat freshness. *International Journal of Biological Macromolecules, 166*, 741–750.

Alonso-Salinas, R., López-Miranda, S., Pérez-López, A. J., & Acosta-Motos, J. R. (2024). Strategies to delay ethylene-mediated ripening in climacteric fruits: Implications for shelf life extension and postharvest quality. *Horticulturae, 10*(8), 840.

Andevari, G. T., & Rezaei, M. (2011). Effect of gelatin coating incorporated with cinnamon oil on the quality of fresh rainbow trout in cold storage. *International Journal of Food Science and Technology, 46*(11), 2305–2311.

Asdagh, A., & Pirsa, S. (2020). Bacterial and oxidative control of local butter with smart/active film based on pectin/nanoclay/Carum copticum essential oils/β-carotene. *International Journal of Biological Macromolecules, 165*, 156–168.

Atta, O. M., Manan, S., Ul-Islam, M., Ahmed, A. A. Q., Ullah, M. W., & Yang, G. (2022). Development and characterization of plant oil-incorporated carboxymethyl cellulose/bacterial cellulose/glycerol-based antimicrobial edible films for food packaging applications. *Advanced Composites and Hybrid Materials*, 1–18.

Bao, H., Wang, Y., Huang, Y., Zhang, Y., & Dai, H. (2025). The beneficial role of polysaccharide hydrocolloids in meat products: A review. *Gels, 11*(1), 55.

Bhavya, E. P., & Raman, M. (2024). Comprehensive review on synthesis of nano starch and its applications in food packaging industry. *Journal of Packaging Technology and Research*, 1–14.

Bi, S., Zhang, L., & Li, C. (2019). Multifunctional films with a highly oriented "nano-brick wall" structure by regulating modified TiO 2@ graphene oxide/poly (vinyl alcohol) nanocomposites. *Nanoscale, 11*(15), 7424–7432.

Chaudhary, S., Kumar, S., Kumar, V., & Sharma, R. (2020). Chitosan nanoemulsions as advanced edible coatings for fruits and vegetables: Composition, fabrication and developments in last decade. *International Journal of Biological Macromolecules, 152*, 154–170.

Chen, S., Wu, M., Lu, P., Gao, L., Yan, S., & Wang, S. (2020). Development of pH indicator and antimicrobial cellulose nanofibre packaging film based on purple sweet potato anthocyanin and oregano essential oil. *International Journal of Biological Macromolecules, 149*, 271–280.

Chen, K., Tian, R., Jiang, J., Xiao, M., Wu, K., Kuang, Y., et al. (2024). Moisture loss inhibition with biopolymer films for preservation of fruits and vegetables: A review. *International Journal of Biological Macromolecules, 263*, 130337.

Cheng, M., Wang, J., Zhang, R., Kong, R., Lu, W., & Wang, X. (2019). Characterization and application of the microencapsulated carvacrol/sodium alginate films as food packaging materials. *International Journal of Biological Macromolecules, 141*, 259–267.

Choulitoudi, E., Ganiari, S., Tsironi, T., Ntzimani, A., Tsimogiannis, D., Taoukis, P., & Oreopoulou, V. (2017). Edible coating enriched with rosemary extracts to enhance oxidative and microbial stability of smoked eel fillets. *Food Packaging and Shelf Life, 12*, 107–113.

Chrysargyris, A., Nikou, A., & Tzortzakis, N. (2016). Effectiveness of Aloe vera gel coating for maintaining tomato fruit quality. *New Zealand Journal of Crop and Horticultural Science, 44*(3), 203–217.

Condés, M. C., Añón, M. C., Dufresne, A., & Mauri, A. N. (2018). Composite and nanocomposite films based on amaranth biopolymers. *Food Hydrocolloids, 74*, 159–167.

Cortés-Rodríguez, M., Villegas-Yépez, C., González, J. H. G., Rodríguez, P. E., & Ortega-Toro, R. (2020). Development and evaluation of edible films based on cassava starch, whey protein, and bees wax. *Heliyon, 6*(9).

Cui, C., et al. (2023). Hydrophobic biopolymer-based films: Strategies, properties, and food applications. *Food Engineering Reviews, 15*(2), 360–379.

D'Almeida, A. P., & de Albuquerque, T. L. (2024). Innovations in food packaging: From bio-based materials to smart packaging systems. *Processes, 12*(10), 2085.

de Oliveira, K. Á. R., da Conceição, M. L., de Oliveira, S. P. A., Lima, M. D. S., de Sousa Galvão, M., Madruga, M. S., et al. (2020). Postharvest quality improvements in mango cultivar Tommy Atkins by chitosan coating with Mentha piperita L. essential oil. *The Journal of Horticultural Science and Biotechnology, 95*(2), 260–272.

Dehghani, S., Hosseini, S. V., & Regenstein, J. M. (2018). Edible films and coatings in seafood preservation: A review. *Food Chemistry, 240*, 505–513.

Delgado Ospina, J., Grande, C. D., Monsalve, L. V., Advíncula, R. C., Herminsul Mina, J., Valencia, M. E., … & Rodrigues, D. (2019). Evaluation of the chitosan films of essential oils from Origanum vulgare L (oregano) and Rosmarinus officinalis L (rosemary).

Devi, L. S., Jaiswal, A. K., & Jaiswal, S. (2024). Lipid incorporated biopolymer based edible films and coatings in food packaging: A review. *Current Research in Food Science, 8*, 100720.

Dinika, I., Verma, D. K., Balia, R., Utama, G. L., & Patel, A. R. (2020). Potential of cheese whey bioactive proteins and peptides in the development of antimicrobial edible film composite: A review of recent trends. *Trends in Food Science & Technology, 103*, 57–67.

Dong, M., Tian, L., Li, J., Jia, J., Dong, Y., Tu, Y., et al. (2022). Improving physicochemical properties of edible wheat gluten protein films with proteins, polysaccharides and organic acid. *LWT—Food Science and Technology, 154*, 112868.

Drago, E., Campardelli, R., Pettinato, M., & Perego, P. (2020). Innovations in smart packaging concepts for food: An extensive review. *Food, 9*(11), 1628.

Duong, N. T. C., Uthairatanakij, A., Laohakunjit, N., Jitareerat, P., & Kaisangsri, N. (2022). An innovative single step of cross-linked alginate-based edible coating for maintaining postharvest quality and reducing chilling injury in rose apple cv. Tabtimchan (Syzygium samarangenese). *Scientia Horticulturae, 292*, 110648.

Duran, A., & Kahve, H. I. (2020). The effect of chitosan coating and vacuum packaging on the microbiological and chemical properties of beef. *Meat Science, 162*, 107961.

Dussán-Sarria, S., Barragán-Coral, S. D., & Buitrago-Dueñas, E. M. (2023). Efecto de un recubrimiento comestible a base de aloe vera en piña (Ananas comosus) Oro Miel mínimamente procesada. *Información tecnológica, 34*(1), 11–20.

Eranda, D. H. U., Chaijan, M., Panpipat, W., Karnjanapratum, S., Cerqueira, M. A., & Castro-Muñoz, R. (2024). Gelatin-chitosan interactions in edible films and coatings doped with plant extracts for biopreservation of fresh tuna fish products: A review. *International Journal of Biological Macromolecules, 280*, 135661.

Etemadipoor, R., Dastjerdi, A. M., Ramezanian, A., & Ehteshami, S. (2020). Ameliorative effect of gum arabic, oleic acid and/or cinnamon essential oil on chilling injury and quality loss of guava fruit. *Scientia Horticulturae, 266*, 109255.

Fernández-Pan, I., Carrión-Granda, X., & Maté, J. I. (2014). Antimicrobial efficiency of edible coatings on the preservation of chicken breast fillets. *Food Control, 36*(1), 69–75.

Forato, L. A., de Britto, D., de Rizzo, J. S., Gastaldi, T. A., & Assis, O. B. (2015). Effect of cashew gum-carboxymethylcellulose edible coatings in extending the shelf-life of fresh and cut guavas. *Food Packaging and Shelf Life, 5*, 68–74.

Formiga, A. S., Junior, J. S. P., Pereira, E. M., Cordeiro, I. N., & Mattiuz, B. H. (2019). Use of edible coatings based on hydroxypropyl methylcellulose and beeswax in the conservation of red guava 'Pedro Sato'. *Food Chemistry, 290*, 144–151.

Gaikwad, K. K., Singh, S., Negi, Y. S., & Lee, Y. S. (2020). The effect of trans-polyisoprene/LDPE based active films on oxidative stability in roasted peanuts. *Journal of Food Measurement and Characterization, 14*, 1857–1864.

Galus, S., Mikus, M., Ciurzyńska, A., Domian, E., Kowalska, J., Marzec, A., & Kowalska, H. (2021). The effect of whey protein-based edible coatings incorporated with lemon and lemongrass essential oils on the quality attributes of fresh-cut pears during storage. *Coatings, 11*(7), 745.

Gheorghita, R., Gutt, G., & Amariei, S. (2020). The use of edible films based on sodium alginate in meat product packaging: An eco-friendly alternative to conventional plastic materials. *Coatings, 10*(2), 166.

Gulati, S., Amar, A., Chhabra, L., Katiyar, R., Sahu, T., & Varma, R. S. (2024). Greener nanobiopolymers and nanoencapsulation: Environmental implications and future prospects. *RSC Sustainability, 2*(10), 2805–2832.

Guo, Y., Chen, X., Yang, F., Wang, T., Ni, M., Chen, Y., et al. (2019). Preparation and characterization of chitosan-based ternary blend edible films with efficient antimicrobial activities for food packaging applications. *Journal of Food Science, 84*(6), 1411–1419.

Gutierrez, J., Barry-Ryan, C., & Bourke, P. (2009). Antimicrobial activity of plant essential oils using food model media: Efficacy, synergistic potential and interactions with food components. *Food Microbiology, 26*(2), 142–150.

Hajebi Seyed, R., Rastegar, S., & Faramarzi, S. (2021). Impact of edible coating derived from a combination of Aloe vera gel, chitosan and calcium chloride on maintain the quality of mango fruit at ambient temperature. *Journal of Food Measurement and Characterization, 15*, 2932–2942.

Hamed, I., Jakobsen, A. N., & Lerfall, J. (2022). Sustainable edible packaging systems based on active compounds from food processing byproducts: A review. *Comprehensive Reviews in Food Science and Food Safety, 21*(1), 198–226.

Hasan, M. M. (2022). Biochemical study of rapid discolouration mechanisms in bison meat.

Hashemi, M., Dastjerdi, A. M., Mirdehghan, S. H., Shakerardekani, A., & Golding, J. B. (2021a). Incorporation of Zataria multiflora Boiss essential oil into gum Arabic edible coating to maintain the quality properties of fresh in-hull pistachio (Pistacia vera L.). *Food Packaging and Shelf Life, 30*, 100724.

Hashemi, M., Dastjerdi, A. M., Shakerardekani, A., & Mirdehghan, S. H. (2021b). Effect of alginate coating enriched with Shirazi thyme essential oil on quality of the fresh pistachio (Pistacia vera L.). *Journal of Food Science and Technology, 58*, 34–43.

He, Y., Li, H., Fei, X., & Peng, L. (2021). Carboxymethyl cellulose/cellulose nanocrystals immobilized silver nanoparticles as an effective coating to improve barrier and antibacterial properties of paper for food packaging applications. *Carbohydrate Polymers, 252*, 117156.

Hedegaard, R. V., & Skibsted, L. H. (2024). Shelf-life of food powders. In *Handbook of food powders* (pp. 335–354). Woodhead Publishing.

Huang, X. L., Wang, Z. W., Hu, C. Y., Zhu, Y., & Wang, J. (2013). Factors affecting migration of contaminants from paper through polymer coating into food simulants. *Packaging Technology and Science, 26*, 23–31.

Jridi, M., Boughriba, S., Abdelhedi, O., Nciri, H., Nasri, R., Kchaou, H., et al. (2019). Investigation of physicochemical and antioxidant properties of gelatin edible film mixed with blood orange (Citrus sinensis) peel extract. *Food Packaging and Shelf Life, 21*, 100342.

Jurić, M., Bandić, L. M., Carullo, D., & Jurić, S. (2024). Technological advancements in edible coatings: Emerging trends and applications in sustainable food preservation. *Food Bioscience, 58*, 103835.

Kandasamy, S., Yoo, J., Yun, J., Kang, H. B., Seol, K. H., Kim, H. W., & Ham, J. S. (2021). Application of whey protein-based edible films and coatings in food industries: An updated overview. *Coatings, 11*(9), 1056.

Kapetanakou, A. E., Nestora, S., Evageliou, V., & Skandamis, P. N. (2019). Sodium alginate–cinnamon essential oil coated apples and pears: Variability of Aspergillus carbonarius growth and ochratoxin A production. *Food Research International, 119*, 876–885.

Karthik, C., Caroline, D. G., & Pandi Prabha, S. (2022). Nanochitosan augmented with essential oils and extracts as an edible antimicrobial coating for the shelf life extension of fresh produce: A review. *Polymer Bulletin, 79*(10), 8009–8032.

Kasaai, M. R. (2025). *Oxidative and hydrolytic deteriorations of lipids and several alternative pathways for their protections: An overview. Food Nutrition Chemistry* 3(1): 238

Kingwascharapong, P., Arisa, K., Karnjanapratum, S., Tanaka, F., & Tanaka, F. (2020). Effect of gelatin-based coating containing frog skin oil on the quality of persimmon and its characteristics. *Scientia Horticulturae, 260*, 108864.

Kocira, A., Kozłowicz, K., Panasiewicz, K., Staniak, M., Szpunar-Krok, E., & Hortyńska, P. (2021). Polysaccharides as edible films and coatings: Characteristics and influence on fruit and vegetable quality—A review. *Agronomy, 11*(5), 813.

Kontominas, M. G., Badeka, A. V., Kosma, I. S., & Nathanailides, C. I. (2021). Recent developments in seafood packaging technologies. *Foods, 10*(5), 940.

Kumar, S., Mudai, A., Roy, B., Basumatary, I. B., Mukherjee, A., & Dutta, J. (2020). Biodegradable hybrid nanocomposite of chitosan/gelatin and green synthesized zinc oxide nanoparticles for food packaging. *Food, 9*(9), 1143.

Kumar, A., Gupta, V., Singh, S., Saini, S., & Gaikwad, K. K. (2021). Pine needles lignocellulosic ethylene scavenging paper impregnated with nanozeolite for active packaging applications. *Industrial Crops and Products, 170*, 113752.

Kumar, L., Ramakanth, D., Akhila, K., & Gaikwad, K. K. (2022). Edible films and coatings for food packaging applications: A review. *Environmental Chemistry Letters, 20(1)*, 826–875.

Kumar, N., Pratibha, Prasad, J., Yadav, A., Upadhyay, A., Neeraj, et al. (2023). Recent trends in edible packaging for food applications-perspective for the future. *Food Engineering Reviews, 15*(4), 718–747.

Kurubas, M. S., Yildirim, I. K., Ali, Q., & Erkan, M. (2022). Maintenance shelf-life quality of cocktail tomatoes by using UV-C illumination and Arabic gum coating. *Journal of the Science of Food and Agriculture, 102*(9), 3897–3907.

Leelaphiwat, P., Pechprankan, C., Siripho, P., Bumbudsanpharoke, N., & Harnkarnsujarit, N. (2022). Effects of nisin and EDTA on morphology and properties of thermoplastic starch and PBAT biodegradable films for meat packaging. *Food Chemistry, 369*, 130956.

Li, J., Li, Q., Lei, X., Tian, W., Cao, J., Jiang, W., & Wang, M. (2018). Effects of wax coating on the moisture loss of cucumbers at different storage temperatures. *Journal of Food Quality, 2018*(1), 9351821.

Li, X. Y., Du, X. L., Liu, Y., Tong, L. J., Wang, Q., & Li, J. L. (2019). Rhubarb extract incorporated into an alginate-based edible coating for peach preservation. *Scientia Horticulturae, 257*, 108685.

Lin, L., Xue, L., Duraiarasan, S., & Haiying, C. (2018). Preparation of ε-polylysine/chitosan nanofibers for food packaging against Salmonella on chicken. *Food Packaging and Shelf Life, 17*, 134–141.

Liu, J., Junejo, S. A., Xiao, Y., Jin, Y., Shi, S., & Zhou, Y. (2024). Effect of camellia oil body-based oleogels on the film-forming properties of soy protein isolate. *Food Chemistry, 458*, 140282.

Luís, Â., Pereira, L., Domingues, F., & Ramos, A. (2019). Development of a carboxymethyl xylan film containing licorice essential oil with antioxidant properties to inhibit the growth of foodborne pathogens. *Lwt, 111*, 218–225.

Luo, Q., Hossen, M. A., Zeng, Y., Dai, J., Li, S., Qin, W., & Liu, Y. (2022). Gelatin-based composite films and their application in food packaging: A review. *Journal of Food Engineering, 313*, 110762.

Marín, A., Atarés, L., & Chiralt, A. (2017). Improving function of biocontrol agents incorporated in antifungal fruit coatings: A review. *Biocontrol Science and Technology, 27*(10), 1220–1241.

Mendy, T. K., Misran, A., Mahmud, T. M. M., & Ismail, S. I. (2019). Application of Aloe vera coating delays ripening and extend the shelf life of papaya fruit. *Scientia Horticulturae, 246*, 769–776.

Merino, D., Bellassi, P., Paul, U. C., Morelli, L., & Athanassiou, A. (2023). Assessment of chitosan/pectin-rich vegetable waste composites for the active packaging of dry foods. *Food Hydrocolloids, 139*, 108580.

Mohammadi, M., Mirabzadeh, S., Shahvalizadeh, R., & Hamishehkar, H. (2020). Development of novel active packaging films based on whey protein isolate incorporated with chitosan nanofiber and nano-formulated cinnamon oil. *International Journal of Biological Macromolecules, 149*, 11–20.

Nandhini, J., Bellarmin, M., Siva Prakash, S., Sowmya Sri, D., & Karthikeyan, E. (2024). Advancements in metal oxide bio-nanocomposites for sustainable food packaging: Fabrication, applications, and future prospectives. *Food. Bioengineering, 3, 438.*

Nastasi, J. R., Kontogiorgos, V., Daygon, V. D., & Fitzgerald, M. A. (2022). Pectin-based films and coatings with plant extracts as natural preservatives: A systematic review. *Trends in Food Science & Technology, 120*, 193–211.

Nehchiri, N., Amiri, S., & Radi, M. (2021). Improving the water barrier properties of alginate packaging films by submicron coating with drying linseed oil. *Packaging Technology and Science, 34*(5), 283–295.

Nicolau-Lapeña, I., Colàs-Medà, P., Alegre, I., Aguiló-Aguayo, I., Muranyi, P., & Viñas, I. (2021). Aloe vera gel: An update on its use as a functional edible coating to preserve fruits and vegetables. *Progress in Organic Coatings, 151*, 106007.

Nongnual, T., Butprom, N., Boonsang, S., & Kaewpirom, S. (2024). Citric acid crosslinked carboxymethyl cellulose edible films: A case study on preserving freshness in bananas. *International Journal of Biological Macromolecules, 267*, 131135.

Oun, A. A., & Rhim, J. W. (2017). Characterization of carboxymethyl cellulose-based nanocomposite films reinforced with oxidized nanocellulose isolated using ammonium persulfate method. *Carbohydrate Polymers, 174*, 484–492.

Panahirad, S., Dadpour, M., Peighambardoust, S. H., Soltanzadeh, M., Gullón, B., Alirezalu, K., & Lorenzo, J. M. (2021). Applications of carboxymethyl cellulose-and pectin-based active edible coatings in preservation of fruits and vegetables: A review. *Trends in Food Science & Technology, 110*, 663–673.

Pashova, S. (2023). Application of plant waxes in edible coatings. *Coatings, 13*(5), 911.

Pavinatto, A., de Almeida Mattos, A. V., Malpass, A. C. G., Okura, M. H., Balogh, D. T., & Sanfelice, R. C. (2020). Coating with chitosan-based edible films for mechanical/biological protection of strawberries. *International Journal of Biological Macromolecules, 151*, 1004–1011.

Perera, K. Y., Jaiswal, S., & Jaiswal, A. K. (2022). A review on nanomaterials and nanohybrids based bio-nanocomposites for food packaging. *Food Chemistry, 376*, 131912.

Perumalla, A. V. S., & Hettiarachchy, N. S. (2011). Green tea and grape seed extracts—Potential applications in food safety and quality. *Food Research International, 44*(4), 827–839.

Prajapati, R. A., & Jadeja, G. C. (2024). Red dragon fruit-soy protein isolate biofilm: UV-blocking, antioxidant & improved mechanical properties for sustainable food packaging. *Journal of Food Science and Technology, 61*(9), 1686–1700.

Priyadarshi, R., Kim, H. J., & Rhim, J. W. (2021). Effect of sulfur nanoparticles on properties of alginate-based films for active food packaging applications. *Food Hydrocolloids, 110*, 106155.

Ribeiro-Santos, R., Sanches-Silva, A., Motta, J. F. G., Andrade, M., de Araújo Neves, I., Teófilo, R. F., et al. (2017). Combined use of essential oils applied to protein base active food packaging: Study in vitro and in a food simulant. *European Polymer Journal, 93*, 75–86.

Roy, S., & Rhim, J. W. (2020). Preparation of antimicrobial and antioxidant gelatin/curcumin composite films for active food packaging application. *Colloids and Surfaces B: Biointerfaces, 188*, 110761.

Roy, S., Priyadarshi, R., Łopusiewicz, Ł., Biswas, D., Chandel, V., & Rhim, J. W. (2023). Recent progress in pectin extraction, characterization, and pectin-based films for active food packaging applications: A review. *International Journal of Biological Macromolecules, 239*, 124248.

Saleh, I., & Abu-Dieyeh, M. (2022). Novel Prosopis juliflora leaf ethanolic extract coating for extending postharvest shelf-life of strawberries. *Food Control, 133*, 108641.

Sani, M. A., Azizi-Lalabadi, M., Tavassoli, M., Mohammadi, K., & McClements, D. J. (2021). Recent advances in the development of smart and active biodegradable packaging materials. *Nanomaterials, 11*(5), 1331.

Sant'anna Andrade, L., & da Silva, E. O. (2024). Nanohybrids with potential barrier property and antimicrobial activity for application in sustainable polymer food packaging: A mini-review. *Polymers for Advanced Technologies, 35*(6), e6455.

Sapper, M., & Chiralt, A. (2018). Starch-based coatings for preservation of fruits and vegetables. *Coatings, 8*(5), 152.

Sathiyaseelan, A., Saravanakumar, K., Mariadoss, A. V. A., Ramachandran, C., Hu, X., Oh, D. H., & Wang, M. H. (2021). Chitosan-tea tree oil nanoemulsion and calcium chloride tailored edible coating increase the shelf life of fresh cut red bell pepper. *Progress in Organic Coatings, 151*, 106010.

Seiwert, K., Kamdem, D. P., Kocabaş, D. S., & Ustunol, Z. (2021). Development and characterization of whey protein isolate and xylan composite films with and without enzymatic crosslinking. *Food Hydrocolloids, 120*, 106847.

Serino, N., Boari, A., Santagata, G., Masi, M., Malinconico, M., Evidente, A., & Vurro, M. (2021). Biodegradable polymers as carriers for tuning the release and improve the herbicidal effectiveness of Dittrichia viscosa plant organic extracts. *Pest Management Science, 77*(2), 646–658.

Sganzerla, W. G., Rosa, G. B., Ferreira, A. L. A., da Rosa, C. G., Beling, P. C., Xavier, L. O., et al. (2020). Bioactive food packaging based on starch, citric pectin and functionalized with Acca sellowiana waste by-product: Characterization and application in the postharvest conservation of apple. *International Journal of Biological Macromolecules, 147*, 295–303.

Shah, Y. A., Bhatia, S., Al-Harrasi, A., Tarahi, M., Almasi, H., Chawla, R., & Ali, A. M. M. (2024). Insights into recent innovations in barrier resistance of edible films for food packaging applications. *International Journal of Biological Macromolecules, 271*, 132354.

Shekhar, S., Sharma, R., Sharma, S., Sharma, B., Sarkar, A., & Jain, P. (2020). An exploration of electrocatalytic analysis and antibacterial efficacy of electrically conductive poly (D-glucosamine)/graphene oxide bionanohybrid. *Carbohydrate Polymers, 240*, 116242.

Shen, X., Sun, X., Xie, Q., Liu, H., Zhao, Y., Pan, Y., et al. (2014). Antimicrobial effect of blueberry (Vaccinium corymbosum L.) extracts against the growth of Listeria monocytogenes and salmonella Enteritidis. *Food Control, 35*(1), 159–165.

Shiekh, K. A., Ngiwngam, K., & Tongdeesoontorn, W. (2021). Polysaccharide-based active coatings incorporated with bioactive compounds for reducing postharvest losses of fresh fruits. *Coatings, 12*(1), 8.

Siciliano, S., Lopresto, C. G., & Lamonaca, F. (2024). From traditional packaging to smart bio-packaging for food safety: A review. *Euro-Mediterranean Journal for Environmental Integration*, 1–16.

Sonseca, A., Madani, S., Rodríguez, G., Hevilla, V., Echeverría, C., Fernández-García, M., et al. (2019). Multifunctional PLA blends containing chitosan mediated silver nanoparticles: Thermal, mechanical, antibacterial, and degradation properties. *Nanomaterials, 10*(1), 22.

Sousa, F. F., Junior, J. S. P., Oliveira, K. T., Rodrigues, E. C., Andrade, J. P., & Mattiuz, B. H. (2021). Conservation of 'Palmer'mango with an edible coating of hydroxypropyl methylcellulose and beeswax. *Food Chemistry, 346*, 128925.

Syahida, S. N., Ismail-Fitry, M. R., Ainun, Z. M. A. A., & Hanani, Z. A. N. (2020). Effects of palm wax on the physical, mechanical and water barrier properties of fish gelatin films for food packaging application. *Food Packaging and Shelf Life, 23*, 100437.

Thakur, R., Pristijono, P., Bowyer, M., Singh, S. P., Scarlett, C. J., Stathopoulos, C. E., & Vuong, Q. V. (2019). A starch edible surface coating delays banana fruit ripening. *Lwt, 100*, 341–347.

Tian, F., Decker, E. A., & Goddard, J. M. (2013). Controlling lipid oxidation via a biomimetic iron chelating active packaging material. *Journal of Agricultural and Food Chemistry, 61*(50), 12397–12404.

Vizzini, P., Beltrame, E., Zanet, V., Vidic, J., & Manzano, M. (2020). Development and evaluation of qPCR detection method and Zn-MgO/alginate active packaging for controlling Listeria monocytogenes contamination in cold-smoked salmon. *Food, 9*(10), 1353.

Wang, Z. C., Qin, C. Q., Zhang, X., Wang, Q., Li, R. X., & Ren, D. F. (2021). Effect of whey protein isolate/chitosan/microcrystalline cellulose/PET multilayer bottles on the shelf life of rosebud beverages. *Food Chemistry, 347*, 129006.

Wu, Y., Wu, H., & Hu, L. (2024). Recent advances of proteins, polysaccharides and lipids-based edible films/coatings for food packaging applications: A review. *Food Biophysics, 19*(1), 29–45.

Xu, J., Manepalli, P. H., Zhu, L., Narayan-Sarathy, S., & Alavi, S. (2020). Morphological and performance characteristics of nanocomposite films based on poly (lactic acid) compounded with nanocrystalline cellulose and chitin whiskers using melt extrusion. *Cellulose, 27*, 7523–7534.

Yang, Z. K., Zou, X. B., Li, Z. H., Huang, X. W., Zhai, X. D., Zhang, W., et al. (2019). Improved postharvest quality of cold stored blueberry by edible coating based on composite gum Arabic/roselle extract. *Food and Bioprocess Technology, 12(9)*, 1537–1547.

Yousuf, B., Sun, Y., & Shimin, W. (2022). Lipid and lipid-containing composite edible coatings and films. *Food Reviews International, 38*(sup1), 574–597.

Zareie, Z., Yazdi, F. T., & Mortazavi, S. A. (2020). Development and characterization of antioxidant and antimicrobial edible films based on chitosan and gamma-aminobutyric acid-rich fermented soy protein. *Carbohydrate Polymers, 244*, 116491.

Zhang, X., Li, Y., Guo, M., Jin, T. Z., Arabi, S. A., He, Q., et al. (2021). Antimicrobial and UV blocking properties of composite chitosan films with curcumin grafted cellulose nanofiber. *Food Hydrocolloids, 112*, 106337.

Zhang, L., Yang, D., Luo, R., Luo, Y., & Hou, Y. (2024). Research progress on the mechanism of the impact of myofibrillar protein oxidation on the flavor of meat products. *Food, 13*(20), 3268.

Zinoviadou, K. G., Koutsoumanis, K. P., & Biliaderis, C. G. (2009). Physico-chemical properties of whey protein isolate films containing oregano oil and their antimicrobial action against spoilage flora of fresh beef. *Meat Science, 82*(3), 338–345.

Chapter 8
Effect of Plasticizers on Properties of Edible Coating and Films

Himanshi Singh, Nikita, Khushbu Kumari, and Anand Kishore

Abstract Plasticizers are vital additives that modify the functional performance of edible films and coatings by improving flexibility, reducing brittleness, and enhancing processing characteristics. Their incorporation into starch-based biopolymer matrixes alters the structural arrangement by disrupting strong intermolecular hydrogen bonds and increasing polymer chain mobility. This restructuring enables improved film-forming ability and mechanical compliance, making the films more suitable for packaging applications. This chapter explores the role of plasticizers in modifying the functional properties of edible films and coatings. By analyzing the effects of different plasticizers on key attributes such as water absorption, solubility, oxygen permeability, water vapor transmission rate (WVTR), thermal behavior, and mechanical properties, we aim to provide a comprehensive understanding of their impact.

Keywords Edible coating and films · Mechanical properties · Plasticizers · Structural properties · Barrier properties

Introduction

Edible films and coatings represent an innovative and sustainable approach to food packaging, offering a biodegradable and eco-friendly alternative to traditional synthetic packaging materials (Kumar & Neeraj, 2019; Janowicz et al., 2023). These

---

H. Singh · Nikita · A. Kishore (✉)
National Institute of Food Technology Entrepreneurship and Management, Kundli, Sonepat, Haryana, India

K. Kumari
National Dairy Research Institute, Karnal, Haryana, India

N. Kumar et al. (eds.), *Trends in Edible Packaging: A sustainable approaches for postharvest management of fruits and vegetables*,
https://doi.org/10.1007/978-3-032-13199-7_8

thin layers of biopolymer-based materials can either be applied as coatings directly onto food surfaces or used as standalone films to protect and preserve food products. Their primary function is to act as a barrier against environmental factors such as moisture, oxygen, and microbial contamination, thereby enhancing the shelf life, quality, and safety of food products (Perez-Vazquez et al., 2023; Kumar et al., 2024). The global shift toward sustainability has increased interest in edible films and coatings, primarily due to their potential to reduce plastic waste and reliance on petroleum-based packaging (Pillai et al., 2024). Derived from natural biopolymers such as polysaccharides (starch, cellulose, chitosan), proteins (gelatin, casein, whey protein), and lipids (wax, oils), these films possess desirable characteristics such as biodegradability, biocompatibility, and food-grade safety (Kumar et al., 2022; Kumar et al., 2023; Gupta et al., 2024). However, despite their advantages, edible films often exhibit limitations such as brittleness, low flexibility, and poor mechanical integrity, which restrict their practical applications (Rostamabadi et al., 2024). To overcome these shortcomings, plasticizers are incorporated into the film formulation to improve film flexibility, durability, and handling properties (Eslami et al., 2023). Plasticizers are low-molecular-weight compounds that interact with polymer chains, reducing intermolecular forces and increasing chain mobility (Sánchez-González et al., 2025). This modification enhances the mechanical flexibility, extensibility, and overall usability of the films. Common plasticizers used in edible films and coatings include glycerol, sorbitol, polyethylene glycol (PEG), and citric acid (Yermagambetova et al., 2024). Each plasticizer has a unique influence on the film's physicochemical properties, including water absorption, solubility, oxygen permeability, water vapor transmission rate (WVTR), and thermal stability (Kumarihami et al., 2023; Eslami et al., 2023).

Understanding the effect of plasticizers on these properties is critical for tailoring edible films to specific applications. For instance, glycerol is widely used for its ability to impart flexibility, though it often increases hydrophilicity and water vapor permeability (Caicedo et al., 2022). On the other hand, citric acid not only acts as a plasticizer but also improves cross-linking in starch-based films, enhancing their mechanical strength and water resistance (Pooja et al., 2024). The choice and concentration of plasticizers must therefore be optimized to balance the trade-offs between flexibility, barrier properties, and durability (Makhathini et al., 2025). This chapter explores the role of plasticizers in modifying the functional properties of edible films and coatings. By analyzing the effects of different plasticizers on key attributes such as water absorption, solubility, oxygen permeability, WVTR, thermal behavior, and mechanical properties, we aim to provide a comprehensive understanding of their impact.

Types of Plasticization

In polymer science, plasticizers are broadly classified into two categories: internal and external plasticizers. These substances are added to enhance polymer flexibility, primarily by lowering the glass transition temperature (Tg) and modifying the mechanical strength.

Internal Plasticizers

Internal plasticizers are chemically integrated into the polymer's backbone during synthesis, either through copolymerization or direct chemical modification of the polymer. This incorporation reduces the structural order of the polymer chains, increases molecular mobility, and results in a softer, more flexible material with a reduced Tg. For example, the use of a second monomer in copolymerization leads to greater chain disorder, making it more difficult for polymer chains to pack closely together (Skelly et al., 2022; SpecialChem Tech Library, 2023). Although internal plasticization effectively softens the polymer, it often results in a relatively narrow working temperature range. Internally plasticized copolymers may soften more abruptly compared to externally plasticized systems or polymer blends. Another mechanism of internal plasticization involves introducing bulky side chainseither as substituents or graftswhich interfere with intermolecular forces. This is welldocumented in polyacrylates and polymethacrylates, where steric hindrance from side chains enhances flexibility but may lead to side chain crystallization if the chains become too long (Godwin, 2024). Additionally, alkylation of polyamides is a technique that reduces crystallinity and Tg, yielding elastic fibers—precursors to commercial products such as Lycra and Vyrene (Special Chem Tech Library, 2023).

External Plasticizers

In contrast, external plasticizers are not chemically bonded to the polymer but are added physically during processing. These are typically low-volatility, high-boiling-point compounds that interact with polymer chains via solvation or swelling mechanisms, especially at elevated temperatures (Banker, 1966; Wilson, 1995). The main advantage of external plasticizers is the wide range of available compounds, allowing for targeted tuning of polymer properties without altering the primary structure. This versatility has made external plasticizers the preferred choice for many commercial applications. External plasticizers can be further classified into primary (solvent-type) and secondary (non-solvent-type) categories. In amorphous polymers, all plasticizers behave like solvents and may dissolve the polymer over time. However, in semicrystalline or crystalline polymers, primary plasticizers are capable of penetrating both the crystalline and amorphous regions, thereby effectively reducing rigidity. Secondary plasticizers, also referred to as softeners, interact only with the amorphous regions and are typically used as adjuncts to primary plasticizers, helping to improve cost-efficiency and fine-tune performance characteristics (Godwin, 2005; Bonifacio et al., 2023). External plasticizers can be categorized into solvent plasticizers and nonsolvent plasticizers, depending on their interaction with the polymer matrix:

- *Solvent Plasticizers:* In amorphous polymers, any plasticizer acts as a solvent plasticizer, meaning that under suitable conditions, the polymer can dissolve in the plasticizer, leading to enhanced chain mobility and flexibility.
- *True Plasticizers (Primary Plasticizers):* In crystalline or semicrystalline polymers, certain compounds can penetrate both amorphous and crystalline regions, acting as true plasticizers that enhance overall polymer flexibility (Rahman & Brazel, 2004).
- *Nonsolvent Plasticizers (Secondary Plasticizers or Softeners):* If a plasticizer interacts only with the amorphous regions of a polymer, it is classified as a nonsolvent plasticizer. These secondary plasticizers sometimes act as diluents for primary plasticizers, modifying mechanical properties without significantly altering the crystalline phase of the polymer (Mark et al., 2017).

Mechanism of Plasticization

Plasticization in polymers has been explained through various theoretical models. Among the earliest explanations are the free volume theory, which focuses on the role of molecular spacing within polymers, and the coiled spring theory, which describes the behavior of entangled polymer chains (Consolati et al., 2022; Kumar et al., 2019; Uneyama & Masubuchi, 2012). Subsequent theories, including the gel theory and lubricity theory, provide additional perspectives on how plasticizers influence polymer properties. The gel theory attributes the stiffness of unplasticized polymers to weaken intermolecular forces creating a gel-like network along the polymer chains (Donhowe & Fennema, 1993; Sears & Darby, 1982). On the other hand, the lubricity theory suggests that plasticizers function by reducing friction between polymer chains, thereby facilitating their movement and decreasing resistance to deformation (Eslami et al., 2023; Sears & Darby, 1982). Plasticizers integrate into these systems by attaching to specific sites on the polymer chains, displacing polymer–polymer interactions and thereby softening the structure. Additionally, excess plasticizers that are not bound to the polymer may aggregate, forming micro-domains that increase free volume and support greater chain mobility (Special Chem Tech Library, 2023). This dual mechanism enhances the overall flexibility and deformability of the polymer. In the context of edible films, the free volume and gel theories are particularly pertinent and will be the focus of a more detailed discussion.

Gel Theory

The gel theory is a foundational framework used to understand the mechanism of plasticization in polymers, particularly those used in food applications, such as edible films and coatings. According to this theory, the introduction of plasticizers alters the internal polymer matrix by modifying its semirigid, three-dimensional network structure (Doolittle, 1954; Cheyney et al., 1979). In hydrocolloid systems, this internal matrix is stabilized by various molecular interactions, including hydrogen bonds, van der Waals attractions, and other cohesive forces like primary valence interactions (Aiken, 1960). Plasticizers function by intercalating themselves between polymer chains, thereby weakening or displacing these internal forces. This process enhances the material's flexibility and enables the formation of mobile plasticizer-rich domains that further promote chain movement.

Furthermore, temperature plays a critical role in determining how plasticized polymers behave. Doolittle (1954) noted that increasing temperature typically enhances the elasticity and extensibility of plasticized systems. At lower temperatures, however, plasticizers and polymer chains tend to interact more rigidly, limiting solvation and leading to increased brittleness. Interestingly, in certain systems like nitrocellulose blended with di-(n-hexyl carbitol) succinate, plasticization is paradoxically more effective at lower temperatures. This is attributed to the plasticizer's improved solvating capacity under cooler conditions, which disrupts polymer associations more efficiently.

Free Volume Theory

The free volume theory explains the role of unoccupied space within a polymer system—space not taken up by polymer molecules themselves in influencing the physical behavior of polymers. This available internal space, known as free volume, is key to understanding how materials respond to plasticization. In the case of edible films and coatings, this theory provides a framework for understanding how plasticizers improve the mechanical flexibility, moisture barrier function, and overall performance of the films by increasing chain mobility and expanding internal free space within the polymer matrix (Caicedo et al., 2022; Eslami et al., 2023). Free volume refers to the difference between a polymer's specific volume at a particular temperature and its volume at a baseline reference point, such as absolute zero. It is mathematically expressed as:

$$V_\mathrm{f} = V_\mathrm{t} - V_\mathrm{r}$$

Where:

V_f is the free volume,
V_t is the specific volume of the polymer at a given temperature T, and
V_r is the reference specific volume, often taken at 0 K

Plasticizers Used in Edible Films and Coatings

Plasticizers are critical additives in edible films and coatings, primarily enhancing flexibility, reducing brittleness, and improving mechanical properties by disrupting polymer–polymer interactions, increasing chain mobility, and lowering the glass transition temperature (T_g). These plasticizers are classified into categories such as polyols, carbohydrates, lipids, and other organic compounds. The choice of plasticizer depends on the film-forming material, targeted mechanical attributes, and intended food preservation applications (Thakur et al., 2021).

Water as a Natural Plasticizer

Water acts as a natural, low-molecular-weight, and highly volatile plasticizer for various biopolymers, including alginate. Its ability to form hydrogen bonds with polymer chains effectively lowers glass transition temperature (Tg) by increasing the free volume within the polymer matrix (Eslami et al., 2023; Gao et al., 2017). Additionally, water serves as a solvent for hydrophilic biopolymers, enhancing chain mobility and flexibility (Chiaoprakobkij et al., 2019; Bocqué et al., 2016). Olivas and Barbosa-Cánovas (2008) demonstrated that water vapor permeability (WVP) increases in alginate films with rising relative humidity (RH), a consequence of water-induced swelling and increased free volume facilitating moisture and gas diffusion. Barbut and Harper (2019) reported that dried alginate films at higher RH exhibited increased elongation and decreased tensile strength and Young's modulus, highlighting water's effect in reducing intermolecular cohesion. Water also acts as a restructuring agent during thermo-mechanical processing; Gao et al. (2017) observed that water combined with glycerol disrupts the alginate polymer network, facilitating better chain rearrangement and improved mechanical behavior. Similarly, Souza and Andrade (2001) reported that increasing water content in starch–alginate blends lowered tensile strength and Young's modulus but increased elongation, indicating a transition from a glassy to a rubbery state. Despite its benefits, water's volatility requires nonvolatile plasticizers like glycerol for long-term flexibility and stability. Excess glycerol can cause uneven plasticizer distribution and segregation, reducing elongation due to weakened polymer–plasticizer interactions (Averous et al., 2000).

Ethylene Glycol and Polyethylene Glycol (PEG)

Ethylene glycol (EG), a low-molecular-weight diol with two hydroxyl groups, is frequently employed in edible film formulations to improve flexibility and reduce brittleness. Its small structure allows it to interface effectively with polymer chains (starch, alginate) through hydrogen bonding, enabling reduced intermolecular stiffness and enhanced processability. Closely related polyols, diethylene glycol (DEG), triethylene glycol, and polyethylene glycol (PEG), are used to fine-tune film characteristics in polysaccharide-based systems. The molecular weight of these plasticizers critically influences their function: low molecular weight plasticizers increase mobility and softness, while higher molecular weight compounds modulate thermal behavior, compatibility, and phase stability. Polyethylene glycol (PEG), or Macrogol, comprises repeated ethylene glycol units $[–CH_2CH_2O–]_n$ and is available in a wide range of molecular weights. PEG 100–700 is liquid at room temperature, while PEG >2000 tends to be a semisolid or crystalline solid. PEG is highly hydrophilic, chemically inert, electrically neutral across pH ranges, and resistant to microbial growth, making it widely used in food, pharmaceutical, and biomedical film systems. Its amphiphilic structure and hydration capacity aid in reducing brittleness, improving solubility, and enhancing softness, while also reducing contact angles on hydrophobic surfaces for better wettability and permeability (D'souza & Shegokar, 2016). However, according to Olivas and Barbosa-Cánovas (2008), films plasticized with PEG-8000 exhibited notable phase separation, producing an opaque "white" surface and significantly reducing tensile strength and elongation compared to glycerol-plasticized films. Moreover, PEG-8000–plasticized films showed the highest water vapor permeability among all plasticizers tested, likely due to incompatibility and phase heterogeneity with the alginate matrix.

Sugar Alcohols as Plasticizers

Sorbitol

Sorbitol ($C_6H_{14}O_6$) is a sugar alcohol produced by the reduction of glucose and is widely used as a plasticizer in biopolymer films such as alginate, starch, and chitosan. Compared to glycerol, sorbitol is less hygroscopic, making it more suitable for films requiring enhanced moisture resistance. Its molecular structure, rich in hydroxyl groups, enables effective integration within the polymer network by forming hydrogen bonds, which improves flexibility while maintaining tensile strength and structural integrity under humid conditions (Santana & Kieckbusch, 2013; Jost et al., 2014; Olivas & Barbosa-Cánovas, 2008). Sorbitol is a small molecule that shows greater resistance to moisture due to its lower affinity for water and its ability to establish stronger molecular interactions with polymer chains. The use of sorbitol, either alone or in combination with glycerol, enhances the mechanical and barrier properties of edible films. Notably, the combination of sorbitol and glycerol, when applied to oxidized starch-based films, was found to significantly improve their physicochemical properties, making them more suitable for food preservation applications (González-Torres et al., 2021).

Glycerol

Glycerol is one of the most widely used plasticizers in edible films and coatings due to its non-toxic nature, high biocompatibility, and food-grade approval. Though not a sugar alcohol, it is frequently used alongside polyols like sorbitol in biopolymer systems. Glycerol is particularly effective in enhancing film flexibility by interfering with the tight packing of polymer chains, such as amylose, thereby increasing chain mobility. Owing to its lower molecular weight compared to sorbitol, glycerol can achieve similar mechanical properties at lower concentrations (Jost et al., 2014; Tong et al., 2013). This effect is consistent with the free volume theory, which posits that small, hydrophilic molecules like glycerol increase the free space between polymer chains, enhancing polymer mobility. However, glycerol's hygroscopic nature also increases the water vapor and oxygen permeability of films, potentially reducing their barrier properties. Its concentration has a substantial influence on film behavior. At levels below 10%, films tend to be brittle; between 10% and 20%, flexibility remains inadequate. For specific starch-based systems such as sago starch, a minimum of 30% glycerol is required to ensure acceptable mechanical resistance and heat sealability. Excessive plasticizer levels (above 50%) further improve ductility but can significantly compromise moisture and gas barrier properties. Thus, glycerol concentration must be optimized depending on the intended application (Basiak et al., 2018).

Mannitol

Mannitol ($C_6H_{14}O_6$), derived from the hydrogenation of fructose, is another sugar alcohol utilized in edible films. Although less efficient as a plasticizer than sorbitol or glycerol, mannitol imparts better water resistance due to its lower hygroscopicity. It is often combined with other polyols in celluloseand protein-based films. Films plasticized with mannitol tend to be thicker and may exhibit phase separation or crystal formation, resulting in a whitish appearance. Such effects can be mitigated by cross-linking treatments (Santana & Kieckbusch, 2013; Parreidt et al., 2018).

Xylitol

Xylitol ($C_5H_{12}O_5$) is a naturally occurring five-carbon sugar alcohol (polyol) found in small amounts in fruits and vegetables and commercially produced predominantly from hemicellulose-rich biomass. Its molecular structure features a hydroxyl group on each carbon atom, contributing to its hygroscopicity and high-water solubility, which support its broad applicability in food, pharmaceutical, and material science sectors (Umai et al. (2022)). With a sweetness level similar to sucrose but a lower caloric value (2.4 kcal/g), xylitol is widely used as a low-energy sweetener in functional foods, diabetic-friendly products, chewing gums, and beverages. Xylitol is also non-cariogenic, meaning it cannot be fermented by Streptococcus mutans, the bacteria responsible for dental caries. This characteristic underpins its widespread use in toothpaste, mouthwash, and dental chewing gums, where it helps reduce plaque formation and promotes oral health (Janakiram et al., 2017).

In addition to its health benefits, xylitol has gained attention in the chemical and materials industry as one of the top value-added products derived from lignocellulosic biomass. It serves as a platform chemical for the synthesis of ethylene glycol, glycerol, xylaric acid, and biodegradable polymers (Isikgor & Becer, 2015). Furthermore, xylitol functions effectively as a plasticizer in the formulation of biodegradable and edible films. It enhances film transparency and uniformity, performing comparably to glycerol in improving flexibility and thickness. This is attributed to xylitol's ability to increase free volume within the polymer matrix, making the film more ductile and hydrophilic. Films plasticized with xylitol often outperform those using mannitol in terms of clarity and structural consistency (Santana & Kieckbusch, 2013).

Sugars

Monosaccharides, disaccharides, and oligosaccharides are the sugars most commonly used as plasticizers in polysaccharide-based polymers, such as starch and alginate, because of the similarity between the molecular structures of the plasticizers and polymers. Monosaccharides are the simplest carbohydrates, and they cannot be hydrolyzed into smaller carbohydrates. Glucose, fructose, sucrose, and lactose are some examples of sugars that come from different sources, such as fruits or milk. The plasticizing efficiency of fructose was investigated by Olivas et al. (2008). According to their results, because of the similarity in the sorbitol and fructose molecular weights, 182.17 and 180.16, respectively, the plasticized films with these plasticizers possessed similar properties. At higher relative humidity (76% and 100%), the WVP of the films containing fructose was the lowest among the plasticizers. They reported that the low WVP of fructose-plasticized alginate films is comparable to that of lipid-plasticized whey protein emulsion films. However, there was no considerable difference between the WVP of fructose and sorbitol-plasticized films at other RHs. They also showed that the films containing fructose and sorbitol have similar tensile strength and elastic modulus. The similar behavior of sorbitol and fructose is also highlighted by Tong et al. (2013): sorbitoland fructoseplasticized blend films exhibited the lowest and similar elongation at break values at any given plasticizer concentration compared with glyceroland xylitol-plasticized films, with the fructoseplasticized film being even more brittle, with higher tensile strength and lower elongation at break. However, the moisture contents of films containing fructose were lower than those of films containing sorbitol, possibly due to their different molecular structures. Zhang and Han (2006) also reported that monosaccharide-plasticized pea starch films were more resistant to water vapor permeation than polyolplasticized films, although their mechanical properties were comparable. Fruit and vegetable purees have good potential to be used in making edible films. Since they are high in sugars such as glucose, fructose, and sucrose, they can have a plasticizing effect on the produced film (Otoni et al., 2017). In the study by Kadzińska et al. (2020), the physical properties of apple puree–sodium alginate edible films incorporated with vegetable oils were investigated. They found that in the samples containing apple puree, the glass transition temperature was lower, indicating the plasticizing effect of natural plasticizers such as fructose and glucose present in the apple puree.

Lipid-Based Plasticizers

Oils

Oils are especially valued as plasticizers in polysaccharide-based films because they are naturally derived, hydrophobic, and safe for food packaging. Their high molecular weights help reduce migration to the surface, which improves their stability as additives. Common oils used include castor oil (Aziz et al., 2018), soybean oil (Gutiérrez-Jara et al., 2020), coconut oil (Marismandani & Husni, 2020), canola oil (Syarifuddin et al., 2017), oregano essential oil, and garlic oil. Although oils are often combined with primary plasticizers like glycerol, they enhance plasticization by increasing flexibility and lowering the tensile strength of films. The dispersed oil droplets act as softening agents within the polymer matrix, improving film deformability. To achieve uniform dispersion of these hydrophobic oils in hydrophilic matrixes such as alginate, surfactants like Tween 80 and Span are usually employed.

Citrate Plasticizers

Citric acid itself, which is obtained from citrus fruits, sugar cane, or beetroots, is also used as a plasticizer (Bocqué et al., 2016; Alhanish & Abu Ghalia, 2021). Citrate-based plasticizers, including tri- and tetra-esters derived from citric acid esterification, represent a group of hydrophobic plasticizers studied in alginate films (Paixão et al., 2019; Sharmin et al., 2021; Chen et al., 2020; Remuñán-López & Bodmeier, 1996). Research indicates that citric acid (CA) primarily acts as a plasticizer at higher concentrations. At lower concentrations (0.5–1% w/v), CA reduces water vapor transmission rate (WVTR) by forming hydrophobic ester bonds with alginate chains. However, at higher concentrations, the plasticizing effect dominates, increasing WVTR. Mechanically, CA significantly improves elongation at break (up to a 475% increase at 2% CA), as it reduces intermolecular bonding between alginate chains through unreacted CA acting as a plasticizer (Paixão et al., 2019; Sharmin et al., 2021).

Fatty Acids as Plasticizers

Fatty acids, characterized by long hydrocarbon chains and a carboxylic acid group, serve as hydrophobic plasticizers in alginate films. Common examples include stearic acid, lauric acid (Chen et al., 2022), and oleic acid (Azucena Castro-Yobal et al., 2021). These fatty acids tend to increase film opacity and create a more heterogeneous structure, weakening intermolecular interactions and reducing tensile strength. Typically, elongation at break decreases due to weaker interactions compared to polar plasticizers. However, at higher concentrations, shorter-chain fatty acids like lauric acid can enhance flexibility due to increased molecular mobility. Fatty acids also impact water vapor permeability (WVP): low concentrations improve barrier properties, while higher concentrations cause discontinuities that raise WVP. Oleic acid, owing to its hydrophobicity, slightly reduces WVP (Azucena Castro-Yobal et al., 2021; Chen et al., 2022).

Tri-butyl Citrate (TC) and Related Plasticizers

Hydrophobic tri-butyl citrate (TC), either alone or combined with hydrophilic glycerol, has been investigated in alginate films (Paixão et al., 2019). TC incorporation results in film opacity and a globular, sponge-like microstructure, as seen by scanning electron microscopy (SEM) imaging. Its hydrophobic character reduces both water solubility and WVP. Mechanically, TC increases tensile stress but decreases elongation at break, likely due to stronger interactions between alginate chains and TC, influenced by TC's high polarity. Dynamic mechanical analysis (DMA) reveals a single glass transition temperature (T_g) that increases with TC content, indicating a homogeneous polymer mixture. Similarly, triacetin and tri-ethyl citrate plasticizers behave differently: triacetin-plasticized films are brittle and thermally stable with increased recrystallization (Chen et al., 2020), whereas tri-ethyl citrate at 30% results in opaque, brittle films (Remuñán-López & Bodmeier, 1996).

Effect of Plasticizers on Properties of Edible Film and Coating

Film Thickness and Density

The addition of plasticizers plays a crucial role in modifying the thickness and density of edible films. Generally, increasing the plasticizer concentration leads to an increase in film thickness while causing a slight reduction in density. In one of the studies by Mohammed et al. (2022), it was observed that the addition of plasticizers significantly influenced the thickness and density of edible films. Plasticizers such as glycerol, sorbitol, polyethylene glycol (PEG 400), urea, and fructose play a crucial role in modifying the structural properties of biopolymer films by reducing intermolecular forces and increasing polymer chain mobility. The incorporation of plasticizers disrupts hydrogen bonding within the polymer matrix, allowing for greater molecular spacing and flexibility. This results in an increase in film thickness, as the polymer network expands due to the plasticizers' ability to enhance free volume and reduce molecular packing density (Galdeano et al., 2009). The extent of this increase varies depending on the plasticizer type and concentration. Glycerol, for instance, has been reported to restructure the polymer chain network by converting free volumes into a more expanded and less compact film structure. This effect is more pronounced at higher concentrations, as more plasticizer molecules integrate into the polymer matrix, thereby increasing the spacing between chains and further enhancing thickness (Razavi et al., 2015).

The density of plasticized films, on the other hand, generally decreases with an increase in plasticizer content due to the reduced compactness of the polymer network. The ability of plasticizers to penetrate the polymer structure and reduce intermolecular interactions leads to a more open network, which results in a lower density. Studies on starch-based films have shown that this reduction in density is a direct consequence of the increase in film volume, as plasticizers create additional free volume by disrupting polymer–polymer interactions and forming new hydrogen bonds with the polymer chains (Sanyang et al., 2016). The extent of this decrease is dependent on the type of plasticizer used, with hydrophilic plasticizers such as glycerol and sorbitol demonstrating a greater tendency to reduce film density due to their strong interactions with the polymer matrix. Conversely, plasticizers with lower water affinity, such as PEG 400, may have a less pronounced effect on density since they interact differently with the polymer network and do not significantly disrupt its compact structure (Sahari et al., 2012).

Water Absorption

Water absorption is a critical parameter in the characterization of edible films, as it directly influences their mechanical integrity, barrier properties, and potential applications. Water, acting as a natural plasticizer, enhances film flexibility; however, excessive absorption is undesirable due to its impact on dimensional stability, leading to swelling and structural weakening (Tarique et al., 2021). The hydrophilic nature of starch-based films makes them particularly susceptible to water uptake, as the hydroxyl groups present in starch molecules readily form hydrogen bonds with water (Souza et al., 2021). The incorporation of plasticizers alters this behavior by modifying intermolecular interactions and influencing the overall structural compactness of the film. Several studies have demonstrated that the water absorption capacity of edible films decreases with increasing plasticizer concentration. This reduction can be attributed to the formation of strong hydrogen bonds between the plasticizer and starch molecules, which reduces the availability of hydroxyl groups to interact with water molecules (Sanyang et al., 2016). For instance, in glycerol-plasticized arrowroot starch films, lower glycerol concentrations (15%) led to higher water absorption, whereas films with higher concentrations (30% and 45%)

exhibited significantly reduced water uptake (Tarique et al., 2021). This phenomenon is explained by the plasticizers' ability to disrupt the existing hydrogen bonds within the starch network and establish new bonds with starch molecules, thereby limiting water penetration and enhancing the film's water resistance. Beyond glycerol, different plasticizers exhibit distinct effects on water absorption depending on their molecular structure, polarity, and hydrophilicity. In wheat starch films plasticized with different plasticizers at 35% concentration, films with fructose showed the lowest water absorption (~140%), while sorbitol and urea films exhibited significantly higher absorption levels of ~234% and ~300%, respectively (Mohammed et al., 2022). These variations are attributed to differences in the solubility and polarity of plasticizers. Highly hydrophilic plasticizers like urea increase water uptake due to their strong affinity for water, whereas less hydrophilic ones like fructose enhance water resistance by forming tighter polymer networks. Moreover, plasticized films often reach a saturation point beyond which no further water absorption occurs, indicating that film composition and matrix interaction govern long-term moisture behavior (Mohammed et al., 2022). These insights are crucial for applications where moisture resistance is vital, such as food packaging. Overall, the chemical nature, concentration, and interaction of plasticizers with the polymer matrix are key determinants in modulating water absorption behavior.

Moisture Content and Water Solubility

Moisture content and water solubility are key factors influencing the functionality of edible films, particularly in applications requiring controlled water retention and solubility. The addition of plasticizers significantly affects these properties, with moisture content generally increasing as plasticizer concentration rises. This is due to the hydrophilic nature of plasticizers, which disrupt intermolecular interactions within the polymer matrix, enhancing water retention (Debeaufort et al., 1998). Higher plasticizer concentrations in film-forming solution lead to increased water solubility by disrupting the biopolymer network and reducing intermolecular forces within the film. This structural modification increases free volume, making the polymer matrix more permeable to water.

Water Vapor Permeability (WVP)

Water vapor permeability is a vital barrier property in starch-based edible films, especially for packaging applications where controlling moisture transfer between food and the external environment is essential for maintaining quality and prolonging shelf life. Plasticizers, while primarily added to enhance flexibility and processability, also significantly influence the permeability characteristics of these films. Hydrophilic plasticizers such as glycerol and sorbitol generally increase WVP by disrupting the polymer network and enhancing water molecule mobility (Hazrati et al., 2021). Glycerol, a small, highly hydrophilic molecule with three hydroxyl groups, weakens the intra- and intermolecular hydrogen bonding between starch chains, increasing the free volume and polymer chain mobility. This loosening of the film structure allows easier water diffusion, resulting in higher WVP. For instance, increasing glycerol concentration in arrowroot starch films from 15% to 45% increased WVP from 5.71×10^{-9} to 10.8×10^{-9} $g \cdot s^{-1} \cdot m^{-1} \cdot Pa^{-1}$. A similar trend was observed in sugar palm starch films, where WVP rose substantially with plasticizer loading (Sanyang et al., 2015). At lower glycerol concentrations, the polymer matrix remains denser and more cohesive, minimizing the diffusion of water vapor (Lintang et al., 2021). As the plasticizer concentration increases, barrier properties deteriorate due to enhanced segmental motion of polymer chains and expansion of free volume. Abdillah et al. (2021) demonstrated that cassava starch–gelatin films showed progressive increases in WVP with rising glycerol content, attributing this to the increased polarity and plasticizing efficiency of glycerol. The type of plasticizer also plays a critical role. Sorbitol, due to its higher molecular weight and lower hygroscopicity, produces films with comparatively lower WVP than glycerol. Sharma et al. (2023) showed that sorbitol-plasticized finger millet starch films exhibited lower WVP values across equivalent plasticizer concentrations. This behavior was attributed to stronger hydrogen bonding and reduced chain mobility, which create a denser, less permeable matrix. Blends of glycerol and sorbitol can result in moderate permeability, balancing flexibility and barrier performance more effectively than either plasticizer alone. In contrast, highly hydrophilic plasticizers like urea promote greater water absorption and diffusion. When used in SPS films at 35% concentration, urea yielded WVP values of 1.88×10^{-10} $g \cdot mm/s \cdot m^2 \cdot Pa$, significantly higher than films plasticized with sorbitol or glycerol alone, underlining the importance of plasticizer selection relative to desired packaging properties (Sanyang et al., 2015). Compared to biopolymer-based films, synthetic materials such as low-density polyethylene (LDPE) exhibit far superior barrier properties. For instance, LDPE's WVP is as low as 0.0036×10^{-10} $g \cdot s^{-1} \cdot m^{-1} \cdot Pa^{-1}$, which in orders of magnitude lower than starch-based systems (Shahbazi et al., 2021). However, with careful optimization of plasticizer type and concentration, starch-based edible films can achieve acceptable WVP levels for a range of food packaging applications. The underlying mechanism of increased WVP in plasticized starch films relates to the water affinity of both the starch matrix and the plasticizer. Glycerol-rich systems exhibit greater permeability due to enhanced water uptake and chain mobility, while less hydrophilic additives like sorbitol or those forming stronger interactions with the starch chains can restrict water transport across the film. Therefore, tailoring the plasticizer system is key to achieving the desired balance between mechanical flexibility and moisture barrier performance.

Mechanical Properties

Tensile strength (TS), Young's modulus (YM), and elongation at break (EAB) are key mechanical properties that determine the integrity and performance of edible and biodegradable films used in food packaging. Tensile strength (TS) indicates the maximum stress a film can endure before rupture, while Young's modulus (YM) reflects its stiffness or resistance to deformation under stress. Unplasticized starch-based films generally exhibit high TS and YM values because of strong intermolecular hydrogen bonding and tightly packed polymer chains, resulting in rigid and brittle materials (Sothornvit & Krochta, 2001). However, the incorporation of plasticizers such as glycerol or sorbitol significantly reduces TS and YM, thereby enhancing flexibility but compromising mechanical strength. This occurs because plasticizers increase free volume and chain mobility by disrupting hydrogen bonds and loosening the film matrix (Tharanathan, 2003). Recent studies have confirmed that the magnitude of this reduction depends on both the type and concentration of plasticizer used. For instance, Hazrati et al. (2021) found that increasing glycerol content in Dioscorea hispida starch films decreased TS from 17.9 MPa to 6.4 MPa, while also reducing YM by nearly 60%, indicating a clear plasticization effect. Similarly, Sharma et al. (2023) reported that in Eleusine coracana (finger millet) starch films, both glycerol and sorbitol addition led to substantial reductions in tensile strength, although sorbitol-plasticized films retained higher stiffness due to stronger intermolecular interactions between sorbitol and starch chains. Elongation at break (EAB) measures how much a film can stretch before breaking, which reflects its flexibility, a critical property for packaging materials expected to deform without tearing. The addition of plasticizers significantly increases elongation at break (EAB). Glycerol-plasticized films typically show higher EAB values than unplasticized films, with increases proportional to glycerol concentration (Han, 2014). Sorbitol-plasticized films also exhibit increased stretchability, although to a slightly lesser extent due to their higher molecular weight and lower hygroscopicity (Hazrati et al., 2021; Nordin et al., 2021). A study on sago starch films demonstrated that the elongation at break (EAB) improved from 4.3% in unplasticized films to over 35% upon addition of 40% glycerol, while TS simultaneously decreased from 22 MPa to 8 MPa, highlighting the classic strength–flexibility trade-off. These findings underscore the importance of optimizing plasticizer selection not only to meet the mechanical requirements of the application but also to maintain a balance between film durability and usability (Lintang et al., 2021). The effect of plasticizers on the mechanical properties of film is shown in Table 8.1.

Table 8.1 Effect of different plasticizers on mechanical and barrier properties of film

Polymer base	Plasticizer	Mechanical property	Barrier property	References
Sugar palm starch	Glycerol	Tensile strength decreases significantly as concentration increases from 15% to 45%. Elongation at break increases from 15% to 30% but drops at 45%	WVP increases with plasticizer content. Highest WVP	Sanyang et al. (2015)
	Sorbitol	Higher tensile strength than glycerol-plasticized. Elongation at break increases from 15% to 30% and remains stable even at 45% plasticizer content	Lowest WVP values among plasticized films	
	Glycerol (GLY) + sorbitol	Tensile strength falls between G- and S-plasticized films (45%). Elongation increases from 15% to 30% but reduces at 45%	Sorbitol based film has slightly lower WVP values than glycerol based plasticized films	
Sodium alginate	0% plasticiser	High tensile strength, but very brittle with low elongation	Low oxygen permeability (OP) and water vapor transmission rate (WVTR)	Jost et al. (2014)
	Glycerol	Tensile strength deceases and elongation at break increases as concentration increases	Significantly increases OP and WVTR	
	Sorbitol	Provides better balance between flexibility and strength	No significant effect on OP and WVTR	
Curdlan	Pure Chinese lacquer (CL) film	High tensile strength, low elongation at break	Lowest water solubility & water vapor permeability	Chen et al. (2024)
	Glycerol	Best plasticizing effect; highest elongation at break and reduced tensile strength	WVP is lower than PEG	
	Polypropylene glycol	Moderate plasticizing effect; lower TS than GLY but higher than most others	Slightly better water resistance than GLY and EG	
	Polyethylene glycol	Reduces TS significantly, moderate EAB increase	Highest water solubility (WS) and water vapor permeability (WVP)	
	Xylitol	Decreases TS and increases EAB moderately; better than PEG but lower than GLY	Enhance water absorption and WVP	

Whey protein concentrate	Glycerol	Lower TS and YM, making films weaker and higher EAB	Highest moisture content and solubility in water	Pérez et al. (2016)
	Trehalose (Tre)	Highest TS and YM, making films the strongest initially & moderate EAB	Lowest MC & solubility	
	Glycerol (GLY) + Trehalos	Lower TS and YM compared to WPC/Tre films	Decreasing moisture content (MC) with increasing Tre content & decreasing solubility	
Soybean polysaccharide	Glycerol	Increases EAB but reduced tensile strength as concentration increases	Increasing plasticizer concentration increases water vapor permeability	Dong et al. (2023)
	Sorbitol	Better TS than glycerol film, moderate EAB	Least effect on WVP	
	Polyethylene glycol (400)	Increased EAB as concentration increases but less than glycerol	Moderate increase in WVP	

Thermal Property

The thermal properties of edible films and coatings are crucial in determining their performance, particularly in food packaging applications where they are exposed to a range of temperature conditions. Plasticizers are essential additives incorporated into these biopolymer films to improve flexibility, reduce brittleness, and enhance processability. Beyond mechanical effects, plasticizers profoundly impact the thermal behavior of edible films by altering degradation pathways, moisture interactions, and polymer network stability. Understanding how different plasticizers and their concentrations affect thermal stability is key to optimizing edible film formulations for improved shelf life, safety, and functionality (Dang & Yoksan, 2015; García et al., 1998; Sanyang et al., 2015). Thermogravimetric analysis (TGA) of starch films typically reveals a characteristic three-stage thermal degradation process. The first stage, occurring below 100 °C (often between 40 and 47 °C), is attributed to the evaporation of moisture, including weakly bound water molecules and low molecular-weight compounds. Studies consistently show that unplasticized (control) films exhibit greater mass loss in this phase compared to plasticized films due to their higher moisture content (Dang & Yoksan, 2015; Sanyang et al., 2016). The presence of plasticizers like glycerol tends to reduce water affinity by occupying sites that otherwise hold water molecules, thus decreasing moisture loss during initial heating (Sanyang et al., 2016). The second degradation phase, typically between 125 and 290 °C, corresponds to the volatilization and decomposition of the plasticizer molecules and bound water. Plasticizers, particularly glycerol, have relatively lower thermal stability compared to the starch polymer, and their degradation in this temperature range can lead to accelerated mass loss in plasticized films (Sanyang et al., 2015; Ilyas et al., 2018). The extent of this effect is dependent on the plasticizer concentration; higher plasticizer content increases the proportion of thermally labile molecules, thus lowering the onset temperature of significant mass loss (Sanyang et al., 2015). The third and most significant stage, commencing around 290–300 °C, involves the thermal breakdown of starch's saccharide rings and depolymerization of the polysaccharide carbon backbone. This phase represents the major structural degradation of the starch matrix, including the cleavage of glycosidic bonds and removal of hydrogen functional groups, leading to polymer chain scission (Nascimento et al., 2012). Although plasticizers tend to reduce the onset temperature of degradation due to their labile nature, interactions between plasticizers and starch molecules can modify the polymer network and influence the thermal decomposition pathway (Gutiérrez et al., 2015; Lintang et al., 2021). The chemical nature and molecular weight of the plasticizer significantly influence thermal behavior. For example, plasticizers with higher molecular weight or different polarity may interact differently with starch chains, altering thermal degradation patterns and film stability. Lower molecular weight plasticizers like glycerol tend to volatilize more readily, which can lead to earlier mass loss in TGA profiles, whereas larger plasticizers might confer different thermal stabilization effects (García et al., 1998; Rajan et al., 2006).

Influence of Plasticizer Type and Concentration on Thermal Properties

An increase in plasticizer concentration generally results in a decrease in the thermal resistance of edible films. For example, glycerol-plasticized arrowroot starch films with glycerol content ranging from 15% to 45% show a higher mass loss at degradation temperatures between 318 and 322 °C compared to unplasticized films, which show comparatively lower weight loss at around 314 °C. This suggests that higher plasticizer loading facilitate faster thermal degradation, likely due to the plasticizer molecules disrupting the hydrogen bonding network and weakening intermolecular forces within the starch polymer matrix (Tarique et al., 2021). Interestingly, despite promoting earlier degradation, plasticizers can also improve the thermal stability of films in terms of the maximum degradation temperature (Tmax). Studies report that the Tmax shifts to higher temperatures in plasticized films compared to controls, indicating stronger plasticizer–starch interactions that enhance polymer compatibility and retard decomposition at higher temperatures (Lintang et al., 2021; Gutiérrez et al., 2015; Nordin et al., 2020). This dual effect-lower onset thermal resistance but increased Tmax reflects the complex role of plasticizers in modifying the polymer network. Furthermore, residual mass after thermal degradation decreases with increasing plasticizer content. Films containing higher glycerol percentages show lower residue at 600 °C compared to controls, attributable to volatilization and decomposition of plasticizers during heating, resulting in reduced solid char formation (Sanyang et al., 2015). The effect of the type and amount of plasticizer in the film-forming solution on the applicability and performance of edible films is illustrated in Table 8.2.

Table 8.2 The effect of the type and amount of plasticizer on the thermal properties of film

Plasticizer (concentration)	Max degradation temp (T_{max}, °C)	Mass loss near T_{max} (%)	Thermal stability	References
None (Control- 0%)	~314	75.09	Moderate thermal stability	Sanyang et al. (2016)
Glycerol (15%)	318–322	78.64	Increased mass loss; Tmax increased	Sanyang et al. (2015)
Glycerol (30%)	318–322	79.46	Higher glycerol further lowers resistance but raises Tmax	Sanyang et al. (2015)
Glycerol (45%)	318–322	57.46	Lowest residue due to plasticizer volatilization	Sanyang et al. (2015)
Sorbitol (Variable)	Higher than glycerol	Lower than glycerol	Sorbitol generally provides higher thermal stability	Ilyas et al. (2018)
PEG (Variable)	Higher than glycerol	Not specified	PEG shows higher Tmax due to better compatibility	García et al. (1998)

Effect of Plasticizers on Surface and Cross-Sectional Morphology of Film

SEM images consistently demonstrate that control films without plasticizers exhibit rough, porous, or cracked surfaces, indicating poor film-forming capability and weak internal bonding. For instance, unplasticized Chinese lacquer (CL) films presented rough and loosely structured surfaces, as well as cracked cross-sections, pointing to a lack of cohesion in the matrix (Liu et al., 2025). Similarly, urea-based films without plasticizer showed coarse textures laden with unmelted starch agglomerates and structural imperfections, such as pores and cracks (Liu et al., 2013). Upon addition of plasticizers, particularly at higher concentrations, SEM micrographs reveal smoother, more continuous surfaces and betterintegrated film structures. This smoothing effect arises because of the plasticizer's ability to reduce internal stress and improve chain mobility, resulting in more uniform matrix formation during film casting and drying. For instance, in CL films plasticized with glycerol (GLY) or polyethylene glycol (PEG), the cross-sections became significantly smoother and devoid of cracks, implying good miscibility and enhanced interaction between the plasticizer and starch matrix (Xie et al., 2025). Likewise, fructose-plasticized wheat starch films at 35% concentration exhibited the smoothest surfaces among the formulations tested, indicating superior film formation (Mohammed et al., 2022). Different plasticizers exhibit varying degrees of compatibility with starch or biopolymer matrixes, which is reflected in the microstructural features. Glycerol, one of the most commonly used plasticizers, showed excellent interaction with arrowroot starch (AS) films. SEM analysis revealed that higher concentrations of glycerol (e.g., 45%) produced films with more homogeneous cross-sections compared to films with lower glycerol content (15% and 30%), suggesting improved dispersion and lower interfacial tension at higher plasticizer loadings (Abera et al., 2020). This morphological uniformity correlates with better mechanical behavior and flexibility, attributed to enhanced plasticization. In contrast, films plasticized with sorbitol or ethylene glycol displayed denser and more irregular cross-sections with ridges and surface patterns, indicating stronger internal cohesion and less compatibility (Xie et al., 2025). These features are likely due to the slower diffusion and limited interaction of bulkier or less polar plasticizers with the starch network, resulting in phase separation or localized stiffness. Importantly, SEM findings are supported by related mechanical data from other studies, where smoother and more homogeneous film structures areassociated with optimal plasticizer type and concentration also yielded better tensile properties and reduced brittleness (Sun et al., 2020). In general, surface homogeneity is not only an aesthetic feature but also a marker of mechanical integrity and functional performance. In summary, SEM provides compelling evidence of the role plasticizers play in modulating the film morphology of starch-based edible coatings. The type and concentration of plasticizer influence the degree of homogeneity, crack formation, and surface smoothness, which in turn affect the functional properties of the films. Glycerol and PEG, in particular, demonstrate superior compatibility with starch matrixes, while other plasticizers like sorbitol or EG may require optimization to prevent structural heterogeneity. These findings underscore the importance of morphological characterization in plasticizer selection for edible film development.

Effect of Plasticizers on the Transparency of Edible Films

Film transparency plays a pivotal role in determining the visual appeal and consumer acceptability of food packaging materials. In particular, high transparency is preferred in edible films intended for applications where product visibility is essential. Transparency, typically evaluated through light transmittance or the inverse measure of opacity, is influenced by various factors, including the type and concentration of plasticizers, film thickness, surface morphology, crystallinity, and moisture content (Wen et al., 2022). The incorporation of plasticizers has been found to significantly enhance the light transmittance of starch-based edible films. In a study on Chinese lacquer (CL) films, all plasticized formulations demonstrated improved transparency compared with the control (unplasticized) film. Among the various plasticizers tested, polyethylene glycol (PEG) yielded the highest light transmittance, followed by ethylene glycol (EG), sorbitol (SOR), xylose (XY), propylene glycol (PRO), and glycerol (GLY), respectively (Corrêa-Filho et al., 2024). This improvement is attributed primarily to modifications in the film's microstructure. For instance, PEG's ability to lower film crystallinity and promote a smoother surface was key to its superior transmittance. Similarly, EG-enhanced films exhibited high transparency, likely due to a combination of elevated moisture content and favorable morphological characteristics, even though EG tends to increase crystallinity. In contrast, plasticizers such as PRO and GLY, despite their ability to increase moisture content, resulted in slightly lower transmittance values. This is likely due to the relatively rougher surfaces observed in these films, which scatter light and reduce transparency. Nevertheless, their performance was still better than that of unplasticized films, indicating the general benefit of plasticizer addition in reducing opacity (Li et al., 2022). A related study on wheat starch-based films corroborated these findings, demonstrating that plasticizer concentration plays a critical role in determining film transparency. Films plasticized with fructose at 25% and 35% (w/w, dry starch basis) showed the lowest opacity values (0.620 and 0.625 nm/mm, respectively), indicating high light transmission and uniform matrix formation. In contrast, films with lower plasticizer content (15%) exhibited higher opacity, likely due to phase separation or incomplete interaction between the starch and the plasticizer. Urea-plasticized films, especially at higher concentrations, showed the highest opacity, suggesting poor dispersion and possibly increased light scattering due to surface roughness or internal discontinuities (Domene-López et al., 2019). These findings highlight the interplay between polymer–plasticizer compatibility and microstructural features in determining film transparency. Plasticizers that promote molecular mobility and reduce crystalline domains tend to enhance transparency. Furthermore, moisture retention associated with certain plasticizers may also improve light transmittance, though excessive moisture may compromise mechanical stability. Overall, plasticizer type and concentration must be optimized not only for mechanical and barrier properties but also for achieving the desired optical clarity in edible films.

Conclusion

Plasticizers are vital additives that modify the functional performance of edible films and coatings by improving flexibility, reducing brittleness, and enhancing processing characteristics. Their incorporation into starch-based biopolymer matrixes alters the structural arrangement by disrupting strong intermolecular hydrogen bonds and increasing polymer chain mobility. This restructuring enables improved film-forming ability and mechanical compliance, making the films more suitable for packaging applications. Among the plasticizers explored, glycerol remains widely used due to its excellent miscibility and ability to impart flexibility, reflected in increased elongation at break. However, at higher concentrations, it compromises tensile strength and barrier properties, such as increased water vapor permeability (WVP), because of its hydrophilic nature. Sorbitol, by contrast, maintains higher tensile strength and provides better moisture resistance, offering an advantage in applications where mechanical strength and water barrier performance are critical. Other plasticizers like PEG, fructose, and urea show varied effects: PEG improves transparency and thermal stability; fructose enhances optical clarity and water resistance; while urea increases WVP, making it less suitable for moisture-sensitive uses. Structural and morphological analyses, such as scanning electron microscopy (SEM), affirm the relationship between plasticizer compatibility and film uniformity. Films with glycerol and PEG exhibit smoother, crack-free surfaces and better internal cohesion, translating to improved mechanical integrity. Similarly, thermal degradation studies reveal that plasticizers influence both the onset and maximum degradation temperatures. While they reduce thermal resistance at initial stages, compatible plasticizer–polymer interactions can increase the maximum degradation temperature (T_{max}), improving film stability at higher temperatures. Overall, the choice of plasticizer must align with the specific performance requirements of the target application—whether flexibility, barrier properties, thermal resistance, or transparency is the priority. A balanced formulation that leverages the synergistic effects of different plasticizers or optimal concentrations can significantly enhance the functional potential of starch-based edible films, paving the way for sustainable alternatives to conventional plastic packaging in the food industry.

Acknowledgments The authors are thankful to the National Institute of Food Technology Entrepreneurship and Management, Kundli, Sonepat (Haryana), India, for providing the infrastructure and other support for this work.

References

Abdillah, M., Nisa, F. A. K., & Arifin, S. (2021). The effect of glycerol concentration on the characteristics ofcassava starch-gelatin edible film. IOP Conference Series: Earth and Environmental Science, 733(1), 012019. https://doi.org/10.1088/1755-1315/733/1/012019

Abera, M. E., Tekele, F. Y., Tadesse, B. K., Tsegay, S. M., Weldu, A. A., &Mebrahtu, A. F. (2020). Optimization ofglycerol concentration on the physical, mechanical, and barrier properties of biodegradable film based on ensetstarch and gelatin. *Heliyon, 6*(12), e05669. https://doi.org/10.1016/j.heliyon.2020.e05669

Aiken, W. H. (1960). In R. L. Whistler & M. L. Wolfrom (Eds.), *Polysaccharide Chemistry* (pp. 535–560). AcademicPress.

Alhanish, A., & Abu Ghalia, M. (2021). Developments of biobased plasticizers for compostable polymers in the green packaging applications: A review. *Biotechnology Progress, 37*(6), e3210.

Averous, L., Moro, L., Dole, P., & Fringant, C. (2000). Properties of thermoplastic blends: Starch–polycaprolactone. *Polymer, 41*(11), 4157–4167.

Aziz, M. S. A., Salama, H. E., & Sabaa, M. W. (2018). Biobased alginate/castor oil edible films for active food packaging. *LWT, 96*, 455–460.

Azucena Castro-Yobal, M., Contreras-Oliva, A., Saucedo-Rivalcoba, V., Rivera-Armenta, J. L., Hernández-Ramírez, G., Salinas-Ruiz, J., & Herrera-Corredor, A. (2021). Evaluation of physicochemical properties of film-based alginate for food packing applications. *E-Polymers, 21*(1), 82–95.

Banker, G. S. (1966). Film coating theory and practice. *Journal of Pharmaceutical Sciences, 55*(1), 81–89.

Barbut, S., & Harper, B. A. (2019). Dried Ca-alginate films: Effects of glycerol, relative humidity, soy fibers, and carrageenan. *LWT, 103*, 260–265.

Basiak, E., Lenart, A., & Debeaufort, F. (2018). How glycerol and water contents affect the structural and functional properties of starch-based edible films. *Polymers, 10*(4), 412.

Bocqué, M., Voirin, C., Lapinte, V., Caillol, S., & Robin, J. J. (2016). Petro-based and bio-based plasticizers: Chemical structures to plasticizing properties. *Journal of Polymer Science Part A: Polymer Chemistry, 54*(1), 11–33.

Bonifacio, A., Bonetti, L., Piantanida, E., & De Nardo, L. (2023). Plasticizer design strategies enabling advanced applications of cellulose acetate. *European Polymer Journal, 197*, 112360.

Caicedo, C., Díaz-Cruz, C. A., Jiménez-Regalado, E. J., & Aguirre-Loredo, R. Y. (2022). Effect of plasticizer content on mechanical and water vapor permeability of maize starch/PVOH/chitosan composite films. *Materials, 15*(4), 1274.

Chen, H., Wu, C., Feng, X., He, M., Zhu, X., Li, Y., & Teng, F. (2022). Effects of two fatty acids on soy protein isolate/sodium alginate edible films: Structures and properties. *LWT, 159*, 113221.

Chen, P., Xie, F., Tang, F., & McNally, T. (2020). Unexpected plasticization effects on the structure and properties of polyelectrolyte complexed chitosan/alginate materials. *ACS Applied Polymer Materials, 2*(7), 2957–2966.

Chen, J., He, J., Fan, Z., & Chen, J. (2024). Effect of different plasticizers on the properties of pure Chinese lacquerfilm. *Journal of Coatings Technology and Research, 21*(1), 163–175. https://doi.org/10.1007/s11998-023-00827-0

Cheyney, A. L., Bair, H. E., & Geyer, C. J. (1979). The effect of plasticizer structure on the glass transitiontemperature of PVC. *Journal of Applied Polymer Science, 23*(11), 3121–3133.

Chiaoprakobkij, N., Seetabhawang, S., Sanchavanakit, N., & Phisalaphong, M. (2019). Fabrication and characterization of novel bacterial cellulose/alginate/gelatin biocomposite film. *Journal of Biomaterials Science, Polymer Edition, 30*(11), 961–982.

Consolati, G., Nichetti, D., & Quasso, F. (2022). Free volume expansion in some polybutadiene–acrylonitrilerubbers: comparison between theory and experiments. *Polymer International, 71*(8), 918–926. https://doi.org/10.1002/pi.6393

Corrêa-Filho, L. C., dos Santos Junior, J. R., Ramos, A. V., Martinazzo, A. P., Habert, A. C., de Carvalho, C. W. P., et al. (2024). Chitosan-based nanocomposite films with carnauba wax, rosin resin, and zinc oxide nanoparticles. *Food Research International, 188*, 114475.

Dang, K. M., & Yoksan, R. (2015). Development of thermoplastic starch blown film by incorporating plasticized chitosan. *Carbohydrate Polymers, 115*, 575–581.

Debeaufort, F., Quezada-Gallo, J. A., & Voilley, A. (1998). Edible films and coatings: Tomorrow's packagings: A review. *Critical Reviews in Food Science and Nutrition, 38*(4), 299–313.

Dong, R., Liu, W., Xu, W., Han, H., & Zhou, Y. (2023). Preparation and characterization of polysaccharide-basededible films plasticized with glycerol: A review. *Food Hydrocolloids, 139*, 108573. https://doi.org/10.1016/j.foodhyd.2023.108573

De Souza, A. C., Akyildiz, A. D., Furtado, D. F., & Bilodeau, M. (2021). Influence of preparation method on theproperties of corn starch films: A review. *Carbohydrate Polymers, 264*, 118012. https://doi.org/10.1016/j.carbpol.2021.118012

Donhowe, I. G., & Fennema, O. (1993). The effects of plasticizers on crystallinity, permeability, and mechanical properties of methylcellulose films. *Journal of Food Processing and Preservation, 17*(4), 247–257.

Domene-López, D., García-Quesada, J. C., Martin-Gullon, I., & Montalbán, M. G. (2019). Influence of starch composition and molecular weight on physicochemical properties of biodegradable films. *Polymers, 11*(7), 1084.

Doolittle, A. K. (1954). *The technology of solvents and plasticizers*. Wiley.

D'souza, A. A., & Shegokar, R. (2016). Polyethylene glycol (PEG): A versatile polymer for pharmaceutical applications. *Expert Opinion on Drug Delivery, 13*(9), 1257–1275.

Eslami, Z., Elkoun, S., Robert, M., & Adjallé, K. (2023). A review of the effect of plasticizers on the physical and mechanical properties of alginate-based films. *Molecules, 28*(18), 6637.

Galdeano, M. C., Grossmann, M. V. E., Mali, S., Bello-Perez, L. A., Garcia, M. A., & Zamudio-Flores, P. B. (2009). Effects of production process and plasticizers on stability of films and sheets of oat starch. *Materials Science and Engineering: C, 29*(2), 492–498.

Gao, C., Pollet, E., & Avérous, L. (2017). Properties of glycerol-plasticized alginate films obtained by thermo-mechanical mixing. *Food Hydrocolloids, 63*, 414–420.

García, M. A., Martino, M. N., & Zaritzky, N. E. (1998). Plasticized starch-based coatings to improve strawberry (Fragaria × ananassa) quality and stability. *Journal of Agricultural and Food Chemistry, 46*(9), 3758–3767.

Godwin, A. D. (2024). Plasticizers. In *Applied plastics engineering handbook* (pp. 595–618). William Andrew Publishing.

González-Torres, B., Robles-García, M. Á., Gutiérrez-Lomelí, M., Padilla-Frausto, J. J., Navarro-Villarruel, C. L., Del-Toro-Sánchez, C. L., et al. (2021). Combination of sorbitol and glycerol, as plasticizers, and oxidized starch improves the physicochemical characteristics of films for food preservation. *Polymers, 13*(19), 3356.

Gupta, D., Lall, A., Kumar, S., Patil, T. D., & Gaikwad, K. K. (2024). Plant-based edible films and coatings for food-packaging applications: Recent advances, applications, and trends. *Sustainable Food Technology, 2*(5), 1428–1455.

Gutiérrez, T. J., Tapia, M. S., & Pérez, E. (2015). Effect of different plasticizers on the thermal and functionalproperties of thermoplastic cassava starch. *Polymer Degradation and Stability, 118*, 26–36. https://doi.org/10.1016/j.polymdegradstab.2015.04.017

Gutiérrez-Jara, C., Bilbao-Sainz, C., McHugh, T., Chiou, B. S., Williams, T., & Villalobos-Carvajal, R. (2020). Physical, mechanical and transport properties of emulsified films based on alginate with soybean oil: Effects of soybean oil concentration, number of passes and degree of surface crosslinking. *Food Hydrocolloids, 109*, 106133.

Han, J. H. (2014). Edible Films and Coatings: A Review. *Comprehensive Reviews in Food Science and FoodSafety, 13*(6), 1162–1172. https://doi.org/10.1111/1541-4337.12071

Hazrati, K. Z., Sapuan, S. M., Zuhri, M. Y. M., & Jumaidin, R. (2021). Effect of plasticizers on physical, thermal, and tensile properties of thermoplastic films based on Dioscorea hispida starch. *International Journal of Biological Macromolecules, 185*, 219–228.

Ilyas, R. A., Sapuan, S. M., Ishak, M. R., & Zainudin, E. S. (2018). Thermogravimetric and mechanical properties of starch-based films plasticized with glycerol. *Polymers, 10*(2), 190.

Isikgor, F. H., & Becer, C. R. (2015). Lignocellulosic biomass: A sustainable platform for the production of bio-based chemicals and polymers. *Polymer Chemistry, 6*(25), 4497–4559.

Janakiram, C., Kumar, C. D., & Joseph, J. (2017). Xylitol in preventing dental caries: A systematic review and meta-analyses. *Journal of Natural Science, Biology, and Medicine, 8*(1), 16.

Janowicz, M., Galus, S., Ciurzyńska, A., & Nowacka, M. (2023). The potential of edible films, sheets, and coatings based on fruits and vegetables in the context of sustainable food packaging development. *Polymers, 15*(21), 4231.

Jost, V., Kobsik, K., Schmid, M., & Noller, K. (2014). Influence of plasticiser on the barrier, mechanical and grease resistance properties of alginate cast films. *Carbohydrate Polymers, 110*, 309–319.

Kadzińska, J., Janowicz, M., & Ciurzyńska, A. (2020). Influence of vegetable oils addition on the selected physicalproperties of apple–sodium alginate edible films. *Polymer Bulletin, 77*(2), 883–900. https://doi.org/10.1007/s00289-019-02840-x

Krauskopf, L. G., & Godwin, A. D. (2005). Plasticizers. In C. E. Wilkes, J. W. Summers, & C. A. Daniels (Eds.),PVC Handbook (pp. 173–193). *Hanser*

Kumar, N., & Neeraj. (2019). Polysaccharide-based component and their relevance in edible film/ coating: A review.*Nutrition & Food Science, 49*(5), 793-823.

Kumar, N., Ojha, A., & Singh, R. (2019). Preparation and characterization of chitosan-pullulan blended edible filmsenrich with pomegranate peel extract. *Reactive and Functional Polymers, 144*, 104350.

Kumar, N., Pratibha, Petkoska, A. T., & Singla, M. (2022). Natural gums for fruits and vegetables preservation: areview. *Gums, resins and latexes of plant origin: chemistry, biological activities and uses*, 81-116.

Kumar, N., Pratibha, Prasad, J., Yadav, A., Upadhyay, A., Neeraj, & Kieliszek, M. (2023). Recent trends in ediblepackaging for food applications—Perspective for the future. *Food Engineering Reviews, 15*(4), 718-747

Kumar, N., Khan, A. A., Pyngrope, D., Alanazi, A. M., Upadhyay, A., & Shukla, S. (2024). Development andcharacterization of novel starch (mango kernel and litchi seed) based active edible coatings and films usingultrasonication: Effects on postharvest shelf life of Khasi mandarins. *Sustainable Chemistry and Pharmacy, 39*,101610.

Li, H., Liu, C., Sun, J., & Lv, S. (2022). Bioactive edible sodium alginate films incorporated with tannic acid as antimicrobial and antioxidative food packaging. *Food, 11*(19), 3044.

Lintang, T. I., Nurnida, J., Fitria, Z., & Anam, A. (2021). The effect of glycerol concentration on the physical andmechanical properties of cassava starch-chitosan film. IOP Conference Series: *Earth and Environmental Science,629*(1), 012015. https://doi.org/10.1088/1755-1315/629/1/012015

Liu, F., Cao, Y., Ren, J., Xie, Y., Xiao, X., Zou, Y., et al. (2025). Optimization of starch foam extrusion through PVA polymerization, moisture content control, and CMCS incorporation for enhanced antibacterial cushioning packaging. *Carbohydrate Polymers, 347*, 122763.

Liu, H., Adhikari, R., Guo, Q., & Adhikari, B. (2013). Preparation and characterization of glycerol plasticized (high-amylose) starch–chitosan films. *Journal of Food Engineering, 116*(2), 588–597.

Makhathini, N., Kumar, N., & Fawole, O. A. (2025). Enhancing circular bioeconomy: Alginate-cellulose nanofibrefilms/coatings functionalized with encapsulated pomegranate peel extract for postharvest preservation ofpomegranate arils. *International Journal of Biological Macromolecules, 309*, 142848.

Marismandani, A. D. P., & Husni, A. (2020). Development and characterization of biobased alginate/glycerol/virgin coconut oil as biodegradable packaging. In *E3S web of conferences* (Vol. 147, p. 03016). EDP Sciences.

Mark, D., Mader, B., & Wypych, G. (2017). *Plasticizers. In G. Wypych (Ed.), Handbook of Polymer Additives* (2nded., pp. 109–165). ChemTec Publishing.

Mohammed, A. A., Hasan, Z., Omran, A. A. B., Elfaghi, A. M., Khattak, M. A., Ilyas, R. A., & Sapuan, S. M. (2022). Effect of various plasticizers in different concentrations on physical, thermal, mechanical, and structural properties of wheat starch-based films. *Polymers, 15*(1), 63.

Moore, D. J., J. M. B., & W. O. H. (2006). The influence of plasticizers on the mechanical and thermal properties ofstarch-based bioplastics. *Journal of Applied Polymer Science, 100*(1), 589–599. https://doi.org/10.1002/app.22910

Nascimento, T. A., L. A. C. S. A. A. L., J. L. R. A., B. N. A., & W. D. L. V. J. (2012). Thermal, mechanical, andbarrier properties of biodegradable films based on cassava starch, alginate, and glycerol. *Journal ofAppliedPolymer Science, 125*(1), E618–E627. https://doi.org/10.1002/app.36625

Nordin, N., Othman, S. H., Rashid, S. A., & Basha, R. K. (2020). Effects of glycerol and thymol on physical, mechanical, and thermal properties of corn starch films. *Food Hydrocolloids, 106*, 105884.

Nordin, N., Sanyang, M. L., Mohd-Setapar, S. H., Yaakob, Z., & Ahmad, N. (2021). Effect of sorbitol and glycerolblend on the physicochemical and barrier properties of sugar palm starch film. *Journal of Polymers and theEnvironment, 29(1)*, 169–181. https://doi.org/10.1007/s10924-020-01869-3

Olivas, G. I., & Barbosa-Cánovas, G. V. (2008). Alginate–calcium films: Water vapor permeability and mechanical properties as affected by plasticizer and relative humidity. *LWT-Food Science and Technology, 41*(2), 359–366.

Olivas, G. I., & Barbosa-Cánovas, G. V. (2008). Alginate films with four different plasticizers and two different M/Gratios. *Journal of Food Engineering, 88*(1), 127–134. https://doi.org/10.1016/j.jfoodeng.2008.02.016

Otoni, C. G., Avena-Bustillos, R. J., Azeredo, H. M., Lorevice, M. V., Moura, M. R., Mattoso, L. H., & McHugh, T. H. (2017). Recent advances on edible films based on fruits and vegetables—A review. *Comprehensive Reviews in Food Science and Food Safety, 16*(5), 1151–1169.

Paixão, L. C., Lopes, I. A., Barros Filho, A. K. D., & Santana, A. A. (2019). Alginate biofilms plasticized with hydrophilic and hydrophobic plasticizers for application in food packaging. *Journal of Applied Polymer Science, 136*(48), 48263.

Parreidt, T. S., Müller, K., & Schmid, M. (2018). Alginate-based edible films and coatings for food packagingapplications. *Foods, 7*(10), 170. https://doi.org/10.3390/foods7100170

Perez-Vazquez, A., Barciela, P., Carpena, M., & Prieto, M. A. (2023). Edible coatings as a natural packaging system to improve fruit and vegetable shelf life and quality. *Food, 12*(19), 3570.

Pérez, L., Ruz, M., Cussons, K., & Nerín, C. (2016). Effect of glycerol, trehalose, and their mixture on themechanical, water barrier and surface properties of whey protein concentrate films. *Food Hydrocolloids, 56*, 280–288. https://doi.org/10.1016/j.foodhyd.2015.12.029

Pillai, A. R., Eapen, A. S., Zhang, W., & Roy, S. (2024). Polysaccharide-based edible biopolymer-based coatings for fruit preservation: A review. *Food, 13*(10), 1529.

Pooja, N., Chakraborty, I., Mal, S. S., Prasad, A. S. B., Mahato, K. K., & Mazumder, N. (2024). Evaluation of physicochemical properties of citric acid crosslinked starch elastomers reinforced with silicon dioxide. *RSC Advances, 14*(1), 139–146.

Rahman, M., & Brazel, C. S. (2004). The plasticizer market: An assessment of traditional plasticizers and researchtrends to meet new challenges. *Progress in Polymer Science, 29*(12), 1223–1248. https://doi.org/10.1016/j.progpolymsci.2004.10.001

Rajan, A. S., S. S., & P. J. (2006). Preparation and characterization of biodegradable films from cassava starch.*Polymer Degradation and Stability, 91*(3), 540–547. https://doi.org/10.1016/j.polymdegradstab.2005.06.017

Razavi, S. M. A., Amini, A. M., & Zahedi, Y. (2015). Characterisation of a new biodegradable edible film based on sage seed gum: Influence of plasticiser type and concentration. *Food Hydrocolloids, 43*, 290–298.

Remuñán-lópez, C., & Bodmeier, R. (1996). Mechanical and water vapor transmission properties of polysaccharide films. *Drug Development and Industrial Pharmacy, 22*(12), 1201–1209.

Rostamabadi, H., Demirkesen, I., Colussi, R., Roy, S., Tabassum, N., de Oliveira Filho, J. G., & Falsafi, S. R. (2024). Recent trends in the application of films and coatings based on starch, cellulose, chitin, chitosan, xanthan, gellan, pullulan, Arabic gum, alginate, pectin, and carrageenan in food packaging. *Food Frontiers, 5*(2), 350–391.

Sahari, J., Sapuan, S. M., Zainudin, E. S., & Maleque, M. A. (2012). A new approach to use Arenga pinnata as sustainable biopolymer: Effects of plasticizers on physical properties. *Procedia Chemistry, 4*, 254–259.

Sánchez-González, L., Caro, A. A., Navarro-Tarazaga, M. L., & González-Martínez, C. (2025). Role of plasticizers in enhancing mechanical and barrier properties of biopolymer-based edible films: Mechanisms and recent advances. *Frontiers in Sustainable Food Systems*, 1569458.

Santana, A. A., & Kieckbusch, T. G. (2013). Physical evaluation of biodegradable films of calcium alginate plasticized with polyols. *Brazilian Journal of Chemical Engineering, 30*, 835–845.

Sanyang, M. L., Sapuan, S. M., Jawaid, M., Ishak, M. R., & Sahari, J. (2015). Degradation and water barrierproperties of sugar palm starch biopolymer film plasticized with glycerol. *BioResources, 10(3)*, 5424–5434. https://doi.org/10.15376/biores.10.3.5424-5434

Sanyang, M. L., Sapuan, S. M., Jawaid, M., Ishak, M. R., & Sahari, J. (2016). Effect of plasticizer type and concentration on physical properties of biodegradable films based on sugar palm (Arenga pinnata) starch for food packaging. *Journal of Food Science and Technology, 53*(1), 326–336.

Sears, J. K., & Darby, J. R. (1982). Mechanism of plasticizer action. In *The technology of plasticizers* (pp. 35–77). Wiley.

Shahbazi, M., Ramezanpour, S., & Ghasempour, Z. (2021). Preparation and characterization of bionanocompositefilms based on corn starch and clay reinforced with cellulose nanocrystals. *Polymer Composites, 42*(12), 6542–6552. https://doi.org/10.1002/pc.26330

Sharma, R., Gautam, N., Giribabu, G., Kumar, R., & Yadav, S. (2023). Synergistic effect of plasticizer and cross-linking agent on the properties of finger millet starch-based film. *Starch-Stärke, 75*(9–10), 2200251.

Sharmin, N., Sone, I., Walsh, J. L., Sivertsvik, M., & Fernandez, E. N. (2021). Effect of citric acid and plasma activated water on the functional properties of sodium alginate for potential food packaging applications. *Food Packaging and Shelf Life, 29*, 100733.

Skelly, P. W., Li, L., & Braslau, R. (2022). Internal plasticization of PVC. *Polymer Reviews, 62*(3), 485–528.

Sothornvit, R., & Krochta, J. M. (2001). Plasticizer effect on mechanical properties of β-lactoglobulin films. *Journal of Food Engineering, 50*(3), 149–155.

Souza, R. C. R., & Andrade, C. T. (2001). Processing and properties of thermoplastic starch and its blends withsodium alginate. *Journal of Applied Polymer Science, 82*(10), 2420–2429. https://doi.org/10.1002/app.2081

SpecialChem Tech Library. (2023). *Polymer plasticization—Theories, types, process & key factors*. Polymer Additives Tech Library.

Sun, Y., Liu, Z., Zhang, L., Wang, X., & Li, L. (2020). Effects of plasticizer type and concentration on rheological,physico-mechanical and structural properties of chitosan/zein film. *International Journal of BiologicalMacromolecules, 143*, 334-340.

Syarifuddin, A., Dirpan, A., & Mahendradatta, M. (2017, December). Physical, mechanical, and barrier properties of sodium alginate/gelatin emulsion based-films incorporated with canola oil. In *IOP conference series: Earth and environmental science* (Vol. 101, No. 1, p. 012019). IOP Publishing.

Tarique, J. S. M. S., Sapuan, S. M., & Khalina, A. (2021). Effect of glycerol plasticizer loading on the physical, mechanical, thermal, and barrier properties of arrowroot (Maranta arundinacea) starch biopolymers. *Scientific Reports, 11*(1), 13900.

Thakur, K., Thakur, N., Sharma, N., & Mahajan, K. (2021). A comprehensive review on plasticizing agent. *Journal of Emerging Technologies and Innovative Research, 8*, a79–a85.

Tharanathan, R. N. (2003). Biodegradable films and composite coatings: Past, present and future. *Trends in Food Science & Technology, 14*(3), 71–78.

Tong, Q., Xiao, Q., & Lim, L. T. (2013). Effects of glycerol, sorbitol, xylitol and fructose plasticisers on mechanical and moisture barrier properties of pullulan–alginate–carboxymethylcellulose blend films. *International Journal of Food Science and Technology, 48*(4), 870–878.

Umai, D., Kayalvizhi, R., Kumar, V., & Jacob, S. (2022). Xylitol: Bioproduction and applications–a review. *Frontiers in Sustainability, 3*, 826190.

Uneyama, T., & Masubuchi, Y. (2012). Multi-chain slip-spring model for entangled polymer dynamics. *The Journal of Chemical Physics, 137*(15), 154902.

Wilson, K. (1995). Plasticizers: Types, uses, and safety. *Polymer Engineering and Science, 35*(9), 732–737.

Xie, F., Wu, Y., Liu, X., Liu, C., & Zhang, J. (2025). The effect of plasticizers on the structure and properties ofChinese lacquer composite films. *Progress in Organic Coatings, 194*, 108605. https://doi.org/10.1016/j.porgcoat.2025.108605

Yermagambetova, A. D., Tazhibayeva, S. M., Tyussyupova, B. B., Musabekov, K. B., & Pastorino, L. (2024). Effect of plasticizers on the rheological properties of xanthan gum-starch biodegradable films. *Heliyon, 10*(14).

Zhang, Y., & Han, J. H. (2006). Plasticization of pea starch films with monosaccharides and polyols. *Journal of Food Science, 71*(6), E253–E261.

Chapter 9
Effects of Essential Oils on Properties of Edible Coating and Films

Mandavi Goswami, Vikrant Bodana, and Vimal Katiyar

Abstract In the world market nowadays, you can buy all kinds of food shipped anywhere in the world. The only problem with fruits and vegetables is that they go bad quickly. This fresh produce usually gets destroyed during shipment because it does not last long, which leads to microbial illness. The edible coating with an essential oil is very significant for stopping this kind of contamination and mechanical damage. The essential oil in edible coatings helps slow the growth of microbes, and the coatings form a very thin film around the fruit's surface that protects it from mechanical damage. Finding the right dose of essential oil is very important because it impacts the taste, smell, and look of any fresh food. Essential oils can improve the taste and smell of the fruit, but this can sometimes make the fruit worse. Too much essential oil is what makes some fruits turn brown. Essential oils have antibacterial and antioxidant properties that improve the quality of the fruit and extend the shelf life of fresh food by slowing down the growth of microbes. This chapter of the book talks about the many ways to get essential oils and their influence on edible coatings for different kinds of fresh produce.

Keywords Edible coatings · Essential oils · Antimicrobial mechanism · Shelf life

Introduction

Ethereal oils, commonly referred to as essential oils, are concentrated, volatile liquids extracted from plant tissues, exhibiting high solubility in oil and low affinity for water (Bakkali et al., 2008). These oils are produced through a range of techniques, including steam distillation, expression, enfleurage, resin distillation, and

M. Goswami · V. Bodana · V. Katiyar (✉)
Centre For Sustainable Polymers, Indian Institute of Technology Guwahati, Guwahati, Assam, India
e-mail: vkatiyar@iitg.ac.in

N. Kumar et al. (eds.), *Trends in Edible Packaging: A sustainable approaches for postharvest management of fruits and vegetables*,
https://doi.org/10.1007/978-3-032-13199-7_9

cold pressing, as well as more specialized methods such as solvent extraction and embedding in wax (Baser & Buchbauer, 2015). As a result, they are now ubiquitous in the formulation of soaps, perfumes, scented candles, personal care items, and surface cleaners. Besides their cosmetic and olfactory uses, the food and beverage industries deploy oils to impart flavor, and the medical and wellness communities employ them in aromatherapy protocols (Hassid et al., 2025). Albeit valued, they also warrant prudence, for excessive application or improper dilution may precipitate contact dermatitis, respiratory sensitivity, and other cutaneous or systemic toxicity (Tisserand & Young, 2013; Ju et al., 2019; Froiio et al., 2019; Fathi et al., 2021; Li et al., 2023).

Nevertheless, to remain harmonious and prevent masking the fruit or vegetable's inherent flavors, the dosage of each essential oil must be carefully calibrated. A concentration that deviates to the high end can dominate the palate, overshadowing the delicate balance that the fresh produce itself provides, which is why finding the right level is essential for maintaining both consumer acceptance and the oil's intended functional benefits (Hassid et al. 2025; Ju et al. 2019). The antioxidant action of essential oils springs largely from the flavonoids and phenolic constituents they harbor. By scavenging free radicals, these compounds curb lipid oxidation, thereby safeguarding the texture, color, and taste of coated produce (Amorati et al., 2013). When incorporated into edible films, the oils deter the formation of off-flavors and the fading of color, ensuring that bell peppers, strawberries, and other fruits and vegetables arrive at consumers without sensory compromise. Furthermore, the films create a semipermeable barrier that curbs moisture evaporation and moderates gas exchange of oxygen and carbon dioxide, fundamentally retarding the metabolic acceleration that ripening entails (Paidari et al., 2021). This metabolic throttling postpones softening, decay, and the visible effects of spoilage, thus extending marketable life (Zhao et al., 2024). From a sensory standpoint, the addition of essential oils offers the dual advantage of pleasant aroma and very subtle flavor. Herbal oils like thyme or oregano contribute soft, savory undertones, while cold-pressed citrus oils lend a bright, refreshing aroma. When perceptible at the consumer level, these subtle enhancements can increase the perceived freshness of the produce (Zhao et al., 2024). Careful calibration is, however, nonnegotiable; a concentration that favors antioxidant performance can, in the same breath, overwhelm the inherent flavor profile of, say, a ripe strawberry (Polito et al., 2021). When the oils exceed a minimum, the coating can impart a bitter or medicinal note that consumers instinctively reject. The dual objective of sensory balance and functional efficiency thus governs the formulation of edible films fortified with essential oils (da Silva et al., 2025a, b).

One of the outstanding merits of embedding essential oils within edible coatings is their ability to provide slow-release profiles. When oils are simply drizzled on the surface of the fruit, vapor losses and oxidation can strip their effectiveness within minutes; however, entrainment in a tailored edible matrix converts oil into timed micro-reservoirs. They defer the release, permitting the biopolymer to gradually let the active volatiles diffuse to the fruit surface (da Silva et al., 2025a, b; Zhao et al., 2024). The lengthened release, in turn, sharpens the antimicrobial punch and tames the volatile aroma. Even the guiltiest pungents are coated, inhaled, and subdued in a controlled cloud rather than a sudden slap. Blending essential oils in a kaleidoscope of edible biopolymers such as whey, gelatin, chitosan, alginate, starch, carnauba wax, and beeswax, wedges tiny droplets within a global matrix that obeys fruit cuticle transport kinetics (Dairywala & Virdi, 2022; Rao & Zhang, 2021). Protein, polysaccharides, and lipids masquerade as invisible buses that stabilize, orient, and make the droplets mobile within the cuticle wax bilayer. The rigid gels, or brittle monolayer waxes, the oils bank into a shared choreography. When integrated homogeneously the rigid gel and brittle monolayer waxes, surface emulsions disperse the actives without layering, gaining a choir that improves film tenacity, aroma endurance, and microbe carding (Mohamed et al., 2020).

The ability of edible coatings infused with essential oils to lower the use of artificial preservatives is one of their biggest benefits. Essential oils provide a safe and sustainable substitute for chemicals in food preservation as consumer demand for natural and chemical-free food products rises (Mohamed et al., 2020). Essential oils used in edible coatings are clove bud, coriander, neem, turmeric, eucalyptus, onion, thyme, oregano, basil (Ocimum), lemon, cinnamon bark, garlic, lime, fennel, clove leaf, and lavender oils, etc. (Ju et al., 2019; Kumar Pandey et al., 2022; Sánchez-González et al., 2011; Shahidi & Hossain, 2022). All things considered, adding essential oils to edible films and coatings offers a safe, all-natural way to increase the shelf life of fresh produce while also improving food safety and sensory qualities. They are useful substitutes for artificial preservatives due to their antimicrobial, antioxidant, and shelf life–extending qualities. Furthermore, their method of gradual release guarantees long-term preservation without sacrificing the fresh produce's organoleptic qualities. Edible coatings containing essential oils have enormous potential to enhance food storage in a sustainable and consumer-friendly way, with continued research and technological developments (Ju et al., 2019; Sánchez-González et al., 2011; Shahidi & Hossain, 2022).

Properties of Essential Oils

Antimicrobial Properties

Properties of essential oils include antimicrobial characteristics. Essential oils have strong antimicrobial properties because they contain bioactive compounds like phenols, aldehydes, and terpenes. These compounds break down the cell membranes of microorganisms and let important ions leak out. When essential oils are added to edible coatings, they slowly release antimicrobial agents, which protect against spoiling while the food is being stored (Hager et al., 2019). This gradual-release mechanism makes them work better by giving them more time to encounter spoilage microorganisms. Studies show that oregano and clove oils significantly reduce fungal contamination in fruits and vegetables, preventing spoilage and extending shelf life. Conversely, oils of cinnamon and thyme efficiently eradicate foodborne pathogens such as *Listeria monocytogenes* and *Escherichia coli*. Studies show that clove and oregano oils significantly reduce fungal contamination in fruits and vegetables, preventing spoiling and extending shelf life, while thyme and cinnamon oils effectively inhibit foodborne pathogens like *Listeria monocytogenes* and *Escherichia coli* (Teixeira et al., 2022). Meat and poultry products are also preserved using edible coatings infused with essential oils. The essential oils of garlic and cinnamon had a strong antimicrobial effect that stabilized the chicken tenders' water retention and stopped protein deterioration (Shen et al., 2023).

Edible coatings composed of natural polymers such as chitosan, alginate, or starch adhere better and last longer when essential oils are added. This produces a semipermeable barrier that prevents moisture from escaping and slows down breathing. In addition to maintaining the firmness and freshness of fruits and vegetables, this prevents biofilm formation by disrupting microbial communication. The essential oils of noni and rosemary pepper were found to be effective in controlling *C. gloeosporioides* that were isolated from papaya fruit (Dias et al., 2020). While rosemary pepper can inhibit conidial germination at 5000 ppm, essential oils extracted from noni can inhibit it at 3000 ppm. Essential oils from noni and rosemary pepper inhibited the mycelial growth of C. gloeosporioides at concentrations of 4000 and 6000 parts per million. The mycelial growth of *C. gloeosporioides* derived from papaya fruit was entirely suppressed by savory oil containing the active chemical compound carvacrol (71.2%) (Sarkhosh et al., 2018). An essential oil based on thymol was successful in controlling *C. gloeosporioides* that were isolated from mangoes (Chillet et al., 2018). By altering membrane permeability, causing cellular constituents to leak out, and dissipating the H+ and K+ ion gradients, essential oils also harm the microbial cell's structure and functionality (Sivakumar et al., 2021).

Antioxidant Properties

Essential oils are extracted from a variety of aromatic plants and are a complex blend of 20–60 distinct phenolic and terpenoid components. Essential oils contain antioxidants called phenols, flavonoids, and terpenoids that help keep fresh produce from going bad. When added to edible coatings, these bioactive compounds protect fresh produce from damage caused by free radicals and stop lipid oxidation. This maintains the nutritional value, flavor, and color of the produce (Kumar et al., 2024; Mehra et al., 2024; Shenbagam et al., 2023). Essential oils inhibits the change in color of produce with minimizing the oxidation process and spoilage. Their antioxidant properties extend the shelf life of fresh produce while also enhancing its flavor and appearance. It is well known that some essential oils have strong antioxidant properties. For example, adding thyme oil to edible coatings has shown promise in meat preservation by dramatically lowering lipid oxidation. Additionally, the oils of cloves and rosemary are effective at protecting fresh produce from oxidative stress, maintaining its quality while being stored. This enhances the meat's flavor and prolongs its shelf life by preserving its color and texture. Essential oils are advantageous natural bioadditives in edible coatings because they prevent oxidation. This is a long-lasting, chemical-free way to maintain the quality and freshness of food.

Preservation of Shelf Life

The encapsulation of essential oils is combined with edible coatings; a barrier is formed between the food and the environment that allows gases such as carbon dioxide and oxygen to flow through while preventing water vapor from escaping. The texture, firmness, and general quality of fruits and vegetables are preserved by this regulated barrier, which also slows down breathing and retains moisture (Das et al., 2021). By preventing excessive drying and reducing oxidative stress, these coatings effectively slow down ripening and spoiling. Fruit quality was preserved by the infusion of bergamot essential oil and natural antioxidant extract into the coating formulation, which demonstrated a good chance of achieving the dual effects of barrier resistance to transpiration and respiration and an antimicrobial effect against microbial growth during shelf life (De Bruno et al., 2023). By protecting food from oxidative deterioration and microbial contamination, essential oils' antimicrobial and antioxidant qualities help it remain fresher for longer. Several studies show that strawberries' shelf life can be significantly increased by adding lemon and oregano oils to edible coatings, which prevents them from spoiling and preserves their texture and color. Clove oil-based tomato coatings also prevent water loss, preventing the fruit from shrinking and prolonging its firmness. (Goswami et al., 2024) reported that the shelf life of Khasi mandarins was extended up to 20 days after the incorporation of cinnamon bark and clove bud essential oil. In addition to reducing food waste, this shelf life extension increases the consumer appeal and ease of sale of fresh produce. Fruits and vegetables can be safely and effectively kept fresh while maintaining their flavor, aroma, and appearance by using essential oils in coatings.

Flavor and Aroma Enhancement

Essential oils improve the sensory qualities of food products by adding natural flavor and aroma to edible coatings, in addition to their antimicrobial and antioxidant properties. Essential oils' volatile compounds give the coated produce pleasing, unique scents that enhance its appeal. Particularly for fruits, desserts, and confections, this extra sensory element can enhance the entire eating experience. Nevertheless, it is important to carefully balance the concentration of essential oils because too much of them can overpower the food's inherent flavor and aroma, possibly producing an unappealing or artificial flavor profile (Kumar Pandey et al., 2022).

Essential oils and edible coating polymers have a lot of potential for food preservation. For instance, oregano oil-enhanced chitosan-based coatings have increased antimicrobial activity, which successfully lowers microbial contamination and prolongs the fresh produce's shelf life. Likewise, it has been shown that gelatin coatings containing clove oil can lessen bacterial contamination on meat products while preserving their safety and quality. Through the controlled release of bioactive compounds, the coatings offer a promising natural solution to improve food preservation and safety, and provide extended protection by embedding essential oils into edible polymers (Yousuf et al., 2021a). Essential oil-based coatings are a helpful way to create healthier, more natural, and consumer-friendly food products because they can improve food safety and quality while using fewer artificial chemicals. For instance, coatings with oils of lemongrass and oregano have been shown to reduce the need for artificial preservatives in fresh-cut fruits by inhibiting microbial growth and maintaining quality. Similarly, coating meat products with cinnamon oil significantly reduces the need for chemical antimicrobials because it preserves the meat's safety and freshness naturally (Yousuf et al., 2021b).

The use of naturally occurring and widely available essential oils in edible coatings offers a promising alternative to synthetic chemical preservatives for food preservation. Essential oils, derived from various plant sources, are recognized for their antimicrobial, antifungal, and antioxidant properties, making them highly effective in protecting fresh produce from spoilage. Their natural origin makes them particularly appealing for biodegradable food packaging, as they not only help extend the shelf life of food products but also add value by promoting clean-label, chemical-free preservation. Essential oils create a barrier that prevents oxidation and microbial growth when added to edible coatings, maintaining the food's freshness, texture, and quality while it is being stored. These coatings are also very appealing in contemporary food preservation techniques since they provide a more environmentally responsible and sustainable substitute for traditional wax and petroleum packaging.

Consumer Preference for Natural Additives

Consumers' awareness of the possible health hazards linked to chemical preservatives and synthetic additives has grown in recent years. The need for natural, plant-based preservation techniques that guarantee food safety without sacrificing health has increased as a result of this increased awareness. Essential oils provide a clean-label substitute that appeals to consumers who are health-conscious because of their natural origin and multipurpose qualities. Food manufacturers can improve product appeal and satisfy consumer preferences for sustainable, chemical-free food preservation by substituting essential oil-infused coatings for artificial preservatives.

Challenges in Real Food Applications

Although essential oils have shown remarkable efficacy in vitro against a range of pathogens, there are some concerns regarding their performance in real food systems. In laboratory settings, essential oils can effectively inhibit microbial growth; however, in real food matrixes, their potency may be diminished. This is because fruits and vegetables contain more nutrients, which can speed up the healing of damaged microbial cells. Therefore, higher concentrations of essential oils are often required in food applications to achieve the same antimicrobial effects observed in vitro. Increasing the concentration of essential oils at the expense of sensory quality may have an impact on the food's flavor, aroma, and overall acceptability (De Bruno et al., 2023). Thus, it is crucial to find the ideal balance between antimicrobial efficacy and sensory preservation. All cells recover faster. Therefore, higher concentrations of essential oils are often required in food applications to achieve the same antimicrobial effects observed in vitro. Increasing the concentration of essential oils at the expense of sensory quality may have an impact on the food's flavor, aroma, and overall acceptability. Thus, it is crucial to find the ideal balance between antimicrobial efficacy and sensory preservation.

Limitations and Formulation Challenges

The use of essential oils in edible coatings poses formulation challenges despite their encouraging potential. Over time, essential oils' stability and efficacy may be limited by their high volatility and poor water solubility. Furthermore, if not adequately managed, their potent flavor profiles and strong odor can overpower the food's inherent flavor and aroma. Emulsifiers or encapsulation techniques are necessary to improve the solubility and controlled release of essential oils to formulate stable and effective coatings. This guarantees that the bioactive ingredients are dispersed uniformly and released gradually, offering extended protection without sacrificing the food's sensory appeal.

Safety and Toxicity Issues

When used in the right amounts, essential oils are generally recognized as safe (GRAS); however, their toxicity at higher concentrations is still a concern. It is crucial to maximize the concentration of certain essential oils in edible coatings because overuse of these oils can have negative effects. To guarantee that essential oils offer efficient food protection without sacrificing consumer safety, careful formulation techniques and comprehensive toxicity evaluations are required.

Future Prospects

The creation and use of edible coatings infused with essential oils are becoming better thanks to continuous research and technical developments. Essential oils' solubility and volatility issues can be resolved with the development of nano-emulsions (Pandey et al. 2022), encapsulation, and polymer-based delivery systems, which will increase their stability and efficacy in food preservation. Additionally, essential oils may work in concert with other natural preservatives or antimicrobials to increase their overall effectiveness. Essential oil-based edible coatings are anticipated to become increasingly important in improving food safety, increasing shelf life, and decreasing food waste as the food industry continues to place a higher priority on sustainable and natural solutions.

Essential Oils in Edible Coatings Derived from Different Plant Sources

The strong antibacterial qualities of essential oils that are extracted from a variety of spices, herbs, and other plant sources have been thoroughly investigated. Because they are high in bioactive substances like phenols, aldehydes, and terpenes, these oils have more potent antimicrobial qualities when they are in their oil form as opposed to when they are in their corresponding ground or raw plant state. This is because the essential oil fraction has a higher concentration of antimicrobial agents, which are particularly good at stopping the growth of bacteria that cause spoiling and harm. As a result, edible coatings that are applied to fresh and fresh-cut fruits and vegetables now contain essential oils, offering a natural and safe means of preserving food quality.

Spice-Derived Essential Oils

Essential oils that have been extracted from a range of spice plants and have shown remarkable antimicrobial activity are commonly found in edible coatings. Clove, thyme, sage, garlic, onion, and cinnamon are the most widely used sources (Goswami et al., 2024; Mondal et al., 2024). Bioactive substances found in these herbs can successfully lower microbial contamination. Coatings frequently use essential oils from the Myrtaceae family, a group of woody dicotyledonous plants, because of their high essential oil content. Two well-known examples of oils with strong antifungal and antibacterial properties are tea tree oil (*Melaleuca alternifolia*) and clove oil (*Syzygium aromaticum*) (Sarengaowa et al., 2022). Moreover, essential oils from the Lauraceae family, a flowering plant family rich in essential oils, are commonly found in food coatings. Cinnamon oil (*Cinnamomum verum*), which has demonstrated broad-spectrum antimicrobial qualities, is one noteworthy example. Cinnamon oil (*Cinnamomum verum*) has been shown to have broad-spectrum antimicrobial activity and highly effective at preserving the freshness of produce that has been coated.

Essential Oils from Herbs and Grasses

Essential oils from various grasses and herbs are also commonly added to edible coatings due to their potent preservative properties. The Lamiaceae family, also known as the mint or sage family, is one of the primary sources of essential oils used in food preservation. This family includes herbs like savory, peppermint, rosemary, oregano, and thyme that have potent antimicrobial and antioxidant properties (Liguori et al., 2024). Additionally, oils from the Asteraceae family, which is mainly composed of annual or perennial herbs, have also been utilized. For example, tarragon oil (*Artemisia dracunculus*) is widely recognized for its ability to inhibit bacterial growth and extend the shelf life of coated foods (Parham et al., 2020). Cumin essential oil is added to modified sweet potato starch, which has been shown to maintain the stomatal densities and internal quality of "early crisp." The Poaceae family, which includes economically significant monocotyledonous flowering plants, also produces essential oils used in edible coatings (Oyom et al., 2022). A typical example is lemongrass oil (*Cymbopogon citratus*), which is valued for its strong antibacterial qualities and pleasant citrusy aroma, both of which enhance the sensory appeal of the coated produce.

Essential Oils from Citrus Sources

Essential oils extracted from citrus fruits are also commonly used in edible coatings for fresh and fresh-cut fruits and vegetables due to their antimicrobial and antioxidant properties (da Silva et al., 2025b). The Rutaceae family, also known as the citrus family, produces essential oils with broad-spectrum antimicrobial qualities. This family includes economically important citrus species such as grapefruit (*Citrus paradisi*), bergamot (*Citrus bergamia*), lemon (*Citrus limon*), and lime (*Citrus aurantifolia*) (Bora et al., 2020). These citrus oils are rich in citral and limonene, which have strong antifungal and antibacterial qualities. In addition to their preservative qualities, citrus essential oils enhance the coated produce's sensory appeal by adding a pleasant flavor and aroma. In addition to extending shelf life, the use of essential oils from the Rutaceae family in edible coatings provides a clean-label option, satisfying consumer demands for natural and chemical-free food preservation techniques (Shenbagam et al., 2023; Li et al., 2024).

Extraction and Characterization of Essential Oils

Many plant parts, including flowers, buds, fruits, leaves, stems, bark, roots, and seeds, are used to make essential oils, which are high-concentration extracts. These oils capture the natural scent and beneficial bioactive compounds of the plant, making them useful for aromatherapy, cosmetics, and food preservation. The extraction process is crucial in determining the quality, purity, and composition of the essential oil since it influences the preservation of its aromatic and bioactive properties (Başer & Buchbauer, 2015).

Extraction Methods of Essential Oils

Numerous techniques, each with its own principles and outcomes, are used to extract essential oils from plant parts. The most widely used methods include:

Steam Distillation

The most popular technique for removing essential oils from plant materials is steam distillation. The plant material is put in a distillation apparatus above boiling water to begin the process. The volatile substances (essence) are drawn out of the plant by the rising steam as the water boils. After passing through a condensation tube, the steam carrying the essential oil cools and transforms back into liquid. The remaining water, called hydrosol, contains traces of the essential oil and is occasionally used for cosmetic or medicinal purposes. The oil is then extracted from the water and collected. For woody plant parts, flowers, and herbs, steam distillation works especially well (Aziz et al., 2018).

Cold Press Extraction

The cold press extraction mainly used for extraction of oils from citrus fruits such as bergamot, lemon, orange, grapefruit, and lime etc.. After being cleaned, the whole fruit is put in a container and either agitated or mechanically pressed. The released oil from fruit peel collecting by spraying water on the fruit, and it is then separated from the water. Citrus oils benefit greatly from cold pressing since it keeps their bright, fresh scent intact without heating them up (Chemat et al., 2012).

Solvent Extraction

This method is applied to plant materials like vanilla or jasmine that are too delicate for steam distillation. The essential oils and other aromatic compounds are dissolved by mixing the plant material with a solvent (such as ethanol or hexane). The extremely concentrated essential oil, also known as an absolute, is then extracted from the mixture by vacuum distillation, which eliminates the solvent (Ghasemy-piranloo et al., 2020).

Water Distillation

This method releases the volatile oils along with the steam by submerging plant material directly in boiling water. The essential oil floats to the top of the mixture after it has been separated and condensed. This technique is frequently applied to delicate flowers that direct steam distillation might harm (Naqvi et al., 2023).

Maceration

In this method, plant material is soaked in warm fat or oil, which gradually absorbs the essential oil. To extract the infused oil from the plant matter, the mixture is subsequently filtered. Usually, delicate flowers like roses or jasmine are macerated (Ahmad et al., 2018).

CO_2 Extraction

This cutting-edge technique extracts essential oils by using carbon dioxide (CO_2) at low temperatures and high pressures. By acting like a liquid under pressure, CO_2 efficiently extracts the oil. Pure essential oil is left behind when the CO_2 is reduced to its gaseous state. This method works well for delicate botanicals like vanilla because it maintains the oil's bioactive qualities and natural scent (Pourmortazavi & Hajimirsadeghi, 2007).

Enfleurage

This traditional extraction method uses animal or vegetable fat to absorb the essential oil from plant material. The fat is then treated with alcohol to release the oil. Although less commonly used today, enfleurage is still applied for delicate flowers such as jasmine and tuberose. Essential oils are derived from various natural plant parts, each contributing unique chemical profiles and aromatic qualities (Başer & Buchbauer, 2015).

Flowers

Essential oils extracted from flowers, such as rose, jasmine, lavender, carnation, clove, mimosa, and rosemary, are highly prized for their fragrance and therapeutic properties. They are frequently used in edible coatings to enhance sensory appeal and provide antimicrobial protection (Tajkarimi et al., 2010; Bashir et al., 2023; Venkatachalam & Lekjing, 2020).

Leaves and Stems

Terpenes and phenolic compounds, which have potent antimicrobial and antioxidant properties, are abundant in essential oils derived from leaves, including mint, Ocimum spp. (basil), lemongrass, and jamrosa (Wigati et al., 2023). A variety of plants, including verbena, petitgrain, geranium, patchouli, and cinnamon, produce essential oils from both their leaves and stems, increasing the oil yield and variety of bioactive substances (Tajkarimi et al., 2010).

Bark and Wood

Known for their strong antimicrobial and antifungal qualities, essential oils derived from bark include those from cinnamon, cassia, and canella. Because of their preservative properties, these oils are frequently used in edible coatings (Mathew & Abraham, 2006). Essential oils with earthy, resinous scents and antimicrobial qualities are produced by woody plants like pine, cedar, and sandalwood. These oils are less common in food applications but are valued in cosmetics and perfumery (Ibrahim et al., 2025).

Roots and Seeds

Essential oils derived from roots, including valerian, saussurea, vetiver, angelica, and sassafras, are prized for their potent, earthy fragrances and therapeutic qualities (Akhoon et al., 2015). Seeds produce essential oils with unique preservative and aromatic properties. Examples include fennel, coriander, caraway, dill, and nutmeg, which are often incorporated into edible coatings for their antimicrobial and flavor-enhancing properties (Fidan et al., 2022; Hussain et al., 2008).

Fruits and Rhizomes

Essential oils are frequently found in citrus fruits like bergamot, orange, lemon, and juniper. Because of their high limonene and citral content, their oils are very effective at warding off foodborne pathogens and can help prolong the shelf life of coated produce (Li et al., 2022). Essential oils derived from rhizomes, such as those of orris, curcuma, calamus, and ginger, are prized for their antioxidant and antimicrobial qualities, which qualify them for application in edible coatings (Jiang et al., 2021).

Gums and Oleoresins

Certain plants secrete gums or oleoresins, which are exudates that contain potent antimicrobial essential oils. Peruvian balsam, Myroxylon balsamum, storax, myrrh, and benzoin are a few examples that are used in food and cosmetic products (Mulyaningsih et al., 2010).

Hydrodistillation Techniques for Essential Oil Production

Water and Steam Distillation

In this method, steam is circulated through plant material that is partially immersed in water. Oil can be extracted more effectively with this hybrid method without harming heat-sensitive substances (Ferhat et al., 2007).

Direct Steam Distillation

This technique extracts the essential oil by injecting steam straight into the plant material. Hardy plant materials like wood and bark, which need higher temperatures for efficient oil release, are frequently treated with it (Ahmad et al., 2018). Essential oil producers can obtain premium oils with retained aromatic and bioactive qualities by using these different extraction methods. Enhancing the shelf life, safety, and sensory quality of fresh and fresh-cut produce can be achieved naturally and effectively by incorporating these oils into edible coatings (Salehi et al., 2019).

Effect of Different Essential Oils on Various Fresh Produce

The physical, mechanical, and functional characteristics of yam starch-based edible coatings infused with lime essential oil showed higher antioxidant activity (Gómez-Contreras et al., 2021). Essential oils like lime, fennel, and lavender have been successfully added to yam starch coatings, improving their capacity to maintain strawberry quality and extend shelf life. The addition of essential oils affected the films' water interaction characteristics, barrier efficacy, and tensile properties. The coatings had tensile strengths ranging from 1.98 MPa to 4 MPa and moduli of elasticity ranging from 76 MPa to 127 MPa. The elongation percentage, a gauge of the film's flexibility, varied from 16% to 130% depending on the type of essential oil used. This variation demonstrates how the plasticizing effect of the essential oils increases the flexibility and adaptability of the coatings (Perdones et al., 2012). The water vapor permeability (WVP) of the films decreased due to the hydrophobic nature of the essential oils, reducing the amount of moisture that was transferred. Conversely, the films became less dense due to the oil content, which led to a slight increase in oxygen permeability (OP). The interactions of films with water was also impacted by the hydrophobic qualities of the oils, which resulted in a slightly higher capacity for water absorption, a lower moisture content, and decreased solubility (Bonilla et al., 2012). The antioxidant properties of the coatings were significantly enhanced by the addition of lavender oil, which had the highest antioxidant capacity of all the essential oils tested. Applying the yam starch-based coating with essential oils to strawberries prevents fungal contamination and successfully decreases water loss. To preserve their freshness and firmness, strawberries coated with the essential oil-infused film lost 22.8 g less water per 100 g of fruit over the course of a 14-day storage period than the uncoated control samples. Furthermore, the coatings showed potent antifungal qualities, totally preventing Aspergillus niger from growing until day 8 of storage. The fungal growth on the coated strawberries was considerably less than that on the uncoated ones, even after day 10. This study demonstrates how edible coatings made from yam starch enhanced with essential oils can prolong the shelf life and preserve the quality of fresh produce by reducing weight loss, preventing fungal contamination, and maintaining antioxidant qualities. These coatings offer a secure and effective means of enhancing the quality of fruits like strawberries after harvest.

Essential Oil Composition and Antimicrobial Activity

The antimicrobial properties of essential oils (EOs) are primarily attributed to their low molecular weight compounds, primarily terpenes and terpenoids, which have aromatic or aliphatic carbon chains (Sharma et al., 2024). Terpenes are hydrocarbons classified according to the number of isoprene units (C5) that comprise them. The two most common types of compounds found in essential oils are monoterpenes (C10) and sesquiterpenes (C15). Terpenoids, on the other hand, are oxygenated terpene derivatives. Monoterpenes and monoterpenoids make up between 80% and 90% of the chemical composition of most essential oils. The aromatic compounds in essential oils are derived from phenylpropane and include a variety of organic functional groups, including alcohols, esters, phenols, ketones, and aldehydes. These bioactive compounds work together to make strong antimicrobial effects by changing the permeability, breaking down the structural integrity, and interfering with cellular metabolic processes. The composition of essential oils can be greatly affected by several factors, such as where the plant comes from, how it is grown, when it is harvested, and how it is extracted. Supercritical fluid extraction with carbon dioxide and water or steam distillation are the two most common ways to get essential oils. Both methods change the oil's final chemical profile. The oil's ability to kill germs can change a lot with even small changes in its chemical makeup, which can happen because of extraction methods or the environment. Thymol and cinnamaldehyde are the main bioactive components of thyme and cinnamon essential oils (EOs), respectively. Both substances are very good at killing bacteria and fungi. However, the overall antimicrobial effectiveness of EOs is contingent upon factors beyond their primary constituents. Instead, the oil's bioactivity is improved by the complex interactions and synergy between its major and minor parts. Isomeric variations can sometimes change the antimicrobial activity of EO compounds (Table 9.1). Carvacrol and thymol, differing solely in the positioning of the hydroxyl group, demonstrate similar antimicrobial properties. In contrast, eugenol and isoeugenol, which are isomers with varying double bond placements, can exhibit significantly divergent bactericidal activities against identical bacterial strains. Figure 9.1 summarizes the antimicrobial mechanism of essential oils. Essential oils are very effective natural preservatives for food and medicine because their overall ability to kill germs depends on the combination of their major and minor bioactive compounds and the way their functional groups are arranged (Marcovich & Ansorena, 2025).

Conclusion

Essential oils considered as effective bioactive agents as an alternative of chemical fungicides for intensifying the effectiveness of edible coatings for prolonging the shelf life of the fresh horticulture produce. The exhibit spectrum of functionalities such as antioxidant and antimicrobial properties of essential oils is highly effective in enchaining the post-harvest shelf life as well as the storability of the fresh produce. Also, when incorporated, essential oils also improve various properties of edible coatings such as gas and moisture barrier, slowing down the action of oxidative enzymes which delays ripening which in turn minimizes spoilage. The antimicrobial activity of essential oils which are primarily ascribed to the terpenes, terpenoids, phenols, and aldehydes present which disrupts microbial cell wall and hinders their metabolic activity. Incorporation of essential oils in the polymeric matrix ensures sustained release of these antimicrobials in food matrix ensuring protection against microbial as well as enzymatic spoilage. However, further investigation needs to be carried out on the formulation of edible coatings incorporated with essential oils covering the aspects of consumer acceptability, off flavour and commercial viability as well.

Table 9.1 Source, functional compound, property, and the effect of various essential oils in edible coatings and their application to various types fresh produce

Source with scientific names	Functional compound	Properties/microorganism	Application	Reference
Cinnamon (*Cinnamomum verum*)	Cinnamaldehyde, cinnamyl ester, salicylaldehyde, eugenol	Antioxidant and antimicrobial against bacteria, yeast, and fungi/ effective against *Listeria monocytogenes*	Incorporated in a chitosan coating solution at different concentrations (0, 0.2, 0.4, and 0.6% v/v) and applied to fresh-cut potatoes to enhance shelf life	Sarengaowa et al. (2022)
Coriander (*Coriandrum sativum*)	Linalool	Antifungal, antibacterial, as well as anti-oxidative activities/effective against *Listeria monocytogenes*	Coriander essential oil (0.01% v/v) was incorporated into a chitosan-based edible coating solution to improve the freshness of the vegetable cakes during storage	Fidan et al. (2022)
Eucalyptus (*Eucalyptus teriticornis*)	Eucalyptol	Anti-inflammatory and antimicrobial activity/effective against *Colletotrichum gloeosporioides* and *Aspergillus niger*	Carboxymethyl cellulose incorporated with eucalyptus essential oil at the concentration of 0.5% used as a coating agent for orange fruits to control postharvest soft rot	Adeogun et al. (2025)
Thyme (*Thymus vulgaris*)	Thymol	Antioxidant, antimicrobial, antitussive, antitumoral, and spasmolytic ingredient/effective against *Aspergillus niger*	Thyme essential oil encapsulated into β–cyclodextrin at the ratio of 8:92 (%, w/w) incorporated into a chitosan-based edible coating applied to kiwi fruits for improved organoleptic quality	Marcovich and Ansorena (2025)
Rosemary (*Salvia rosmarinus*)	Carnosic acid, carnosol, and rosmarinic acid	Antifungal and antibacterial/effective against *Penicillium expansum*, blue mold	Water chestnut-extracted starch loaded with rosemary essential oil at different concentrations (0.1, 0.25, and 0.5%) for the storage stability of apples at ambient temperature	Bashir et al. (2023)

Lavender (*Lavandula angustifolia*)	Linalool, linalyl acetate, 1,8-cineole, terpenen-4-ol etc.	Antibacterial, antioxidant, anti-inflammatory, analgesic and antifungal activities/effective against *Staphylococcus aureus, Escherichia coli* and *listeria monocytogenes*	Lavender essential oil in the form of nanoemulsion incorporated into a gelatin-based edible film solution at 50, 100, 200, and 300 μL/10 mL concentration to enhance the quality of cherry tomatoes	Parham et al. (2020)
Turmeric (*Curcuma longa*)	Curcumin, curcuminoids	Antioxidant, anticancerous, antibrowning, and texture enhancer/ effective against *Penicillium* sp. and *Cladosporium* sp.	Turmeric oil (TO) was studied for its potential as an antioxidant in starch edible coating for fresh-cut apples and its degree of oxidation was analyzed	Ahmad et al. (2018)
Garlic (*Allium sativum*)	Diallyl trisulfide, diallyl disulfide, and diallyl sulfide	Suppressing oxidative rancidity and microbial growth/effective against mesophilic and psychrotrophic microorganisms	Chitosan-based edible coating enriched with garlic essential oil extract, at 0.1 and 0.5% concentration, improved microbial and chemical properties of rainbow trout fillet stored at 4 °C	Foromandi and Khani (2023)
Neem (*Azadirachta indica*)	Limonoids-azadirachtin, nimbin, etc.	Antibacterial, antifungal, and anti-inflammatory	Application of neem essential oil incorporated in chitosan-based edible film to improve the shelf life of cut mangoes was investigated	Passafiume et al. (2022)
Clove (*Syzygium aromaticum*)	Eugenol and Eugenol acetate	Antioxidant and antimicrobial/ effective against *Enterobacteriaceae* and LAB	Incorporated into chitosan-based edible film solution at the concentration of 6400 μg ml^{-1} to improve quality and shelf life of pork patties	Venkatachalam, and Lekjing (2020)
Fennel seeds (*Foeniculum vulgare*)	Ketones, olefins, and anethole	Antibacterial, anti-inflammatory, antifungal, and especially analgesic and antioxidant properties/ effective against *Escherichia coli* and *Staphylococcus aureus*	Fennel seed oil incorporated into a cyclodextrin nanoemulsion-based coating solution showed better shelf life of blueberry	Gu et al. (2024)

(continued)

Table 9.1 (continued)

Source with scientific names	Functional compound	Properties/microorganism	Application	Reference
Basil (*Ocimum basilicum*)	Neryl acetate, 1,8-cineole, p-allylanisole, geraniol, methyl eugenol, methyl chavicol, and trans-α-bergamotene	Antioxidant, antifungal, easily accessible, safe source of lipids	Mandarin orange was coated with pregelatized starch, basil essential oil (2–3%), and cellulose nano-fiber for the prevention of postharvest loss	Wigati et al. (2023)
Oregano (*Origanum vulgare*)	Thymol, carvacrol	Antioxidant, antimicrobial and antifungal properties/effective against *Listeria monocytogenes* and *Staphylococcus aureus*	Two percent of oregano essential oil was incorporated into a gellan-gum based edible coating solution for coating of mandarin to protect it from mold infection	Liguori et al. (2024)
Lemon (*Citrus limon*)	Limonene, ocimene	Antibacterial and antioxidant properties/ effective against *Penicullium* sp.	Lemon essential oil (1% v/v) chitosan-enriched coating solution was applied to citrus fruits to improve the postharvest quality	Zhang et al. (2022)
Orange (*Citrus sinensis*)	Vitamin C, flavonoids, and carotenoids	Antioxidant, anticancerous, and flavor enhancer	Aloe vera-based edible coating solution enriched with orange essential oil at the concentration of 500, 1000 and 1500 μL/L, improved the storage stability of mushrooms at 4 °C	Shenbagam et al. (2023)
Tea tree seed (*Camellia sinensis*)	Lignan, saponin	Antioxidant, antibacterial, anti-inflammatory, and antifungal/ effective against *Botrytis cinerea*	Edible-coatings solution was prepared with 1% (w/v) chitosan (CH) incorporated into 0.05%, 0.25%, or 0.5% (w/v) tea seed oil for the enhancement of Japanese pear shelf life	Kingwascharapong et al. (2021)

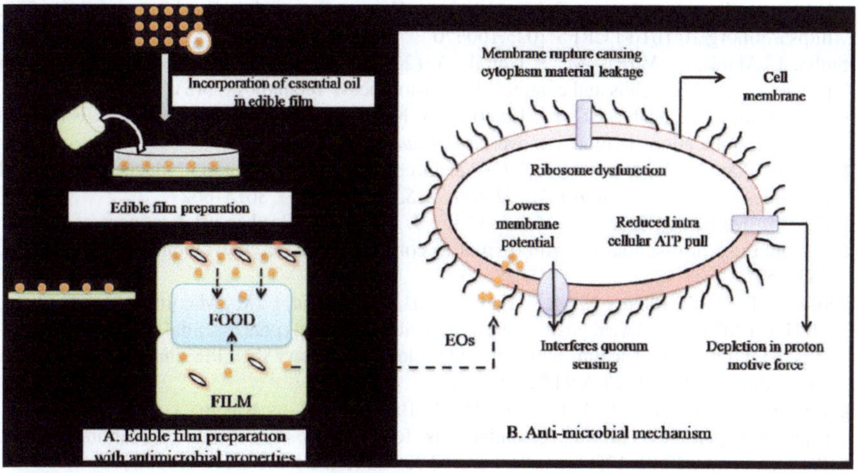

Fig. 9.1 The antimicrobial mechanism of edible film coating with essential oil. (Li et al., 2024; Mahendra Shah et al., 2021)

Acknowledgments The authors acknowledge the CoE-SusPol, Indian Institute of Technology Guwahati, India, for their continuous support and encouragement throughout the writing of this book chapter.

References

Adeogun, O. O., Oluwa, O. K., Nejo, A. O., Salami, S. O., Egwu, P. C., & Adekunle, A. A. (2025). Assessment of eucalyptus oil integrated with carboxymethyl cellulose (CMC) for controlling postharvest soft rot in Citrus sinensis fruits induced by Colletotrichum gloeosporioides and Aspergillus niger. *Bulletin of the National Research Centre, 49*(1), 1–19. https://doi.org/10.1186/S42269-025-01307-5

Ahmad, N. A., Sharif, Z. I. M., Jai, J., Yusof, N. M., & Mustapha, F. A. (2018). Tumeric oil as the antioxidation agent in edible coating film. In *IOP conference series: materials science and engineering* (Vol. 334, No. 1) (p. 012064). IOP Publishing.

Akhoon, B. A., Singh, K. P., Karmakar, M., Smita, S., Pandey, R., & Gupta, S. K. (2015). Virtual screening and prediction of the molecular mechanism of bioactive compounds in silico. In *Biotechnology of bioactive compounds: sources and applications* (pp. 371–394). Wiley.

Amorati, R., Foti, M. C., & Valgimigli, L. (2013). Antioxidant activity of essential oils. *Journal of Agricultural and Food Chemistry, 61*(46), 10835–10847. https://doi.org/10.1021/JF403496K

Aziz, Z. A., Ahmad, A., Setapar, S. H. M., Karakucuk, A., Azim, M. M., Lokhat, D., et al. (2018). Essential oils: Extraction techniques, pharmaceutical and therapeutic potential-a review. *Current Drug Metabolism, 19*(13), 1100–1110.

Bakkali, F., Averbeck, S., Averbeck, D., & Idaomar, M. (2008). Biological effects of essential oils–a review. *Food and Chemical Toxicology, 46*(2), 446–475.

Baser, K. H. C., & Buchbauer, G. (Eds.). (2015). Handbook of essential oils: Science, technology, and applications (2nd ed.). CRC Press.

Bashir, O., Amin, T., Hussain, S. Z., Naik, H. R., Goksen, G., Wani, A. W., Manzoor, S., Malik, A. R., Wani, F. J., & Proestos, C. (2023). Development, characterization and use of rosemary essential oil loaded water-chestnut starch based nanoemulsion coatings for enhancing postharvest quality of apples var. Golden delicious. *Current Research in Food Science, 7*, 100570. https://doi.org/10.1016/J.CRFS.2023.100570

Bonilla, J., Atarés, L., Vargas, M., & Chiralt, A. (2012). Effect of chitosan and essential oils on properties of edible films and coatings. Food Bioprocess Technology, 5(5), 1834–1848.

Bora, H., Kamle, M., Mahato, D. K., Tiwari, P., & Kumar, P. (2020). Citrus essential oils (CEOs) and their applications in food: An overview. *Plants, 9*(3), 357.

Chemat, F., Vian, M. A., & Cravotto, G. (2012). Green extraction of natural products: Concept and principles. *International Journal of Molecular Sciences, 13*(7), 8615–8627.

Chillet, M., Minier, J., Ducrocq, M., & Meile, J. C. (2018). Postharvest treatment of mango: Potential use of essential oil with thymol to control anthracnose development. Fruits, 73(3), 153–157.

da Silva, P. P. M., de Toledo, N. M. V., de Oliveira, J., da Gloria, E. M., Maximo, F. B., & Spoto, M. H. F. (2025a). Biobased postharvest treatment using eucalyptus essential oils in edible coatings to inhibit Colletotrichum acutatum and prolong strawberry shelf life. *Plants, 14*(16), 2565. https://doi.org/10.3390/PLANTS14162565

da Silva, E. F., dos Santos, F. A. L., Pires, H. M., Bastos, L. M., & Ribeiro, L. N. d. M. (2025b). Lipid nanoparticles carrying essential oils for multiple applications as antimicrobials. *Pharmaceutics, 17*(2), 178. https://doi.org/10.3390/PHARMACEUTICS17020178

Dairywala, A., & Virdi, R. (2022). Controlled-release edible films: Mechanisms and application of essential oils in biopolymer coatings. *Comprehensive Reviews in Food Science and Food Safety, 21*(4), 1234–1256. https://doi.org/10.1111/1541-4337.1298

Das, S., Vishakha, K., Das, S., Chakravarty, M., Ganguli, A., & Adhikary, J. (2021). Carboxymethyl cellulose and tapioca starch based edible coating with geraniol for fruit shelf-life extension. International Journal of Biological Macromolecules, 188, 837–850.

De Bruno, A., Gattuso, A., Ritorto, D., Piscopo, A., & Poiana, M. (2023). Effect of edible coating enriched with natural antioxidant extract and bergamot essential oil on the shelf life of strawberries. *Food, 12*(3), 488.

Dias, B. L., dos Santos, G. R., de Castro, R. S., de Oliveira, S. F., & Ferreira, T. P. S. (2020). Control of papaya fruits anthracnose by essential oils of medicinal plants associated to different coatings. Journal of Medicinal Plants Research, 14(6), 239–246.

Fathi, M., Vinceković, M., Jurić, S., Viskić, M., Režek Jambrak, A., & Donsì, F. (2021). Food-grade colloidal systems for the delivery of essential oils. *Food Reviews International, 37*(1), 1–45.

Ferhat, M. A., Meklati, B. Y., & Chemat, F. (2007). Comparison of different isolation methods of essential oil from citrus fruits: Cold pressing, hydrodistillation and microwave 'dry'distillation. *Flavour and Fragrance Journal, 22*(6), 494–504.

Fidan, H., Stankov, S., Georgieva, D., Gandova, V., Stoyanova, A., & Ercisli, S. (2022). Antimicrobial effect of edible coating with coriander essential oil in vegetarian cake during storage. *Acta Microbiologica Bulgarica, 4*, 315–323.

Foromandi, M., & Khani, M. (2023). The effect of chitosan coatings enriched with garlic extract and coriander essential oil on the shelf life of rainbow trout (Oncorhynchus mykiss) fillet stored at 4 °C. *International Journal of Food Science and Technology, 58*(7), 3618–3627.

Froiio, F., Mosaddik, A., Morshed, M. T., Paolino, D., Fessi, H., & Elaissari, A. (2019). Edible polymers for essential oils encapsulation: Application in food preservation. *Industrial & Engineering Chemistry Research, 58*(46), 20932–20945.

Ghasemy-piranloo, F., Moameri, M., Farajpour, M., & Khalighi-Sigaroodi, F. (2020). Comparison of various extraction methods on the essential oil composition and antioxidant activity of Rosa damascena Mill. from Iran. Journal of Applied Research on Medicinal and Aromatic Plants, 18, 100256.

Gómez-Contreras, P., Figueroa-Lopez, K. J., Hernández-Fernández, J., Cortés Rodríguez, M., & Ortega-Toro, R. (2021). Effect of different essential oils on the properties of edible coatings based on yam (Dioscorea rotundata L.) starch and its application in strawberry (Fragaria vesca L.) preservation. Applied Sciences, 11(22), 11057.

Goswami, M., Mondal, K., Prasannavenkadesan, V., Bodana, V., & Katiyar, V. (2024). Effect of guar gum-chitosan composites edible coating functionalized with essential oils on the post-harvest shelf life of Khasi mandarin at ambient condition. *International Journal of Biological Macromolecules, 254*, 127489. https://doi.org/10.1016/J.IJBIOMAC.2023.127489

Gu, X., Li, J., Yang, L., Liu, L., Li, T., Zhang, H., et al. (2024). Comparative study on the different edible coatings loaded with fennel essential oil/β-cyclodextrin microcapsules for blueberry preservation. *The Journal of Horticultural Science and Biotechnology, 99*(5), 584–596.

Hager, J. V., Rawles, S. D., Xiong, Y. L., Newman, M. C., & Webster, C. D. (2019). Edible corn-zein-based coating incorporated with nisin or lemongrass essential oil inhibits Listeria monocytogenes on cultured hybrid striped bass, Morone chrysops× Morone saxatilis, fillets during refrigerated and frozen storage. Journal of the World Aquaculture Society, 50(1), 204-218.

Hassid, A., Salla, M., Krayem, M., Khaled, S., Hassan, H. F., & El Khatib, S. (2025). A review on the versatile applications of plant-based essential oils in food flavoring, culinary uses and health benefits. *Discover Food, 5*(1), 130.

Hussain, A. I., Anwar, F., Sherazi, S. T. H., & Przybylski, R. (2008). Chemical composition, antioxidant and antimicrobial activities of basil (Ocimum basilicum) essential oils depends on seasonal variations. *Food Chemistry, 108*(3), 986–995.

Ibrahim, A. A., Abd-Ellatif, S., Razik, E. S. S. A., & Salem, K. F. (2025). Advances in Cinnamon (Cinnamomum verum L.) Breeding Strategies. In Biodiversity and Genetic Improvement of Herbs and Spices (pp. 21-59). Cham: Springer Nature Switzerland.

Jiang, T., Ghosh, R., & Charcosset, C. (2021). Extraction, purification and applications of curcumin from plant materials-a comprehensive review. *Trends in Food Science & Technology, 112*, 419–430.

Ju, J., Xie, Y., Guo, Y., Cheng, Y., Qian, H., & Yao, W. (2019). Application of edible coating with essential oil in food preservation. *Critical Reviews in Food Science and Nutrition, 59*(15), 2467–2480.

Kingwascharapong, P., Tanaka, F., & Tanaka, F. (2021). Effect of edible coatings developed from chitosan incorporated with tea seed oil on Japanese pear. *Scientia Horticulturae, 288*, 110314.

Kumar, N. (2019). Polysaccharide-based component and their relevance in edible film/coating: A review. Nutrition & Food Science, 49(5), 793-823.

Kumar, N., Petkoska, A. T., Al-Hilifi, S. A., & Fawole, O. A. (2021a). Effect of chitosan–pullulan composite edible coating functionalized with pomegranate peel extract on the shelf life of mango (Mangifera indica). Coatings, 11(7), 764.

Kumar, N., Neeraj, Pratibha, & Trajkovska Petkoska, A. (2021b). Improved shelf life and quality of tomato (Solanum lycopersicum L.) by using chitosan-pullulan composite edible coating enriched with pomegranate peel extract. ACS Food Science & Technology, 1(4), 500-510.

Kumar, N., Pratibha, Trajkovska Petkoska, A., Khojah, E., Sami, R., & Al-Mushhin, A. A. (2021c). Chitosan edible films enhanced with pomegranate peel extract: Study on physical, biological, thermal, and barrier properties. Materials, 14(12), 3305.

Kumar, N., Daniloski, D., D'cunha, N. M., Naumovski, N., & Petkoska, A. T. (2022). Pomegranate peel extract–A natural bioactive addition to novel active edible packaging. Food Research International, 156, 111378.

Kumar Pandey, V., Shams, R., Singh, R., Dar, A. H., Pandiselvam, R., Rusu, A. V., & Trif, M. (2022). A comprehensive review on clove (Caryophyllus aromaticus L.) essential oil and its significance in the formulation of edible coatings for potential food applications. *Frontiers in Nutrition, 9*, 987674.

Kumar, N., Pratibha, Prasad, J., Yadav, A., Upadhyay, A., Neeraj,... & Kieliszek, M. (2023a). Recent trends in edible packaging for food applications—Perspective for the future. Food Engineering Reviews, 15(4), 718-747.

Kumar, N., Pratibha, Upadhyay, A., Trajkovska Petkoska, A., Gniewosz, M., & Kieliszek, M. (2023b). Extending the shelf life of mango (Mangifera indica L.) fruits by using edible coating based on xanthan gum and pomegranate peel extract. Journal of Food Measurement and Characterization, 17(2), 1300-1308.

Kumar, N., Khan, A. A., Pyngrope, D., Alanazi, A. M., Upadhyay, A., & Shukla, S. (2024). Development and characterization of novel starch (mango kernel and litchi seed) based active edible coatings and films using ultrasonication: Effects on postharvest shelf life of Khasi mandarins. Sustainable Chemistry and Pharmacy, 39, 101610.

Kumar, N., Upadhyay, A., & Shukla, S. (2024). Effects of essential oils and ultrasonic treatments on properties of edible coatings and their application on citrus fruits. *Starch, 76*(11–12). https://doi.org/10.1002/STAR.202300104

Kumarihami, H. P. C., Kumar, N., Pratibha, Petkoska, A. T., & Neeraj. (2023). Effect of Natural Materials on Thermal Properties of Food Packaging Film: An Overview. Natural Materials for Food Packaging Application, 255-274.

Li, Y., Liu, S., Zhao, C., Zhang, Z., Nie, D., Tang, W., & Li, Y. (2022). The chemical composition and antibacterial and antioxidant activities of five citrus essential oils. *Molecules, 27*(20), 7044.

Li, Y., Zhang, J., Ji, N., Zuo, X., Ru, X., Jin, P., et al. (2023). PpMYB44 positively affects salicylic acid biosynthesis in Pichia guilliermondii-induced peach fruit resistance against Rhizopus stolonifer. *Postharvest Biology and Technology, 202*, 112395.

Li, X., Li, F., Zhang, X., Tang, W., Huang, M., Huang, Q., & Tu, Z. (2024). Interaction mechanisms of edible film ingredients and their effects on food quality. *Current Research in Food Science, 8*, 100696.

Liguori, G., Greco, G., Salsi, G., Garofalo, G., Gaglio, R., Barbera, M., et al. (2024). Effect of the gellan-based edible coating enriched with oregano essential oil on the preservation of the 'Tardivo di Ciaculli'mandarin (Citrus reticulata Blanco cv. Tardivo di Ciaculli). *Frontiers in Sustainable Food Systems, 8*, 1334030.

Mahendra Shah, M. A., Bosco, S. J. D., & Mir, S. A. (2021). Plant-based edible coatings and films with essential oils: A review on their antimicrobial and antioxidant activities and quality retention of fresh-cut fruits and vegetables. Food Packaging and Shelf Life, 28, 100658.

Marcovich, N. E., & Ansorena, M. R. (2025). Development and characterization of sustainable active films incorporated with free and microencapsulated thyme essential oil for kiwi-fruit preservation. *Frontiers in Sustainable Food Systems, 9*, 1571114.

Mathew, S., & Abraham, T. E. (2006). In vitro antioxidant activity and scavenging effects of Cinnamomum verum leaf extract, polyphenols and ethyl acetate fractions. Food Chemistry, 94(4), 520–528.

Mehra, M., Kumar, N., Ojha, A., & Upadhyay, A. (2024). Extending the postharvest shelf life of sweet orange (Citrus sinensis L.) fruits using muskmelon seeds flour based edible coatings functionalized with oregano essential oil. *Journal of Food Measurement and Characterization, 18*(10), 8472–8488.

Mohamed, S. A., El-Sakhawy, M., & El-Sakhawy, M. A. M. (2020). Polysaccharides, protein and lipid-based natural edible films in food packaging: A review. *Carbohydrate Polymers, 238*, 116178.

Mondal, K., Goswami, M., Goud, V. V., & Katiyar, V. (2024). Optimization of guar gum-based anti-browning coating for prolonging the shelf life of cut potatoes. *International Journal of Biological Macromolecules, 269*, 132051. https://doi.org/10.1016/J.IJBIOMAC.2024.132051

Mulyaningsih, S., Sporer, F., Zimmermann, S., Reichling, J., & Wink, M. (2010). Synergistic properties of the terpenoids aromadendrene and 1,8-cineole from the essential oil of Eucalyptus globulus against antibiotic-susceptible and antibiotic-resistant pathogens. Phytomedicine, 17(13), 1061–1066

Naqvi, S. R. A., Shahid, H., Zahoor, A. F., Saeed, M., Usman, M., Abbas, A.,... & Hussain, T. (2023). Classical Methods for Obtaining Essential Oils. Essential Oils: Extraction Methods and Applications, 565-582.

Oyom, W., Xu, H., Liu, Z., Bi, Y., Li, Z., Sun, P.,... & Prusky, D. (2022). Effects of modified sweet potato starch coating enriched with cumin essential oil on postharvest quality and decay of Early Crisp pear. LWT - Food Science and Technology, 154, 112746.

Paidari, S., Zamindar, N., Tahergorabi, R., Kargar, M., Ezzati, S., shirani, N., & Musavi, S. H. (2021). Edible coating and films as promising packaging: A mini review. *Journal of Food Measurement and Characterization, 15*(5), 4205–4214. https://doi.org/10.1007/S11694-021-00979-7

Pandey, V. K., Islam, R. U., Shams, R., & Dar, A. H. (2022). A comprehensive review on the application of essential oils as bioactive compounds in Nano-emulsion based edible coatings of fruits and vegetables. *Applied Food Research, 2*(1), 100042.
Parham, S., Kharazi, A. Z., Bakhsheshi-Rad, H. R., Nur, H., Ismail, A. F., Sharif, S., Ramakrishna, S., & Berto, F. (2020). Antioxidant, antimicrobial and antiviral properties of herbal materials. *Antioxidants, 9*(12), 1309. https://doi.org/10.3390/ANTIOX9121309
Passafiume, R., Tinebra, I., Gaglio, R., Settanni, L., Sortino, G., Allegra, A., & Farina, V. (2022). Fresh-cut mangoes: How to increase shelf life by using neem oil edible coating. *Coatings, 12*(5), 664.
Perdones, A., Sánchez-González, L., Chiralt, A., & Vargas, M. (2012). Effect of chitosan–lemon essential oil coatings on storage-keeping quality of strawberry. Postharvest biology and technology, 70, 32-41.
Polito, R. A., Pasqualotto, L., Dysarz, R., Cinelli, R., Heck, T., & Nunes, A. L. (2021). Interaction of acetyl-CoA carboxylase enzyme inhibiting herbicides with auxin herbicides on ryegrass. *Ciência Rural, 51*(4), e20200462.
Pourmortazavi, S. M., & Hajimirsadeghi, S. S. (2007). Supercritical fluid extraction in plant essential and volatile oil analysis. *Journal of Chromatography A, 1163*(1–2), 2–24. https://doi.org/10.1016/J.CHROMA.2007.06.021
Rao, M., & Zhang, L. (2021). Smart composite edible films for controlled release of essential oils in post-harvest preservation. *Journal of Agricultural and Food Packaging, 14*(2), 78–92. https://doi.org/10.1016/j.jafp.2021.04.003
Salehi, B., Sharifi-Rad, J., Quispe, C., Llaique, H., Villalobos, M., Smeriglio, A., et al. (2019). Insights into Eucalyptus genus chemical constituents, biological activities and health-promoting effects. *Trends in Food Science & Technology, 91*, 609–624. https://doi.org/10.1016/J.TIFS.2019.08.003
Sánchez-González, L., Vargas, M., González-Martínez, C., Chiralt, A., & Chafer, M. (2011). Use of essential oils in bioactive edible coatings: A review. *Food Engineering Reviews, 3*(1), 1–16.
Sarengaowa, Wang, L., Liu, Y., Yang, C., Feng, K., & Hu, W. (2022). Screening of essential oils and effect of a chitosan-based edible coating containing cinnamon oil on the quality and microbial safety of fresh-cut potatoes. *Coatings, 12*(10), 1492. https://doi.org/10.3390/COATINGS12101492
Sarkhosh, A., Schaffer, B., Vargas, A. I., Palmateer, A. J., Lopez, P., & Soleymani, A. (2018). In vitro evaluation of eight plant essential oils for controlling Colletotrichum, Botryosphaeria, Fusarium and Phytophthora fruit rots of avocado, mango and papaya. Plant Protection Science, 54(3), 153–162.
Shahidi, F., & Hossain, A. (2022). Preservation of aquatic food using edible films and coatings containing essential oils: A review. *Critical Reviews in Food Science and Nutrition, 62*(1), 66–105.
Sharma, R., Nath, P. C., Das, P., Rustagi, S., Sharma, M., Sridhar, N., et al. (2024). Essential oil-nanoemulsion based edible coating: Innovative sustainable preservation method for fresh/fresh-cut fruits and vegetables. *Food Chemistry, 460*, 140545.
Shen, C., Chen, W., Li, C. Z., Lin, L., & Cui, H. (2023). Quality and shelf life assessment of chicken meat products incorporated with garlic and cinnamon essential oil nanoemulsions. Food Control, 145, 109437.
Shenbagam, A., Kumar, N., Rahul, K., Upadhyay, A., Gniewosz, M., & Kieliszek, M. (2023). Characterization of aloe vera gel-based edible coating with orange peel essential oil and its preservation effects on button mushroom (Agaricus bisporus). *Food and Bioprocess Technology, 16*(12), 2877–2897. https://doi.org/10.1007/S11947-023-03107-Z
Sivakumar, D., Bautista-Baños, S., & Romanazzi, G. (2021). A comprehensive review on the impact of edible coatings, essential oils, and their nano formulations on postharvest decay anthracnose of avocados, mangoes, and papayas. Frontiers in Microbiology, 12, 711092.
Tajkarimi, M. M., Ibrahim, S. A., & Cliver, D. O. (2010). Antimicrobial herb and spice compounds in food. *Food Control, 21*(9), 1199–1218. https://doi.org/10.1016/J.FOODCONT.2010.02.003

Teixeira, R. F., Balbinot Filho, C. A., & Borges, C. D. (2022). Essential oils as natural antimicrobials for application in edible coatings for minimally processed apple and melon: A review on antimicrobial activity and characteristics of food models. Food Packaging and Shelf Life, 31, 100781.

Tisserand, R., & Young, R. (2013). *Essential oil safety: A guide for health care professionals.* Elsevier Health Sciences.

Venkatachalam, K., & Lekjing, S. (2020). *A chitosan-based edible film with clove essential oil and nisin for improving the quality and shelf life of pork patties in cold storage*. RSC Advances. https://doi.org/10.1039/d0ra02986f

Wigati, L. P., Wardana, A. A., Tanaka, F., & Tanaka, F. (2023). Application of pregelatinized corn starch and basil essential oil edible coating with cellulose nanofiber as Pickering emulsion agent to prevent quality-quantity loss of mandarin orange. *Food Packaging and Shelf Life, 35*, 101010. https://doi.org/10.1016/J.FPSL.2022.101010

Yousuf, B., Gul, K., Wani, A. A., & Singh, P. (2021a). Health benefits of bioactive compounds from fruit and vegetable wastes and their potential applications in the food industry. Critical Reviews in Food Science and Nutrition, 61(11), 1757–1777.

Yousuf, B., Wu, S., & Siddiqui, M. W. (2021b). Incorporating essential oils into edible coatings: Recent advances in localized delivery and Antimicrobial effects. Trends in Food Science & Technology, 108, 204–215.

Zhang, W., Lin, M., Feng, X., Yao, Z., Wang, T., & Xu, C. (2022). Effect of lemon essential oil-enriched coating on the postharvest storage quality of citrus fruits. *Food Science and Technology, 42*, e125421.

Zhao, P., Wang, J., Wang, K., Wang, Y., Wang, Y., Cheng, M., & Wang, X. (2024). Evaluation of postharvest physiological parameters and shelf life of Agaricus bisporus by addition of anise essential oil Pickering emulsion to corn starch/cassia gum. *Journal of Food Measurement and Characterization, 18*(6), 5103–5117. https://doi.org/10.1007/S11694-024-02559-X

Chapter 10
Effects of Natural Plant Extracts on the Properties of Edible Coatings and Films

A. K. A. N. W. M. R. K. Thamarsha, Pratibha, Nishant Kumar, Rushali Jaiswal, and Nutan Kaushik

Abstract Edible films and coatings have gained popularity in food preservation and protection because of their many benefits compared to synthetic substances, such as being biodegradable and eco-friendly. The functional properties of these biopolymer-based materials can be enhanced by incorporating plant extracts. This chapter provides the recent advancements in the addition of plant extracts into edible films and coatings, emphasizing their impact on techno-functional properties. Moreover, the chapter explores the impact of plant extracts on the mechanical, barrier, functional, antioxidant, and antimicrobial properties of biopolymer-based coatings. Furthermore, the insights of the chapter illustrate how the improved coatings and films are effectively prolonging the shelf life and enhancing the food quality through detailed examples and real-world applications. This chapter also underscores the potential of plant extract–infused edible films to revolutionize food preservation by promoting sustainability and enhancing food safety.

Keywords Edible coatings/films · Food quality · Mechanical properties · Plant extracts · Shelf life

A. K. A. N. W. M. R. K. Thamarsha
Graduate School of Horticulture, Chiba University, Chiba, Japan

Pratibha
Centre of Management and Humanities (CMH), Punjab Engineering College (Deemed to be University), Chandigarh, India

N. Kumar (✉) · N. Kaushik
Amity Food and Agriculture Foundation (AFAF), Amity University Noida, Noida, Uttar Pradesh, India

R. Jaiswal
National Institute of Food Technology Entrepreneurship and Management, Sonepat, Haryana, India

N. Kumar et al. (eds.), *Trends in Edible Packaging: A sustainable approaches for postharvest management of fruits and vegetables*,
https://doi.org/10.1007/978-3-032-13199-7_10

Introduction

Edible films and coatings have become increasingly popular in food protection and preservation due to their numerous advantages over synthetic materials, including biodegradability and environmental friendliness (Ponce et al., 2008; Trajkovska Petkoska et al., 2021). These are consumable layers designed to act as barriers against water content, gases, and solute diffusion in food products. These films and coatings offer an excellent substitute for traditional packaging materials because of their remarkable properties, including edibility, biological compatibility, and extensive applicability (Maan et al., 2021; Flores-Martínez et al., 2017). Edible coatings serve as active packaging, enhancing the lifespan of various foods by reducing dehydration, surface browning, oxidative rancidity, microbial growth, and oil diffusion (Silva-Weiss et al., 2013b). These coatings form a protective layer on the food surface by minimizing respiration, moisture loss, microbial growth, and enzymatic reactions through systematic releasing of active biomolecules (Yousuf et al., 2018; Seididamyeh et al., 2024).

Edible coatings and films come from multiple sources; they are primarily categorized into polysaccharides, lipids, and proteins. By combining these materials, the distinct properties of each can be utilized if they are compatible. The film development process involves a film-creating material, a solvent, an emulsifier, and a plasticizer. Sometimes, modifiers, such as antimicrobial agents and antioxidants, are added to enhance the properties of the films (Saxena et al., 2020; Trajkovska Petkoska et al., 2021). Films created by incorporating plant extracts into polymers typically exhibit altered physicochemical, mechanical, and barrier properties, as well as enhanced antioxidant and antimicrobial characteristics, when compared to films made from the individual components. These films have found diverse applications across various polymer-based functions (Mir et al., 2018).

Addition of plant extracts to edible coatings has recently attracted significant interest (Hasan et al., 2022; Antonino et al., 2024). A large variety of natural extracts have the potential to be utilized to improve the features of edible coatings (Zam, 2019). Plant extracts could be sourced from various parts of the plant, such as roots, petals, leaves, fruits, peels, stems, and xylems (Erkan, 2015). Currently, there is a growing trend toward the usage of unconventional and renewable substances, particularly by-products from fruits and vegetables (Elsayed et al., 2022; Rashwan et al., 2023; Ali et al., 2024; Mohan et al., 2024). Plant extracts are recognized as natural food additives within the food industry because of their potent antibacterial, antifungal, and antioxidant attributes, making them effective for preserving both raw and processed foods (Erkan, 2015). This chapter explores the key properties of plant extract–incorporated edible coatings and their future scope.

Role of Natural Extracts in Edible Coatings and Films

Plant materials are rich in natural compounds, particularly antioxidants, that are utilized in different sectors of the food industry. Recent trends have driven the exploration of plant extracts in edible coatings and films (Table 10.1). These extracts, obtained from multiple sources, including leaves, fruits, and seeds, have been incorporated into biodegradable films to achieve desirable properties (Mir et al., 2018). Plant extracts contain higher amounts of phenolic compounds and are known for their antioxidant and antimicrobial properties (Lopes et al., 2023). Incorporating phenolic compounds and extracts into active packaging helps prevent the oxidation of food and enhances their direct intake by humans, contributing to improved health (Saberi et al., 2017).

Plant extracts are effective at low concentrations, cost-efficient, and easy to apply; maintain high stability during processing; and/or do not alter the sensory attributes of food products (Liu et al., 2021). Plant extracts have gained significant attention from the food industry because of their natural origin and diverse functional attributes. Their potential to be sources of antioxidants, antimicrobial agents, and immune response mediators has sparked interest in their application in food and health products (Abigail et al., 2016; Liu et al., 2021).

Table 10.1 Common plant extracts used for edible coatings

Plant extracts	Scientific name	Reference
Tea leaf	*Camellia sinensis*	Das et al. (2013)
Moringa leaf extract (MLE)	*Moringa oleifera* Lam.	Tesfay and Magwaza (2017)
Grapefruit seed extract (GSE)	*Vitis vinifera*	Xu et al. (2007)
Rosemary	*Rosmarinus officinalis*	Kahya et al. (2022)
Sage	*Salvia officinalis*	Kahya et al. (2022)
Pomegranate peel extract (PPE)	*Punica granatum* L.	Kumar et al. (2022)
Aloe vera (AV)	*Aloe vera*	Valverde et al. (2005)
Murta extract	*Ugni molinae* Turcz.	Silva-Weiss et al. (2013a)
Chinese chive	*Allium tuberosum*	Riaz et al. (2020)
Black cumin	*Nigella sativa* L.	Karkar et al. (2023)
Sweet basil extract	*Ocimum basilicum* L.	Synowiec et al. (2014)
Mango peel extract (MPE)	*Mangifera indica* L.	Elsayed et al. (2022)
Lotus leaf extract (LLE)	*Nelumbo nucifera* Gaertn.	Fan et al. (2019)
Peppermint extract	*Mentha piperita* L.	Morachis-Valdez et al. (2021)
Musk lime extract	*Garcinia atroviridis*	Wai et al. (2019)
Senduduk	*Melastoma Malabathricum*	Chan et al. (2020)
Ginseng extract	*Panax ginseng*	Norajit et al. (2010)
Red raspberry	*Rubus strigosus*	Wang et al. (2012)
Ginger	*Zingiber officinale*	Mir et al. (2018)

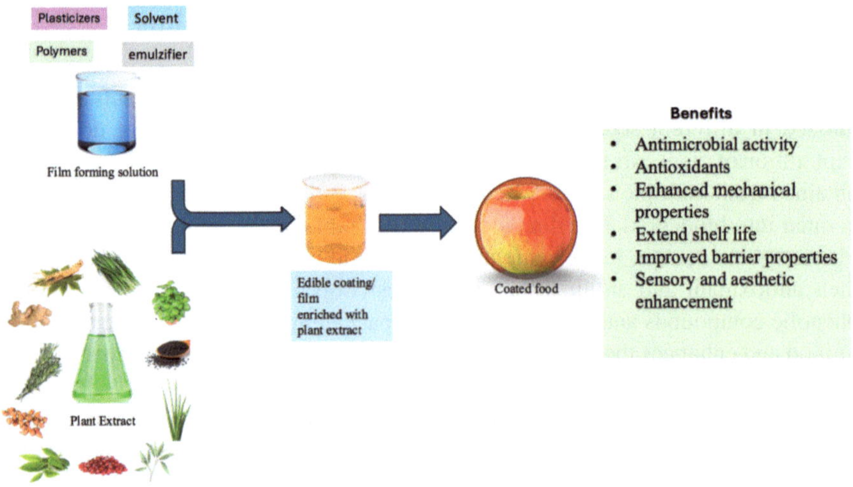

Fig. 10.1 Benefits of incorporating plant extracts into edible coatings

Effect of Plant Extracts on the Properties of the Film

Incorporation of plant extracts may alter the properties of the edible films and coatings. The alterations in the characteristics rely on both the type of extract and the extract concentration (Mir et al., 2018). The structure and performance of biopolymers and extracts are influenced by their physical or chemical interactions. These interactions between biopolymers and plant extracts depend on properties such as the nature, chemical properties, and polymer concentration and the extract type. Furthermore, the role of structural characteristics, such as stereochemistry and conformational flexibility of the active components, is crucial in determining their functionality (Kanmani & Rhim, 2014). The intermolecular interactions within the polymer structure by forming hydrogen bonds between the phenolic compounds in plant extracts and the molecules of polysaccharides or proteins may improving the functionality of edible film and coatings (Wang et al., 2012). Incorporating plant extracts into the polymer can influence various physicochemical and techno-functional attributes, such as color, thickness, tensile strength (TS), solubility, water vapor permeability (WVP), and antimicrobial and antioxidant properties (Mir et al., 2018). Figure 10.1 shows the benefits of incorporating plant extracts into edible coatings.

Thickness

The thickness of the film is a key factor in its effectiveness as a coating to prevent food spoilage and distortion (Karkar et al., 2023). Furthermore, it is a key property that immediately influences the mechanical strength, WVP, light transmittance, and opacity of films (Yong et al., 2019). Thickness is an important factor for food packaging films because it influences various physical attributes, such as TS, WVP, and transparency (Yong et al., 2019). Numerous studies have shown that most plant extracts lead to an enhanced thickness of the film (Kola & Carvalho, 2023). The addition of mango kernel extracts (MKEs) to edible films significantly enhanced their thickness. The greater film thickness was caused by the higher solid content added to the film matrix, together with the disturbance of its organized structure caused by extract droplets (Maryam Adilah et al., 2018). He et al. (2019) reported that adding pomegranate peel extract (PPE) to polyvinyl alcohol (PVA)–mixed films led to an increased thickness in the film. Similarly, the incorporation of PPE into chitosan (CS)-based coatings showed enhanced thickness (Zeng et al., 2021). The film enhancement is due to the molecular interactions between the functional groups of chitosan (CS) and the phenolic hydroxyl (OH) groups found in the bioactive compounds of pomegranate peels (Riaz et al., 2018; Zeng et al., 2021).

Kanmani and Rhim (2014) found that the addition of grapefruit seed extract (GSE) influenced the thickness of carrageenan films. As the concentration of the extract increased from 0.6 to 13.3 μg/mL, the film thickness increased from 36.66 to 43.33 μm. Furthermore, polyphenol-rich purple rice extract (PRE) and black rice extract (BRE) were added to CS to create pH-sensitive and antioxidant packaging films, and the results showed that CS-PRE and CS-BRE films were thicker than CS films (Yong et al., 2019). Moreover, the inclusion of extracts from *Pistacia terebinthus* to chitosan-based films significantly thickened them compared to the control (50 ± 9.4 μm), with values of 71.75 ± 5.31 μm for stem extract, 104.75 ± 14.3 μm for leaf extract, and 70.8 ± 1.7 μm for seed extract (Kaya et al., 2018a). Besides, it was found that the incorporation of GSE enhanced the thickness of films derived from agarose, attributed to the higher molecular weight of the phenolic compounds present in the extract (Lim et al., 2010). A study focused on creating a CS-based functional film enriched with Chinese chive (*Allium tuberosum*) root extract, and the resulting film showed an increased thickness (Riaz et al., 2020). Additionally, the incorporation of yellow onion peel extract (YOPE) into funoran extracted from *Gloiopeltis furcata* improved the thickness of the resulting film (Ju & Song, 2020). Furthermore, Bigi et al. (2021) demonstrated that the addition of sage leaf extract increased the film thickness, whereas the incorporation of nettle leaf extract resulted in a decrease in thickness in chitosan/hydroxypropyl methylcellulose blend films.

Barrier Properties

Films used in food packaging that have effective barrier properties can improve packaging conditions by minimizing the movement of water vapor and gases between the food and the surroundings (Shan et al., 2023). Barrier properties in edible packaging are crucial for preventing adverse modifications in the foods, such as textural alterations (hardening or softening), physical issues (aggregation, caking, shrinkage, swelling, and breakage), and a decrease in solubility or the capacity to hold water (Kumar et al., 2023b). Plant extracts have a notable impact on the barrier properties of edible packaging (Bifani et al., 2007; Nouri & Nafchi, 2014). Key barrier properties of polymers relevant to food packaging applications are oxygen permeability and WVP (Kumar et al., 2023b). The interaction of moisture, oxygen, other gases, or liquids and flavorings with packaged goods can greatly influence the lifespan of the food (Kola & Carvalho, 2023). Therefore, determining the barrier properties of a coating is essential for estimating the lifespan of the packaged food (Zhang et al., 2010).

One of the main purposes of food packaging films is to minimize the moisture transfer between the food and its surroundings. WVP is an essential barrier property that indicates the effectiveness of the film against moisture (Kola & Carvalho, 2023). A low WVP indicates limited moisture movement between the food and its environment, which helps in prolonging its shelf life (Tong et al., 2020). The addition of 5% MKE to soy protein isolate (SPI) films enhanced their WVP by up to 10%. This enhancement could be due to the hydrophobic quality of the extract, which enhanced the overall hydrophobicity of the film (Maryam Adilah et al., 2018). Edible films made from mango peel extract (MPE) exhibited excellent barrier properties, with WVP ranging from 0.88×10^{-10} to 1.00×10^{-10} g m^{-1} s^{-1} Pa^{-1} (Torres-León et al., 2018).

In addition, the addition of *Sonneratia caseolaris* leaf extract enhanced the moisture barriers of CS-based films (Nguyen et al., 2020). Furthermore, according to Nouri and Nafchi (2014), sago starch–based films containing betel leaf extract showed reduced moisture and oxygen barrier properties. Also, incorporating GSE with varying concentrations into carrageenan-based films enhanced the WVP of the polymer (Kanmani & Rhim, 2014). The addition of PPE and GSE reduced the WVP coefficient of silver carp surimi–based edible films. This reduced WVP in films containing phenolic compounds is likely due to the increased presence of hydroxyl (OH) groups and the enhanced cross-linking between proteins and phenolic compounds (Munir et al., 2019). Furthermore, incorporating watermelon rind extract into CS– guar gum (GG) biopolymers significantly reduced the WVP. The lowest value (3.22×10^{-7} g m^{-1} h^{-1} Pa^{-1}, down from $4.57 \pm 0.06 \times 10^{-7}$ g m^{-1} h^{-1} Pa^{-1}) was shown when 4 wt% of the extract was added. That reduction is attributed to two factors: The watermelon rind extract is abundant in aromatic ring structures, specifically flavones, that form hydrogen bonds with CS and GG, thereby decreasing the hydrophilicity of the film and improving its density, and the enhanced hydrogen bond interactions between CS and watermelon rind extract lead to the formation of a tight cross-linking network that reduces the binding sites for water molecules and compresses the space for water vapor movement; thus water vapor emission is reduced (Wang et al., 2023).

The light barrier properties are crucial when analyzing edible films and coatings, especially if the film will be used on food that is sensitive to light-induced degradation (Torres-León et al., 2018). The light barrier property is important to protect the food against deterioration caused by light (Seididamyeh et al., 2024). According to Nouri and Nafchi (2014), the incorporation of betel leaf extract into sago starch–based films reduced the UV barrier properties. Furthermore, a study on bioactive pectin/pullulan films infused with GSE demonstrated improved UV-blocking capabilities in peanuts (Priyadarshi et al., 2022). Carrageenan-based antimicrobial films, created by incorporating varying concentrations of GSE, exhibited enhanced UV barrier properties of the film (Kanmani & Rhim, 2014). Moreover, CS films containing the leaf extract of *Sonneratia caseolaris* (L.) Engl. exhibited an excellent light barrier property, effectively protecting the packaged products from visible light (Nguyen et al., 2020).

Mechanical Properties

Biodegradable films should possess sufficient strength to secure the stability and protection of packaged foods. Physical and mechanical attributes of biopolymers, including tensile strength (TS), elongation at break (EB), and Young's modulus, are influenced by the interactions between the polymers and the additives used. The addition of plant extracts affects the mechanical properties of these films through their interaction with polysaccharides, proteins, lipids, and the extracts themselves (Li et al., 2014; Mir et al., 2018). On the other hand, the reduction in mechanical strength can be attributed to the increased free volume and molecular mobility within these systems (He et al., 2019; Kumar et al., 2022). TS and EB are significant factors that reflect the mechanical properties and chemical structure of a film. TS indicates the maximum strength, representing the film's resistance to breaking. In contrast, EB gauges the film's stretching ability or flexibility before it breaks (Silva-Weiss et al., 2013b).

Additionally, physical characteristics of SPI films with the addition of GSE (1% w/w), nisin (10,000 IU/g), ethylenediaminetetraacetic acid (EDTA 0.16% w/w), and their combinations revealed that GSE significantly improved the TS and puncture strength of the film (Sivarooban et al., 2008). Also, mulberry leaf extract–incorporated sodium alginate (NaAlg) films display enhanced TS and decreased EB (Kuan et al., 2020). Besides, mechanical tests demonstrated that incorporating MKE enhanced the TS of both SPI and fish gelatin films, with fish gelatin films exhibiting higher TS compared to SPI (Maryam Adilah et al., 2018). Furthermore, the addition of *Melastoma malabathricum* into sodium alginate edible films significantly alters the mechanical properties, including reducing the water activity and EB of the film (Chan et al., 2020). CS-based edible films were improved by incorporating *Berberis crataegina* DC. fruit extract into the chitosan matrix, resulting in a 55% increase in EB against the CS film (Kaya et al., 2018b).

Incorporating murta fruit extract greatly enhanced the mechanical attributes of methyl cellulose films, significantly increasing their elongation at break values (López de Dicastillo et al., 2016). This enhancement was probably due to the improved mobility of polymer chains, aided by substances from the murta extract that inserted themselves between the polymeric chains, thus improving flexibility (López de Dicastillo et al., 2016). Also, the addition of green tea water extract (GTE) marginally enhances both the TS and EB of chitosan-based biopolymers (Siripatrawan & Harte, 2010). However, both TS and EB were reduced due to the incorporation of green tea extract into the agar–gelatin film (Giménez et al., 2013). Also, the addition of yellow onion peel extract into funoran extracted from *Gloiopeltis furcata* showed a decreased TS in the film (Ju & Song, 2020). Moreover, incorporating ginseng extract into alginate films reduced TS and elastic modulus while increasing the percentage of EB (Norajit et al., 2010).

Biodegradability

The term "biodegradable" refers to materials that can be broken down via the enzymatic activity of microbes, such as bacteria, yeasts, and fungi. Under aerobic conditions, this degradation process results in end products like CO_2, H_2O, and biomass. In anaerobic conditions, the degradation yields hydrocarbons, methane, and biomass (Kola & Carvalho, 2023). The properties of CS films to which rosemary and sage extracts were added were assessed by Kahya et al. (2022), and the films with rosemary extract showed a higher biodegradability compared to the films with sage extract. In particular, the films with rosemary extract reached around 99% degradability within 5 days, while those with sage extract exhibited a 95% degradation rate.

The solubility of a film is a key functional property that reflects its biodegradability (Norajit et al., 2010). Edible films were synthesized using a CS matrix enriched with seed oil and fruit extract derived from *Berberis crataegina* DC. The solubility of the chitosan film was measured at 24% over 48 h. The film solubility was significantly increased to 35.27%, 45.31%, and 50.7%, respectively, for leaf-, stem-, and seed-incorporated CS films. The film showed a higher water solubility because of poor intermolecular interactions with the plant extracts and the CS matrix. Furthermore, the soil degradation rate over 15 days was observed to be 89.04% for the CS film, and it was 98.76% for the *Berberis crataegina* DC. leaf–incorporated film. In contrast, soil degradation was lower than the control in the CS–stem film (78.65%) and the CS–seed film (56.35%) (Kaya et al., 2018a). The addition of ginseng extract enhanced the water solubility of the alginate film (Norajit et al., 2010). The incorporation of nettle leaf extract led to an increased water solubility in CS/hydroxypropyl methylcellulose blend films (Bigi et al., 2021). The water solubility has been improved by applying the leaf extract of *Sonneratia caseolaris* (L.) Engl. to an edible film (Nguyen et al., 2020). Furthermore, the inclusion of plant extracts to edible coatings can decrease the water solubility. Saberi et al. (2017) developed functional food packaging using pea starch (PS) and guar gum (GG) films enriched with natural antioxidants, including epigallocatechin gallate, macadamia peel extract, blueberry ash fruit extract, and banana peel extract, and the resulting film exhibited a lower water solubility. Furthermore, the incorporation of MKE into SPI- and fish gelatin–based film reduced the water solubility (Maryam Adilah et al., 2018).

Antioxidant and Antimicrobial Properties

The inclusion of natural extracts to edible coatings can increase their antimicrobial and antioxidant properties, thereby improving their effectiveness in preserving food products (Ruíz-Cruz et al., 2019). Different substances were suggested and evaluated for their potential to prevent microbial growth in food packaging. These substances include organic acids, such as sorbate, propionate, and benzoate; enzymes like lysozyme; fungicides, including benomyl and imazalil; and naturally derived antimicrobial compounds (Tharanathan, 2003). Plant extracts are rich in various substances known as phytochemicals, which are naturally occurring components in plants. These phytochemicals are recognized for their ability to reduce the growth or eliminate foodborne pathogens (Synowiec et al., 2014). These compounds exhibit both antimicrobial and antioxidant properties (Zúñiga et al., 2012). The application of naturally occurring antimicrobial agents in packaging materials to inhibit microbial growth presents a viable alternative to synthetic fungicides. Plant spice extracts offer a promising solution due to their natural composition, alignment with consumer preferences for natural products, and reduced regulatory issues (Ferreira Soares et al., 2016). Furthermore, plant-based extracts enhance the activity of a wide range of antioxidant enzymes, such as ascorbate peroxidase, glutathione peroxidase, and guaiacol peroxidase, which detoxify free radicals of hydrogen peroxide by converting them into water. Furthermore, glutathione reductase restores lowered glutathione by transferring hydrogen to its oxidized form, whereas superoxide dismutase facilitates the conversion of superoxide anions into hydrogen peroxide and molecular oxygen through dismutation. These extracts also offer numerous nutraceutical benefits (Bajaj et al., 2023). Incorporating plant extracts as natural antimicrobials into edible coating films can satisfy the increasing consumer desire for safe and fresh foods (Nguyen et al., 2020). Seididamyeh et al. (2024) studied the synergistic impacts of gum arabic–based coatings incorporated with extracts from *Diploglottis bracteata, Syzygium aqueum*, and *Tasmannia lanceolata* on the characteristics of edible coatings and films. The formulated coating suspensions exhibited robust antimicrobial activity against *Pseudomonas viridiflava*, a microorganism responsible for causing soft rot. Besides, Gómez-Estaca et al. (2009) studied the antioxidant characteristics of gelatin-based edible films containing oregano or rosemary aqueous extracts and concluded that both extracts improved the film properties, with oregano extract displaying elevated total phenolic content

(TPC) and antioxidant activity. Adding basil extract to pullulan coating enhanced its effectiveness on fresh apples, completely preventing mold growth (Synowiec et al., 2014). Bazie et al. (2014) indicated that using aqueous plant extracts, particularly from *Acacia albida* and *Prosopis juliflora* at 50 °C, can effectively reduce anthracnose (*Colletotrichum musae*) incidence and severity in banana fruits without compromising their physicochemical properties, highlighting a potential method to improve postharvest quality.

Incorporating tomato plant extract into CS-based coatings has the potential to improve the physicochemical, microbiological, sensory, and antioxidant qualities of poultry meat, indicating its efficacy as a natural preservation technique (Ruíz-Cruz et al., 2019). Furthermore, the study by Ruíz-Cruz et al. (2019). Demonstrated that CS–tomato plant extract edible coating significantly improved the physicochemical, microbiological, antioxidant, and sensory properties of pork during storage, highlighting its potential as an alternative preservation method. Moreover, the application tomato oily extract–added gelatin-based coatings has been shown to significantly enhance the preservation of garambullo (*Myrtillocactus geometrizans*) fruits, maintaining their physicochemical properties, such as pH, total soluble soilds (TSS), titratable acidity (TA) and color, and their bioactive compound levels (ascorbic acid, betalains, total phenols, and flavonoids) while stored at 5 °C (López-Palestina et al., 2018). The incorporation of *Justicia adhatoda* extract into edible coating solutions significantly enhances their antimicrobial and antioxidant properties while optimizing the balance of sucrose and glycerol to improve their physicochemical and rheological characteristics (Hanumantharaju et al., 2020). Norajit et al. (2010) have investigated the physical and antioxidant characteristics of alginate-based films with red, white, and extruded white ginseng extracts. The findings indicate that extruded white ginseng extract, especially at higher extrusion temperatures, markedly improves free-radical scavenging activity, making it a promising candidate for antioxidant biodegradable films or coatings in food applications. Furthermore, the addition of aqueous green tea extracts to chitosan films resulted in increased antioxidant properties and polyphenol content (Siripatrawan & Harte, 2010). Moreover, the inclusion of *Quillaja saponaria* extract to calcium caseinate and whey protein isolate edible coating reduced the growth of *Botrytis cineria* in strawberries (Zúñiga et al., 2012). Additionally, a study showed that pullulan films with aqueous *Satureja hortensis* extract significantly reduced the growth of both gram-positive and gram-negative bacteria, as well as *Penicillium expansum* (Kraśniewska et al., 2014). According to a study, the addition of loquat leaf extract to alginate-based composite coating showed notable effects against *Penicillium digitatum* in Nanfeng tangerine fruits. The results showed a minimum inhibitory concentration (MIC) of 0.625 mg/mL and a minimum fungicidal concentration (MFC) of 1.25 mg/mL (Zhang et al., 2022). A study examined the use of fungal CS derived from *Aspergillus niger* and PPE in coating films to preserve Nile tilapia (*Oreochromis niloticus*) fillets, maintaining their microbiological quality while stored at 4 °C for 30 days, and it exhibited a substantial decrease in microbial counts during storage (Alsaggaf et al., 2017).

Research evaluated the influence of adding PPE to CS and locust bean gum (LBG) edible coatings on the quality of white shrimp (*Parapenaeus longirostris*) under low temperature. According to this study, CS combined with PPE effectively reduced microbial spoilage. Shrimp coated with CS containing PPE showed a reduction of approximately 2 log units in *Pseudomonas* spp. count, and both CS alone and CS with PPE maintained a psychrotrophic microbial load lower than 7 CFU/g for 6 days. Although LBG alone did not exhibit an inhibitory effect, adding PPE to LBG reduced the psychrotrophic load by 1 log unit (Licciardello et al., 2018).

Color Properties

Color is a key sensory factor that influences the initial acceptance or rejection of a food product. In fact, studies have shown that color accounts for 62–90% of how consumers evaluate a product (Teixeira et al., 2022). Plant extracts often change the color properties of polymers. According to the literature, incorporating plant extracts can modify the original color of the films to varying degrees, with the extent of change depending on the type of plant extract and its concentration (Silva-Weiss et al., 2013b). Aqueous beetroot peel extract (BPE) treatment influences the color of the films based on gelatin–sodium alginate (NaAlg). As the concentration of BPE increased, the lightness (L*) values decreased. However, the yellowness (b*) and redness (a*) values showed a significant increase, resulting in a reddish color in the film due to BPE incorporation. In fact, higher BPE concentrations enhanced the intensity of the red color (Chaari et al., 2022). Furthermore, in SPI films, the addition of raspberry extract led to increased darkness, redness, and yellowness, as indicated by lower L* values, higher positive a* values, and elevated b* values compared to the control film (Wang et al., 2012). According to a study conducted by Kanmani and Rhim (2014), the control film exhibited higher lightness (L*) compared to the carrageenan/grapefruit seed extract composite films. The incorporation of the extract into the carrageenan film led to a reduction in L* values, decreasing from 93.25 to 91.77 as the extract concentration increased. In contrast, a* and b* values showed a significant increase with rising extract concentrations, ranging from 0.6 to 13.3 μg/mL. The addition of a beetroot and carrot extract blend to hydroxypropyl methylcellulose films altered their color properties. A reduction in L* values and an increase in a* values were observed, indicating a noticeable change in film coloration. The significant rise in redness (a*) highlights the blend's strong coloring potential (Mir et al., 2018).

Shelf Life and Quality Attributes

Edible coatings and films incorporating natural plant extracts are increasingly utilized to improve the quality and lengthen the storage life of fruits and vegetables (Nair et al., 2018). Plant extract–incorporated edible coatings improve the antioxidant activity in fruits and reduce the free radicals generated during the storage period. Such coatings create a successful oxygen barrier on the surface of fruit and inhibit enzymes such as polyphenol oxidase (PPO) and peroxidase (POD), preventing the oxidation of phenolic compounds and extending the shelf life of the produce (Bajaj et al., 2023). Plant extract–incorporated edible coatings are effectively utilized to enhance the quality and prolong the lifespan of numerous fruits (Fan et al., 2019). A recent study explored the effects of CS (1% w/v) and alginate (2% w/v) coatings, combined with PPE (1% w/v), on the quality of guavas. The results demonstrated that these coatings effectively limited changes in ripening index, respiration rate, and instrumental color values over 20 days of storage at 10 °C compared to uncoated control samples. Furthermore, coated fruits exhibited limited losses in ascorbic acid (29%), total flavonoids (12%), total phenolics (8%), and antioxidant activity, with reductions of 12% for DPPH (2,2 diphenyl, 1 picrylhydrazyl) and 9% for FRAP(Ferric reducing antioxidant power). Also, a strong correlation was observed between the phytochemicals and antioxidant activity (Nair et al., 2018). Besides that, the incorporation of PPE into xanthan gum–based edible coatings has shown promising results in improving the postharvest storage period of mango (*Mangifera indica* L.) fruits by preserving their quality attributes such as respiration rate, reducing weight loss, ethylene production, maintained texture property, total soluble solids (TSSs), acidity, pH, ascorbic acid, phenols, and antioxidant activity, and reducing postharvest losses (Kumar et al., 2023a).

A study evaluated the effects of ginger extract, garlic extract, gum arabic, and aloe vera gel coatings (in combinations) on fruits stored at 25 ± 3 °C for 15 days, and it showed that the combination of ginger extract and gum arabic resulted in elevated total carotenoid levels. Additionally, the garlic extract and gum arabic combination resulted in higher ascorbic acid content, higher total flavonoid content, and lower total sugars in guavas. Moreover, the combination of garlic extracts and gum arabic minimized skin browning, weight loss, and disease severity of guavas significantly (Anjum et al., 2020). Furthermore, Hasan et al. (2022) conducted a study to examine the effects of alginate-based edible coatings incorporated with black cumin extract on extending the storage life of guava fruits over a 16-day period under specific storage conditions of 11 ± 1 °C temperature and 85 ± 2% relative humidity. It showed significantly enhanced physicochemical attributes, nutritional quality, and biological functions of guava fruits, suggesting that this treatment has the possibility of being a better way to extend the postharvest life and maintain the quality of guava fruits. The garlic extract and gum arabic combination also extended the lifespan of guava fruits (Anjum et al., 2020).

Zam (2019) found that coatings made from CS and alginate, enriched with olive leaves extract, can significantly delay the ripening process and maintain the nutritional quality of sweet cherries, thus extending their shelf life and boosting their antioxidant activity. Recent research has shown that a novel 1.5% chitosan coating incorporated with hairy fig (*Ficus hirta* Vahl.) fruit extract significantly improves the preservation of "Newhall" navel oranges by minimizing decay and weight loss, delaying ripening and senescence, and maintaining fruit quality during cold storage through respiration inhibition and the enhancement of protective enzyme activities (Chen et al., 2018). According to Tesfay and Magwaza (2017), the incorporation of moringa leaf extract (MLE; *Moringa oleifera*) into commercially available edible coatings improved the quality and the postharvest storage life of 'Fuerte' and 'Hass' avocado fruits, resulting in lower respiration rates and higher firmness compared to untreated fruits. The incorporation of MLE into the hydroxypropyl methyl cellulose (HPMC)–based edible coating lengthened the lifespan of *Citrus × sinensis* (Adetunji et al., 2012). The incorporation of lotus leaf extract (LLE) composite coating into fresh goji fruit during storage at room temperature found that LLE coating significantly reduced weight loss, decay rate, and malondialdehyde (MDA) content while maintaining higher levels of ascorbic acid, titratable acidity, soluble solids, and enzyme activities, thereby prolonging the storage life of goji berries by about 4 days (Fan et al., 2019). The addition of pullulan coating with *Satureja hortensis* water extract to peppers and apples reduced weight loss, prolonged storage time, and prevented excessive wilting and wrinkling, thereby maintaining the freshness and consumer preference of the fruits.

Shafique et al. (2023) found that the firmness of strawberries increased by 60% after being coated with a 20% aloe vera (AV) gel and stored at room temperature for 5 days. Additionally, applying a 6% MLE to strawberries showed a 50.7% lower respiration rate and a 49.7% reduction in ethylene production (Shafique et al., 2023). Furthermore, Pagliarulo et al. (2016) demonstrated that an active edible coating comprising peony (*Paeonia rockii*) extracts and chitosan effectively lengthened the lifespan of strawberries by inhibiting fungal growth while retaining their nutritional and sensory attributes. Besides, another novel edible coating composed of basil seed (*Ocimum basilicum* L.) gum (BSG) and incorporated with echinacea extract (EE) revealed that the 3% BSG + 3% EE formulation greatly enhanced the strawberry fruit quality and lifespan by minimizing microbial growth, weight loss, softening, and nutrient degradation while simultaneously boosting antioxidant activity, superoxide dismutase activity, and sensory qualities (Moradi et al., 2019).

The incorporation of GSE into edible coating based on salep solution (SS) improved the quality and shelf life of strawberries by decreasing the weight loss, softening, and degradation of ascorbic acid, phenols, and anthocyanins. Also, the resulting coating demonstrated a higher superoxide dismutase activity and minimal peroxidase activity (Emamifar et al., 2019). Sodium alginate (NaAlg)–based biodegradable coatings containing GSE effectively reduced the weight loss and maintained firmness in grapes while stored. Coatings with GSE were successful in preserving the antioxidant activity and reducing decay incidence in inoculated fruits (Aloui et al., 2014). A research investigated the influence of edible cassava starch–CS coatings, infused with PPE, on the shelf life of tomatoes stored at 25 °C for 12 days. The findings revealed that a concentration of 20 mL L^{-1} of PPE significantly minimized weight loss and decreased the TSS content compared to controlled tomatoes (Araújo et al., 2018). The addition of pineapple core extract to a konjac glucomannan–based edible coating has been shown to effectively inhibit browning, reduce weight loss, and maintain the whiteness and phenolic content of fresh-cut Taaptimjaan rose apple during refrigerated conditions (Supapvanich et al., 2012). Another study assessed the effect of edible coatings made from whey protein concentrate (WPC) and MPE on the shelf life of minimally processed broccoli. This coating appeared to positively impact the quality retention of minimally processed broccoli by slowing changes in weight loss, respiration rate, bioactive compounds such as ascorbic acid, total phenolic content (TPC), total antioxidants, and the yellowing process (Elsayed et al., 2022). Incorporating *Fagonia indica* plant extract into *Aloe vera* gel coatings shows extended shelf life and maintains the quality attributes of the sapodilla fruit while stored (Khaliq et al., 2019). Gelatin coatings enriched with *Aloe vera* gel and green tea extracts have been reported as a prospective technique to enhance the quality and extend the shelf life of fresh-cut oranges, as evidenced by their improved physicochemical properties, reduced microbial growth, and favorable sensory attributes during cold storage (Radi et al., 2017). Also, another study investigated the application of *Aloe vera* and green tea extracts in gelatin-based coatings to maintain the freshness and quality of apple slices while stored. These coatings enhanced sensory attributes, reduced microbial counts, and better preserved quality parameters compared to control samples, though they did not entirely prevent chemical and biochemical reactions (Amiri et al., 2018).

The addition of tomato plant extract to a CS-based edible coating significantly improved the quality and extended the shelf life of sierra fish fillets while stored at cold conditions (Ramírez-Guerra et al., 2018). Furthermore, results from another study showed that CS coating infused with GSE extends the shelf life of trout fillets by 3–5 days during storage at 4 °C (Hassanzadeh et al., 2018). Moreover, a study investigating the application of peppermint extract and CS-based edible coatings on common carp fillets revealed improved storage quality by preventing microbial growth and minimizing lipid and protein oxidation, thus extending the shelf life during refrigerated storage (Morachis-Valdez et al., 2021).

Future Scope

A vast range of natural plant sources is available, and they are rich in bioactive attributes that are being explored for developing biopolymer-based edible films to enhance food preservation and value. Despite significant research interest in films derived from plant extracts, their industrial adoption is currently constrained. Some materials are produced on a large scale and are commercially available. Efforts to advance this innovative technology commercially require concerted actions to connect research progress with broader industrial implementation. The market for plant-based biodegradable films is limited due to challenges in replacing synthetic polymers with eco-friendly alternatives. Thus, extensive research is needed to enhance the characteristics of these films and coatings for industrial food packaging applications. Improving certain characteristics of these materials often involves trade-offs with others, necessitating thorough investigation and optimization.

Future research should focus on determining the optimal concentrations or ratios of plant extracts and exploring their interactions with various edible matrices. There is also a need to explore the use of commercial by-products in industrial-scale applications for food coatings or other food processing purposes, such as enhancing the health benefits of functional foods. Additional research is necessary to evaluate the efficacy of plant extracts in edible coatings and films formulated with diverse polysaccharides, proteins, and lipids. Furthermore, it is essential to investigate more about their effects on various fruits and vegetables. Additionally, some plant extracts might exhibit toxicity at high doses in humans. Therefore, additional studies are required to examine the toxicity levels of plant extracts in both animal and human tissues. Nevertheless, additional experiments are required to examine the cytotoxic and genotoxic characteristics of these coatings. It is also crucial to apply the coatings to food matrices and assess their bioactivity. Moreover, further studies are essential to confirm and establish the impacts of the edible coating on sensory characteristics and consumer acceptance after incorporating plant extracts.

Conclusion

Nowadays, interest is rising in formulating edible coatings that include plant extracts to enrich the characteristics of the coatings and thereby improve the food quality and the extension of the shelf life of foods. Plant extract–incorporated edible coatings offer enhanced techno-functional properties. Significant findings highlight the important role of plant extracts in edible films, suggesting their potential as natural additives. Plant extracts enhance the functionality of coatings by leveraging their antioxidant, antimicrobial, and other bioactive properties, thus significantly contributing to the preservation and safety of perishable foods. However, further research is essential to fine-tune formulations, validate effectiveness across various food types, and ensure consumer acceptance and compliance with regulatory standards.

References

Abigail, R. M., Luisa, C. I., Candy, C. Á., Pimentel-González, D. J., & Brenda, A. S. (2016). Antioxidant activity, antimicrobial and effects in the immune system of plants and fruits extracts. *Life Science, 9*(2), 90–98.

Adetunji, C. O., Fawole, O. B., Arowora, K. A., Nwaubani, S. I., Ajayi, E. S., Oloke, J. K., Majolagbe, O. N., Ogundele, B. A., Aina, J. A., & Adetunji, J. (2012). Quality and safety of *Citrus Sinensis* coated with Hydroxypropyl methylcellulose edible coatings containing *Moringa oleifera* extract stored at ambient temperature. *Global Journal of Scientific Frontier Research, 12*, 530–540.

Ali, M. Q., Ahmad, N., Azhar, M. A., Munaim, M. S. A., Hussain, A., & Mahdi, A. A. (2024). An overview: Exploring the potential of fruit and vegetable waste and by-products in food biodegradable packaging. *Discover Food, 4*(1). https://doi.org/10.1007/s44187-024-00117-4

Aloui, H., Khwaldia, K., Sánchez-González, L., Muneret, L., Jeandel, C., Hamdi, M., & Desobry, S. (2014). Alginate coatings containing grapefruit essential oil or grapefruit seed extract for grapes preservation. *International Journal of Food Science and Technology, 49*(4), 952–959. https://doi.org/10.1111/ijfs.12387

Alsaggaf, M. S., Moussa, S. H., & Tayel, A. A. (2017). Application of fungal chitosan incorporated with pomegranate peel extract as edible coating for microbiological, chemical and sensorial quality enhancement of Nile tilapia fillets. *International Journal of Biological Macromolecules, 99*, 499–505. https://doi.org/10.1016/j.ijbiomac.2017.03.017

Amiri, S., Akhavan, H. R., Zare, N., & Radi, M. (2018). Effect of gelatin-based edible coatings incorporated with aloe vera and green tea extracts on the shelf-life of fresh-cut apple. *Italian Journal of Food Science, 30*(1), 61–74.

Anjum, M. A., Akram, H., Zaidi, M., & Ali, S. (2020). Effect of gum arabic and *Aloe vera* gel based edible coatings in combination with plant extracts on postharvest quality and storability of 'Gola' guava fruits. *Scientia Horticulturae, 271*, 109506. https://doi.org/10.1016/j.scienta.2020.109506

Antonino, C., Difonzo, G., Faccia, M., & Caponio, F. (2024). Effect of edible coatings and films enriched with plant extracts and essential oils on the preservation of animal-derived foods. *Journal of Food Science, 89*(2), 748–772. https://doi.org/10.1111/1750-3841.16894

Araújo, J. M. S., de Siqueira, A. C. P., Blank, A. F., Narain, N., & de Aquino Santana, L. C. L. (2018). A Cassava starch–chitosan edible coating enriched with *Lippia sidoides* Cham. Essential oil and pomegranate peel extract for preservation of Italian Tomatoes (*Lycopersicon esculentum* Mill.) stored at room temperature. *Food and Bioprocess Technology, 11*(9), 1750–1760. https://doi.org/10.1007/s11947-018-2139-9

Bajaj, K., Adhikary, T., Gill, P. P. S., & Kumar, A. (2023). Edible coatings enriched with plant-based extracts preserve postharvest quality of fruits: A review. *Progress in Organic Coatings, 182*, 107669. https://doi.org/10.1016/j.porgcoat.2023.107669

Bazie, S., Ayalew, A., & Woldetsadik, K. (2014). Integrated management of postharvest banana anthracnose (*Colletotrichum musae*) through plant extracts and hot water treatment. *Crop Protection, 66*, 14–18. https://doi.org/10.1016/j.cropro.2014.08.011

Bifani, V., Ramírez, C., Ihl, M., Rubilar, M., García, A., & Zaritzky, N. (2007). Effects of murta (*Ugni molinae* Turcz) extract on gas and water vapor permeability of carboxymethylcellulose-based edible films. *LWT-Food Science and Technology, 40*(8), 1473–1481. https://doi.org/10.1016/j.lwt.2006.03.011

Bigi, F., Haghighi, H., Siesler, H. W., Licciardello, F., & Pulvirenti, A. (2021). Characterization of chitosan-hydroxypropyl methylcellulose blend films enriched with nettle or sage leaf extract for active food packaging applications. *Food Hydrocolloids, 120*, 106979. https://doi.org/10.1016/j.foodhyd.2021.106979

Chaari, M., Elhadef, K., Akermi, S., Ben Akacha, B., Fourati, M., Chakchouk Mtibaa, A., Ennouri, M., Sarkar, T., Shariati, M. A., Rebezov, M., Abdelkafi, S., Mellouli, L., & Smaoui, S. (2022). Novel active food packaging films based on gelatin-sodium alginate containing beetroot peel extract. *Antioxidants, 11*(11), 1–25. https://doi.org/10.3390/antiox11112095

Chan, H. M., Nyam, K. L., Yusof, Y. A., & Pui, L. P. (2020). Investigation of properties of polysaccharide-based edible film incorporated with functional melastoma malabathricum extract. *Carpathian Journal of Food Science and Technology, 12*(1), 120–134. https://doi.org/10.34302/crpjfst/2020.12.1.12

Chen, C., Cai, N., Chen, J., Peng, X., & Wan, C. (2018). Chitosan-based coating enriched with hairy fig (*Ficus hirta* Vahl.) fruit extract for "Newhall" navel orange preservation. *Coatings, 8*(12). https://doi.org/10.3390/coatings8120445

Das, D. K., Dutta, H., & Mahanta, C. L. (2013). Development of a rice starch-based coating with antioxidant and microbe-barrier properties and study of its effect on tomatoes stored at room temperature. *LWT-Food Science and Technology, 50*(1), 272–278. https://doi.org/10.1016/j.lwt.2012.05.018

Elsayed, N., Hassan, A. A. M., Abdelaziz, S. M., Abdeldaym, E. A., & Darwish, O. S. (2022). Effect of whey protein edible coating incorporated with mango peel extract on postharvest quality, bioactive compounds and shelf life of broccoli. *Horticulturae, 8*(9). https://doi.org/10.3390/horticulturae8090770

Emamifar, A., Ghaderi, Z., & Ghaderi, N. (2019). Effect of salep-based edible coating enriched with grape seed extract on postharvest shelf life of fresh strawberries. *Journal of Food Safety, 39*(6), 1–12. https://doi.org/10.1111/jfs.12710

Erkan, N. (2015). Use of natural preservatives in seafood: Plant extracts, edible film and coating. *Journal of Food and Health Science, 1*(1), 33–49. https://doi.org/10.3153/jfhs15004

Fan, X. J., Zhang, B., Yan, H., Feng, J. T., Ma, Z. Q., & Zhang, X. (2019). Effect of lotus leaf extract incorporated composite coating on the postharvest quality of fresh goji (*Lycium barbarum* L.) fruit. *Postharvest Biology and Technology, 148*(October 2018), 132–140. https://doi.org/10.1016/j.postharvbio.2018.10.020

Ferreira Soares, N. d. F., de Sá Silva, C. A., Santiago-Silva, P., Espitia, P. J. P., Conceição Gonçalves, M. P. J., Lopez, M. J. G., & Botrel, D. A. (2016). Active and intelligent packaging for milk and milk products. In *Engineering aspects of milk and dairy products* (pp. 175–199). https://doi.org/10.1201/9781420090390-c8

Flores-Martínez, N. L., Valdez-Fragoso, A., Jiménez-Islas, H., & Pérez-Pérez, M. C. (2017). Physical, barrier, mechanical and microstructural properties of Aloe vera-gelatin-glycerol edible films incorporated with *Pimenta dioica* L. Merrill essential oil. *Revista Mexicana de Ingeniería Química, 16*(1), 109–119. https://doi.org/10.24275/rmiq/alim750

Giménez, B., López de Lacey, A., Pérez-Santín, E., López-Caballero, M. E., & Montero, P. (2013). Release of active compounds from agar and agar-gelatin films with green tea extract. *Food Hydrocolloids, 30*(1), 264–271. https://doi.org/10.1016/j.foodhyd.2012.05.014

Gómez-Estaca, J., Bravo, L., Gómez-Guillén, M. C., Alemán, A., & Montero, P. (2009). Antioxidant properties of tuna-skin and bovine-hide gelatin films induced by the addition of oregano and rosemary extracts. *Food Chemistry, 112*(1), 18–25. https://doi.org/10.1016/j.foodchem.2008.05.034

Hanumantharaju, K. N., Thangavel, K., & Ganapathy, S. (2020). Rheological and surface properties of edible coating solution prepared from vasa plant (*Justicia Adhatoda* L.) leaf extract. *Journal of Agricultural Engineering, 57*(3), 236–247.

Hasan, K., Islam, R., Hasan, M., Sarker, S. H., & Biswas, M. H. (2022). Effect of alginate edible coatings enriched with black cumin extract for improving postharvest quality characteristics of guava (*Psidium guajava* L.) Fruit. *Food and Bioprocess Technology, 15*(9), 2050–2064. https://doi.org/10.1007/s11947-022-02869-2

Hassanzadeh, P., Moradi, M., Vaezi, N., Moosavy, M. H., & Mahmoudi, R. (2018). Effects of chitosan edible coating containing grape seed extract on the shelf-life of refrigerated rainbow trout fillet. *Veterinary Research Forum, 9*(1), 73–79.

He, L., Lan, W., Ahmed, S., Qin, W., & Liu, Y. (2019). Electrospun polyvinyl alcohol film containing pomegranate peel extract and sodium dehydroacetate for use as food packaging. *Food Packaging and Shelf Life, 22*(May). https://doi.org/10.1016/j.fpsl.2019.100390

Ju, A., & Song, K. B. (2020). Incorporation of yellow onion peel extract into the funoran-based biodegradable films as an antioxidant packaging material. *International Journal of Food Science and Technology, 55*(4), 1671–1678. https://doi.org/10.1111/ijfs.14436

Kahya, N., Kestir, S. M., Öztürk, S., Yolaç, A., Torlak, E., Kalaycıoğlu, Z., & Erim, F. B. (2022). Antioxidant and antimicrobial chitosan films enriched with aqueous sage and rosemary extracts as food coating materials: Characterization of the films and detection of rosmarinic acid release. *International Journal of Biological Macromolecules, 217*(May), 470–480. https://doi.org/10.1016/j.ijbiomac.2022.07.073

Kanmani, P., & Rhim, J. W. (2014). Development and characterization of carrageenan/grapefruit seed extract composite films for active packaging. *International Journal of Biological Macromolecules, 68*, 258–266. https://doi.org/10.1016/j.ijbiomac.2014.05.011

Karkar, B., Patır, İ., Eyüboğlu, S., & Şahin, S. (2023). Development of an edible active chitosan film loaded with *Nigella sativa* L. extract to extend the shelf life of grapes. *Biocatalysis and Agricultural Biotechnology, 50*. https://doi.org/10.1016/j.bcab.2023.102708

Kaya, M., Khadem, S., Cakmak, Y. S., Mujtaba, M., Ilk, S., Akyuz, L., & Deligöz, E. (2018a). Antioxidative and antimicrobial edible chitosan films blended with stem, leaf and seed extracts of *Pistacia terebinthus* for active food packaging. *RSC Advances, 8*(8), 3941–3950. https://doi.org/10.1039/c7ra12070b

Kaya, M., Ravikumar, P., Ilk, S., Mujtaba, M., Akyuz, L., Labidi, J., & Erkul, S. K. (2018b). Production and characterization of chitosan based edible films from Berberis crataegina's fruit extract and seed oil. *Innovative Food Science and Emerging Technologies, 45*, 287–297. https://doi.org/10.1016/j.ifset.2017.11.013

Khaliq, G., Ramzan, M., & Baloch, A. H. (2019). Effect of Aloe vera gel coating enriched with *Fagonia indica* plant extract on physicochemical and antioxidant activity of sapodilla fruit during postharvest storage. *Food Chemistry, 286*, 346–353. https://doi.org/10.1016/j.foodchem.2019.01.135

Kola, V., & Carvalho, I. S. (2023). Plant extracts as additives in biodegradable films and coatings in active food packaging. *Food Bioscience, 54*, 102860. https://doi.org/10.1016/j.fbio.2023.102860

Kraśniewska, K., Gniewosz, M., Synowiec, A., Przybył, J. L., Baczek, K., & Weglarz, Z. (2014). The use of pullulan coating enriched with plant extracts from *Satureja hortensis* L. to maintain pepper and apple quality and safety. *Postharvest Biology and Technology, 90*, 63–72. https://doi.org/10.1016/j.postharvbio.2013.12.010

Kuan, Y. L., Navin Sivanasvaran, S., Pui, L. P., Yusof, Y. A., & Senphan, T. (2020). Physicochemical properties of sodium alginate edible film incorporated with mulberry (*Morus australis*) leaf extract. *Pertanika Journal of Tropical Agricultural Science, 43*(3), 357–376.

Kumar, N., Daniloski, D., Pratibha, Neeraj, D'Cunha, N. M., Naumovski, N., & Petkoska, A. T. (2022). Pomegranate peel extract – A natural bioactive addition to novel active edible packaging. *Food Research International, 156*, 111378. https://doi.org/10.1016/j.foodres.2022.111378

Kumar, N., Pratibha, Upadhyay, A., Trajkovska Petkoska, A., Gniewosz, M., & Kieliszek, M. (2023a). Extending the shelf life of mango (*Mangifera indica* L.) fruits by using edible coating based on xanthan gum and pomegranate peel extract. *Journal of Food Measurement and Characterization, 17*(2), 1300–1308. https://doi.org/10.1007/s11694-022-01706-6

Kumar, N., Tripathi, S., & Mehra, M. (2023b). Whey protein based edible coatings: Recent trends. In *whey valorization* (pp. 187–209). https://doi.org/10.1007/978-981-99-5459-9

Li, J., Wu, Z., Huang, C., Liu, H., Huang, R., & Li, L. (2014). Mechanical properties of cyanate ester/epoxy nanocomposites modified with plasma functionalized MWCNTs. *Composites Science and Technology, 90*, 166–173. https://doi.org/10.1016/j.compscitech.2013.11.009

Licciardello, F., Kharchoufi, S., Muratore, G., & Restuccia, C. (2018). Effect of edible coating combined with pomegranate peel extract on the quality maintenance of white shrimps (*Parapenaeus longirostris*) during refrigerated storage. *Food Packaging and Shelf Life, 17*, 114–119. https://doi.org/10.1016/j.fpsl.2018.06.009

Lim, G. O., Jang, S. A., & Song, K. B. (2010). Physical and antimicrobial properties of *Gelidium corneum*/nano-clay composite film containing grapefruit seed extract or thymol. *Journal of Food Engineering, 98*(4), 415–420. https://doi.org/10.1016/j.jfoodeng.2010.01.021

Liu, T., Liu, L., Gong, X., Chi, F., & Ma, Z. (2021). Fabrication and comparison of active films from chitosan incorporating different spice extracts for shelf life extension of refrigerated pork. *LWT-Food Science and Technology, 135*, 110181. https://doi.org/10.1016/j.lwt.2020.110181

Lopes, A. I., Melo, A., Caleja, C., Pereira, E., Finimundy, T. C., Afonso, T. B., & Pintado, M. (2023). Evaluation of antimicrobial and antioxidant activities of alginate edible coatings incorporated with plant extracts. *Coatings, 13*(9), 1–24. https://doi.org/10.3390/coatings13091487

López de Dicastillo, C., Bustos, F., Guarda, A., & Galotto, M. J. (2016). Cross-linked methyl cellulose films with murta fruit extract for antioxidant and antimicrobial active food packaging. *Food Hydrocolloids, 60*, 335–344. https://doi.org/10.1016/j.foodhyd.2016.03.020

López-Palestina, C. U., Aguirre-Mancilla, C. L., Raya-Pérez, J. C., Ramírez-Pimentel, J. G., Gutiérrez-Tlahque, J., & Hernández-Fuentes, A. D. (2018). The effect of an edible coating with tomato oily extract on the physicochemical and antioxidant properties of garambullo

(*Myrtillocactus geometrizans*) fruits. *Agronomy, 8*(11), 1–14. https://doi.org/10.3390/agronomy8110248

Maan, A. A., Reiad Ahmed, Z. F., Iqbal Khan, M. K., Riaz, A., & Nazir, A. (2021). *Aloe vera* gel, an excellent base material for edible films and coatings. *Trends in Food Science and Technology, 116*, 329–341. https://doi.org/10.1016/j.tifs.2021.07.035

Maryam Adilah, Z. A., Jamilah, B., & Nur Hanani, Z. A. (2018). Functional and antioxidant properties of protein-based films incorporated with mango kernel extract for active packaging. *Food Hydrocolloids, 74*, 207–218. https://doi.org/10.1016/j.foodhyd.2017.08.017

Mir, S. A., Dar, B. N., Wani, A. A., & Shah, M. A. (2018). Effect of plant extracts on the techno-functional properties of biodegradable packaging films. *Trends in Food Science & Technology, 80*(August 2017), 141–154. https://doi.org/10.1016/j.tifs.2018.08.004

Mohan, H., Muthukumar Sathya, P., Lee, S. W., Jeong, H. J., Ryu, S., Lee, A. H., & Oh, B. T. (2024). Enhanced health benefits of Psidium guajava peel fermented with Lactiplantibacillus from Myeolchi-jeot: Antioxidant, anti-inflammatory, and anti-cancer properties. *Food Bioscience, 59*(March), 104211. https://doi.org/10.1016/j.fbio.2024.104211

Morachis-Valdez, A. G., Santillán-álvarez, Á., Gómez-Oliván, L. M., García-Argueta, I., Islas-Flores, H., & Dublán-García, O. (2021). Effects of peppermint extract and chitosan-based edible coating on storage quality of common carp (*Cyprinus carpio*) fillets. *Polymers, 13*(19), 1–16. https://doi.org/10.3390/polym13193243

Moradi, F., Emamifar, A., & Ghaderi, N. (2019). Effect of basil seed gum based edible coating enriched with echinacea extract on the postharvest shelf life of fresh strawberries. *Journal of Food Measurement and Characterization, 13*(3), 1852–1863. https://doi.org/10.1007/s11694-019-00104-9

Munir, S., Hu, Y., Liu, Y., & Xiong, S. (2019). Enhanced properties of silver carp surimi-based edible films incorporated with pomegranate peel and grape seed extracts under acidic condition. *Food Packaging and Shelf Life, 19*, 114–120. https://doi.org/10.1016/j.fpsl.2018.12.001

Nair, M. S., Saxena, A., & Kaur, C. (2018). Effect of chitosan and alginate based coatings enriched with pomegranate peel extract to extend the postharvest quality of guava (*Psidium guajava* L.). *Food Chemistry, 240*, 245–252. https://doi.org/10.1016/j.foodchem.2017.07.122

Nguyen, T. T., Thi Dao, U. T., Thi Bui, Q. P., Bach, G. L., Ha Thuc, C. N., & Ha Thuc, H. (2020). Enhanced antimicrobial activities and physiochemical properties of edible film based on chitosan incorporated with *Sonneratia caseolaris* (L.) Engl. leaf extract. *Progress in Organic Coatings, 140*, 105487. https://doi.org/10.1016/j.porgcoat.2019.105487

Norajit, K., Kim, K. M., & Ryu, G. H. (2010). Comparative studies on the characterization and antioxidant properties of biodegradable alginate films containing ginseng extract. *Journal of Food Engineering, 98*(3), 377–384. https://doi.org/10.1016/j.jfoodeng.2010.01.015

Nouri, L., & Nafchi, A. M. (2014). Antibacterial, mechanical, and barrier properties of sago starch film incorporated with betel leaves extract. *International Journal of Biological Macromolecules, 66*, 254–259. https://doi.org/10.1016/j.ijbiomac.2014.02.044

Pagliarulo, C., Sansone, F., Moccia, S., Russo, G. L., Aquino, R. P., Salvatore, P., & Volpe, M. G. (2016). Preservation of strawberries with an antifungal edible coating using Peony extracts in chitosan. *Food and Bioprocess Technology, 9*(11), 1951–1960. https://doi.org/10.1007/s11947-016-1779-x

Ponce, A. G., Roura, S. I., del Valle, C. E., & Moreira, M. R. (2008). Antimicrobial and antioxidant activities of edible coatings enriched with natural plant extracts: In vitro and in vivo studies. *Postharvest Biology and Technology, 49*(2), 294–300. https://doi.org/10.1016/j.postharvbio.2008.02.013

Priyadarshi, R., Riahi, Z., & Rhim, J. W. (2022). Antioxidant pectin/pullulan edible coating incorporated with Vitis vinifera grape seed extract for extending the shelf life of peanuts. *Postharvest Biology and Technology, 183*, 111740. https://doi.org/10.1016/j.postharvbio.2021.111740

Radi, M., Firouzi, E., Akhavan, H., & Amiri, S. (2017). Effect of gelatin-based edible coatings incorporated with *Aloe vera* and black and green tea extracts on the shelf life of fresh-cut oranges. *Journal of Food Quality, 2017*. https://doi.org/10.1155/2017/9764650

Ramírez-Guerra, H. E., Castillo-Yañez, F. J., Montaño-Cota, E. A., Ruíz-Cruz, S., Márquez-Ríos, E., Canizales-Rodríguez, D. F., & Ocaño-Higuera, V. M. (2018). Protective effect of an edible tomato plant extract/chitosan coating on the quality and shelf life of sierra fish fillets. *Journal of Chemistry, 2018*. https://doi.org/10.1155/2018/2436045

Rashwan, A. K., Bai, H., Osman, A. I., Eltohamy, K. M., Chen, Z., Younis, H. A., & Yap, P. S. (2023). Recycling food and agriculture by-products to mitigate climate change: A review. *Environmental Chemistry Letters, 21*(6), 3351–3375. https://doi.org/10.1007/s10311-023-01639-6

Riaz, A., Lei, S., Akhtar, H. M. S., Wan, P., Chen, D., Jabbar, S., & Zeng, X. (2018). Preparation and characterization of chitosan-based antimicrobial active food packaging film incorporated with apple peel polyphenols. *International Journal of Biological Macromolecules, 114*, 547–555. https://doi.org/10.1016/j.ijbiomac.2018.03.126

Riaz, A., Lagnika, C., Luo, H., Dai, Z., Nie, M., Hashim, M. M., & Li, D. (2020). Chitosan-based biodegradable active food packaging film containing Chinese chive (*Allium tuberosum*) root extract for food application. *International Journal of Biological Macromolecules, 150*, 595–604. https://doi.org/10.1016/j.ijbiomac.2020.02.078

Ruíz-Cruz, S., Valenzuela-López, C. C., Chaparro-Hernández, S., Ornelas-Paz, J. d. J., Del Toro-Sánchez, C. L., Márquez-Ríos, E., & Valdez-Hurtado, S. (2019). Effects of chitosan-tomato plant extract edible coatings on the quality and shelf life of chicken fillets during refrigerated storage. *Food Science and Technology, 39*(1), 103–111. https://doi.org/10.1590/fst.23117

Saberi, B., Vuong, Q. V., Chockchaisawasdee, S., Golding, J. B., Scarlett, C. J., & Stathopoulos, C. E. (2017). Physical, barrier, and antioxidant properties of pea starch-guar gum biocomposite edible films by incorporation of natural plant extracts. *Food and Bioprocess Technology, 10*(12), 2240–2250. https://doi.org/10.1007/s11947-017-1995-z

Saxena, A., Sharma, L., & Maity, T. (2020). Enrichment of edible coatings and films with plant extracts or essential oils for the preservation of fruits and vegetables. In *Biopolymer-Based Formulations: Biomedical and Food Applications*. Elsevier. https://doi.org/10.1016/B978-0-12-816897-4.00034-5

Seididamyeh, M., Mantilla, S. M. O., Netzel, M. E., Mereddy, R., & Sultanbawa, Y. (2024). Gum Arabic edible coating embedded aqueous plant extracts: Interactive effects of partaking components and its effectiveness on cold storage of fresh-cut capsicum. *Food Control, 159*, 110267. https://doi.org/10.1016/j.foodcont.2023.110267

Shafique, M., Rashid, M., Ullah, S., Rajwana, I. A., Naz, A., Razzaq, K., & Jbawi, E. A. (2023). Quality and shelf life of strawberry fruit as affected by edible coating by moringa leaf extract, aloe vera gel, oxalic acid, and ascorbic acid. *International Journal of Food Properties, 26*(2), 2995–3012. https://doi.org/10.1080/10942912.2023.2267794

Shan, P., Wang, K., Yu, F., Yi, L., Sun, L., & Li, H. (2023). Gelatin/sodium alginate multilayer composite film crosslinked with green tea extract for active food packaging application. *Colloids and Surfaces A: Physicochemical and Engineering Aspects, 662*, 131013. https://doi.org/10.1016/j.colsurfa.2023.131013

Silva-Weiss, A., Bifani, V., Ihl, M., Sobral, P. J. A., & Gómez-Guillén, M. C. (2013a). Structural properties of films and rheology of film-forming solutions based on chitosan and chitosan-starch blend enriched with murta leaf extract. *Food Hydrocolloids, 31*(2), 458–466. https://doi.org/10.1016/j.foodhyd.2012.11.028

Silva-Weiss, A., Ihl, M., Sobral, P. J. A., Gómez-Guillén, M. C., & Bifani, V. (2013b). Natural additives in bioactive edible films and coatings: Functionality and applications in foods. *Food Engineering Reviews, 5*(4), 200–216. https://doi.org/10.1007/s12393-013-9072-5

Siripatrawan, U., & Harte, B. R. (2010). Physical properties and antioxidant activity of an active film from chitosan incorporated with green tea extract. *Food Hydrocolloids, 24*(8), 770–775. https://doi.org/10.1016/j.foodhyd.2010.04.003

Sivarooban, T., Hettiarachchy, N. S., & Johnson, M. G. (2008). Physical and antimicrobial properties of grape seed extract, nisin, and EDTA incorporated soy protein edible films. *Food Research International, 41*(8), 781–785. https://doi.org/10.1016/j.foodres.2008.04.007

Supapvanich, S., Prathaan, P., & Tepsorn, R. (2012). Browning inhibition in fresh-cut rose apple fruit cv. Taaptimjaan using konjac glucomannan coating incorporated with pineapple fruit extract. *Postharvest Biology and Technology, 73*, 46–49. https://doi.org/10.1016/j.postharvbio.2012.05.013

Synowiec, A., Gniewosz, M., Kraśniewska, K., Przybył, J. L., Bamczek, K., & Węglarz, Z. (2014). Antimicrobial and antioxidant properties of pullulan film containing sweet basil extract and an evaluation of coating effectiveness in the prolongation of the shelf life of apples stored in refrigeration conditions. *Innovative Food Science and Emerging Technologies, 23*, 171–181. https://doi.org/10.1016/j.ifset.2014.03.006

Teixeira, V. M. C., da Silva, R. F. G., Gonçalves, O. H., Pereira, C., Barros, L., Ferreira, I. C. F. R., Bona, E., & Leimann, F. V. (2022). Chemometric approaches to evaluate the substitution of synthetic food dyes by natural compounds: The case of nanoencapsulated curcumin, spirulina, and hibiscus extracts. *LWT-Food Science and Technology, 154*. https://doi.org/10.1016/j.lwt.2021.112786

Tesfay, S. Z., & Magwaza, L. S. (2017). Evaluating the efficacy of moringa leaf extract, chitosan and carboxymethyl cellulose as edible coatings for enhancing quality and extending postharvest life of avocado (*Persea americana* Mill.). Fruit. *Food Packaging and Shelf Life, 11*, 40–48. https://doi.org/10.1016/j.fpsl.2016.12.001

Tharanathan, R. N. (2003). Biodegradable films and composite coatings: Past, present and future. *Trends in Food Science and Technology, 14*(3), 71–78. https://doi.org/10.1016/S0924-2244(02)00280-7

Tong, C., Wu, Z., Sun, J., Lin, L., Wang, L., Guo, Y., & Pang, J. (2020). Effect of carboxylation cellulose nanocrystal and grape peel extracts on the physical, mechanical and antioxidant properties of konjac glucomannan films. *International Journal of Biological Macromolecules, 156*, 874–884. https://doi.org/10.1016/j.ijbiomac.2020.04.051

Torres-León, C., Vicente, A. A., Flores-López, M. L., Rojas, R., Serna-Cock, L., Alvarez-Pérez, O. B., & Aguilar, C. N. (2018). Edible films and coatings based on mango (var. Ataulfo) by-products to improve gas transfer rate of peach. *LWT-Food Science and Technology, 97*(July), 624–631. https://doi.org/10.1016/j.lwt.2018.07.057

Trajkovska Petkoska, A., Daniloski, D., D'Cunha, N. M., Naumovski, N., & Broach, A. T. (2021). Edible packaging: Sustainable solutions and novel trends in food packaging. *Food Research International, 140*(November), 109981. https://doi.org/10.1016/j.foodres.2020.109981

Valverde, J. M., Valero, D., Martínez-Romero, D., Guillén, F., Castillo, S., & Serrano, M. (2005). Novel edible coating based on Aloe vera gel to maintain table grape quality and safety. *Journal of Agricultural and Food Chemistry, 53*(20), 7807–7813. https://doi.org/10.1021/jf050962v

Wai, C. K., Bond, T. Y., Lin, N. K., & Phing, P. L. (2019). Antibacterial properties of chitosan edible films incorporated with musk lime extracts for the preservation of squids. *Malaysian Journal of Analytical Sciences, 23*(6), 914–925. https://doi.org/10.17576/mjas-2019-2306-01

Wang, S., Marcone, M., Barbut, S., & Lim, L. T. (2012). The impact of anthocyanin-rich red raspberry extract (ARRE) on the properties of edible soy protein isolate (SPI) films. *Journal of Food Science, 77*(4), 497–505. https://doi.org/10.1111/j.1750-3841.2012.02655.x

Wang, F., Xie, C., Ye, R., Tang, H., Jiang, L., & Liu, Y. (2023). Development of active packaging with chitosan, guar gum and watermelon rind extract: Characterization, application and performance improvement mechanism. *International Journal of Biological Macromolecules, 227*, 711–725. https://doi.org/10.1016/j.ijbiomac.2022.12.210

Xu, W., Qu, W., Huang, K., Guo, F., Yang, J., Zhao, H., & Luo, Y. B. (2007). Antibacterial effect of grapefruit seed extract on food-borne pathogens and its application in the preservation of minimally processed vegetables. *Postharvest Biology and Technology, 45*(1), 126–133. https://doi.org/10.1016/j.postharvbio.2006.11.019

Yong, H., Liu, J., Qin, Y., Bai, R., Zhang, X., & Liu, J. (2019). Antioxidant and pH-sensitive films developed by incorporating purple and black rice extracts into chitosan matrix. *International Journal of Biological Macromolecules, 137*, 307–316. https://doi.org/10.1016/j.ijbiomac.2019.07.009

Yousuf, B., Qadri, O. S., & Srivastava, A. K. (2018). Recent developments in shelf-life extension of fresh-cut fruits and vegetables by application of different edible coatings: A review. *Lwt-Food Science and Technology, 89*, 198–209. https://doi.org/10.1016/j.lwt.2017.10.051

Zam, W. (2019). Effect of alginate and chitosan edible coating enriched with olive leaves extract on the shelf life of sweet cherries (*Prunus avium* L.). *Journal of Food Quality, 2019*. https://doi.org/10.1155/2019/8192964

Zeng, J., Ren, X., Zhu, S., & Gao, Y. (2021). Fabrication and characterization of an economical active packaging film based on chitosan incorporated with pomegranate peel. *International Journal of Biological Macromolecules, 192*, 1160–1168. https://doi.org/10.1016/j.ijbiomac.2021.10.064

Zhang, C., Guo, K., Ma, Y., Ma, D., Li, X., & Zhao, X. (2010). Incorporations of blueberry extracts into soybean-protein-isolate film preserve qualities of packaged lard. *International Journal of Food Science and Technology, 45*(9), 1801–1806. https://doi.org/10.1111/j.1365-2621.2010.02331.x

Zhang, Y., Yuting, S., Chen, C., Gan, Z., Chen, J., & Wan, C. (2022). Loquat leaf extract and alginate based green composite edible coating for preserving the postharvest quality of *Nanfeng tangerines*. *Sustainable Chemistry and Pharmacy, 27*, 100674. https://doi.org/10.1016/j.scp.2022.100674

Zúñiga, G. E., Junqueira-Gonçalves, M. P., Pizarro, M., Contreras, R., Tapia, A., & Silva, S. (2012). Effect of ionizing energy on extracts of *Quillaja saponaria* to be used as an antimicrobial agent on irradiated edible coating for fresh strawberries. *Radiation Physics and Chemistry, 81*(1), 64–69. https://doi.org/10.1016/j.radphyschem.2011.08.008

Chapter 11
Effects of Probiotics on the Properties of Edible Coatings and Films

Urmila Choudhary, Amrita Poonia, and P. Nandagopal

Abstract In recent years, edible films and coatings have gained a lot of attention due to their low cost and superiority over synthetic films in reducing environmental pollution. The creation of bioactive food products with positive health effects is a result of the growing consumer understanding of nutrition. Recent scientific literature has proposed the use of probiotics in edible films/coatings as a simple, affordable bio-protection technique with a lot of potential for application in the food sector. Applying edible films/coatings loaded with probiotics may enhance or maintain aspects of food products' quality, safety, functionality, and stability during storage. In addition to improving the survivability of probiotic strains and increasing their colonization in the intestinal tract, the simultaneous application of probiotics in edible films or coatings also improves human health and adds value to food products. Knowing the features of the various film and coating production processes enables the selection of the best probiotics that will be added. Important considerations need to be made while choosing the best production methods for probiotics. They have an impact on the food products' techno-functional and physicochemical characteristics. They also help extend the shelf life of food products by preventing the growth of unwanted microbes. Probiotic incorporation in edible films, probiotic selection criteria in edible film/coating formulation, probiotic interactions with other bioactive compounds, and probiotic stability/viability mechanisms in edible films and coatings are the main topics of this chapter. In addition, regulatory

U. Choudhary
Department of Dairy Technology, College of Dairy Science and Technology, Sri Karan Narendra Agricultural University, Jobner, Jaipur, Rajasthan, India

A. Poonia (✉)
Department of Dairy Science and Food Technology, Institute of Agricultural Sciences, Banaras Hindu University, Varanasi, Uttar Pradesh, India
e-mail: amrita12@bhu.ac.in

P. Nandagopal
Department of Dairy Technology, College of Dairy Science and Technology, Kolahalamedu Kerala Veterinary and Animal Sciences University, Kolahalamedu, Kerala, India

N. Kumar et al. (eds.), *Trends in Edible Packaging: A sustainable approaches for postharvest management of fruits and vegetables*,
https://doi.org/10.1007/978-3-032-13199-7_11

considerations in the commercialization of probiotics included in edible coatings/films are also discussed.

Keywords Probiotics · Viability · Edible coatings · Applications · Antimicrobial properties

Introduction

Consumer demand for eco-friendly packaging options has fueled interest in edible films and coatings in recent decades (Choudhary et al., 2022; Galus et al., 2020; Poonia and Mishra, 2021; Yan et al., 2022). Due to their synthetic nature, lack of biodegradability, and lack of environmental friendliness, traditional packaging materials produce a significant quantity of trash (Rai and Poonia, 2019; Moeini et al., 2021; Santhosh et al., 2021). Edible films and coatings have gained widespread acceptability as a substitute for plastics in food packaging (Choudhary et al., 2022). To enhance or alter the fundamental properties of these polymers, they can be made from food-grade polysaccharides, proteins, lipids, and other ingredients, such as plasticizers, emulsifiers, and other additives (Choudhary et al., 2022; Mishra et al., 2022; Sharma et al., 2022a, b). Excellent sensory qualities, high barrier qualities, high mechanical strength, high microbial stability, lack of toxins, human health safety, ease of production, and cost-effectiveness are some of the qualities that the perfect edible film or coating should have (Jeevahan et al., 2020; Sun et al., 2024). Furthermore, probiotic-like microbes can be transported by edible films and coatings to enhance the gut microbiota of consumers. Thus, adding lactic acid bacteria (LAB) with probiotic or antimicrobial properties can be regarded as a cutting-edge technology that has a big influence on enhancing food safety, prolonging food shelf life, and boosting consumer health benefits. Probiotics'potential to improve food safety and preservation has drawn a lot of attention to their integration into edible films and coatings. Probiotic edible films offer practical advantages like antioxidant qualities and enhanced sensory qualities, in addition to acting as barriers against microbial spoiling. Because of fermentation processes, probiotic edible films frequently have changed physicochemical characteristics, including pH and color. When combined with dragon fruit powder, for example, films made from fermented whey solutions exhibited increased acidity and noticeable color changes, but their mechanical qualities remained mostly stable, enabling their efficient use in food products without sacrificing structural integrity (Espinosa-Martinez et al., 2025). According to Asan-Ozusaglam and Celik (2024), probiotic strains added to edible films had strong antimicrobial effects against a variety of infections. For example, films containing *Limosilactobacillus fermentum* showed a considerable inhibition against *Candida albicans*. It has been demonstrated that adding *Lactobacillus* bacteria to xanthan gum coatings improves their antimicrobial qualities and successfully prolongs the shelf life of fresh-cut fruits (Chikhala et al., 2024). For probiotics to have their functional advantages, they must be viable in edible films. Research shows that even after storage, probiotic counts can remain above 1×10^7 CFU/g (Espinosa-Martinez et al., 2025). As demonstrated by edible films enhanced with clove extract and different probiotic strains, adding probiotics to food products not only increases their nutritional profile but also improves their safety (Saleena et al., 2024). Adverse circumstances like heat, acidity, osmotic stress, and mechanical stress pose serious challenges to the survival of probiotics during food preparation and storage. Probiotic cell viability may be improved by recent developments in encapsulation techniques, especially the use of edible films and different wall materials (Adeel et al., 2023; Azeem et al., 2023; Kareem et al., 2023; Wang and Zhong, 2024). The survivability of probiotic strains was also examined using edible films and coatings based on polymeric matrix (Silva et al., 2023c). The impacts of probiotics on the physicochemical properties and antibacterial activity and the general impact on the food quality of edible films and coatings will be examined in this chapter.

Criteria for the Selection of Probiotics in the Formulation of Edible Films/Coatings

The food matrix, including its physicochemical characteristics, physical state, and processing involved, is the sole determinant of the design of the edible packaging formulation and the choice of film-forming method. Additionally, the probiotic edible packaging needs to meet customer expectations about sensory quality, shelf life, ease of handling, cost, sustainability and eco-friendliness, usage of clean label and minimally processed materials, etc. A variety of extrinsic and intrinsic factors can influence how a microbe behaves in various feeding conditions. According to Soukoulis et al. (2014), typical examples include the type of culture that was chosen, the physiological condition of the cells, the feeding matrix, the pH, the temperature, the manufacturing procedures, and the storage conditions. In addition to these variables, probiotic viability is assessed by how well they travel through the gastrointestinal (GI) tract, where bile and low pH are the main obstacles. In addition to exerting antimicrobial effects against pathogenic microorganisms through the production of antimicrobial substances and/or competition for growth factors, nutrients, and/or binding sites, probiotic strains' primary characteristics include the resistance to gastrointestinal conditions, the ability to adhere to human epithelial cells or mucus, the improvement of the intestinal barrier integrity, the enhancement of immunological responses, and the reduction of blood cholesterol levels in the host (FAO, 2002; Pop et al., 2020; Espitia et al., 2016; Poonia, 2018; FSS (Health supplements, nutraceuticals, food for Special dietary Use, Food for Special Medical Purpose and Prebiotic and Probiotic Food) Regulation, 2022). For a probiotic strain, it is relevant to include the history of safe use in foods at the species level as pertinent evidence of safety (Guimarães et al., 2018). Before edible films and coatings can be used as food packaging and ingredients, they must fulfill specific standards. They should offer an efficient barrier against gases or vapors, protect the food from microbiological development and chemical reactions, and offer robust structural protection to prevent mechanical damage (Poonia, 2018).

Suitable Methods for the Incorporation of Probiotics in Edible Film/Coating Formulations

The main components of edible packaging typically include (i) a macromolecule-based material, such as biopolymers; (ii) additives, such as plasticizers, emulsifiers, cross-linkers, and nano-reinforcements; and (iii) precursors, such as proteins, polysaccharides, lipids, or resins. When dissolved in a solvent (often water), the macromolecule-based material serves as the base for a cohesive construction. To increase the package's mechanical qualities (such as its elasticity, toughness, and resistance to ripping), the plasticizer is added. This gives the package more flexibility and stability (Valencia-Chamorro et al., 2011). When the package contains proteins and polysaccharides, plasticizers such as sorbitol, polyethylene glycol, glycerol, and sucrose are frequently required. Emulsifiers are sometimes employed in place of plasticizers to improve the durability of films and coatings composed of polysaccharides and lipids (Ramos et al., 2012). Because of the materials and/or active molecules that are incorporated, edible packaging is intended to either protect the food or simply serve as a carrier for the active compounds, reduce contamination, improve or preserve the natural appearance of food products, improve the mechanical properties of fragile food products, or enhance the appearance and flavor of food products. Recently, the use of probiotic microorganisms in edible films and coatings has become a novel way to offer consumers health benefits. One of the expanding applications for edible coatings and films is the addition of active ingredients such as antioxidants, antimicrobials, and antifungals. When making edible films and coatings, the following steps are taken (Paulo et al., 2021).

A suitable solvent, usually water, is used to dissolve the biopolymer; the pH is then adjusted as necessary; other ingredients, such as plasticizers like glycerol, are added; another biopolymer is mixed if necessary; the mixture is heated and stirred to create a homogenous and viscous solution; the solution is cooled; and functional ingredients, like probiotics or antimicrobials, are added. Factors such as the film's composition, production process, temperature, and storage conditions must be considered when adding probiotics to edible films or coatings because they may affect the probiotics' viability. Figure 11.1 illustrates the use of appropriate methods to incorporate probiotics into edible films or coatings, including direct casting for the creation of edible films (Jeevahan et al., 2020; Paulo et al., 2021; S'aez-Orviz et al., 2022), spraying (Barbosa et al., 2021; Oduro, 2021), dipping (Barbosa et al., 2021; Paulo et al., 2021), and brushing (Barbosa et al., 2021). One interesting field that

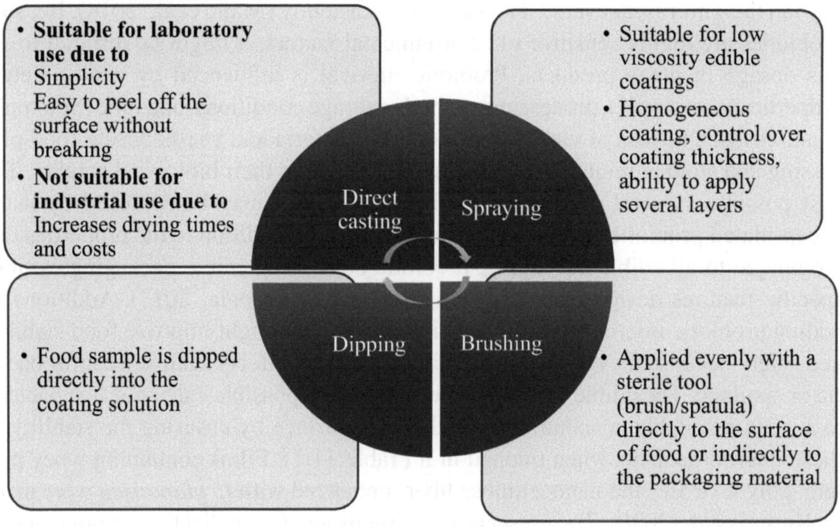

Fig. 11.1 Schematic diagram of suitable methods/techniques for the incorporation of probiotics into edible film/coating formulations

could increase the usage of probiotic edible packaging is nanotechnology and the application of nanomaterials. Because of the novel and distinctive bioactive qualities that can be obtained, the formulation of non-nanomaterials in nanosized structures can have significant advantages (Torres-Giner et al., 2016). An appropriate method for the nanoscale restructuration of biopolymers is the use of electrospinning in the creation of edible packaging materials.

Viability of Probiotics in Probiotic-Incorporated Edible Films/Coatings

One could consider the edible coating or films to be probiotic carriers. The main obstacle that probiotic microbes must overcome is their resistance to entrapment, which is a necessary condition for their survival in the finished product. The colon can only be successfully colonized by probiotics that are alive and have enough cells. Probiotic viability in edible coatings and films is influenced by composition and storage temperature. Probiotics'antibacterial activity and viability are influenced by the kind of substance that makes up the edible package. The Food Safety and Standards (FSS) (health supplements, nutraceuticals, food for special dietary use, food for special medical purpose, and prebiotic and probiotic food) Regulation, 2022, recommends a daily dose of $10^{8–9}$ live cells to maximize the benefits of probiotic use. Foods that are coated with probiotic-containing coatings or films might be considered functional foods, a unique class of foods that provide health benefits beyond their nutritional value if consumed consistently (Majid et al., 2018). Because probiotics are highly sensitive to environmental factors, it might be difficult to get this dosage in many products. Probiotic survival is influenced by strain, dietary properties (such as pH), processing methods, storage conditions, and time (Soukoulis et al., 2014). The loss of viability of probiotic bacteria and yeasts during food processing and storage might have a detrimental effect on their biological activity. The best possible survival of beneficial microbes in food may be achieved by using encapsulated probiotics in edible films or coatings. In addition to the properties that are unique to all edible packaging, probiotic-containing edible packaging exhibits specific features designed to keep the host healthy (Poonia, 2017). Additionally, adding probiotic microorganisms to edible packaging might improve food stability and safety because they frequently exhibit inhibitory effects against harmful bacteria or spoilage. The edible matrix has been used as a possible carrier of live bacteria to sustain a suitable concentration on the food surface by ensuring the stability of these microorganisms when trapped in it (Table 11.1). Films containing whey protein, polydextrose, and nanocellulose fibers enhanced with *L. plantarum* were made by Karimi et al. (2020). The presence of nutrients and free radical scavenging agents and the interaction between polydextrose and the phospholipids of the bacterial membrane were credited by the authors to the probiotic's good survivability inside the matrix. Given that synbiotic films had higher bacterial survivability than probiotic films, Oliveira-Alcântara et al. (2020) found that fructooligosaccharide (FOS) had a protective impact on probiotic bacteria during film drying. Protein-based edible films and coatings combined with plasticizers typically produce materials with appropriate mechanical and physical characteristics (Mihalca et al., 2021; Kaur et al., 2022). Synbiotics have been tested in edible films and may help preserve the probiotics'cell viability. Probiotics *Lactobacillus fermentum ATCC 9398* and *Bifidobacterium infantis ATCC 15697* were more likely to survive in chitosan-based films containing inulin, galacto-oligosaccharide, and FOS. In addition to the prebiotic impact, the oligosaccharides made the film more extensible than a film without prebiotics (Tang et al., 2015).

Mechanisms of Probiotic Stability in Edible Films/Coatings

Numerous factors that improve the viability and usefulness of probiotics also affect their stability in edible films and coatings. Prebiotic inclusion, protective agents, and biopolymer matrices are critical for preserving probiotic stability throughout application and storage, according to recent studies. Coatings made of chitosan have demonstrated great promise in bolstering the stability and regulated release of probiotics while shielding them from environmental stressors (Edo et al., 2025; Sharma, et al., 2022a, b). Because the residues contain antioxidant properties, probiotic viability declined more slowly in starch-based films enhanced with agrifood residues (Coimbra et al., 2023). Chia mucilage's function as an efficient protectant was demonstrated by the fact that it dramatically increased the probiotics'survival under

Table 11.1 Effect of the incorporation of probiotics on the physicochemical properties and techno-functional properties of edible films/coatings

Probiotics	Polymeric matrix for edible films/coatings	Characteristics	Viability	Application	Reference
Lactobacillus rhamnosus Gorbach and Goldin (GG) (LGG)	Probiotic mucilage (quince, flax, and basil)–based edible films	High moisture retention More thermally stable and mechanically robust, with higher elastic moduli and elongation at break Effectively protected against UV light Maintained the probiotic viability and inactivation rate during storage	Quince mucilage–based edible film showed the highest LGG viability after 5 weeks of storage: ~71% at 4 °C and 61% at 25 °C	To preserve fruits and vegetables	Davachi et al. (2021)
L. acidophilus CH -2, *L. casei* B-1922, and *B. lactis* BB12	Chitosan, sodium alginate, and carboxymethyl cellulose	Antimicrobial effect against all tested microbes was observed, but the highest microbial effect was observed for edible films with chitosan and sodium alginate	Probiotic counts in all films were more than 8.0 log CFU/g after 45 days	Cheese	El-Sayed et al. (2021)
B. Animalis Bb-12 or *L. casei*-01	Whey protein isolate	Prevented the growth of *Staphylococcus* spp., *Pseudomonas* spp., *Enterobacteriaceae*, and yeasts/molds	10^6 CFU/g	Sliced ham	Pereira et al. (2018)
L. curvatus 54M16	Whey protein/inulin/gelatin	Films in the presence of nutrients showed a good antimicrobial activity against this pathogen on cooked ham	Whey protein–based films were able to ensure a high viability during 28 days of storage at 4 °C	Cooked ham	La Storia et al. (2020)

(continued)

Table 11.1 (continued)

Probiotics	Polymeric matrix for edible films/coatings	Characteristics	Viability	Application	Reference
L. casei	Whey protein isolate	–	10^6 CFU/g	Cherry tomatoes and Thompson grapes	Dianina et al. (2019)
Bacteriocin-like substances (BLSs) from *L. curvatus* P99	Cassava starch	Effective in controlling *L. monocytogenes* growth	–	"Prato" cheese	de Lima Marques et al. (2017)
Cell-Free Supernatant (CFS) of *L. sakei*	Whey protein isolate	Antimicrobial films significantly reduced the growth of *E. coli*, *L. monocytogenes*	–	Beef cubes	del Carmen Beristain-Bauza et al. (2017)
Lc. lactis ATCC 11454	Sodium alginate, sodium carboxymethyl cellulose, and collagen	Inhibited the growth of *S. aureus*	Highest viability was found in films composed of sodium alginate and methylcellulose	–	Ma et al. (2019)
Lc. lactis ATCC 11454	Polyvinyl alcohol with gelatin, sodium caseinate, gelatin or casein hydrolysates, or yeast extract	The activity of the films was higher when protein hydrolysate– and sodium caseinate–incorporated films were able to inhibit the growth of *L. monocytogenes*	–	Pasteurized milk	Settier-Ramírez et al. (2020)
L. paracasei TEP6	Chitosan with aloe vera and glucose	Highest inhibition of the fungus with the formulation with 100% aloe vera gel	Maintained bacterial viability for 14 days of film storage, with a loss to values of 9.5 log $CFU \cdot g^{-1}$	Aloe vera gel	Barrag'an-Men'endez et al. (2020)
L. rhamnosus CECT 8361	Alginate with inulin and oligofructose	The bioactive coating showed a strong antilisterial effect	–	Blueberries	Bambace et al. (2019)
L. plantarum CECT 9567	Egg yolk proteins and lactobionic acid as prebiotics	–	10^7 CFU/g	–	S'aez-Orviz et al. (2021)

L. casei ATCC 393	Duck feet gelatin, dextrin, polydextrose, gum arabic, and sago starch as prebiotics	–	Increased viability by 40% with gum arabic		Abedinia et al. (2021)
B. animalis Bb-12	Whey protein isolate with inulin and fructooligosaccharides as prebiotics	–	10^7 CFU/g with inulin		Odila Pereira et al. (2019)
Enterocin whey solution (EWS) of *E. faecalis* L2B21K3, L3A21K6	Gelatin/glycerol with EWS	Reduction of *L. monocytogenes* contamination to undetected levels after 20–30 days	–	Cheese	Silva et al. (2023a, b)
Levilactobacillus brevis SJC120	Gelatin/glycerol with fermented cheese whey	Reduced fungal contamination by 55–76%	–		
Lc. lactis L3A21M1 and *Lactococcus garvieae* SJM17	Alginate with maltodextrin and glycerol	Reduction of *L. monocytogenes* on cheese surface, no migration to the cheese	–	Fresh cheese	Silva et al. (2022)
L. rhamnosus CECT 8361 and *B. lactis* CECT 7347	Alginate with fructooligosaccharides and inulin	Probiotics exerted bactericidal effects against *L. innocua* inoculated on apple cubes	–	Apple pieces	Alvarez et al. (2021)
L. acidophilus PTCC 1643, *L. reuteri* PTCC 1655, *L. casei* PTCC 1608, *L. rhamnosus* PTCC 1637, *B. bifidum* PTCC 1644	Carboxymethyl cellulose and sodium caseinate	A delay of microbial growth on trout fillets was observed	–	Trout fillets	Mozaffarzogh et al. (2020)

stress when added to sodium caseinate films (Semwal et al., 2022). Although their impact on probiotic viability can vary, prebiotics like inulin can improve the structural integrity of films (Hashemi et al., 2022). In addition to protecting probiotics, Prebiotic loaded films provide antibacterial and antioxidant qualities, which are essential for maintaining minimally processed fruits and vegetables (Chang Et al., 2023). The ability of the film to deliver viable probiotic bacteria to the gastrointestinal (GI) tract is crucial, regardless of the technological process used to create edible films containing probiotics. Several probiotic strains' viability, stability, and survival under varied conditions have been thoroughly examined and reviewed, but there are limited studies that specifically evaluate probiotic strains' viability in films to determine their suitability as probiotic carriers (Espitia et al., 2016). Because it is directly linked to bacterial survival after processing and after ingestion, the film's chemistry and the film-forming procedure are crucial (Soukoulis et al., 2017). Although the incorporation of probiotics into edible films has encouraging advantages, there are still issues with formulation optimization to guarantee optimal viability and functionality throughout application and storage.

Interactions Between Probiotics and Other Components Present in Edible Films/Coatings

Probiotic interactions with other ingredients in edible films and coatings are essential for improving food safety and preservation. Probiotics can be added to biopolymer-based edible films, which not only increase the shelf life but also enhance the nutritional value of minimally processed fruits and vegetables, according to recent studies. Edible coatings composed of polysaccharides like xanthan gum and alginate can successfully include probiotics, such as strains of *Lactobacillus* and *Bifidobacterium* (Chang et al., 2023; Chikhala et al., 2024). According to research, these coatings significantly preserve probiotic viable counts (such as 7 log CFU/g) in coated fruits, ensuring their viability throughout storage (Chikhala et al., 2024). Food safety is improved and microbiological deterioration is decreased by probiotic-loaded films' antibacterial qualities (Silva et al., 2023c). According to Chang et al. (2023) and Chikhala et al. (2024), the addition of probiotics also enhances antioxidant activity, protecting bioactive chemicals in fruits and vegetables, which are crucial for consumer health. Probiotics are shielded from environmental deterioration by encapsulation techniques, which enable controlled release during digestion (Ramanathan et al., 2024). By increasing the bioavailability of probiotics and other bioactive substances, this technique guarantees that they will effectively support health advantages. Although there are encouraging results from the incorporation of probiotics into edible films, there are still issues with the films' durability in different storage settings and the possibility of interactions with other food ingredients that could compromise the viability of the probiotics.

Effect of the Incorporation of Probiotics on the Physicochemical Properties and Techno-functional Properties of Edible Films/Coatings

While some studies (Soukoulis et al., 2016; Khodaei et al., 2020) found probiotics had no influence on the physicochemical and mechanical properties of films/coatings, others (Shahrampour et al., 2020; Akman et al., 2021; Li et al., 2020) observed significant modifications. For instance, by creating hydrogen bonds with the matrix, *L. plantarum* decreased the permeability of water vapor in alginate/pectin films. Encapsulated probiotics improved water vapor permeability and tensile strength in alginate films. The structure and composition of the biopolymer determine how probiotics work (Soukoulis et al., 2016). Probiotic viability can be increased by adding prebiotics to films, which can either reinforce or plasticize them (Paulo et al., 2021). *Bacillus coagulans* and fructooligosaccharide (FOS) in synbiotic films resulted in plasticizing effects that increased the elongation and lowered the strength, while FOS also enhanced the water vapor permeability and deteriorated the structure of the film (Oliveira-Alcântara et al., 2020). *Lactobacillus plantarum*, cellulose nanofiber (CNF), and inulin were combined to create carboxymethyl cellulose (CMC)–based nanocomposite films by Zabihollahi et al. (2020). While inulin improved the compactness, CNF decreased the film roughness. The film completely contained the probiotic cells. Whereas the thinnest film had 5% CNF and no inulin, the thickest film had 2.5% CNF and 20% inulin. CNF and 10% inulin had no effect on the tensile strength, whereas 20% inulin decreased it. Because of the stiff network that was created between CMC, CNF, and inulin, elongation to break decreased when 5% CNF and 20% inulin were combined. The effects of adding probiotics to edible films and coatings on their physicochemical and techno-functional characteristics are shown in Table 11.1.

Regulatory Aspects in the Commercialization of Probiotic-Incorporated Edible Films/Coatings/Packaging

Adopting probiotics into edible films, coatings, and packaging is subject to a number of important regulatory requirements. To guarantee the safety of the probiotic strains utilized, it is crucial to first secure the Generally Recognized as Safe (GRAS) accreditation. Additionally, since regional variations exist in food safety rules, adherence to them is essential to avoiding product recalls and safeguarding brand reputation. Additionally, manufacturers must follow labeling and claims laws to prevent customers from being misinformed by false nutrition information or health claims. Furthermore, when new or modified microbes are involved, novel food regulations apply, which means that a thorough understanding of these regulations is necessary to assure consumer safety. Finally, maintaining a clean manufacturing environment lowers the danger of contamination and ensures product quality, which is made possible by putting Good Manufacturing Practices (GMPs) into practice. These regulatory elements work together to provide a thorough framework for the safe marketing of probiotic goods.

Application of Probiotic-Incorporated Edible Films/Coatings in the Food Sector

Numerous studies have examined the use of edible films and coatings containing probiotics on meat and fish products (Pavli et al., 2017; Poonia and Mishra, 2021; Ma et al., 2019; La Storia et al., 2020; Zabihollahi et al., 2020; Ceylan and Atasoy, 2022), baked goods (García-Argueta et al., 2016; Soukoulis et al., 2016, 2017), dairy products (Costa et al., 2018; Moghanjougi et al., 2020; S'aez-Orviz et al., 2020; Settier-Ramírez et al., 2020; El-Sayed et al., 2021; Silva et al., 2022), and fruits (Bambace et al., 2019; Dianina et al., 2019; Alvarez et al., 2021; Silva et al., 2022). Figure 11.2 illustrates the key features of probiotic-encapsulated edible films or coatings for food product preservation.

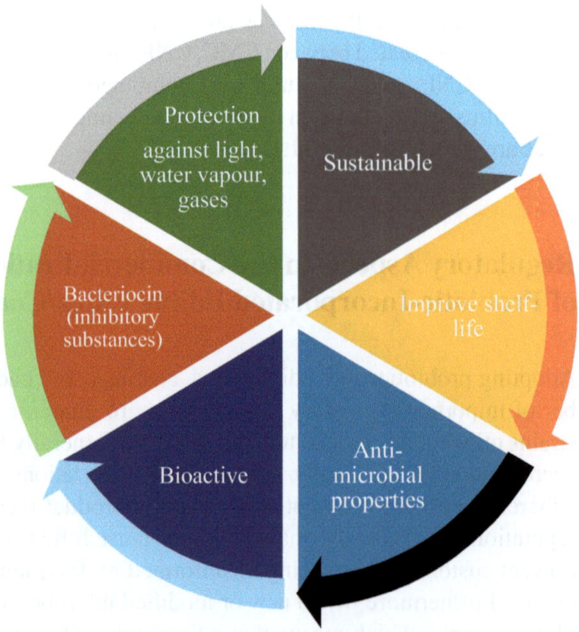

Fig. 11.2 Characteristics of probiotic-encapsulated edible films/coatings for the preservation of food products

Conclusion

Food shelf life, safety, and nutritional value are all improved by edible films and coatings, which are becoming a viable substitute for traditional food packaging. Probiotics and antibacterial agents are added to increase their efficacy and provide customers with extra health benefits. By offering an environmentally sustainable way to preserve food safety and quality, this invention has the potential to completely transform the food sector. The market for foods high in probiotics is being driven by consumer awareness of the connection between nutrition and health, and the use of probiotics in new food products is being expanded via edible packaging. However, laboratory research is ahead of industry applications. Finding the appropriate materials, technology, and probiotic strains at a reasonable price is one of the primary obstacles. Another is preserving a high probiotic cell density throughout the manufacturing, storage, and post-ingestion phases to guarantee health benefits. For adoption to be more widespread, these obstacles must be removed. In addition, the inclusion of probiotics in the edible coating and film may also help improve the gut microbiota and provide health benefits. Further research is required to explore the potential roles of edible coatings functionalized with probiotics on the gut microbiota health.

References

Abedinia, A., Alimohammadi, F., Teymori, F., Razgardani, N., Saeidi Asl, M. R., Ariffin, F., Mohammadi Nafchi, A., Huda, N., & Roslan, J. (2021). Characterization and cell viability of probiotic/prebiotics film based on duck feet gelatin: A novel poultry gelatin as a suitable matrix for probiotics. *Foods, 10*, 1761.

Adeel, M., Afzaal, M., Saeed, F., Ahmed, A., Mahmood, K., Abbas Shah, Y., et al. (2023). Encapsulation of probiotic bacteria using polyelectrolytes stabilized nanoliposomes for improved viability under hostile conditions. *Journal of Food Science, 88*(9), 3839–3848.

Akman, P. K., Bozkurt, F., Dogan, K., Tornuk, F., & Tamturk, F. (2021). Fabrication and characterization of probiotic lactobacillus plantarum loaded sodium alginate edible films. *Journal of Food Measurement and Characterization, 15*, 84–92.

Alvarez, M. V., Bambace, M. F., Quintana, G., Gomez-Zavaglia, A., & del Rosario Moreira, M. (2021). Prebiotic-alginate edible coating on fresh-cut apple as a new carrier for probiotic *lactobacilli* and *bifidobacteria*. *LWT – Food Science and Technology, 137*, Article 110483.

Aşan Özüsağlam, M., & Celik, I. (2024). Formulating edible films with red pitahaya extract and probiotic. *Foods and Raw Materials.*

Azeem, M., Saeed, F., Afzaal, M., Ateeq, H., Ahmad, A., Liaqat, A., et al. (2023). Encapsulation of probiotics in solid lipid micro particle for improved viability and stability under stressed conditions. *International Journal of Food Properties, 26*(1), 1612–1623.

Bambace, M. F., Alvarez, M. V., & del Rosario Moreira, M. (2019). Novel functional blueberries: Fructo-oligosaccharides and probiotic lactobacilli incorporated into alginate edible coatings. *Food Research International, 122*, 653–660.

Barbosa, C. H., Andrade, M. A., Vilarinho, F., Fernando, A. L., & Silva, A. S. (2021). *Active edible packaging. Encyclopedia, 1*, 360–370.

Barrag'an-Men'endez, C., G'alvez-L'opez, D., Rosas-Quijano, R., Salvador-Figueroa, M., Ovando-Medina, I., & V'azquez-Ovando, A. (2020). Films of chitosan and *Aloe vera* for maintaining the viability and antifungal activity of *Lactobacillus paracasei* TEP6. *Coatings, 10*, 259.

Ceylan, H. G., & Atasoy, A. F. (2022). Optimization and characterization of prebiotic concentration of edible films containing *Bifidobacterium animalis* subsp. *lactis* BB-12® and its application to block type processed cheese. *International Dairy Journal, 134*, Article 105443.

Chang, S., Guo, Q., Du, G., Tang, J., Liu, B., Shao, K., & Zhao, X. (2023). Probiotic-loaded edible films made from proteins, polysaccharides, and prebiotics as a quality factor for minimally processed fruits and vegetables: A review. *International Journal of Biological Macromolecules, 253*, 127226.

Chikhala, T., Seke, F., Slabbert, R. M., Sultanbawa, Y., & Sivakumar, D. (2024). Utilizing xanthan gum coatings as probiotic bacteria carriers to enhance postharvest quality and antioxidants in fresh-cut cantaloupe and honeydew (*Cucumis melo* L.) melons. *Foods, 13*(6), 940.

Choudhary, U., Bhinchhar, B. K., Yadav, S. P., & Singh, P. (2022). Utilization of Agro-Industrial Wastes as EdibleCoating and. *A glance at food processing applications, 93*

Coimbra, P., Alarico, S., Empadinhas, N., Braga, M. E., & Gaspar, M. C. (2023). Sustainable starch-based edible films with agrifood residues as potential carriers for the probiotic Lactobacillus rhamnosus. *Innovative Food Science & Emerging Technologies, 88*, 103452.

Costa, M. J., Maciel, L. C., Teixeira, J. A., Vicente, A. A., & Cerqueira, M. A. (2018). Use of edible films and coatings in cheese preservation: Opportunities and challenges. *Food Research International, 107*, 84–92.

Davachi, S. M., Pottackal, N., Torabi, H., & Abbaspourrad, A. (2021). Development and characterization of probiotic mucilage based edible films for the preservation of fruits and vegetables. *Scientific reports, 11*, 16608.

de J Espinosa-Martínez, D., Aparicio-Fernández, X., Ramírez-López, C., Hernández-Carranza, P., Ruiz-López, I. I., & Ochoa-Velasco, C. E. (2025). Application of functional edible films with antioxidant, probiotic, and antimicrobial properties on Mexican sweet potato candy. *International Journal of Food Science and Technology, 60*, vvae083.

de Lima Marques, J., Funck, G. D., da Silva Dannenberg, G., dos Santos Cruxen, C. E., El Halal, S. L. M., Dias, A. R. G., Fiorentini, Â. M., & da Silva, W. P. (2017). Bacteriocin like substances of *Lactobacillus curvatus* P99: Characterization and application in biodegradable films for control of *Listeria monocytogenes* in cheese. *Food Microbiology, 63*, 159–163.

del Carmen Beristain-Bauza, S., Mani-López, E., Palou, E., & López-Malo, A. (2017). Antimicrobial activity of whey protein films supplemented with Lactobacillus sakei cell-free supernatant on fresh beef. *Food microbiology, 62*, 207–211.

Dianina, I. B., Jrb, A. G. O., Pimentelc, T. C., Hernandesa, N. F., & Costaa, G. N. (2019). Edible biofilms formulated with whey protein isolate and *L. casei* probiotic culture: Characterization and application in tomatoes and grapes. *Chemical Engineering Transactions, 75*, 469–474.

Edo, G. I., Mafe, A. N., Razooqi, N. F., Umelo, E. C., Gaaz, T. S., Isoje, E. F., et al. (2025). Advances in bio-polymer coatings for probiotic microencapsulation: chitosan and beyond for enhanced stability and controlled release. *Designed Monomers and Polymers, 28*(1), 1–34.

El-Sayed, H. S., El-Sayed, S. M., Mabrouk, A. M., Nawwar, G. A., & Youssef, A. M. (2021). Development of eco-friendly probiotic edible coatings based on chitosan, alginate and carboxymethyl cellulose for improving the shelf life of soft cheese. *Journal of Polymers and the Environment, 29*, 1941–1953.

Espitia, P. J. P., Batista, R. A., Azeredo, H. M., & Otoni, C. G. (2016). Probiotics and their potential applications in active edible films and coatings. *Food Research International, 90*, 42–52.

Food and Agriculture Organization of the United Nations/World Health Organization (FAO/WHO). (2002). *Guidelines for the evaluation of probiotics in food. Report of a Joint FAO/WHO Working Group on Drafting Guidelines for the Evaluation of Probiotics in Food.* WHO.

FSS (Health supplements, nutraceuticals, food for Special dietary Use, Food for Special Medical Purpose and Prebiotic and Probiotic Food) Regulation. (2022).

Galus, S., Arik Kibar, E. A., Gniewosz, M., & Kra'sniewska, K. (2020). Novel materials in the preparation of edible films and coatings—a review. *Coatings, 10*, 674.

García-Argueta, I., Quintero-Salazar, B., Dominguez-Lopez, A., G'omez-Oliv'an, L., Díaz-Bandera, D., & Dubl'an-García, O. (2016). Effect of edible coating based on whey, inulin and gelatine with *Lactobacillus casei* on the textural and sensorial properties of a cracker cookie. *Journal of Probiotics & Health, 4*, 153.

Guimarães, A., Abrunhosa, L., Pastrana, L. M., & Cerqueira, M. A. (2018). Edible films and coatings as carriers of living microorganisms: A new strategy towards biopreservation and healthier foods. *Comprehensive Reviews in Food Science and Food Safety, 17*, 594–614.

Hashemi, S. S., Khorshidian, N., & Mohammadi, M. (2022). An insight to potential application of synbiotic edible films and coatings in food products. *Frontiers in Nutrition, 9*, 875368.

Jeevahan, J. J., Chandrasekaran, M., Venkatesan, S., Sriram, V., Joseph, G. B., Mageshwaran, G., & Durairaj, R. (2020). Scaling up difficulties and commercial aspects of edible films for food packaging: A review. *Trends in Food Science & Technology, 100*, 210–222.

Kareem, R.A., Razavi, S.H., Mousavi, Z.E. (2023). Effect of two types of wall matrix on viability of lactobacillus plantarum under simulated gastrointestinal fluids.

Karimi, N., Alizadeh, A., Almasi, H., & Hanifian, S. (2020). Preparation and characterization of whey protein isolate/polydextrose-based nanocomposite film incorporated with cellulose nanofiber and L. plantarum: A new probiotic active packaging system. *LWT, 121*, 108978.

Kaur, J., Gunjal, M., Rasane, P., & Poonia, A. (2022). Edible packaging: An overview. In *Edible food packaging, applications, innovations and sustainability*. Springer. https://doi.org/10.1007/978-981-16-2383-7_1

Khodaei, D., Hamidi-Esfahani, Z., & Lacroix, M. (2020). Gelatin and low methoxyl pectin films containing probiotics: film characterization and cell viability. *Food Bioscience, 36*, 100660.

La Storia, A., Di Giuseppe, F. A., Volpe, S., Oliviero, V., Villani, F., & Torrieri, E. (2020). Physical properties and antimicrobial activity of bioactive film based on whey protein and *Lactobacillus curvatus* 54M16 producer of bacteriocins. *Food Hydrocolloids, 108*, Article 105959.

Li, S., Ma, Y., Ji, T., Sameen, D. E., Ahmed, S., Qin, W., et al. (2020). Cassava starch/ carboxymethylcellulose edible films embedded with lactic acid bacteria to extend the shelf life of banana. *Carbohydrate Polymers, 248*, 116805.

Ma, D., Jiang, Y., Ahmed, S., Qin, W., & Liu, Y. (2019). Physical and antimicrobial properties of edible films containing *Lactococcus lactis*. *International Journal of Biological Macromolecules, 141*, 378–386.

Majid, I., Ahmad Nayik, G., Mohammad Dar, S., & Nanda, V. (2018). Novel food packaging technologies: Innovations and future prospective. *Journal of the Saudi Society of Agricultural Sciences, 17*, 454–462.

Mihalca, V., Kerezsi, A. D., Weber, A., Gruber-Traub, C., Schmucker, J., Vodnar, D. C., Dulf, F. V., Socaci, S. A., F˘arcaş, A., & Mureşan, C. I. (2021). Protein-based films and coatings for food industry applications. *Polymers, 13*, 769.

Mishra, A., Mann, B., Poonia, A., & Rai, D. C. (2022). Milk protein-based edible coatings: Properties and applications. In A. Poonia & T. Dhewa (Eds.), *Edible food packaging*. Springer. https://doi.org/10.1007/978-981-16-2383-7_11

Moeini, A., Germann, N., Malinconico, M., & Santagata, G. (2021). Formulation of secondary compounds as additives of biopolymer-based food packaging: A review. *Trends in Food Science & Technology, 114*, 342–354.

Moghanjougi, Z. M., Bari, M. R., Khaledabad, M. A., Almasi, H., & Amiri, S. (2020). Biopreservation of white brined cheese (feta) by using probiotic bacteria immobilized in bacterial cellulose: Optimization by response surface method and characterization. *LWT – Food Science and Technology, 117*, Article 108603.

Mozaffarzogh, M., Misaghi, A., Shahbazi, Y., & Kamkar, A. (2020). Evaluation of probiotic carboxymethyl cellulose-sodium caseinate films and their application in extending shelf life quality of fresh trout fillets. *LWT – Food Science and Technology, 126*, 109305.

Odila Pereira, J., Soares, J., Costa, E., Silva, S., Gomes, A., & Pintado, M. (2019). Characterization of edible films based on alginate or whey protein incorporated with *Bifidobacterium animalis* subsp. *lactis BB-12* and prebiotics. *Coatings, 9*, 493.

Oduro, K. O. A. (2021). Edible coating. In *Postharvest technology-recent advances, new perspectives and applications*. IntechOpen. 978-1-83969-925-2.

Oliveira-Alcântara, A. V., Abreu, A. A. S., Gonçalves, C., Fuciños, P., Cerqueira, M. A., FMP, G., et al. (2020). Bacterial cellulose/cashew gum films as probiotic carriers. *LWT – Food Science and Technology, 130*, 109699.

Paulo, A. F. S., Baú, T. R., Ida, E. I., & Shirai, M. A. (2021). Edible coatings and films with incorporation of prebiotics—A review. *Food Research International, 148*, 110629.

Pavli, F., Kovaiou, I., Apostolakopoulou, G., Kapetanakou, A., Skandamis, P., Nychas, G. J. E., Tassou, C., & Chorianopoulos, N. (2017). Alginate-based edible films delivering probiotic bacteria to sliced ham pretreated with high pressure processing. *International Journal of Molecular Sciences, 18*, 1867.

Pereira, J. O., Soares, J., Monteiro, M. J., Gomes, A., & Pintado, M. (2018). Impact of whey protein coating incorporated with *Bifidobacterium* and *Lactobacillus* on sliced ham properties. *Meat Science, 139*, 125–133.

Poonia A. (2017). Prebiotics, probiotics and synbiotic foods. In D. Mudgil & S. Barak (Eds.), *Functional foods: Sources and health benefits*. Scientific Publishers. ISBN: 978-93-86237-00-2 eISBN: 978-93-87869-63-9.

Poonia, A. (2018). Antimicrobial edible films and coatings for fruits and vegetables. In *Food science and nutrition: Breakthroughs in research and practice* (pp. 177–195). IGI Global Scientific Publishing. https://doi.org/10.4018/978-1-5225-5207-9.ch009

Poonia, A., & Mishra, A. (2021). Edible nanocoatings: potential food applications, challenges and safety regulations. *Nutrition & Food Science, 52*(3), 497–514. https://doi.org/10.1108/NFS-07-2021-0222

Pop, O. L., Pop, C. R., Dufrechou, M., Vodnar, D. C., Socaci, S. A., Dulf, F. V., Minervini, F., & Suharoschi, R. (2020). Edible films and coatings functionalization by probiotic incorporation: A review. *Polymers, 12*, 12.

Rai, S., & Poonia, A. (2019). Formulation and characterization of edible films from pea starch and casein. *Journal of Pharmacognosy and Phytochemistry, 8*(2), 317–321.

Ramanathan, S., Leya, B., Freeda Blessie, R., & Pragalyaashree, M. M. (2024). Role of bioactive carrier in edible films: Encapsulation theory. In *Food Coatings and Preservation Technologies* (pp. 403–446).

Ramos, O. L., Fernandes, J. C., Silva, S. I., Pintado, M. E., & Malcata, F. X. (2012). Edible films and coatings from whey proteins: A review on formulation, and on mechanical and bioactive properties. *Critical Reviews in Food Science and Nutrition, 52*, 533–552.

S'aez-Orviz, S., Puertas, C., Marcet, I., Rendueles, M., & Díaz, M. (2020). Bioactive synbiotic coatings with lactobionic acid and *Lactobacillus plantarum* CECT 9567 in the production and characterization of a new functional dairy product. *Journal of Functional Foods, 75*, 104263.

S'aez-Orviz, S., Marcet, I., Rendueles, M., & Díaz, M. (2021). Bioactive packaging based on delipidated egg yolk protein edible films with lactobionic acid and *Lactobacillus plantarum* CECT 9567: Characterization and use as coating in a food model. *Food Hydrocolloids, 119*, 106849.

S'aez-Orviz, S., Rendueles, M., & Díaz, M. (2022). Impact of adding prebiotics and probiotics on the characteristics of edible films and coatings – A review. *Food Research International*, 112381.

Saleena, L. A. K., Nyam, K. L., & Pui, L. P. (2024). Probiotic edible film added with aqueous clove extract milk powder. *Kafkas Üniversitesi Veteriner Fakültesi Dergisi, 30*(3).

Santhosh, R., Nath, D., & Sarkar, P. (2021). Novel food packaging materials including plant-based byproducts: A review. *Trends in Food Science & Technology, 118*, 471–489.

Semwal, A., Ambatipudi, K., & Navani, N. K. (2022). Development and characterization of sodium caseinate based probiotic edible film with chia mucilage as a protectant for the safe delivery of probiotics in functional bakery. *Food Hydrocolloids for Health, 2*, 100065.

Settier-Ramírez, L., L'opez-Carballo, G., Gavara, R., & Hern'andez-Mu˜noz, P. (2020). Pvoh/protein blend films embedded with lactic acid bacteria and their antilisterial activity in pasteurized milk. *International Journal of Food Microbiology, 322*, 108545.

Shahrampour, D., Khomeiri, M., Razavi, S. M. A., & Kashiri, M. (2020). Development and characterization of alginate/pectin edible films containing lactobacillus plantarum KMC 45. *LWT – Food Science and Technology, 118*, 108758.

Sharma, K., Babaei, A., Oberoi, K., Aayush, K., Sharma, R., & Sharma, S. (2022a). Essential oil nanoemulsion edible coating in food industry: A review. *Food and Bioprocess Technology, 15*, 2375–2395.

Sharma, N., Bhardwaj, A., Sid, P., Sahu, J. K., & Poonia, A. (2022b). Scope, functions, and novelty of packaging edibles. In *Edible food packaging, applications, innovations and sustainability*. https://doi.org/10.1007/978-981-16-2383-7_4

Silva, S. P., Ribeiro, S. C., Teixeira, J. A., & Silva, C. C. (2022). Application of an alginate based edible coating with bacteriocin-producing *Lactococcus* strains in fresh cheese preservation. *LWT – Food Science and Technology, 153*, 112486.

Silva, S. P., Teixeira, J. A., & Silva, C. C. (2023a). Application of enterocin-whey films to reduce *Listeria monocytogenes* contamination on ripened cheese. *Food Microbiology, 109*, 104134.

Silva, S. P., Teixeira, J. A., & Silva, C. C. (2023b). Prevention of fungal contamination in semi-hard cheeses by whey–gelatin film incorporated with *Levilactobacillus brevis* SJC120. *Foods, 12*, 1396.

Silva, S. P., Teixeira, J. A., & Silva, C. C. (2023c). Recent advances in the use of edible films and coatings with probiotic and bacteriocin-producing lactic acid bacteria. *Food Bioscience, 56*, 103196.

Soukoulis, C., Behboudi-Jobbehdar, S., Yonekura, L., Parmenter, C., & Fisk, I. D. (2014). Stability of Lactobacillus rhamnosus GG in prebiotic edible films. *Food Chemistry, 159*, 302–308.

Soukoulis, C., Singh, P., Macnaughtan, W., Parmenter, C., & Fisk, I. D. (2016). Compositional and physicochemical factors governing the viability of *Lactobacillus rhamnosus GG* embedded in starch-protein based edible films. *Food Hydrocolloids, 52*, 876–887.

Soukoulis, C., Behboudi-Jobbehdar, S., Macnaughtan, W., Parmenter, C., & Fisk, I. D. (2017). Stability of *Lactobacillus rhamnosus* GG incorporated in edible films: Impact of anionic biopolymers and whey protein concentrate. *Food Hydrocolloids, 70*, 345–355.

Sun, J., Leng, X., Zang, J., & Zhao, G. (2024). Bio-based antibacterial food packaging films and coatings containing cinnamaldehyde: A review. *Critical Reviews in Food Science and Nutrition, 64*, 140–152.

Tang, Y., Xie, F., Zhang, D., Zhu, M., Liu, L., Liu, P., & Gu, C. (2015). Physical properties and prebiotic activity of maize starch-based functional films. *Starch – Stärke, 67*, 124–131.

Torres-Giner, S., Perez-Masia, R., & Lagaron, J. M. (2016). A review on electrospun polymer nanostructures as advanced bioactive platforms. *Polymer Engineering and Science, 56*, 500–527.

Valencia-Chamorro, S. A., Palou, L., Del Rio, M. A., & Perez-Gago, M. B. (2011). Antimicrobial edible films and coatings for fresh and minimally processed fruits and vegetables: A review. *Critical Reviews in Food Science and Nutrition, 51*, 872–900.

Wang, A., & Zhong, Q. (2024). Drying of probiotics to enhance the viability during preparation, storage, food application, and digestion: A review. *Comprehensive Reviews in Food Science and Food Safety, 23*(1), e13287.

Yan, M. R., Hsieh, S., & Ricacho, N. (2022). Innovative food packaging, food quality and safety, and consumer perspectives. *Processes, 10*, 747.

Zabihollahi, N., Alizadeh, A., Almasi, H., Hanifian, S., & Hamishekar, H. (2020). Development and characterization of carboxymethyl cellulose based probiotic nanocomposite film containing cellulose nanofiber and inulin for chicken fillet shelf life extension. *International Journal of Biological Macromolecules, 160*, 409–417.

Chapter 12
Effects of Nanomaterials on the Properties of Edible Coatings and Films

Aditya, Neeraj, J. N. Bhatia, and Ajar Nath Yadav

Abstract The integration of nanomaterials into edible films and coatings has emerged as a transformative strategy to enhance food preservation, safety, and quality while aligning with global sustainability goals. Recent advancements demonstrate that incorporating metallic nanoparticles (NPs), biopolymer-based nanostructures, nanoclays, and nanoemulsions can significantly improve mechanical strength, barrier properties, and antimicrobial functionality. These enhancements address long-standing limitations of conventional edible films, such as limited moisture resistance and mechanical fragility, while also enabling novel functionalities like active and intelligent packaging. Despite these promising benefits, the field faces critical challenges, including safety and toxicity concerns related to nanoparticle migration, regulatory gaps in defining safe use parameters, and technical barriers in achieving uniform dispersion and large-scale production. Additionally, consumer perception and labeling requirements add complexity to commercialization pathways. Emerging trends, including the use of green synthesis methods, biodegradable nanomaterials, and artificial intelligence (AI)–assisted material design, offer potential pathways to mitigate these challenges while supporting sustainable and scalable innovations. Further, multidisciplinary research bridging materials science, food technology, toxicology, and policy will be essential to fully realize the potential of nanomaterial-enhanced edible films and coatings. This chapter provides a review of the current nanomaterials used, their effects on functional and physico-

Aditya (✉) · Neeraj
Department of Agriculture and Environmental Sciences, National Institute of Food Technology Entrepreneurship and Management, Kundli (An Institute of National Importance of India, NIFTEM-K), Sonipat, Haryana, India

J. N. Bhatia
Department of Plant Pathology, Chaudhary Charan Singh Haryana Agricultural University (CCS HAU), Hisar, Haryana, India

A. N. Yadav
Department of Biotechnology, Dr. Khem Singh Gill Akal College of Agriculture, Eternal University, Baru Sahib, Sirmour, Himachal Pradesh, India

N. Kumar et al. (eds.), *Trends in Edible Packaging: A sustainable approaches for postharvest management of fruits and vegetables*,
https://doi.org/10.1007/978-3-032-13199-7_12

chemical properties, recent technological trends, safety considerations, and the practical barriers to market adoption, with a special emphasis on the integration of nanotechnology.

Keywords Antimicrobial activity · Barrier properties · Edible coatings · Edible films · Nanomaterials

Introduction

The global food industry is increasingly facing the dual challenge of ensuring high-quality food products while adhering to environmental sustainability. As consumer awareness regarding the detrimental effects of plastic pollution grows, industry stakeholders are compelled to seek alternatives to traditional synthetic packaging solutions, predominantly derived from petroleum-based polymers. These synthetic materials, characterized by their durability, are unfortunately nonbiodegradable, and their unchecked proliferation contributes significantly to environmental degradation. According to a report, approximately 300 million tons of plastic is produced each year, of which a significant fraction contributes to marine and terrestrial pollution, thereby posing a threat to ecosystems and human health (Adeyeye & Ashaolu, 2021). In this context, edible films and coatings, which are biodegradable and derived from natural biopolymers, emerge as a viable alternative. These materials are not only safe for consumption but also offer unique functionalities that enhance food quality and extend shelf life (Dewi & Khoerunnisa, 2024). The concept of edible films has historical roots, tracing back to ancient civilizations that utilized natural products like leaves, wax, or animal skins to preserve food. These practices underscore an early understanding of the protective capabilities of naturally occurring materials. The evolution of edible films into their modern form can be attributed to advances in polymer science and food technology, particularly throughout the twentieth century, during which the demand for sustainable food packaging solutions began to escalate. Contemporary edible films are typically composed of biopolymers such as proteins (gelatin, casein, and whey), polysaccharides (starch, alginate, cellulose, and chitosan), and lipids (waxes and fatty acids). Each of these biopolymers offers distinct physicochemical properties that contribute to food preservation (Nikhanj, 2023; Gurjar et al., 2024). Edible films and coatings serve multiple functions, primarily acting as barriers to gases, moisture, and contaminants, which helps maintain food quality by reducing spoilage (Fig. 12.1). Films made from sago starch have been shown to effectively inhibit gas exchange, thereby prolonging the freshness of packaged produce (Dewi & Khoerunnisa, 2024). Furthermore, these films can be engineered to incorporate active agents like antimicrobials, antioxidants, and flavor encapsulants. This functionality not only enhances their preservative role but also opens avenues for creating active packaging that

Fig. 12.1 Schematic illustration of the effect of nanomaterial-incorporated edible coatings/films on the preservation and shelf life of the apple fruit. The figure compares the apple fruit with (left) and without (right) an edible nanocomposite coating/film. The coated apple demonstrates multiple functional benefits. The incorporation of nanomaterials within the biopolymer-based edible film imparts antimicrobial activity, reducing microbial proliferation on the fruit surface; enhanced mechanical properties increase structural integrity and resistance to physical damage; the coating acts as a water vapor barrier, minimizing transpiration and moisture loss; and reduced gas permeability lowers the respiration rate and oxidative reactions, further delaying senescence. On the other hand, the uncoated apple is vulnerable to microbial attack, exhibits higher moisture loss and oxidation, and, consequently, suffers from reduced shelf life. This schematic highlights the multifunctional role of nanomaterial-enhanced edible coatings/films in improving the postharvest quality, storability, and safety of fresh produce through synergistic barrier and bioactive properties

interacts with food to further inhibit spoilage (Muizniece-Brasava & Kirse, 2018; Suvarna et al., 2022). Nanomaterial-based films and coatings offer advanced protection against microbial contamination by incorporating antimicrobial nanoparticles (NPs), such as silver (Ag)-, zinc oxide (ZnO)–, or mushroom-derived nanostructures. These smart coatings act as active barriers, effectively preventing the spread of pathogens and enhancing the safety and shelf life of food and biomedical products (Xing et al., 2019; Pant et al., 2023; Maurya et al., 2023). The environmental benefits of these materials also cannot be overstated; they are typically biodegradable, nontoxic, and crafted from renewable resources, aligning with global sustainability targets and consumer preferences regarding eco-friendly packaging (Suvarna et al., 2022).

Despite their potential, the commercial viability of conventional edible films is hindered by several limitations. For example, biopolymer-based films often exhibit subpar mechanical strength, high permeability to water vapor, and limited inherent antioxidant and antimicrobial properties. These weaknesses present challenges, particularly when packaging moist or highly perishable foods, leading to compromised functionality (Gurjar et al., 2024). Additionally, issues such as susceptibility to microbial degradation and variability in structural uniformity pose hurdles for widespread adoption. Therefore, innovative strategies are imperative to enhance the durability, barrier properties, and bioactivity of these materials (Muizniece-Brasava

& Kirse, 2018). Nanotechnology has emerged as a promising avenue for addressing the limitations of edible films. Defined as the manipulation of materials at the nanoscale (1–100 nm), nanotechnology introduces unique properties inherent to nanoscale materials that differ significantly from their bulk counterparts. These properties include an enhanced surface area, increased reactivity, and improved mechanical reinforcement, which can be leveraged to optimize the performance of edible films (Omerovic et al., 2021). Integrating nanomaterials such as nanoclays, metal nanoparticles (silver and zinc oxide), nanocellulose, and carbon nanotubes into film matrices has demonstrated significant improvements in film performance. For example, the addition of silver nanoparticles imparts potent antimicrobial properties that effectively suppress a broad spectrum of foodborne pathogens (Han et al., 2018; Awwaly et al., 2024). Additionally, nanoemulsions containing essential oils or bioactive compounds have been successfully incorporated into edible coatings to facilitate controlled release mechanisms, thus maintaining the sensory quality of food products during storage. Innovations in this area have led to the development of smart and active packaging solutions that not only serve conventional protective roles but also actively monitor and respond to changes in the food condition (Nikita & Agnihotri, 2025). Also, nanomaterials derived from oyster mushrooms have emerged as promising candidates for sustainable food packaging and preservation applications. These bio-based nanostructures exhibit antioxidant and antimicrobial properties, enabling the development of eco-friendly smart packaging systems that enhance shelf life while minimizing synthetic inputs (Aditya et al., 2024a, 2025a). Nanosensors embedded within edible films can be employed to detect alterations in pH, microbial load, or gas composition, thus offering real-time insights into food quality (Firdaus et al., 2023; Kondle et al., 2023). Implementing such smart materials can significantly reduce food waste and enhance consumer trust in food quality and safety. The shift toward incorporating nanotechnology in edible packaging does not come without its own set of challenges. Safety remains a critical concern, as public health bodies grapple with the potential long-term effects of nanoscale materials on human health. While many nanomaterials are classified as Generally Recognized as Safe (GRAS), the unique properties associated with their nanoscale form may give rise to unforeseen health risks (Lestari & Wilujeng, 2022). Public perception of nanoscale materials in food applications also presents a challenge, as consumer acceptance is often colored by skepticism and concerns regarding the invisibility of engineered nanoparticles in food products. Although consumers may recognize the benefits associated with sustainable and technologically advanced packaging, their willingness to embrace such innovations is contingent upon transparency, labeling, and rigorous safety assurance (Misir et al., 2014). In light of these challenges, this chapter aims to provide a comprehensive assessment of edible films and coatings, with a special emphasis on the integration of nanotechnology.

Overview of Edible Films and Coatings

Edible films and coatings have emerged as innovative packaging solutions that offer a promising alternative to traditional synthetic polymers for food preservation. These thin layers made from natural and edible substances can be applied directly onto food items or produced as stand-alone films serving a dual purpose, protecting food from spoilage while also being safe for consumption. The historic use of edible barriers can be traced back to ancient traditions where materials like leaves, wax, and animal intestines were employed to wrap and shield food. In contemporary society, the push toward developing edible films has been propelled by growing concerns surrounding environmental sustainability, particularly concerning plastic waste (Reichert et al., 2020; Tavassoli et al., 2022). The ability of these films to act as semipermeable barriers primarily controls the movement of gases such as oxygen and carbon dioxide, alongside moisture and solutes, which effectively delays oxidative deterioration, moisture loss, and microbial spoilage (Suput et al., 2019). Beyond their primary function as barriers, edible films and coatings are increasingly engineered to deliver additional benefits. These can include mechanical support to prevent physical damage during handling and transportation, as well as the incorporation of bioactive compounds such as antimicrobial agents, antioxidants, and flavoring agents (Khezerlou et al., 2021). The study on chitosan-based coatings has illustrated that, when enriched with essential oils, they significantly reduce microbial loads on fresh produce, thereby extending their shelf life (Diaz-Montes & Castro-Munoz, 2021). Similarly, whey protein films are recognized for their exceptional oxygen barrier properties, which effectively curb oxidation in meat products, thus improving their durability and maintaining freshness (Giannakas et al., 2023). The adaptability of these edible coatings enables them to cater to a broad array of food categories, including fresh fruits and vegetables, dairy products, meats, and bakery items (Juikar & Warkar, 2022).

The effectiveness of edible films and coatings hinges upon the selection of biopolymers used in their formulation, which typically fall into three main types: proteins, polysaccharides, and lipids. Each category exhibits distinct characteristics and advantages, contributing to how these edible films interact with food products and environmental conditions. Protein-based films often exhibit excellent mechanical strength and barrier properties. For example, gelatin films are widely used in applications involving meat and dairy due to their flexibility and transparency (Ali & Ahmed, 2018). On the other hand, soy protein isolate (SPI) showcases exceptional strength and thermal stability but typically requires plasticizers to mitigate brittleness (Giannakas et al., 2023). Corn-derived zein forms hydrophobic films that are particularly suitable for coating snacks like candies and nuts (Cherpinski et al., 2017). Whey protein, recognized for its clarity and ability to lower oxygen permeability, is particularly beneficial in protecting oxygen-sensitive products, thus enhancing shelf life (Balan & Kadeppagari, 2020). Polysaccharide-based films attract interest due to their biodegradability, cost-effectiveness, and capacity to form flexible films. Starch, for instance, has garnered attention for its abundance and affordability, though native starch films often exhibit brittleness and require the addition of plasticizers like glycerol for improved flexibility (Jost & Stramm, 2015). Another notable polysaccharide (chitosan) is celebrated not only for its film-forming capabilities but also for its intrinsic antimicrobial properties, making it particularly valuable for food coatings (Alboofetileh et al., 2018). Furthermore, alginate, derived from brown seaweed, forms gel-like coatings in the presence of calcium ions, making it suitable for fresh-cut fruit preservation (Juikar & Warkar, 2022). Pectin is often leveraged for its adhesive properties and efficacy in retaining moisture in high-moisture foods like fruits (Vartiainen et al., 2015). Cellulose derivatives, including carboxymethyl cellulose (CMC) and hydroxypropyl methylcellulose (HPMC), are employed to create oxygen-resistant coatings and multilayered films with enhanced performance characteristics (Suput et al., 2019). Lipid-based films, including waxes, fatty acids, and mono- or diglycerides, provide superior moisture barrier properties owing to their hydrophobic nature. Common lipids used in this context include beeswax, carnauba wax, and paraffin. Despite their advantages, lipid films often lack mechanical strength and can be brittle, highlighting the necessity for composite films that blend proteins or polysaccharides to enhance flexibility and adhesion (Suput et al., 2019). Furthermore, edible biomass films developed from oyster mushrooms have shown potential as functional packaging materials enriched with bioactives like vitamin D_2 and lovastatin. These films not only offer biodegradable and sustainable alternatives to conventional plastics but also contribute added health benefits through the controlled release of nutraceutical compounds during storage (Aditya et al., 2024b, 2025b). These innovations have facilitated the development of composite or multilayered films that utilize different types of biopolymers to optimize the benefits of each. For example, starch–chitosan composites exhibit significant improvements in mechanical strength and water resistance, while combinations of protein and lipid forms harness both oxygen and moisture barrier functionalities (Juikar & Warkar, 2022).

Despite their considerable benefits, conventional edible films and coatings face several limitations that hinder their large-scale application. A notable challenge is their mechanical weakness and propensity toward brittleness, particularly within polysaccharide- and protein-based films. While the addition of plasticizers can enhance flexibility, it often results in increased water vapor permeability, which compromises the effectiveness of films intended for moisture-sensitive applications (Jost & Stramm, 2015; Alboofetileh et al., 2018). Moreover, many biopolymers exhibit hydrophilic properties, which can deter their efficacy in high-humidity environments and limit their protective capabilities for moisture-sensitive foods (Mostafa et al., 2020). Another significant drawback lies in the limited inherent antimicrobial and antioxidant activities of several natural biopolymers. Excluding chitosan, most natural biopolymers lack bioactivity and necessitate the incorporation of functional agents to impart these properties. The introduction of essential oils, plant extracts, or synthetic preservatives can negatively affect the mechanical or sensory characteristics of the films (Krasniewska et al., 2020). Additionally, uneven migration of these agents within the film matrix or their degradation over time can lead to inconsistent protection and reduced shelf life. Structural inconsistencies associated with variations in molecular weight, moisture content, and extraction methods pose challenges in achieving reproducible and reliable film characteristics.

Consumer acceptance plays a pivotal role in the wider adoption of edible coatings. Changes to the visual aesthetics or the texture of food products may diminish consumer appeal, and concerns about the safety of ingesting certain coatings, especially those containing nanomaterials or chemical additives, present further barriers. While edible films are generally classified as Generally Recognized as Safe (GRAS), providing regulatory clarity and enhancing consumer education are critical for promoting broader acceptance. The absence of unified global food contact regulations complicates the commercial deployment of edible coatings, particularly in the international marketplace. In response to these obstacles, recent research has turned toward nanotechnology as a means of enhancing the functionality of edible films and coatings. By operating at the nanoscale, nanotechnology has the potential to significantly improve the mechanical properties, barrier efficacy, and bioactivity of edible films without compromising their safety or biodegradability. The incorporation of nanoparticles, nanoclays, nanoemulsions, and biopolymer nanocomposites has shown promising results in bolstering these features. Studies have indicated that the integration of silver nanoparticles, zinc oxide, and chitosan nanoparticles into biopolymer films can lead to heightened antimicrobial efficacy and improved mechanical strength and thermal stability (Alboofetileh et al., 2018; Suput et al., 2019; Krasniewska et al., 2020). Furthermore, advancements in active and intelligent packaging systems capable of responding to spoilage or releasing functional compounds in relation to environmental stimuli represent a significant step forward in the evolution of edible films and coatings.

Nanomaterials Used in Edible Films and Coatings

Edible films made from natural materials, such as proteins, polysaccharides, and lipids, combined with nanomaterials, have shown promising results in improving the performance and sustainability of food packaging solutions, mitigating the reliance on synthetic polymers that pose environmental challenges (Hassan et al., 2018; Naskar et al., 2023). This chapter elaborates on the various categories of nanomaterials used in edible films and coatings, including metallic nanoparticles, biopolymer-based nanomaterials, nanoclays, lipid-based structures, and nanoemulsions (Fig. 12.2; Table 12.1).

Metallic and Metal Oxide Nanoparticles

Metallic and metal oxide nanoparticles, particularly silver (Ag), zinc oxide (ZnO), titanium dioxide (TiO_2), and copper oxide (CuO), have been extensively explored for improving the antimicrobial properties of edible films. The incorporation of nanoparticles like Ag has demonstrated significant antimicrobial efficacy against a

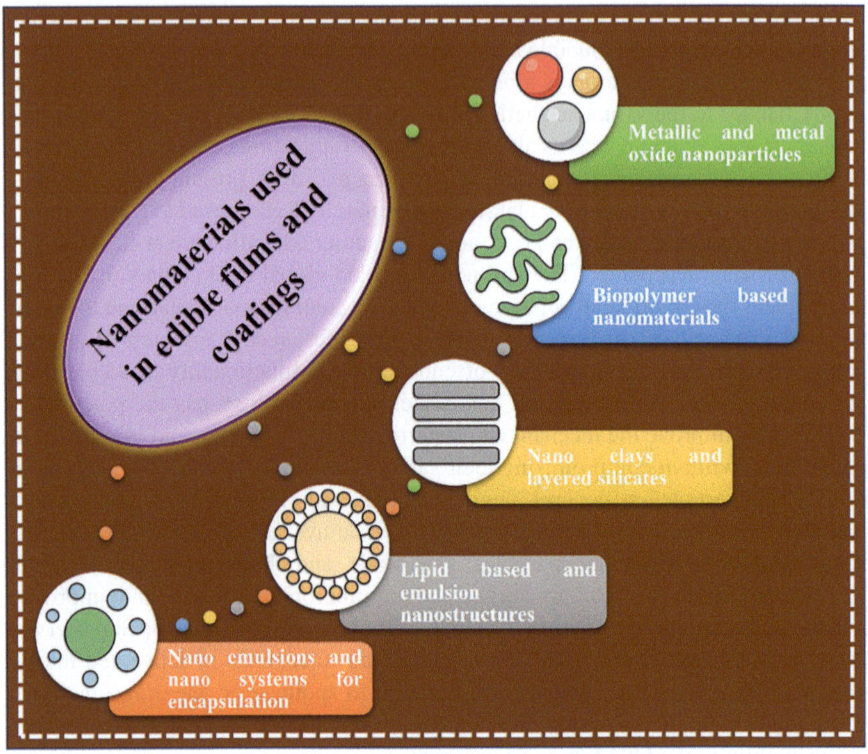

Fig. 12.2 Five different categories of nanomaterials used to improve the structural, barrier, and active functionalities of edible films and coatings

variety of foodborne pathogens, providing enhanced food safety and quality during storage (Ahari et al., 2021). Additionally, ZnO particles are known to improve the mechanical strength and UV barrier properties of films while also having a role in inhibiting microbial growth (Ahari et al., 2021). Studies have shown that the controlled release of these nanoparticles into the food environment can lead to prolonged shelf life without compromising the organoleptic properties of the food (Orsuwan & Sothornvit, 2018; Ahari et al., 2021).

Biopolymer-Based Nanomaterials

Biopolymer-based nanomaterials, such as chitosan nanoparticles and nanocellulose, are gaining traction in the field of edible films due to their biodegradability and non-toxicity. Chitosan, a biopolymer derived from chitin, is known for its excellent film-forming properties and ability to form antibacterial nanocomposites when

Table 12.1 Different types of nanomaterials used in edible films and coatings

Sr. No.	Nanomaterial type	Example	Structure/form	Source	Mode of incorporation	Functionality	References
1.	Silver nanoparticles	Ag nanosuspension	Colloidal dispersion	Synthetic	Film embedding	Antimicrobial activity	Ebrahimi et al. (2019)
2.	Zinc oxide	ZnO nanoparticles	Nanoparticle	Synthetic	Coating	Antimicrobial and UV blocking	Ebrahimi et al. (2019)
3.	Graphene oxide (GO)	GO nanocomposite films	Layered nanosheets	Synthetic	Blending	Mechanical reinforcement, barrier properties	Das et al. (2023)
4.	Chitosan	Chitosan nanoparticles	Polymer matrix	Natural	Coating	Biodegradability, antimicrobial activity	Suvarna et al. (2022)
5.	Cellulose	Cellulose nanofibers	Fiber	Natural	Fiber reinforcement	Mechanical strength and moisture control	Shi et al. (2021)
6.	Starch	Nanocrystalline starch	Hydrogel	Natural	Coating	Barrier properties, moisture retention	Nandiyanto and Dewi (2022)
7.	Carboxymethyl cellulose	CMC nanocomposites	Film	Synthetic	Film casting	Biodegradable, film-forming	Ebrahimi et al. (2019)
8.	Iron oxide (Fe_3O_4)	Fe_3O_4 nanoparticles	Colloidal form	Synthetic	Functional coating	Antimicrobial activity, food preservation	Al-Shabib et al. (2018)
9.	Molybdenum disulfide (MoS_2)	Molybdenum disulfide–layered film	Layered structure	Synthetic	Blending	Barrier properties, mechanical strength	Shi et al. (2021)
10.	Nanoclays	Montmorillonite (MMT)	Layered silicate minerals	Natural	Blending	Gas barrier, moisture control	Garcia (2022)
11.	Nanosilver	Ag incorporated in films	Particulate dispersion	Synthetic	Incorporation	Antibacterial and antifungal activities	Al-Shabib et al. (2018)
12.	Polyvinyl alcohol (PVA)	PVA nanocomposite coatings	Film	Synthetic	Coating	Biodegradability, moisture barrier	Garcia (2022)
13.	Nanoparticles	Curcumin nanoparticles	Nanocrystalline	Natural	Film integration	Antioxidant properties	Mishra et al. (2024)

(continued)

Table 12.1 (continued)

Sr. No.	Nanomaterial type	Example	Structure/form	Source	Mode of incorporation	Functionality	References
14.	Nanocomposite films	Antibacterial cellulose nanocomposite	Film	Synthetic	Film casting	Mechanical strength, antimicrobial activity	Ebrahimi et al. (2019)
15.	Biodegradable polymers	Polylactic acid (PLA) nanocomposites	Biopolymer film	Synthetic	Coating	Biocompatibility, sustainable packaging	Mishra et al. (2024)
16.	Titanium dioxide (TiO_2)	TiO_2 nanomaterials	Nanoparticle	Synthetic	Coating	Antimicrobial activity, stabilizing	Al-Shabib et al. (2018)
17.	Calcium carbonate	Nanocalcite particles	Nano-sized mineral	Natural	Blending	Reinforcement, moisture management	Mishra et al. (2024)
18.	Protein nanoparticles	Nanofibrillar proteins	Protein film	Natural	Emulsion	Edible coating, functionality	Mishra et al. (2024)
19.	Biogenic nanoparticles	Green synthesized gold nanoparticles	Nanoparticle	Natural	Coating	Antimicrobial activity, safe for food contact	Bahrulolum et al. (2021)
20.	Polycaprolactone (PCL)	PCL-based films	Aliphatic polyester	Synthetic	Coating	Biodegradable, mechanical properties	Ebrahimi et al. (2019)

combined with metal nanoparticles (Naskar et al., 2023). Additionally, the use of nanocellulose reinforces the mechanical properties of biopolymer films, enhancing their flexibility and water barrier characteristics (Munoz-Gimena et al., 2023). Recent advancements have shown that incorporating nanochitosan into starch-based matrices can significantly improve the physical and mechanical properties of the edible films, thereby increasing their practical usage in food preservation (Mohammadzadeh-Vazifeh et al., 2020; Fadilla et al., 2023).

Nanoclays and Layered Silicates

Nanoclays, including montmorillonite (MMT) and halloysite, have emerged as effective nanofillers in biopolymer films, where they enhance mechanical strength and barrier properties by creating a more tortuous path for moisture and gases. Their unique layered structure contributes to improved thermal stability and reduces the water vapor transmission rate, making them ideal for food packaging applications (Orsuwan & Sothornvit, 2018; Salmas et al., 2023). Furthermore, studies have demonstrated that the addition of nanoclays can also provide antimicrobial properties to the biocomposite films, promoting food safety (Mohammadzadeh-Vazifeh et al., 2020). The incorporation of various concentrations of nanoclays into edible films has shown synergistic effects on their physical and chemical properties, leading to optimized performance for food coatings (Tozluoglu & Fidan, 2023).

Lipid-Based and Emulsion Nanostructures

Lipid-based nanostructures and emulsions, such as those containing fish oil, have been investigated for their roles in enhancing the moisture barrier properties of edible films while also providing essential fatty acids and desirable flavor (Fadilla et al., 2023). The integration of lipid-based materials into biopolymer films can significantly alter their physical characteristics, such as thickness and solubility, thus influencing their efficacy as food coatings. The studies have also shown that films enriched with lipids demonstrate lower water vapor transmission rates and enhanced mechanical properties, resulting in longer preservation times for perishable food products (Fadilla et al., 2023; Hassan et al., 2018).

Nanoemulsions and Nanosystems for Encapsulation

Nanoemulsions and nanosystems are employed for the encapsulation of bioactive compounds and essential oils within edible films, offering controlled release and enhanced functionality. By using nanoemulsion technology, it is possible to stabilize volatile compounds, protecting them from degradation while providing antimicrobial and antioxidant properties to the food coating (Alorabi et al., 2021; Salmas et al., 2023). Recent studies have highlighted the potential of incorporating these nano-formulations into edible films, resulting in significant improvements in their shelf life and overall quality (Ahari et al., 2021; Shahrbabaki et al., 2025). The versatility of nanoemulsions also enables the incorporation of a variety of natural bioactive compounds, enhancing the nutritional and sensory qualities of the final food products (Shahrbabaki et al., 2025). The integration of various nanomaterials into edible films and coatings presents a viable solution for enhancing food preservation and safety. Latest research and findings in these areas continue to evolve, exploring novel combinations and applications that aim to address both consumer demands for sustainable packaging solutions and the challenges posed by food spoilage and waste.

Effects of Nanomaterials on Functional Properties

The integration of nanomaterials into various systems has been shown to enhance functional properties significantly, with a broad range of applications across diverse fields such as materials science, medicine, and food technology. Recent literature highlights how nanoparticles improve mechanical strength, barrier efficacy, antimicrobial action, and optical properties while also addressing their environmental impact (Table 12.2). This part of the chapter provides the information from numerous studies that collectively illustrate the transformative role of nanomaterials in modulating these functional properties, leading to the development of innovative solutions tailored to specific needs in various industrial applications.

Mechanical Properties (Tensile Strength, Elongation, and Flexibility)

Mechanical properties are crucial in determining the performance of materials across applications. The incorporation of nanomaterials, such as carbon nanotubes and nanoclay, into polymer matrices has been shown to enhance tensile strength and elongation. For example, a study revealed that blending thermoplastic starch with sugar palm nanocellulose led to improved tensile properties, demonstrating enhanced elongation and flexibility (Nazrin et al., 2020). Similarly, another study involving two-dimensional (2D) materials highlighted their contributions to the mechanical enhancement of composite materials, indicating significant improvements in toughness and tensile strength compared to traditional materials (Ji et al., 2020). Moreover, integrating nanomaterials into biocomposite matrices can lead to

Table 12.2 Effects of nanomaterials used in edible films and coatings on mechanical, barrier, antioxidant, and antimicrobial properties

Sr. No.	Edible film/ coating type	Active substance incorporated	Commodity targeted	Mechanical property	Barrier property	Antioxidant property	Antimicrobial property	References
1.	Chitosan-based film	Cellulose nanocrystals and beta-cyclodextrin	Fresh produce	Enhanced tensile strength	Improved moisture barrier	Yes	Strong antibacterial activity	Bizymis et al. (2022)
2.	Gelatin film	Oregano essential oil	Fish (mullet)	Enhanced flexibility	Reduced oxygen permeability	Yes	Strong antimicrobial effectiveness	Ma et al. (2023)
3.	Whey protein isolate film	Microbial transglutaminase	Dairy products	High tensile strength	Excellent barrier against moisture	Yes	Strong antimicrobial activity	Kouravand et al. (2021)
4.	Chitosan-based nanocomposite	Alpha-tocopherol and zein	Edible films	Enhanced mechanical properties	Effective barrier	High antioxidant capacity	Inhibition of microbial growth	Zhang et al. (2020)
5.	Hemicellulose/ chitosan film	Cellulose nanofibers	Fresh produce	Improved mechanical strength	Enhanced moisture barrier	Minimal antioxidant activity	Effective against certain bacteria	Liu et al. (2023)
6.	Protein-based films	Antioxidants and essential oils	Bakery products	Improved elongation at break	Strong barrier properties	High antioxidant effectiveness	Antimicrobial activities	Shah et al. (2023)
7.	Biopolymer-based composite	Metal oxide nanostructures	Multiple commodities	Mechanical stability	High barrier to gases	Minimal effect	Antimicrobial activity	Galstyan et al. (2018)
8.	Chitosan composite	Silver nanoparticles	Prepared foods	Enhanced flexibility	Good moisture barrier	Moderate antioxidant activity	Significant antibacterial properties	Ranjha et al. (2022)
9.	Edible coating (corn starch)	Antioxidants	Fruits and vegetables	Improved mechanical integrity	Effective moisture retention	High antioxidant levels	Antimicrobial effectiveness	Nile et al. (2020)

(continued)

Table 12.2 (continued)

Sr. No.	Edible film/ coating type	Active substance incorporated	Commodity targeted	Mechanical property	Barrier property	Antioxidant property	Antimicrobial property	References
10.	Casein film	Nanoemulsions	Dairy products	Notable tensile strength	Strong barrier	Moderate antioxidant activity	Effective against pathogens	Nile et al. (2020)
11.	Composite film	Calcium nanoparticles	Meat products	High tensile strength	Good oxygen barrier	High antioxidant property	Strong antimicrobial properties	Nile et al. (2020)
12.	Protein–fiber composite	Plant extracts (natural bioactives)	General food	Enhanced strength and elasticity	Reduced moisture absorption	Strong antioxidant activity	Effective against bacteria	Shah et al. (2023)

a synergistic effect, whereby mechanical reinforcement is achieved without compromising flexibility, which is essential in applications like bio-based plastics (Roy et al., 2024).

Barrier Properties (Water Vapor, Oxygen, and Light Transmission)

Barrier properties are vital in packaging and protective coatings, particularly concerning moisture, oxygen, and light transmission. Nanomaterials can markedly enhance these barrier characteristics. Recent studies have demonstrated that incorporating nanoclays or metal organic frameworks (MOFs) into polymer matrices results in significantly lower moisture permeability and improved oxygen barrier properties (Lu et al., 2021; Roy et al., 2024). The capability of nanomaterials to create a dense, tortuous path that hinders molecular transport underlies these improvements. For instance, the addition of silica nanoparticles in composite films has notably enhanced their resistance to water vapor transmission, thereby extending the shelf life of food products (Gomez-Gomez et al., 2023). This improvement is attributed to the nanoscale structures, which provide a more effective barrier due to their high surface area and aspect ratio.

Antimicrobial and Antioxidant Properties

The antimicrobial and antioxidant properties imparted by nanomaterials are critical for applications in food packaging and biomedical devices. The unique properties of metal nanoparticles, such as silver and copper, have been extensively studied for their antimicrobial efficacy, demonstrating a broad spectrum of activity against bacteria, viruses, and fungi (Zhou & McClements, 2022; Tammaro, 2023). Additionally, incorporating nanomaterials into food packaging films has been shown to exhibit significant antioxidant properties, thus preventing oxidative spoilage and extending the shelf life of perishable products (Zhou & McClements, 2022). Specifically, studies indicate that plasticized composite films with biogenic nanoparticles not only display antimicrobial activity but also retain antioxidant ability, effectively enhancing the overall quality of packaged food (Roy et al., 2024).

Optical and Sensory Attributes

Nanomaterials also impact the optical and sensory properties of materials, which is particularly relevant in consumer goods and electronics. The incorporation of metal nanoparticles into polymers can lead to significant changes in optical characteristics, such as enhanced light absorption and scattering due to localized surface plasmon resonance effects (Khan et al., 2021). This property is exploited in various applications, from sensors to coatings, designed for enhanced visibility and aesthetic appeal. For instance, studies have shown that organic–inorganic hybrid materials exhibit improved photonic properties, which can be tailored for specific light transmission applications (Chen et al., 2023). Sensing technologies have also benefited from nanomaterials due to their high surface-to-volume ratio, which amplifies signal transduction efficiency in biosensors (Lee et al., 2018).

Biodegradability and Environmental Impact

Despite the numerous advantages offered by nanomaterials, their environmental implications are a growing concern. Studies indicate that nanomaterials' biodegradability can be highly variable, depending on their chemical composition and structural properties. Biodegradable polymer nanocomposites show promise in reducing environmental impact. The blending of polylactic acid with natural nanomaterials has been shown to yield materials that degrade under composting conditions while maintaining their functional integrity during use (Roy et al., 2024). However, the potential toxicity of certain nanomaterials remains a critical area of investigation, as improper disposal may lead to adverse environmental effects, necessitating thorough assessments of their life cycle impacts and regulatory frameworks (Jeevanandam et al., 2018; Primozic et al., 2021). Understanding these dynamics is essential for developing environmentally sustainable nanotechnology applications that do not compromise ecological integrity. Therefore, the functional enhancement achieved through the incorporation of nanomaterials is profound, influencing mechanical, barrier, antimicrobial, optical, and environmental properties significantly.

Safety, Toxicity, and Regulatory Aspects of Nanoparticles

The rapid growth and integration of nanotechnology into various domains necessitate an assessment of the safety, toxicity, and regulatory aspects associated with nanoparticle applications (Table 12.3). The unique properties of nanoparticles, such as their size and surface chemistry, often contribute to their biological activity but also present potential hazards for human health and the environment. A number of

Table 12.3 Regulatory status of nanomaterials in food packaging (global overview)

Sr. No.	Country/region	Regulatory body	Nanomaterials approved	Labeling requirement	Guidelines/year	References
1.	USA	Food and Drug Administration (FDA)	Nanoclays, graphene oxide	Mandatory	FDA GRAS, Nanotech Guidance (2022)	Galstyan et al. (2018), Shafiq et al. (2020), Adeyeye (2019)
2.	EU	European Food Safety Authority (EFSA)	Silver nanoparticles, nanozeolite	Not mandatory	EFSA opinion on nanomaterials (2018)	Shafiq et al. (2020), Saikanth et al. (2023)
3.	Canada	Health Canada	SiO_2 nanoparticles	Optional	Guidelines for food additives (2021)	Saikanth et al. (2023), Adeyeye (2019)
4.	Australia	Food Standards Australia New Zealand (FSANZ)	Nano–encapsulated antimicrobial substances	Not mandatory	FSANZ guidance (2020)	Shafiq et al. (2020), Adeyeye (2019)
5.	Japan	Food Safety Commission	Titanium dioxide nanoparticles	Not mandatory	New food additive regulations (2019)	Saikanth et al. (2023)
6.	South Korea	Korea Food and Drug Administration (KFDA)	Gold and silver nanoparticles	Mandatory	KFDA food safety guidelines (2022)	Galstyan et al. (2018)
7.	China	National Health Commission	Graphene nanomaterials	Mandatory	Standards for food safety (2023)	Das et al. (2023), Adeyeye (2019)
8.	Brazil	Agencia Nacional de Vigilancia Sanitaria (ANVISA)	Biodegradable polymer nanocomposites	Optional	Regulatory framework for Genetically Modified Organisms (GMOs) (2022)	Saikanth et al. (2023)
9.	UK	Food Standards Agency (FSA)	Carbon nanotubes	Not mandatory	FSA nanotechnology guidelines (2023)	Shafiq et al. (2020), Adeyeye (2019)
10.	New Zealand	FSANZ	Nanostructured lipid carriers	Optional	Guidelines for novel foods (2021)	Shafiq et al. (2020), Adeyeye (2019)

(continued)

Table 12.3 (continued)

Sr. No.	Country/region	Regulatory body	Nanomaterials approved	Labeling requirement	Guidelines/year	References
11.	India	Food Safety and Standards Authority of India (FSSAI)	Silver and zinc oxide nanoparticles	Mandatory	FSSAI draft guidelines (2020)	Saikanth et al. (2023), Shafiq et al. (2020)
12.	Germany	Bundesinstitut fur Risikobewertung (BfR)	Polymeric nanocomposites	Not mandatory	BfR recommendations (2021)	Shafiq et al. (2020), Adeyeye (2019)
13.	France	Agence Nationale de Securite Sanitaire de L'alimentation, de L'environment et du travail (ANSES)	Iron and zinc oxide nanoparticles	Not mandatory	ANSES report on food safety (2019)	Shafiq et al. (2020), Suvarna et al. (2022)
14.	Italy	Italian Ministry of Health	Bio-nanocomposite films	Mandatory	National guidelines (2022)	Saikanth et al. (2023)
15.	Russia	Rospotrebnadzor	Active nanomaterials	Not mandatory	National standards (2020)	Shafiq et al. (2020)
16.	Mexico	Comision Federal para la Proteccion contra Riesgos Sanitarios (COFEPRIS)	Antimicrobial nanomaterials	Mandatory	Food additive guidelines (2021)	Das et al. (2023), Adeyeye (2019)
17.	Singapore	Singapore Food Agency (SFA)	Nanosilica	Not mandatory	SFA food safety standards (2022)	Shafiq et al. (2020), Adeyeye (2019)
18.	South Africa	Foodstuffs Control Act	Carbon black nanoparticles	Not mandatory	Food safety regulations (2020)	Shafiq et al. (2020), Suvarna et al. (2022)
19.	Thailand	FDA	Nano-sized calcium carbonate	Optional	Thai food additive standards (2023)	Saikanth et al. (2023), Adeyeye (2019)
20.	UAE	Ministry of Climate Change and Environment (MOCCAE)	Organic nanomaterials	Mandatory	Ministry guidelines on food safety (2023)	Shafiq et al. (2020)

studies have consistently highlighted that nanoparticles can exhibit different toxicological profiles than their bulk counterparts, potentially leading to unintended exposure risks in various settings, including consumer products, food packaging, and industrial applications (Ganguly et al., 2018; Ljunggren et al., 2019; Russo et al., 2020). Thorough evaluations that encompass both in vitro and in vivo studies are vital for elucidating the interaction dynamics between nanoparticles and biological systems (Suthar et al., 2023). Moreover, regulatory frameworks are still evolving to address these implications, with a growing emphasis on the need for dynamic oversight that balances innovation with safety (Franz et al., 2020).

Migration of Nanoparticles and Exposure Risks

The migration of nanoparticles from materials into the environment or biological systems raises significant exposure risk concerns. Chen and Zhou (2022) found that titanium dioxide nanoparticles (TiO2 NPs) could migrate through biological membranes, accumulating in various tissues, including the brain, leading to neurotoxic effects. Similarly, Choi et al. (2018) investigated silver nanoparticles' (AgNPs) migration from consumer products into food simulants and highlighted the potential health risks posed by accidental migration into consumables. Furthermore, Tang et al. (2020) acknowledged that nanoparticles embedded in food packaging materials could leach into food during storage or heating, creating safety concerns stemming from potential ingestion. The regulatory approach to managing these risks must adapt to the complexity of nanoparticle behavior in diverse environments, and the current strategies should evolve to include robust testing and monitoring for migration and exposure (Labouta et al., 2019).

In Vitro and In Vivo Toxicological Studies

In vitro and in vivo toxicological studies play a pivotal role in understanding the adverse effects of nanoparticle exposure. Valentini et al. (2018) highlighted the necessity of such studies in elucidating the impact of nanoparticles on health and well-being, particularly regarding neurotoxicological outcomes. They also demonstrated that exposure to TiO_2 NPs could induce inflammatory responses and histopathological changes in kidney and liver tissues. Albalawi et al. (2020) evaluated various nanoparticles, revealing significant toxic effects, ranging from cellular alterations to inflammatory responses, in animal models. Such studies provide critical insights into the mechanisms of nanoparticle toxicity, underscoring the need for a comprehensive understanding of their biological interactions and the pathways that lead to health risks.

Global Regulations and Legal Frameworks

Regulatory systems remain paramount in developing safe nanotechnology applications. The current gaps in global regulations pertaining specifically to nanoparticles require urgent attention to ensure public safety (Sousa et al., 2021). Many existing frameworks primarily address chemical substances rather than the unique challenges posed by engineered nanomaterials. The European Union Registration, Evaluation, Authorisation and Restriction of Chemicals (REACH) regulation is one of the frameworks that attempts to incorporate nanoscale materials. However, further specificity is needed regarding the assessment of risks associated with nanoparticle exposure (Labouta et al., 2019). Studies indicate that regulatory guidelines must facilitate proactive risk assessments that account for the unique behaviors and characteristics of nanoparticles, particularly regarding their potential to evade traditional toxicity evaluations (Choi et al., 2018; Labouta et al., 2019). Collaborative efforts among various national and international regulatory bodies are essential to create unified guidelines that promote safe practices in nanoparticle usage across industries (Ganguly et al., 2018).

Consumer Acceptance and Labeling Challenges

Consumer acceptance of products containing nanoparticles is heavily influenced by perceived risks and benefits, as well as transparency regarding their use. The complexity of nanoparticle characteristics complicates labeling initiatives, as many consumers lack an understanding of nanotechnology and its implications. Clear and comprehensible information regarding the presence and role of nanoparticles in consumer products is crucial for fostering trust and acceptance (Kumari & Kumar, 2021). In food packaging, efforts are made to educate consumers about the safety and functional benefits of nanoparticles while also addressing concerns over potential migration and exposure, which are vital components in building consumer confidence (Choi et al., 2018; Tang et al., 2020). As the landscape of consumer products increasingly integrates nanotechnology, regulatory frameworks must adapt to include guidelines that ensure proper labeling and transparent communication, thus addressing consumer hesitations while promoting the responsible use of nanotechnology (Kumari & Kumar, 2021). Therefore, there is a need for thorough and ongoing assessments of safety, toxicity, and regulatory practices. A closer examination of nanoparticle migration and exposure risks, alongside the results of in vitro and in vivo studies, will enhance our understanding of potential hazards. The development of comprehensive regulatory frameworks, paired with effective consumer communication strategies, can ensure the safe integration of nanoparticles into products. These measures will not only safeguard public health but also foster consumer confidence in nanotechnology's benefits, leading to more widespread acceptance and utilization.

Recent Advances and Emerging Trends in Food Packaging Technologies

The food packaging industry has witnessed extraordinary advancements in recent years, driven by the demands for sustainability, safety, and enhanced food quality. Recent developments in smart and active packaging emphasize the integration of technology to ensure that food products remain safe and retain their nutritional value throughout their life cycle (Table 12.4). The incorporation of natural pigments and anthocyanins into packaging materials has resulted in pH-sensitive films that exhibit colorimetric changes, enabling real-time monitoring of food quality and freshness (Kanha et al., 2022; Shiri, 2025). This emerging trend leverages biocompatibility and environmental sustainability, aligning with contemporary consumer preferences, which are increasingly inclined toward eco-friendly solutions. A review by Wongphan et al. (2023) elucidated the potential of biodegradable active packaging, such as thermoplastic starch (TPS) films containing preservatives, showcasing enhancements in the mechanical and barrier properties essential for long-lasting food preservation. Additionally, the utilization of artificial intelligence (AI) in the design of edible coatings represents a transformative shift toward highly personalized food safety solutions, responding adaptively to various spoilage indicators (Siddiqui et al., 2021). The convergence of these technologies promises a substantial improvement in quality assurance throughout the food supply chain, meeting the escalating demands for food safety due to global population growth and resource limitations (Yousefi et al., 2019).

Smart and Active Packaging Integration

Smart packaging integrates electronic and biochemical sensors to provide dynamic feedback regarding food quality. Poyatos-Racionero et al. (2018) revealed that intelligent packaging could significantly reduce food waste while ensuring consumer safety by changing color or giving alerts when food is spoiled or contaminated. Recent studies highlight the effectiveness of films made with natural ingredients, particularly those that contain health-beneficial compounds from sources like sumac (Shiri, 2025) and red cabbage (Abedi-Firoozjah et al., 2022). These films not only offer barriers to gas and moisture but also serve as indicators of food quality deterioration, thus seamlessly combining functionality and environmental sustainability. Moreover, active packaging systems designed to release preservatives or absorb ethylene gas can extend shelf life by mitigating oxidative reactions in food products (Sohail et al., 2018). Such packaging technologies can enhance the organoleptic properties of food, ensuring better market appeal while addressing food safety concerns in agricultural and distribution practices (Firouz et al., 2021). Mushroom spawn, particularly from oyster mushrooms, serves as a sustainable biofactory for nanomaterial synthesis, utilizing its enzymes and metabolites to reduce metal ions

Table 12.4 Recent innovations and trends in nano–edible films

Sr. No.	Year	Nanomaterial matrix/ polymer	Innovation	Result	Highlights	References
1.	2020	Alginate	Active edible films	Enhanced shelf life	Development of alginate/ pectin films containing *Lactobacillus plantarum* that improved the preservation of fruits and vegetables	Shahrampour et al. (2020)
2.	2021	Alginate	Nanoscale modification	Improved mechanical properties	The incorporation of sulfur nanoparticles into alginate films significantly enhanced their mechanical properties and gas barrier effectiveness	Priyadarshi et al. (2021)
3.	2021	Starch/chitosan	Bio-nanomaterials	Higher stability	Starch/chitosan films modified with graphene oxide exhibited enhanced physical and antimicrobial properties, facilitating food preservation	Krystyjan et al. (2021)
4.	2022	Nano-sized starch particles	Biodegradable films	Enhanced mechanical properties	Starch films with nanoparticles demonstrated lower moisture content and improved thermal properties, making them suitable for food packaging applications	Podgorbunskikh et al. (2022)

(continued)

5.	2023	Ulvan	Smart edible films	Enhanced performance	Ulvan-based films showed responsive properties in different environments, improving mechanical and barrier characteristics relevant for food preservation	Guidara et al. (2020)
6.	2023	Cellulose/ulvan	Antioxidant films	Good barrier properties	Composite films of cellulose and ulvan demonstrated significant antioxidant characteristics and improved UV barrier protection	Gomaa et al. (2022)
7.	2023	Pectin	Probiotic edible films	Improved microbial stability	Development of pectin-based films with *Lactobacillus* enhanced storage life and microbial stability of perishable foods	Hashemi et al. (2023)
8.	2023	Plant-derived bioactives	Functional films	Antimicrobial activity	Films mixed with plant-derived bioactive compounds displayed significant antimicrobial activities, increasing their safety for food applications	Nogueira et al. (2020)

(continued)

Table 12.4 (continued)

Sr. No.	Year	Nanomaterial matrix/ polymer	Innovation	Result	Highlights	References
9.	2023	Alginate/chitosan	Multilayered edible films	Enhanced shelf life	Multilayered edible films combining alginate and chitosan provided robust preservation characteristics against microbial spoilage	Liyanapathiranage et al. (2023)
10.	2023	Starch nanoparticles	Novel biodegradable films	Improved mechanical strength	The utilization of starch nanoparticles led to a decrease in permeability while increasing the overall mechanical integrity of the resulting films	Podgorbunskikh et al. (2022)
11.	2024	Plant-based materials	Environmentally safe films	Eco-friendly packaging	Recent advancements in plant-based materials for edible films make them suitable alternatives to synthetic plastics in food packaging	Olawade et al. (2024)
12.	2024	Chitosan with lipid systems	Bioactive films	Delay in spoilage	Chitosan films enhanced with lipid-based systems provided an effective barrier against spoilage and retained nutritional quality in meat products	Hashemi et al. (2023)

13.	2024	Hydroxypropyl methylcellulose	Antioxidant films	Effective preservation	Developing HPMC-based films integrated with antioxidants resulted in significant improvements in the quality and shelf life of food items	Poovai et al. (2023)
14.	2025	Various polymer blends	Ultrathin smart films	Enhanced mechanical and barrier properties	Blending various biopolymers resulted in the creation of ultrathin films with advanced barrier properties for improved food preservation	Thakur and Raposo (2023)

into stable nanoparticles. This green nanotechnology approach supports the eco-friendly production of functional nanostructures for use in food, agriculture, and biomedical applications (Aditya et al., 2022b, 2023b).

Responsive Films (pH, Temperature, and Spoilage Indicators)

Responsive films that react to environmental stimuli like pH and temperature have emerged as crucial players in food safety technology. The pH-sensitive packaging films based on natural pigments have been developed to change color in response to alterations in the pH of the packaged product, indicating freshness or spoilage (Abedi-Firoozjah et al., 2022; Mohammadian et al., 2020). These innovations not only provide visual cues for consumers but also reduce reliance on harmful synthetic indicator dyes that were traditionally used. In addition, films that respond to temperature variations have garnered interest for their potential to monitor and manage the integrity of temperature-sensitive foods during transport (Mohammadian et al., 2020; Qin et al., 2023). The properties of mushroom-based films and biomass are significantly influenced by the choice of growth media and physiological parameters such as pH, temperature, and incubation time. Variations in these conditions affect the mycelial composition, mechanical strength, and functional attributes of the resulting films, thereby determining their suitability for specific food packaging or biomedical applications (Aditya et al., 2022a, 2023a). The integration of these responsive features aligns with enhanced regulatory measures and consumer expectations for transparency in food sourcing and safety.

Sustainable and Green Synthesis of Nanomaterials

The need for sustainable practices has led to the innovative production of nanomaterials derived from renewable sources. Recent studies highlight biopolymer-based functional films that incorporate copper-based nanoparticles, which not only contribute to improved mechanical properties but also possess antimicrobial qualities essential for food safety (Zhang et al., 2023). The studies also draw attention to the green synthesis pathway that harnesses the natural abilities of plants and microorganisms, leading to environmentally friendly production processes that minimize toxic by-products (Jancikova et al., 2020). This approach aligns with contemporary sustainability goals and consumer demands for less environmentally harmful materials in food contact applications (Zhou et al., 2022). The sustainable synthesis methods also facilitate the development of biodegradable packaging options, positing nanomaterials as a viable alternative to traditional plastics, which contribute significantly to environmental pollution. This shift toward green practices will likely characterize the future of food packaging innovation. The sustainable synthesis of nanomaterials from mushrooms leverages their rich biochemical matrix as a natural reducing and stabilizing agent, eliminating the need for hazardous chemicals. This green approach not only aligns with eco-friendly principles but also yields biofunctional nanostructures with potential applications in food safety, biomedicine, and environmental remediation (Aditya et al., 2024c, 2025c).

Artificial Intelligence (AI) and Nano-informatics in Edible Coating Design

The application of AI in the design of edible coatings has revolutionized conventional packaging practices. By employing machine learning algorithms to analyze large datasets, researchers can optimize the design parameters of coatings for specific applications, leading to improved functionality and performance (Siddiqui et al., 2021). These advancements enable the development of coatings that can adaptively respond to environmental changes in acidity, moisture, and temperature, ensuring optimal food preservation. These developments highlight the potential for intelligent edible coatings that not only enhance food safety but also create new opportunities in waste reduction by introducing entirely edible packaging solutions.

Safety, Toxicity, and Regulatory Aspects

The incorporation of novel materials and technologies into food packaging raises critical questions regarding safety and regulatory compliance. As new materials, such as biopolymer films infused with nanoparticles, become more prevalent, there is a growing urgency to understand their life cycle and interaction with food products (Firouz et al., 2021; Zhang et al., 2023). The potential for toxicological effects from the migration of nanoparticles into food necessitates comprehensive regulatory frameworks to ensure consumer safety. Moreover, packaging materials that include active components must comply with FDA guidelines concerning food contact substances. The shifting regulations emphasize the need for extensive testing to ascertain the safety of newly developed materials (Sohail et al., 2018). Addressing consumer concerns about food safety requires consistent vigilance, monitoring, and transparency about the potential risks associated with modern packaging solutions, ensuring that innovations do not compromise public health (Chung et al., 2020). Therefore, advancements in food packaging technologies hold significant promise for improving food quality, safety, and sustainability; a holistic approach involving rigorous testing and regulatory oversight will be essential for harnessing the potential of these innovations without jeopardizing consumer safety.

Challenges and Future Prospects

The integration of nanomaterials into edible films and coatings holds the transformative potential to enhance barrier properties, mechanical strength, optical clarity, and functional bioactivity. However, it simultaneously introduces a complex web of scientific, technical, and regulatory challenges that must be systematically addressed to unlock large-scale applications. Central to this is the need for comprehensive toxicological evaluation and migration studies, as the nanoscale dimensions and high surface reactivity of these materials may facilitate penetration into food matrices and, by extension, biological systems, raising legitimate concerns regarding bioaccumulation, cytotoxicity, and unintended interactions with the human microbiome. In parallel, ensuring reproducibility and consistency during industrial manufacturing remains a formidable hurdle, as the physicochemical stability and functional performance of nanomaterials can be highly sensitive to the processing conditions, such as temperature, shear, pH, and drying techniques. The challenge of achieving homogeneous dispersion within biopolymer matrices, avoiding nanoparticle aggregation, and maintaining the intended antimicrobial, antioxidant, or active-release functions necessitates sophisticated formulation strategies and process optimization. Moreover, scalability and economic viability are critical constraints. The cost of high-purity nanomaterials, coupled with specialized equipment and process controls required to preserve nanoscale functionality, can impede adoption in price-sensitive food sectors. Regulatory frameworks, currently based largely on conventional additives, must evolve to define clear guidelines for safety assessment, permissible migration limits, and labeling, all while balancing innovation with consumer protection. Consumer perception, shaped by broader societal debates around nanotechnology, further underscores the need for transparent risk–benefit communication and evidence-driven policy. Looking ahead, the future of nanomaterial-enhanced edible films and coatings will likely hinge on interdisciplinary collaboration spanning food science, materials engineering, toxicology, and regulatory science, combined with advances in sustainable synthesis, bio-based nanomaterials, and green manufacturing technologies. This convergence may enable next-generation systems capable not only of passive protection but also of intelligent, responsive functionalities such as real-time spoilage detection, controlled release of nutraceuticals, or dynamic adaptation to environmental conditions, thus redefining the role of edible films and coatings within modern food preservation and packaging. Ultimately, addressing the intertwined challenges of safety, scalability, functionality, and sustainability will be pivotal to translating the remarkable promise of nanotechnology from the laboratory to globally impactful, commercially viable solutions that support both food security and environmental stewardship.

Conclusions

The integration of nanotechnology into edible films and coatings marks a transformative step in the evolution of sustainable food packaging, bridging fundamental scientific insight with applied innovation. Edible films and coatings derived from biopolymers such as proteins, polysaccharides and lipids provide biodegradable and consumer-friendly alternatives to conventional plastic packaging. However, their widespread adoption has been constrained by inherent limitations in mechanical strength, barrier performance and sensitivity to moisture etc. The incorporation of diverse nanomaterials, including metallic and metal oxide nanoparticles, nanoclays, nanocellulose, and advanced lipid-based nanosystems, has redefined these biopolymeric matrices, imparting significant enhancements in tensile strength, flexibility, and optical clarity while markedly reducing permeability to water vapor and gases, key parameters for extending shelf life and preserving food quality. Equally transformative has been the emergence of bioactive functionalities; antimicrobial and antioxidant nanostructures embedded within films now actively contribute to food safety, moving these materials from passive barriers to proactive preservation systems. Beyond these immediate functional gains, recent advances highlight a rapidly expanding frontier where edible films become intelligent interfaces capable of detecting spoilage, responding dynamically to environmental cues, and delivering encapsulated bioactive compounds in controlled release profiles tailored to product needs. Innovations in green synthesis and biodegradable nanomaterials align these technical breakthroughs with broader sustainability and circular economy goals, while computational tools such as AI and nano-informatics accelerate the design and optimization of new formulations. Collectively, these developments underscore a paradigm shift; edible films and coatings are evolving from simple, single-purpose layers into multifunctional, responsive, and environmentally attuned systems that address modern demands for food safety, quality, and sustainability. This convergence of materials science, nanotechnology, and food engineering not only overcomes long-standing limitations of traditional edible films but also positions nanomaterial-enhanced coatings as transformative tools in modern food systems. As research and development continue to refine performance, safety, and scalability, the future promises next-generation edible packaging solutions that actively safeguard food, reduce waste, and contribute meaningfully to global efforts for sustainable and resilient food supply chains.

References

Abedi-Firoozjah, R., Yousefi, S., Heydari, M., Seyedfatehi, F., Jafarzadeh, S., Mohammadi, R., & Garavand, F. (2022). Application of red cabbage anthocyanins as pH-sensitive pigments in smart food packaging and sensors. *Polymers, 14*(8), 1629. https://doi.org/10.3390/polym14081629

Adeyeye, S. (2019). Food packaging and nanotechnology: Safeguarding consumer health and safety. *Nutrition & Food Science, 49*(6), 1164–1179. https://doi.org/10.1108/nfs-01-2019-0020

Adeyeye, S., & Ashaolu, T. (2021). Applications of nano-materials in food packaging: A review. *Journal of Food Process Engineering, 44*(7). https://doi.org/10.1111/jfpe.13708

Aditya, Jarial, R. S., & Jarial, K. (2022a). Laboratory screening of different nutrient media for mycelial growth and cultural characteristics of blue oyster mushroom [*Hypsizygus ulmarius* (Bull.: Fr.) Redhead]. *Indian Journal of Ecology, 49*(03), 826–830. https://doi.org/10.55362/IJE/2022/3602

Aditya, Jarial, R. S., & Jarial, K. (2022b). Evaluation of spawn quality and doses on yield and biological efficiency of blue oyster mushroom [*Hypsizygus ulmarius* (Bull.: Fr.) Redhead]. *Journal of Eco-friendly Agriculture, 17*(02), 393–398. https://doi.org/10.5958/2582-2683.2022.00076.4

Aditya, Jarial, R. S., & Jarial, K. (2023a). Influence of physiological parameters on mycelial growth and cultural characteristics of blue oyster mushroom [*Hypsizygus ulmarius* (Bull.: Fr.) Redhead]. *Indian Journal of Ecology, 50*(01), 151–156. https://doi.org/10.55362/IJE/2023/3869

Aditya, Jarial, R. S., & Jarial, K. (2023b). Evaluation of different grain substrates for spawn production and yield performance of blue oyster mushroom [*Hypsizygus ulmarius* (Bull.: Fr.) Redhead] through bio-conversion of agri/industrial wastes. *Indian Journal of Ecology, 49*(06), 2389–2394. https://doi.org/10.55362/IJE/2022/3838

Aditya, Neeraj, Jarial, R. S., Jarial, K., & Bhatia, J. N. (2024a). Comprehensive review on oyster mushroom species (Agaricomycetes): Morphology, nutrition, cultivation and future aspects. *Heliyon, 10*(5), e26539. https://doi.org/10.1016/j.heliyon.2024.e26539

Aditya, Neeraj, & Bhatia, J. N. (2024b). Escalating vitamin D2 content in fresh sporocarps of elm oyster mushroom, *Hypsizygus ulmarius* (Bull.) Redhead (Agaricomycetes) using ultraviolet irradiation and assessment of its stability, antioxidant capacity, color, and textural properties. *ACS Food Science & Technology, 4*(11), 2739–2751. https://doi.org/10.1021/acsfoodscitech.4c00701

Aditya, Neeraj, Bhatia, J. N., & Yadav, A. N. (2024c). Characterization and yield performance of spawn prepared from *Hypsizygus ulmarius* (Bull.) Redhead and some *Pleurotus* species (Agaricomycetes). *Biocatalysis and Agricultural Biotechnology, 56*, 103047, 1–12. https://doi.org/10.1016/j.bcab.2024.103047

Aditya, Neeraj, Bhatia, J. N., & Yadav, A. N. (2025a). A comprehensive review on multifunctional bioactive properties of elm oyster mushroom *Hypsizygus ulmarius* (Bull.) Redhead (Agaricomycetes): Current research, challenges and future trends. *Heliyon, 11*(2), e41418. https://doi.org/10.1016/j.heliyon.2024.e41418

Aditya, Neeraj, & Bhatia, J. N. (2025b). Amplification and quantification of lovastatin from elm oyster mushroom *Hypsizygus ulmarius* (Bull.) Redhead (Agaricomycetes) through HPLC analysis and determination of its mycochemical constituents. *ACS Food Science & Technology, 5*(1), 368–377. https://doi.org/10.1021/acsfoodscitech.4c00876

Aditya, Neeraj, Bhatia, J. N., Jarial, R. S., & Jarial, K. (2025c). Cultivation technology optimization and identification of secondary metabolites from elm oyster mushroom *Hypsizygus ulmarius* (Bull.) Redhead (Agaricomycetes) through GC-MS metabolomic profiling from India. *Mycologia, 117*(2), 1–27. https://doi.org/10.1080/00275514.2025.2452307

Ahari, H., Anvar, S., Ataee, M., & Naeimabadi, M. (2021). Employing nanosilver, nanocopper, and nanoclays in food packaging production: A systematic review. *Coatings, 11*(5), 509. https://doi.org/10.3390/coatings11050509

Albalawi, A., Alanazi, A., Baharvand, P., Sepahvand, M., & Mahmoudvand, H. (2020). High potency of organic and inorganic nanoparticles to treat cystic echinococcosis: An evidence-based review. *Nanomaterials (Basel), 10*(12), 2538. https://doi.org/10.3390/nano10122538

Alboofetileh, M., Rezaei, M., Hosseini, H., & Abdollahi, M. (2018). Morphological, physico-mechanical, and antimicrobial properties of sodium alginate-montmorillonite nanocomposite films incorporated with marjoram essential oil. *Journal of Food Processing and Preservation, 42*(5), e13596. https://doi.org/10.1111/jfpp.13596

Ali, A., & Ahmed, S. (2018). Recent advances in edible polymer-based hydrogels as a sustainable alternative to conventional polymers. *Journal of Agricultural and Food Chemistry, 66*(27), 6940–6967. https://doi.org/10.1021/acs.jafc.8b0105

Alorabi, A., Hassan, M., Alam, M., Zabin, S., Alsenani, N., & Baghdadi, N. (2021). Natural clay as a low-cost adsorbent for crystal violet dye removal and antimicrobial activity. *Nanomaterials (Basel), 11*(11), 2789. https://doi.org/10.3390/nano11112789

Al-Shabib, N., Husain, F., Ahmed, F., Khan, R., Khan, M., Ansari, F., & Ahmad, I. (2018). Low temperature synthesis of superparamagnetic iron oxide (Fe3O4) nanoparticles and their ROS-mediated inhibition of biofilm formed by food-associated bacteria. *Frontiers in Microbiology, 9*, 1–11. https://doi.org/10.3389/fmicb.2018.02567

Awwaly, K., Widyastuti, E., Apriliyani, M., Thohari, I., Zul'adhar, U., & Rachmawati, A. (2024). The effect of casein-chitosan edible coating with garlic essential oil addition on pH, Haugh unit and egg yolk index fresh chicken egg on different storage. *Bio Web of Conferences, 88*, 00041. https://doi.org/10.1051/bioconf/20248800041

Bahrulolum, H., Nooraei, S., Javanshir, N., Tarrahimofrad, H., Mirbagheri, V., Easton, A., & Ahmadian, G. (2021). Green synthesis of metal nanoparticles using microorganisms and their application in the agrifood sector. *Journal of Nanobiotechnology, 19*(1), 1–15. https://doi.org/10.1186/s12951-021-00834-3

Balan, A., & Kadeppagari, R. (2020). Biopolymers for nano-enabled packaging of foods. In *Handbook of polymer and ceramic nanotechnology* (pp. 839–854). Springer International Publishing. https://doi.org/10.1007/978-3-030-10614-0_57-1

Bizymis, A., Giannou, V., & Tzia, C. (2022). Improved properties of composite edible films based on chitosan by using cellulose nanocrystals and beta-cyclodextrin. *Applied Sciences, 12*(17), 8729. https://doi.org/10.3390/app12178729

Chen, J., & Zhou, C. (2022). Neurotoxicity study with titanium dioxide nanoparticles on rat. *Key Engineering Materials, 905*, 288–293. https://doi.org/10.4028/www.scientific.net/kem.905.288

Chen, X., Wang, K., Yin, J., & Yue, K. (2023). Application of nanomaterials in lithium-sulfur batteries and lithium-air batteries. *Applied and Computational Engineering, 23*(1), 30–38. https://doi.org/10.54254/2755-2721/23/20230605

Cherpinski, A., Torres-Giner, S., Cabedo, L., Méndez, J., & Lagarón, J. (2017). Multilayer structures based on annealed electrospun biopolymer coatings of interest in water and aroma barrier fiber-based food packaging applications. *Journal of Applied Polymer Science, 135*(24), 45501. https://doi.org/10.1002/app.45501

Choi, J., Chae, S., Kim, J., Choi, J., Park, S., Choi, H., & Park, H. (2018). Potential silver nanoparticles migration from commercially available polymeric baby products into food simulants. *Food Additives & Contaminants Part A, 35*(5), 996–1005. https://doi.org/10.1080/19440049.2017.1411611

Chung, E., Kee, D., Chan, J., Tiong, S., Choke, Y., Low, J., & Motwani, H. (2020). Improving food safety and food quality: The case of nestle. *International Journal of Tourism and Hospitality in Asia Pacific, 3*(1), 57–67. https://doi.org/10.32535/ijthap.v3i1.724

Das, M., Panda, R., Ray, L., Sethy, C., Kundu, C., & Tripathy, J. (2023). Fabrication of graphene oxide reinforced bio-nanocomposite films with antibacterial potential. *Macromolecular Symposia, 407*(1), 1–9. https://doi.org/10.1002/masy.202100416

Dewi, S., & Khoerunnisa, Y. (2024). Effect of sago starch (*Metroxylon rumphii*) edible coating on the shelf life of papaya (*Carica papaya* L.). *IOP Conference Series Earth and Environmental Science, 1417*(1), 012028. https://doi.org/10.1088/1755-1315/1417/1/012028

Díaz-Montes, E., & Castro-Muñoz, R. (2021). Edible films and coatings as food-quality preservers: An overview. *Foods, 10*(2), 249. https://doi.org/10.3390/foods10020249

Ebrahimi, Y., Peighambardoust, S., Peighambardoust, S., & Karkaj, S. (2019). Development of antibacterial carboxymethyl cellulose-based nanobiocomposite films containing various metallic nanoparticles for food packaging applications. *Journal of Food Science, 84*(9), 2537–2548. https://doi.org/10.1111/1750-3841.14744

Fadilla, F., Rochima, E., Pratama, R., & Rostini, I. (2023). Physical characteristics of biocomposite edible film based on starch and nanochitosan with the addition of oil and acids: A review. *Asian Journal of Fisheries and Aquatic Research, 22*(1), 1–9. https://doi.org/10.9734/ajfar/2023/v22i1560

Firdaus, M., Rahmana, D., Carolina, D., Firdausi, N., Afiifah, Z., & Sugiarto, B. (2023). Antibacterial edible coating from mandarin orange peel (*Citrus reticulata*) and moringa leaf (*Moringa oleifera*) extract for fish preservation. *Jurnal Riset Kimia, 14*(1), 61–69. https://doi.org/10.25077/jrk.v14i1.563

Firouz, M., Mohi-Alden, K., & Omid, M. (2021). A critical review on intelligent and active packaging in the food industry: Research and development. *Food Research International, 141*, 110113. https://doi.org/10.1016/j.foodres.2021.110113

Franz, R., Bott, J., & Störmer, Á. (2020). Considerations for and guidance to testing and evaluating migration/release of nanoparticles from polymer-based nanocomposites. *Nanomaterials (Basel), 10*(6), 1113. https://doi.org/10.3390/nano10061113

Galstyan, V., Bhandari, M., Sberveglieri, V., Sberveglieri, G., & Comini, E. (2018). Metal oxide nanostructures in food applications: Quality control and packaging. *Chemosensors, 6*(2), 16. https://doi.org/10.3390/chemosensors6020016

Ganguly, P., Breen, A., & Pillai, S. (2018). Toxicity of nanomaterials: Exposure, pathways, assessment, and recent advances. *ACS Biomaterials Science & Engineering, 4*(7), 2237–2275. https://doi.org/10.1021/acsbiomaterials.8b00068

Garcia, C. (2022). Editorial: Nanomaterials in food packaging. *Frontiers in Nutrition, 9*, 1–17. https://doi.org/10.3389/fnut.2022.1083185

Giannakas, A., Salmas, C., Moschovas, D., Karabagias, V., Karabagias, I., Baikousi, M., & Avgeropoulos, A. (2023). Development, characterization and evaluation as food active packaging of low-density-polyethylene-based films incorporated with rich in thymol halloysite nanohybrid for fresh "scaloppini" type pork meat fillets preservation. *Polymers, 15*(2), 282. https://doi.org/10.3390/polym15020282

Gomaa, M., Al-Badaani, A., Hifney, A., & Adam, M. (2022). Utilization of cellulose and ulvan from the green seaweed ulva lactuca in the development of composite edible films with natural antioxidant properties. *Journal of Applied Phycology, 34*(5), 2615–2626. https://doi.org/10.1007/s10811-022-02786-z

Gómez-Gómez, A., Martínez-Ayala, A., Moguel-Concha, D., Borges-Martínez, J., Perea-Flores, M., & Dávila-Ortiz, G. (2023). Relationship of nanomaterials' structure based on their application in the food industry: Physicochemical and techno-functional characteristics. *Applied Sciences, 13*(12), 7167. https://doi.org/10.3390/app13127167

Guidara, M., Yaich, H., Benelhadj, S., Adjouman, Y., Richel, A., Blecker, C., & Garna, H. (2020). Smart ulvan films responsive to stimuli of plasticizer and extraction condition in physico-chemical, optical, barrier and mechanical properties. *International Journal of Biological Macromolecules, 150*, 714–726. https://doi.org/10.1016/j.ijbiomac.2020.02.111

Gurjar, K., Gurjar, P., Gurjar, J., & Choudhary, R. (2024). Assessing the influence of different edible coating methods on the shelf life and biochemical characteristics of tomato. *Journal of Advances in Biology and Biotechnology, 27*(3), 178–185. https://doi.org/10.9734/jabb/2024/v27i3732

Han, J., Ruiz-García, L., Qian, J., & Yang, X. (2018). Food packaging: A comprehensive review and future trends. *Comprehensive Reviews in Food Science and Food Safety, 17*(4), 860–877. https://doi.org/10.1111/1541-4337.12343

Hashemi, S., Kaveh, S., Abedi, E., & Phimolsiripol, Y. (2023). Polysaccharide-based edible films/coatings for the preservation of meat and fish products: Emphasis on incorporation of lipid-based nanosystems loaded with bioactive compounds. *Foods, 12*(17), 3268. https://doi.org/10.3390/foods12173268

Hassan, B., Chatha, S., Hussain, A., Zia, K., & Akhtar, N. (2018). Recent advances on polysaccharides, lipids and protein-based edible films and coatings: A review. *International Journal of Biological Macromolecules, 109*, 1095–1107. https://doi.org/10.1016/j.ijbiomac.2017.11.097

Jančíková, S., Đorđević, Đ., Jamróz, E., Běhalová, H., & Tremlová, B. (2020). Chemical and physical characteristics of edible films, based on κ- and ι-carrageenans with the addition of lapacho tea extract. *Food, 9*(3), 357. https://doi.org/10.3390/foods9030357

Jeevanandam, J., Barhoum, A., Chan, Y., Dufresne, A., & Danquah, M. (2018). Review on nanoparticles and nanostructured materials: History, sources, toxicity and regulations. *Beilstein Journal of Nanotechnology, 9*, 1050–1074. https://doi.org/10.3762/bjnano.9.98

Ji, Z., Zhang, L., Xie, G., Xu, W., Guo, D., Luo, J., & Prakash, B. (2020). Mechanical and tribological properties of nanocomposites incorporated with two-dimensional materials. *Friction, 8*(5), 813–846. https://doi.org/10.1007/s40544-020-0401-4

Jost, V., & Stramm, C. (2015). Influence of plasticizers on the mechanical and barrier properties of cast biopolymer films. *Journal of Applied Polymer Science, 133*(2), 1–5. https://doi.org/10.1002/app.42513

Juikar, S., & Warkar, S. (2022). Biopolymers for packaging applications: An overview. *Packaging Technology and Science, 36*(4), 229–251. https://doi.org/10.1002/pts.2707

Kanha, N., Osiriphun, S., Rakariyatham, K., Klangpetch, W., & Laokuldilok, T. (2022). On-package indicator films based on natural pigments and polysaccharides for monitoring food quality: A review. *Journal of the Science of Food and Agriculture, 102*(15), 6804–6823. https://doi.org/10.1002/jsfa.12076

Khan, M., Imam, M., Irshad, K., Ali, H., Hasan, M., & Islam, S. (2021). Comparative overview of the performance of cementitious and non-cementitious nanomaterials in mortar at normal and elevated temperatures. *Nanomaterials (Basel), 11*(4), 911. https://doi.org/10.3390/nano11040911

Khezerlou, A., Tavassoli, M., Sani, M., Mohammadi, K., Ehsani, A., & McClements, D. (2021). Application of nanotechnology to improve the performance of biodegradable biopolymer-based packaging materials. *Polymers, 13*(24), 4399. https://doi.org/10.3390/polym13244399

Kondle, R., Sharma, K., Singh, G., & Kotiyal, A. (2023). Using nanotechnology for enhancing the shelf life of fruits. In *Food Processing and Packaging Technologies*. IntechOpen. https://doi.org/10.5772/intechopen.108724

Kouravand, F., Jooyandeh, H., Barzegar, H., & Hojjati, M. (2021). Mechanical, barrier and structural properties of whey protein isolate-based films treated by microbial transglutaminase. *Journal of Microbiology Biotechnology and Food Sciences, 9*(5), 960–964. https://doi.org/10.15414/jmbfs.2020.9.5.960-964

Kraśniewska, K., Galus, S., & Gniewosz, M. (2020). Biopolymers-based materials containing silver nanoparticles as active packaging for food applications–A review. *International Journal of Molecular Sciences, 21*(3), 698. https://doi.org/10.3390/ijms21030698

Krystyjan, M., Khachatryan, G., Grabacka, M., Krzan, M., Witczak, M., Grzyb, J., & Woszczak, L. (2021). Physicochemical, bacteriostatic and biological properties of starch/chitosan polymer composites modified by graphene oxide, designed as new bionanomaterials. *Polymers, 13*(14), 2327. https://doi.org/10.3390/polym13142327

Kumari, M., & Kumar, A. (2021). Estimating combined health risks of nanomaterials and antibiotics from natural water: A proposed framework. *Environmental Science and Pollution Research, 29*(10), 13845–13856. https://doi.org/10.1007/s11356-021-16795-x

Labouta, H., Asgarian, N., Rinker, K., & Cramb, D. (2019). Meta-analysis of nanoparticle cytotoxicity via data mining the literature. *ACS Nano, 13*(2), 1583–1594. https://doi.org/10.1021/acsnano.8b07562

Lee, J., Adegoke, O., & Park, E. (2018). High-performance biosensing systems based on various nanomaterials as signal transducers. *Biotechnology Journal, 14*(1). https://doi.org/10.1002/biot.201800249

Lestari, D., & Wilujeng, İ. (2022). Edible coating starch making for preserving fruit as an alternative for junior high school science practicum on additives substance material. *Jurnal Pendidikan Sains (JPS), 10*(1), 1. https://doi.org/10.26714/jps.10.1.2022.1-8

Liu, G., Shi, K., & Sun, H. (2023). Research progress in hemicellulose-based nanocomposite film as food packaging. *Polymers, 15*(4), 979. https://doi.org/10.3390/polym15040979

Liyanapathiranage, A., Dassanayake, R., Gamage, A., Karri, R., Manamperi, A., Evon, P., & Merah, O. (2023). Recent developments in edible films and coatings for fruits and vegetables. *Coatings, 13*(7), 1177. https://doi.org/10.3390/coatings13071177

Ljunggren, S., Karlsson, H., Ståhlbom, B., Krapi, B., Fornander, L., Karlsson, L., & Graff, P. (2019). Biomonitoring of metal exposure during additive manufacturing (3D printing). *Safety & Health at Work, 10*(4), 518–526. https://doi.org/10.1016/j.shaw.2019.07.006

Lü, C., Xiao, H., & Chen, X. (2021). Mofs/pva hybrid membranes with enhanced mechanical and ion-conductive properties. *E-Polymers, 21*(1), 160–165. https://doi.org/10.1515/epoly-2021-0010

Ma, Y., Chen, S., Liu, P., He, Y., Chen, F., Cai, Y., & Yang, X. (2023). Gelatin improves the performance of oregano essential oil nanoparticle composite films—Application to the preservation of mullet. *Foods, 12*(13), 2542. https://doi.org/10.3390/foods12132542

Maurya, A. K., Aditya, John, V., Pant, H., Sharma, S. P., El-Refaey, D. Z., Sami, R., & Ahmed, N. (2023). Unveiling oil seed cakes ability to suppress fusarium wilt (*Fusarium udum* Butler) in Pigeonpea (*Cajanus cajan* L. Millsp.). *Journal of Biobased Materials and Bioenergy, 17*(6), 790–796. https://doi.org/10.1166/jbmb.2023.2319

Mishra, M., Mittal, A., Negi, G., Singh, P., Kaur, S., Prasanna, J., & Sharma, P. (2024). Biodegradable nanomaterials for sustainable food packaging applications. *E3s Web of Conferences, 547*, 01014. https://doi.org/10.1051/e3sconf/202454701014

Misir, J., Brishti, F., & Hoque, M. (2014). Aloe vera gel as a novel edible coating for fresh fruits: A review. *American Journal of Food Science and Technology, 2*(3), 93–97. https://doi.org/10.12691/ajfst-2-3-3

Mohammadian, E., Sani, M., & Jafari, S. (2020). Smart monitoring of gas/temperature changes within food packaging based on natural colorants. *Comprehensive Reviews in Food Science and Food Safety, 19*(6), 2885–2931. https://doi.org/10.1111/1541-4337.12635

Mohammadzadeh-Vazifeh, M., Hosseini, S., Mohammadi, A., Jahanfar, M., & Maleki, H. (2020). Investigation of the antimicrobial properties of nanoclay and chitosan-based nanocomposite on the microbial characteristics of Gouda cheese. *Iranian Journal of Microbiology, 12*(2), 121. https://doi.org/10.18502/ijm.v12i2.2617

Mostafa, Y., Alrumman, S., Alamri, S., Otaif, K., Mostafa, M., & Alfaify, A. (2020). Bioplastic (poly-3-hydroxybutyrate) production by the marine bacterium *Pseudodonghicola xiamenensis* through date syrup valorization and structural assessment of the biopolymer. *Scientific Reports, 10*(1), 1–10. https://doi.org/10.1038/s41598-020-65858-5

Muižniece-Brasava, S., & Ķirse, A. (2018). Attitudes of Latvian consumers to traditional and eco-friendly food packaging materials: Comparison of 2007 and 2017. In proceedings of 17th International Scientific Conference Engineering for Rural Development, Jelgava, 23–25.05.2018. https://doi.org/10.22616/ERDev2018.17.N560

Muñoz-Gimena, P., Oliver-Cuenca, V., Peponi, L., & López, D. (2023). A review on reinforcements and additives in starch-based composites for food packaging. *Polymers, 15*(13), 2972. https://doi.org/10.3390/polym15132972

Nandiyanto, A., & Dewi, A. (2022). Bibliometric mapping analysis of nanocrystalline starch in food packaging application using VOS viewer. *Advance Sustainable Science Engineering and Technology, 4*(2), 0220202. https://doi.org/10.26877/asset.v4i2.13332

Naskar, A., Sanyal, I., Nahar, N., Ghosh, D., & Chakraborty, S. (2023). Bio nanocomposites films applied as active and smart food packaging: A review. *Polymer Engineering and Science, 63*(9), 2675–2699. https://doi.org/10.1002/pen.26415

Nazrin, A., Sapuan, S., & Zuhri, M. (2020). Mechanical, physical and thermal properties of sugar palm nanocellulose reinforced thermoplastic starch (TPS)/poly (lactic acid) (PLA) blend bionanocomposites. *Polymers, 12*(10), 2216. https://doi.org/10.3390/polym12102216

Nikhanj, P. (2023). Edible coatings: A novel technique for shelf-life extension of fresh-cut vegetables. *World Journal of Food Nutrition, 3*(1), 1–2. https://doi.org/10.54026/wjfn/1015

Nikita, & Agnihotri, S. (2025). Nanoemulsions as advanced delivery systems of bioactive compounds for sustainable food preservation applications. *Biocatalysis and Agricultural Biotechnology*, 103702. https://doi.org/10.1016/j.bcab.2025.103702

Nile, S., Venkidasamy, B., Selvaraj, D., Nile, A., Xiao, J., & Kai, G. (2020). Nanotechnologies in food science: Applications, recent trends, and future perspectives. *Nano-Micro Letters, 12*(1), 1–6. https://doi.org/10.1007/s40820-020-0383-9

Nogueira, G., Oliveira, R., Velasco, J., & Fakhouri, F. (2020). Methods of incorporating plant-derived bioactive compounds into films made with agro-based polymers for application as food packaging: A brief review. *Polymers, 12*(11), 2518. https://doi.org/10.3390/polym12112518

Olawade, D., Wada, O., & Ige, A. (2024). Advances and recent trends in plant-based materials and edible films: A mini-review. *Frontiers in Chemistry, 12*, 1–11. https://doi.org/10.3389/fchem.2024.1441650

Omerović, N., Djisalov, M., Živojević, K., Mladenović, M., Vunduk, J., Milenković, I., & Vidić, J. (2021). Antimicrobial nanoparticles and biodegradable polymer composites for active food packaging applications. *Comprehensive Reviews in Food Science and Food Safety, 20*(3), 2428–2454. https://doi.org/10.1111/1541-4337.12727

Orsuwan, A., & Sothornvit, R. (2018). Reinforcement of banana flour biocomposite film with beeswax and montmorillonite and effects on water barrier and physical properties. *International Journal of Food Science and Technology, 53*(12), 2642–2649. https://doi.org/10.1111/ijfs.13859

Pant, H., Maurya, A. K., Aditya, Singh, M. K., John, V., Mehra, M., Sami, R., & Helal, M. (2023). Ecofriendly management of root knot nematode (*Meloidogyne incognita*) in okra (*Abelmoschus esculentus* (L.) Moench). *Journal of Biobased Materials and Bioenergy, 17*(3), 311–317. https://doi.org/10.1166/jbmb.2023.2286

Podgorbunskikh, E., Сапожников, A., Kuskov, T., Gurova, D., Kopylova, A., Бычков, A., & Lomovsky, O. (2022). Comprehensive enzymatic conversion of starch for the food industry. *Polymers, 14*(21), 4575. https://doi.org/10.3390/polym14214575

Poovai, P., Kumaran, N., Iyengar, A., Kalpana, P., & Ramasubramaniyan, M. (2023). A study on coating of hydroxypropyl methylcellulose incorporated with a nano-emulsion of piper betel leaf essential oil to enhance shelf-life and improve postharvest quality of tomato (*Solanum lycopersicum* L.). *Journal of Applied and Natural Science, 15*(1), 252–261. https://doi.org/10.31018/jans.v15i1.4005

Poyatos-Racionero, E., Ros-Lis, J., Vivancos, J., & Martínez-Máñez, R. (2018). Recent advances on intelligent packaging as tools to reduce food waste. *Journal of Cleaner Production, 172*, 3398–3409. https://doi.org/10.1016/j.jclepro.2017.11.075

Primožič, M., Knez, Ž., & Leitgeb, M. (2021). Bio-nanotechnology in food science—Food packaging. *Nanomaterials (Basel), 11*(2), 292. https://doi.org/10.3390/nano11020292

Priyadarshi, R., Kim, H., & Rhim, J. (2021). Effect of sulfur nanoparticles on properties of alginate-based films for active food packaging applications. *Food Hydrocolloids, 110*, 106155. https://doi.org/10.1016/j.foodhyd.2020.106155

Qin, S., Sun, H., Wan, X., Wu, Y., Lin, X., Kan, H., & Liu, C. (2023). Carboxymethylcellulose reinforced starch films and rapid detection of spoiled beverages. *Frontiers in Bioengineering and Biotechnology, 10*, 1–10. https://doi.org/10.3389/fbioe.2022.1099118

Ranjha, M., Shafique, B., Rehman, A., Mehmood, A., Ali, A., Zahra, S., & Siddiqui, S. (2022). Biocompatible nanomaterials in food science, technology, and nutrient drug delivery: Recent developments and applications. *Frontiers in Nutrition, 8*, 1–10. https://doi.org/10.3389/fnut.2021.778155

Reichert, C., Bugnicourt, E., Coltelli, M., Cinelli, P., Lazzeri, A., Canesi, I., & Schmid, M. (2020). Bio-based packaging: Materials, modifications, industrial applications and sustainability. *Polymers, 12*(7), 1558. https://doi.org/10.3390/polym12071558

Roy, S., Singh, K., & Kumar, A. (2024). Biodegradable polymer nanocomposites for active food packaging. *E3S Web of Conferences, 509*, 01013. https://doi.org/10.1051/e3sconf/202450901013

Russo, D., Yan, X., Shende, S., Huang, H., Yan, B., & Zhu, H. (2020). Virtual molecular projections and convolutional neural networks for the end-to-end modelling of nanoparticle activities and properties. *Analytical Chemistry, 92*(20), 13971–13979. https://doi.org/10.1021/acs.analchem.0c02878

Saikanth, D., Gawande, V., Chaithanya, G., Mishra, S., Swami, S., Singh, B., & MP, H. (2023). Development and current state of nanotechnology in the agricultural and food sectors: A review. *International Journal of Environment and Climate Change, 13*(9), 1602–1612. https://doi.org/10.9734/ijecc/2023/v13i92392

Salmas, C., Giannakas, A., Karabagias, V., Moschovas, D., Karabagias, I., Gioti, C., & Proestos, C. (2023). Development and evaluation of a novel-thymol natural-zeolite/low-density-polyethylene active packaging film: Applications for pork fillets preservation. *Antioxidants, 12*(2), 523. https://doi.org/10.3390/antiox12020523

Shafiq, M., Anjum, S., Hano, C., Anjum, I., & Abbasi, B. (2020). An overview of the applications of nanomaterials and nanodevices in the food industry. *Foods, 9*(2), 148. https://doi.org/10.3390/foods9020148

Shah, Y., Bhatia, S., Al-Harrasi, A., Afzaal, M., Saeed, F., Anwer, M., & Faisal, Z. (2023). Mechanical properties of protein-based food packaging materials. *Polymers, 15*(7), 1724. https://doi.org/10.3390/polym15071724

Shahrampour, D., Khomeiri, M., Razavi, S., & Kashiri, M. (2020). Development and characterization of alginate/pectin edible films containing *Lactobacillus plantarum* KMC 45. *LWT - Food Science and Technology, 118*, 108758. https://doi.org/10.1016/j.lwt.2019.108758

Shahrbabaki, M., Shahab Lavasani, A. N., Nateghi, L., & Eshaghi, M. R. (2025). Designing an optimized nanocomposite film comprising hydroxypropyl methylcellulose, nanoclay, and dill essential oil to extend the shelf life of eggs. *International Journal of Food Science, 1*, 6633069. https://doi.org/10.1155/ijfo/6633069

Shi, S., Mandal, P., & Chen, T. (2021). Mechanical properties and tribological behavior of MoS2-enhanced cellulose-based biocomposites for food packaging. *Polymers, 13*(11), 1838. https://doi.org/10.3390/polym13111838

Shiri, A. (2025). Eco-friendly and smart electrospun food packaging films based on polyvinyl alcohol and sumac extract: Physicochemical, mechanical, antibacterial, and antioxidant properties. *Food Science & Nutrition, 13*(4), 1–7. https://doi.org/10.1002/fsn3.70190

Siddiqui, J., Taheri, M., Alam, A., & Deen, M. (2021). Nanomaterials in smart packaging applications: A review. *Small, 18*(1), 1–8. https://doi.org/10.1002/smll.202101171

Sohail, M., Sun, D., & Zhu, Z. (2018). Recent developments in intelligent packaging for enhancing food quality and safety. *Critical Reviews in Food Science and Nutrition, 58*(15), 2650–2662. https://doi.org/10.1080/10408398.2018.1449731

Sousa, M., Arezes, P., & Silva, F. (2021). Occupational exposure to ultrafine particles in metal additive manufacturing: A qualitative and quantitative risk assessment. *International Journal of Environmental Research and Public Health, 18*(18), 9788. https://doi.org/10.3390/ijerph18189788

Šuput, D., Popović, S., Hromiš, N., Bulut, S., & Lazić, V. (2019). Biopolymer films properties change affected by essential oils addition. *Journal on Processing and Energy in Agriculture, 23*(2), 61–65. https://doi.org/10.5937/jpea1902061s

Suthar, J., Rakesh, B., Vaidya, A., & Ravindran, S. (2023). Comprehensive analysis of titanium oxide nanoparticle size and surface properties on neuronal PC-12 cells: Unraveling cytotoxicity, dopaminergic gene expression, and acetylcholinesterase inhibition. *Journal of Xenobiotics, 13*(4), 662–684. https://doi.org/10.3390/jox13040043

Suvarna, V., Nair, A., Mallya, R., Khan, T., & Omri, A. (2022). Antimicrobial nanomaterials for food packaging. *Antibiotics, 11*(6), 729. https://doi.org/10.3390/antibiotics11060729

Tammaro, L. (2023). Antibacterial nanostructured coatings. *Nanomaterials (Basel), 13*(23), 2982. https://doi.org/10.3390/nano13232982

Tang, Z., Fan, F., Fan, C., Jiang, K., & Qin, Y. (2020). The performance changes and migration behavior of PLA/nano-TiO2 composite film by high-pressure treatment in ethanol solution. *Polymers, 12*(2), 471. https://doi.org/10.3390/polym12020471

Tavassoli, M., Sani, M., Khezerlou, A., Ehsani, A., Khaniki, G., & McClements, D. (2022). Smart biopolymer-based nanocomposite materials containing pH-sensing colorimetric indicators for food freshness monitoring. *Molecules, 27*(10), 3168. https://doi.org/10.3390/molecules27103168

Thakur, N., & Raposo, A. (2023). Development and application of fruit and vegetable-based green films with natural bio-actives in meat and dairy products: A review. *Journal of the Science of Food and Agriculture, 103*(13), 6167–6179. https://doi.org/10.1002/jsfa.12686

Tozluoğlu, A., & Fidan, H. (2023). Effect of size press coating of cationic starch/ nanofibrillated cellulose on physical and mechanical properties of recycled papersheets. *BioResources, 18*(3), 5993–6012. https://doi.org/10.15376/biores.18.3.5993-6012

Valentini, X., Deneufbourg, P., Paci, P., Rugira, P., Laurent, S., Frau, A., & Nonclercq, D. (2018). Morphological alterations induced by the exposure to TiO2 nanoparticles in primary cortical neuron cultures and in the brain of rats. *Toxicology Reports, 5*, 878–889. https://doi.org/10.1016/j.toxrep.2018.08.006

Vartiainen, J., Shen, Y., Kaljunen, T., Malm, T., Vähä-Nissi, M., Putkonen, M., & Harlin, A. (2015). Bio-based multilayer barrier films by extrusion, dispersion coating and atomic layer deposition. *Journal of Applied Polymer Science, 133*(2), 1–8. https://doi.org/10.1002/app.42260

Wongphan, P., Harnkarnsujarit, N., Estacho, P., Aguareles, E., & Puerta, C. (2023). Blown-extrusion of biodegradable active packaging with food preservatives: Packaging properties, food quality and safety. *Jornadas De Jóvenes Investigadores Del, 3*(11), 1–5. https://doi.org/10.26754/jjii3a.20238995

Xing, Y., Li, W., Wang, Q., Li, X., Xu, Q., Guo, X., & Yang, H. (2019). Antimicrobial nanoparticles incorporated in edible coatings and films for the preservation of fruits and vegetables. *Molecules, 24*(9), 1695. https://doi.org/10.3390/molecules24091695

Yousefi, H., Su, H., Imani, S., Alkhaldi, K., Filipe, C., & Didar, T. (2019). Intelligent food packaging: A review of smart sensing technologies for monitoring food quality. *ACS Sensors, 4*(4), 808–821. https://doi.org/10.1021/acssensors.9b00440

Zhang, L., Liu, Z., Sun, Y., Wang, X., & Li, L. (2020). Effect of α-tocopherol antioxidant on rheological and physicochemical properties of chitosan/zein edible films. *LWT - Food Science and Technology, 118*, 108799. https://doi.org/10.1016/j.lwt.2019.108799

Zhang, W., Roy, S., & Rhim, J. (2023). Copper-based nanoparticles for biopolymer-based functional films in food packaging applications. *Comprehensive Reviews in Food Science and Food Safety, 22*(3), 1933–1952. https://doi.org/10.1111/1541-4337.13136

Zhou, H., & McClements, D. (2022). Recent advances in the gastrointestinal fate of organic and inorganic nanoparticles in foods. *Nanomaterials (Basel), 12*(7), 1099. https://doi.org/10.3390/nano12071099

Zhou, S., Wang, L., Wang, S., Ma, H., Yan, B., Guo, K., & Xiang, J. (2022). Sustainable indicators based on furfural-derived colorant-doped biobased polyurethane to improve food safety. *ACS Sustainable Chemistry & Engineering, 10*(26), 8624–8630. https://doi.org/10.1021/acssuschemeng.2c02357

Chapter 13
Effects of Nutrients on Edible Coatings and Films

Waseem Ahmed, Abdul Qayyum, Aniqa Anam, Mehnaz Mehboob, Rafia Azmat, and Muhammad Liaquat

Abstract The increasing demand for sustainable food packaging has led to significant advancements in edible coatings and films, which offer an eco-friendly alternative to conventional plastic packaging. Edible coatings and films are biodegradable layers derived from natural materials, which serve as protective barriers against moisture, oxygen, and microbial contamination, effectively extending the shelf life of various food products. Beyond their role in preservation, edible coatings can be enriched with essential nutrients, bioactive compounds, and functional ingredients to enhance the nutritional value and health benefits of foods. This chapter explores the composition, mechanisms of nutrient incorporation, and structural properties of edible coatings; factors affecting the coating; and case studies emphasizing their role in food quality enhancement and nutrient stability. Additionally, it examines the impact of nutrients on the effectiveness of these coatings in food preservation, as well as consumer acceptance factors. Through a combination of scientific research and practical applications, this chapter provides valuable insights into the potential of nutrient-enriched edible coatings in addressing global nutritional challenges while promoting sustainability in food packaging.

W. Ahmed (✉)
Department of Horticulture, The University of Haripur, Haripur, Khyber Pakhtunkhwa, Pakistan

A. Qayyum
Department of Agronomy, The University of Haripur, Haripur, Khyber Pakhtunkhwa, Pakistan

A. Anam · M. Mehboob
Department of Botany, The University of Haripur, Haripur, Khyber Pakhtunkhwa, Pakistan

R. Azmat
Department of Chemistry, University of Karachi, Karachi, Pakistan

M. Liaquat
Department of Food Science and Technology, The University of Haripur, Haripur, Khyber Pakhtunkhwa, Pakistan

N. Kumar et al. (eds.), *Trends in Edible Packaging: A sustainable approaches for postharvest management of fruits and vegetables*,
https://doi.org/10.1007/978-3-032-13199-7_13

Keywords Biodegradable packaging · Sustainability · Biopolymers · Encapsulation · Migration

Introduction

In recent years, the demand for sustainable food packaging solutions has grown significantly due to increasing environmental concerns and consumer preferences for natural products. One promising innovation in this field is the development of edible coatings and films. These coatings are applied to food surfaces and are made from natural biodegradable materials, providing an eco-friendly alternative to traditional plastic packaging. Beyond their sustainability benefits, edible coatings also play a vital role in preserving the quality and nutritional value of foods (Perez-Vazquez et al., 2023).

The application of edible coatings and films extends far beyond their environmental advantages. These materials serve as a barrier against moisture, oxygen, and microbial contamination, which are key factors contributing to the spoilage of fresh foods (Silue & Fawole, 2024). By providing protection against these elements, edible coatings can significantly extend the shelf life of a variety of food products, including fruits, vegetables, meats, and baked goods (Fig. 13.1). Furthermore, the composition of these coatings allows for the incorporation of functional ingredients such as nutrients, bioactive compounds, and preservatives, which can enhance the nutritional profile and health benefits of the food (Quirós-Sauceda et al., 2014).

The potential of edible coatings and films as tools for both food preservation and quality enhancement is vast. These coatings not only enhance the texture, taste, and visual appeal of the food but also provide valuable nutritional benefits. By incorporating vitamins, antioxidants, essential fatty acids, and dietary fibers, edible coatings can transform food into functional products that support overall health and well-being. As research in the field of edible coatings continues to evolve, the incorporation of nutrients into these coatings has become a key area of focus. Nutrient-enriched edible coatings can help address global health concerns related to nutrient deficiencies, particularly in populations with limited access to fresh, nutrient-rich foods. Additionally, these coatings can act as a carrier for functional ingredients that promote specific health outcomes, such as improved digestion, strengthened immunity, or heart health (Díaz-Montes & Castro-Muñoz, 2021). This chapter provides an in-depth exploration of the role of nutrients in edible coatings and films, focusing on their composition, mechanisms of incorporation, nutritional benefits, and practical applications. It also examines the structural properties of edible coatings, as well as the effects of different nutrients on their performance and stability. By integrating scientific research with practical insights, this chapter aims to contribute to the growing body of knowledge on edible coatings and their potential in food packaging and preservation (Gupta et al., 2024).

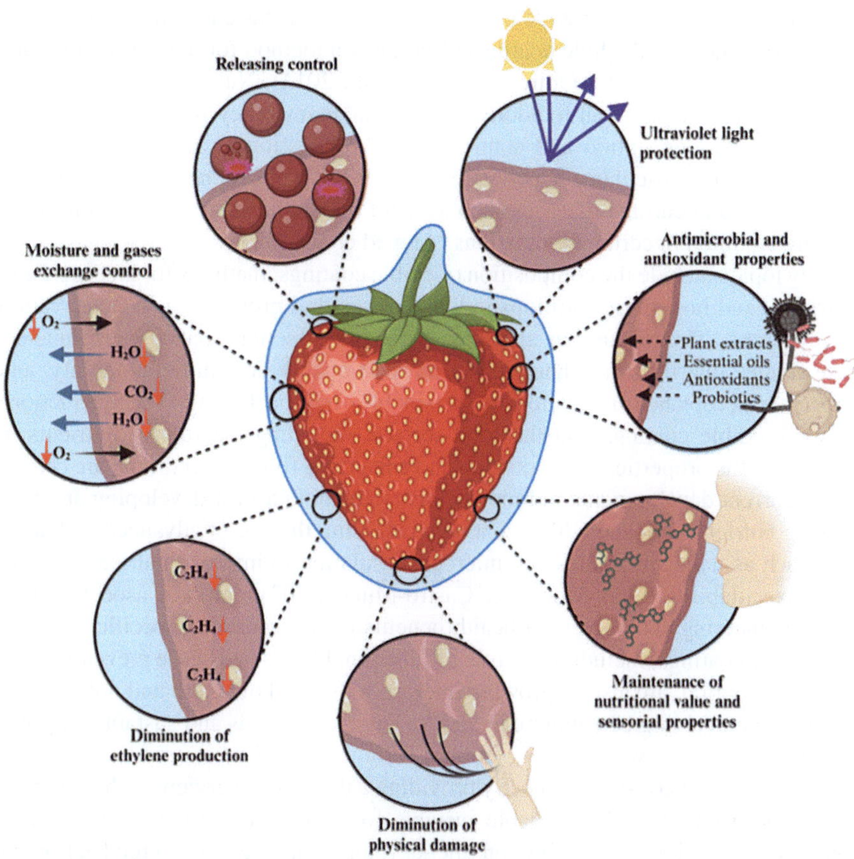

Fig. 13.1 Mechanism and functions of edible coatings

Background and Scope

The need for sustainable food packaging solutions has never been more critical. Traditional plastic packaging, which is widely used in the food industry, poses significant environmental challenges due to its nonbiodegradable nature and the growing accumulation of plastic waste. In response, there has been a shift toward the development of biodegradable and edible alternatives that can reduce waste and promote environmental sustainability. Edible coatings and films offer a promising solution by providing a functional, biodegradable option for packaging that can be safely consumed or decomposed without causing harm to the environment's health (Díaz-Montes & Castro-Muñoz, 2021).

In addition to sustainability, consumers are increasingly concerned with the health benefits of the products they consume. The growing trend toward healthier eating habits has led to the incorporation of functional ingredients in food products.

Functional foods, which provide health benefits beyond basic nutrition, are becoming more popular, and edible coatings offer a novel method for delivering nutrients directly to food products (Quirós-Sauceda et al., 2014). Nutrient-enriched edible coatings have the potential to address these challenges: They provide a more sustainable packaging option and contribute to the nutritional value of foods. This development is particularly beneficial in addressing global concerns about dietary deficiencies and chronic diseases such as heart disease, diabetes, and malnutrition that are common in certain populations (Ahmed et al., 2023).

Key topics include the composition of edible coatings, methods for incorporating nutrients, and how these additions influence structural properties and preservation effectiveness. The discussion also covers factors affecting coating performance, case studies that illustrate their impact on food quality and nutrient stability, and consumer acceptance in relation to sustainable food packaging. The interactions between edible coatings and the food surface, including the role of nutrients in modifying the properties of the coating and enhancing food preservation, are central to the current developments in food chemistry. Techniques for developing and processing nutrient-enriched edible coatings, including the use of advanced technologies such as nanocomposites and microencapsulation to improve nutrient stability and bioavailability (Díaz-Montes & Castro-Muñoz, 2021), are key aspects within food engineering. The potential health benefits of incorporating specific nutrients into edible coatings, including the role of functional foods in disease prevention and health promotion, are also of growing interest in the field of health and nutrition. In parallel, the evolving consumer demand for functional foods and sustainable packaging solutions, as well as how edible coatings fit into these trends, highlights important market considerations. By providing a thorough overview of these topics, the chapter aims to offer valuable insights for researchers, food scientists, and industry professionals working on the development and commercialization of nutrient-enriched edible coatings and films (Perez-Vazquez et al., 2023).

Edible Coatings and Films

Edible coatings and films are materials made from natural or synthetic substances that are designed to be applied to the surface of food products. These coatings serve various functions, such as preserving freshness, extending shelf life, and enhancing the nutritional profile of the food. The composition of edible coatings and films is critical in determining their effectiveness as well as their compatibility with specific food types. The materials used in edible coatings and films are diverse, ranging from polysaccharides and proteins to lipids. This section explores the key components of edible coatings and films, their characteristics, and the factors that influence their selection and performance (Silue & Fawole, 2024).

Composition of Edible Coatings and Films

Edible film or coating is a material with a thickness of less than 0.3 mm, formed from a combination of biopolymers and various additives dispersed in an aqueous medium. While some researchers use the terms "edible film" and "edible coating" interchangeably, a distinction exists between the two: *Edible coatings are directly formed on the surface of food products*, whereas *edible films are preformed sheets that are later applied to the product*. Despite this difference, both create rigid matrices with similar structural characteristics, contributing to food preservation and quality enhancement (Díaz-Montes & Castro-Muñoz, 2021).

The incorporation of nutrients into edible coatings plays a crucial role in enhancing their structural and functional properties. The impact of different nutrients on the mechanical strength, moisture barrier effectiveness, and overall functionality of coatings will be examined, including how these nutrients interact with the base materials and influence the coating performance. Various methods for incorporating nutrients, such as encapsulation, microencapsulation, and other advanced techniques, can be utilized to improve nutrient stability and bioavailability (Ahmed et al., 2023). Additionally, a focus will be placed on the types of nutrients suitable for integration, including vitamins, minerals, antioxidants, and essential fatty acids, and how these additions contribute to the nutritional value of the coated food product (Priya et al., 2023). Beyond structural and nutritional contributions, the inclusion of nutrients can enhance the effectiveness of edible coatings in food preservation, particularly in extending shelf life, preventing spoilage, and reducing food waste. Finally, consumer acceptance and the market potential of nutrient-enriched edible coatings are also key considerations, as sensory attributes, consumer perceptions, and health-related benefits strongly influence purchasing decisions and the broader adoption of these innovative food technologies (Karnwal et al., 2025).

Polysaccharides

Polysaccharides are among the most widely used materials in the formulation of edible coatings and films. These natural polymers are derived from plant sources, and they offer several desirable properties such as biodegradability, non-toxicity, and water solubility. Polysaccharides form the backbone of many edible coatings due to their ability to form films and their role in improving moisture retention and texture (Gupta et al., 2024).

Chitosan

Chitosan biopolymer, derived from chitin (found in the exoskeletons of crustaceans), is one of the most used polysaccharides for edible coatings. It has excellent antimicrobial properties, which help prevent microbial growth on food surfaces, extending shelf life. Chitosan is also biodegradable and nontoxic and can easily be modified to improve its film-forming properties. It has been used in coatings for fruits, vegetables, meat, and seafood (Wijesekara & Xu, 2024).

Alginate

Alginate is a natural polysaccharide extracted from brown seaweed and is known for its ability to form gels in the presence of divalent cations like calcium. This property is beneficial for creating edible coatings that are both flexible and durable. Alginate-based coatings help in moisture retention, and they can act as a barrier to oxygen, thus preserving the freshness of the food. Alginate is often used for coating fruits and vegetables to prevent dehydration and maintain their texture (Ahmed et al., 2023).

Carrageenan

Carrageenan is another polysaccharide used in edible coatings. It is often used in combination with other polysaccharides to form films with good mechanical properties. Carrageenan coatings are commonly used in dairy products, meat, and processed food applications, as they can improve the texture, water retention, and overall shelf life.

Gellan and Agar

Both gellan and agar are polysaccharides derived from bacteria and seaweed. These materials are capable of forming strong and elastic films that can provide a protective barrier against moisture loss. They are often used in applications that require food preservation under refrigeration, such as in the case of fruits, vegetables, and ready-to-eat meals (Hashemi et al., 2023).

Proteins

Proteins are another class of materials used in edible coatings, known for their excellent film-forming properties, water resistance, and ability to form strong bonds with other ingredients. Protein-based coatings are often used for their functional properties in enhancing the texture, acting as a barrier against moisture loss, and providing nutritional benefits (Gupta et al., 2024).

Gelatin

Gelatin is a natural protein derived from collagen in animal tissues, such as the skin and bones. It is widely used for creating edible coatings and films due to its excellent film-forming ability. Gelatin-based coatings are often used in confectionery, meat, and dairy applications, where they help improve the texture and prevent moisture loss (Ahmed et al., 2023).

Whey Protein

Whey protein is derived from milk during the cheese-making process. It is highly soluble in water, and it forms flexible and transparent films when dried. Whey protein–based coatings are often used for fruit and vegetable preservation, as they provide a barrier to oxygen and moisture while improving the shelf life of the product (Shah et al., 2023).

Soy Protein

Soy protein is a plant-based protein that is commonly used in edible coatings for its excellent film-forming properties and nutritional value. Soy protein coatings have been used in a variety of applications, including coatings for fruits, vegetables, and bakery products. These coatings provide protection from dehydration and spoilage while enhancing the nutritional content of the food (Shah et al., 2023).

Corn Zein

Zein is a protein extracted from corn, and it is particularly useful for creating edible coatings that provide a barrier to moisture and oxygen. It is a hydrophobic protein that can produce flexible, durable coatings for food products like nuts, fruits, and confectionery items (Shah et al., 2023).

Lipids

Lipid-based coatings offer excellent moisture barriers and can extend the shelf life of perishable foods by reducing water loss and preventing oxidation (Gupta et al., 2024).

Beeswax

Beeswax is a natural lipid derived from the hive of honeybees. It has been used for centuries as a food coating due to its ability to provide a protective barrier against moisture loss and spoilage. Beeswax coatings are often applied to fruits, vegetables, and cheeses to preserve freshness. Its film-forming ability can help reduce the risk of oxidation and microbial growth (Ahmed et al., 2023).

Carnauba Wax

Carnauba wax, derived from the leaves of the Brazilian carnauba palm tree, is commonly used in the food industry as a glazing agent and edible coating. It has excellent moisture-resistance properties and is often used in conjunction with other ingredients to enhance the texture, appearance, and shelf life of coated food products, particularly confectionery and fruits (Gupta et al., 2024).

Shellac

Shellac is a resin obtained from the secretion of the lac insect, and it is used as an edible coating for fruits and vegetables. Shellac is valued for its glossy finish and ability to act as a barrier against moisture loss. It is commonly applied to citrus fruits and other perishable items to improve their appearance and extend their shelf life (Ahmed et al., 2022).

Plant-Based Oils

Plant oils, such as soybean oil, sunflower oil, and olive oil, are often used in edible coatings to provide a hydrophobic barrier to moisture loss and oxidation. These oils can be incorporated into polysaccharide- or protein-based coatings to enhance their properties. Additionally, the use of plant oils in coatings allows for the incorporation of essential fatty acids, which contribute to the nutritional value of the coated food.

Composite Materials

Composite edible coatings are made by combining two or more types of materials (polysaccharides, proteins, lipids, etc.) to take advantage of their complementary properties. These composite systems can provide improved mechanical strength, flexibility, and barrier properties, making them ideal for a wide range of applications (Qasim et al., 2024).

Polysaccharide–Protein Composite

Combining polysaccharides and proteins can enhance the film-forming properties of edible coatings. The interaction between polysaccharides, such as chitosan or alginate, and proteins, such as whey or soy protein, creates coatings with improved mechanical strength and moisture resistance. These composite coatings are commonly used for coating fruits, vegetables, and meats, as they offer enhanced preservation benefits, such as extended shelf life and reduced microbial growth (Babalola et al., 2022).

Polysaccharide–Lipid Composite

Lipids can be blended with polysaccharides to form composite coatings that have both hydrophobic and hydrophilic properties. These coatings are ideal for use in food products that require both moisture retention and oxidation protection. For example, the combination of alginate and beeswax can be used to create coatings for fruits and vegetables, where the polysaccharide provides flexibility and the lipid provides a moisture barrier (Gallegos-Cedillo et al., 2024).

Protein–Lipid Composite

Protein–lipid composite coatings combine the film-forming properties of proteins with the moisture resistance of lipids. This combination is especially beneficial for meat products and snacks, where the coating needs to offer both flexibility and a strong barrier to moisture and oxygen. For instance, whey protein combined with beeswax or carnauba wax is a common choice for coatings that need to balance texture, moisture retention, and shelf life extension (Dhumal & Sarkar, 2018).

Nanocomposites

Incorporating nanomaterials, such as nanoparticles or nanofibers, into edible coatings has gained attention due to the ability of these nanomaterials to enhance the mechanical, barrier, and antimicrobial properties of the coatings. Nanocellulose, for example, is a biodegradable material derived from plant cellulose that can be used in combination with polysaccharides and proteins to improve the strength and flexibility of edible films and coatings (Matloob et al., 2023).

Plasticizers and Additives

Plasticizers are compounds added to edible coatings to improve flexibility, reduce brittleness, and facilitate film formation. They help create a more pliable film that can better adhere to food surfaces. Common plasticizers include glycerol, sorbitol, and polyethylene glycol (Ahmed, 2021).

Glycerol

Glycerol is one of the most widely used plasticizers in edible coatings. It is a hygroscopic substance that helps reduce the water vapor permeability of films and improves their flexibility. Glycerol is often used in combination with polysaccharides like chitosan and alginate to produce coatings that are flexible and effective in moisture retention (Gallegos-Cedillo et al., 2024).

Sorbitol

Sorbitol, another common plasticizer, is used to improve the flexibility and durability of protein-based films. It is less hygroscopic than glycerol and is often used in edible coatings for dried or preserved foods (Ahmed et al., 2022).

Additives

Additives such as antimicrobial agents, antioxidants, and colorants are often incorporated into edible coatings to improve their functionality and appearance. Antimicrobial agents, such as essential oils (e.g., thyme or oregano oil), can be added to enhance the coating's ability to prevent microbial growth. Antioxidants like vitamin E or green tea extract are also used to improve the stability of the coating and protect sensitive nutrients from oxidation (Shukla et al., 2019).

Nutritional Characteristics of Edible Coatings

Edible coatings are designed not only to serve as barriers that protect food from environmental factors but also to enhance the nutritional profile of the food product. These coatings can incorporate a range of nutrients, bioactive compounds, and functional ingredients that contribute to the overall health benefits of the food. The nutritional characteristics of edible coatings depend on the composition of the coating materials and the nutrients incorporated within them. This section discusses the various types of nutrients, their roles in enhancing the nutritional value of edible coatings, and the potential benefits they offer to consumers (Miteluţ et al., 2021).

Types of Nutrients Incorporated into Edible Coatings

Edible coatings can be enriched with a variety of nutrients, ranging from vitamins and minerals to antioxidants, essential fatty acids, and fiber. The selection of these nutrients depends on the desired functional properties and the specific benefits they offer to the food product and the consumer (Ahmed et al., 2023).

Vitamins

Vitamins are essential for numerous bodily functions, including immune support, skin health, and energy metabolism. Common vitamins incorporated into edible coatings include the following: (i) Vitamin C (ascorbic acid), which is known for its antioxidant properties, is often added to edible coatings to prevent oxidation and preserve the color, texture, and nutritional value of fruits and vegetables. It can also enhance the coating's ability to scavenge free radicals and improve the shelf life of fresh produce. (ii) Vitamin A (retinol) plays a vital role in maintaining vision, immune function, and skin health. Retinol is commonly added to edible coatings to enhance the nutritional value of food products, particularly in dairy- and fruit-based coatings (Miteluţ et al., 2021). (iii) Vitamin E (tocopherols) is often used to improve the oxidative stability of edible coatings, particularly in lipid-based formulations. It helps protect sensitive nutrients and extends the shelf life of coated food items. (iv) Vitamin B, including B1 (thiamine), B2 (riboflavin), B6 (pyridoxine), B12, and folate, is crucial for energy metabolism and neurological health. It can be incorporated into coatings to provide additional nutritional benefits, particularly for consumers with dietary restrictions.

Minerals

Minerals play essential roles in maintaining health, supporting bone strength, and facilitating enzymatic reactions in the body. Edible coatings can be enriched with the following minerals: (i) Calcium is essential for bone health and muscle function. It is often added to edible coatings made from polysaccharides like alginate and chitosan to fortify the coating with a bioavailable form of calcium. (ii) Magnesium supports hundreds of biochemical reactions in the body and is involved in muscle function, nerve transmission, and bone health. Magnesium can be included in edible coatings to improve the nutritional profile of the food and provide additional health benefits. (iii) Iron is vital for the transportation of oxygen in the blood and the formation of red blood cells. Iron-fortified edible coatings can improve the nutritional content of foods, especially in populations prone to iron deficiency.

Antioxidants

Antioxidants help protect food from oxidative degradation and can contribute to the nutritional profile of edible coatings. They are often derived from plant-based compounds and can provide numerous health benefits, including anti-inflammatory and antiaging effects. Common antioxidants in edible coatings include the following: (i) Polyphenols: These plant compounds, found in fruits, vegetables, and herbs, are known for their antioxidant and anti-inflammatory properties. Polyphenols, such as flavonoids, resveratrol, and tannins, can be incorporated into edible coatings to provide a protective layer while offering health benefits such as reducing the risk of chronic diseases. (ii) Carotenoids: Carotenoids, such as beta-carotene, lutein, and zeaxanthin, are antioxidants that contribute to the pigmentation of fruits and vegetables. They offer health benefits such as promoting eye health and reducing the risk of certain cancers. (iii) Essential Fatty Acids: Essential fatty acids, including omega-3 (e.g., alpha (α)–linolenic acid) and omega-6 (e.g., linoleic acid), are crucial for maintaining cardiovascular health, cognitive function, and reducing inflammation. These fatty acids are often incorporated into edible coatings made from lipid-based materials to provide additional nutritional value. Flaxseed oil, fish oil, and chia oil are examples of nutrient-rich oils used for coating enrichment. (iv) Dietary Fiber: Fiber is an essential part of a healthy diet, contributing to digestive health, blood sugar regulation, and cholesterol management. Dietary fiber can be incorporated into edible coatings by using fiber-rich ingredients like psyllium, inulin, pectin, or cellulose. These fibers can improve the texture of the coating and provide additional health benefits, particularly in enhancing the satiety and nutritional content of foods.

Nutrient Retention in Edible Coatings

One of the key challenges in incorporating nutrients into edible coatings is ensuring their stability and retention throughout the shelf life of the coated product. Nutrients can be sensitive to various environmental factors such as temperature, light, oxygen, and moisture, leading to degradation over time. Strategies must be employed to protect these nutrients and maintain their bioavailability (Teixeira-Costa & Andrade, 2021).

Encapsulation and Microencapsulation

One of the most effective methods for preserving the stability of sensitive nutrients is encapsulation. Nutrients can be enclosed within a protective carrier system (such as a lipid or polymer) that shields them from external factors that could cause degradation. Microencapsulation is particularly useful for protecting vitamins, polyphenols, and essential fatty acids from oxidation and thermal degradation. These techniques enable the controlled release of nutrients, preserving their functionality over time and improving the nutrient absorption by consumers (Xu et al., 2024).

Use of Antioxidants

Antioxidants are often included in edible coatings to prevent the oxidative degradation of sensitive nutrients. Vitamins like vitamin C and vitamin E are commonly used as antioxidants within coatings to protect other vitamins, fatty acids, and bioactive compounds from oxidative damage. Additionally, plant-based antioxidants such as polyphenols and carotenoids help stabilize the overall nutrient content of the coating (Gupta et al., 2024).

Storage Conditions and Shelf Life

The storage conditions of nutrient-enriched coated products play a significant role in the retention of nutrients. Factors such as temperature, humidity, and light exposure can affect the stability of the nutrients within the coating. Coatings that provide better protection from moisture and oxygen, such as those made from hydrophobic or lipid-based materials, can help retain the nutrient content for longer periods. Furthermore, refrigerated or controlled atmosphere storage can prolong the shelf life of nutrient-enriched foods (Gupta et al., 2024).

Impact of Food Processing on Nutrient Stability

Food processing methods such as baking, frying, or steaming can impact the stability of both the edible coating and the nutrients incorporated into it. High temperatures and prolonged heat exposure may cause the degradation of heat-sensitive nutrients like vitamins C and B, as well as polyunsaturated fatty acids. Therefore, it is crucial to choose processing methods and coating formulations that minimize the nutrient loss (Ahmed et al., 2022).

Bioavailability of Nutrients in Edible Coatings

Bioavailability refers to the extent to which the body can absorb and utilize the nutrients provided by the edible coatings. The bioavailability of nutrients depends on several factors, including the type of nutrient, its chemical form, the food matrix, and the method of incorporation (Hu et al., 2023).

Interaction with Food Matrix

The bioavailability of nutrients in edible coatings can be influenced by their interaction with the food matrix to which they are applied. For example, fat-soluble nutrients, such as vitamin E or omega-3 fatty acids, may be better absorbed when the coating is applied to foods that naturally contain fats (such as meats or dairy products). On the other hand, water-soluble vitamins may be more bioavailable when applied to fruits and vegetables, which have a high water content (Joglar et al., 2023).

Controlled Release

Nutrients that are incorporated into edible coatings through encapsulation or other controlled release mechanisms are often more bioavailable to consumers. By providing a steady and sustained release of nutrients over time, these techniques ensure that the nutrients remain available for absorption during digestion. Encapsulation also protects sensitive nutrients from the digestive environment, enhancing their bioavailability.

Bioavailability of Nutrients

The chemical form of the nutrient plays a significant role in determining its bioavailability. For example, the bioavailability of calcium can vary depending on whether it is in its inorganic (e.g., calcium carbonate) or organic (e.g., calcium citrate) form. Similarly, the bioavailability of carotenoids like beta-carotene can be influenced by the presence of fat in the food, as carotenoids are fat-soluble (Miteluț et al., 2021).

Digestibility and Absorption of Nutrients

The digestion and absorption of nutrients from edible coatings are influenced by the properties of the coating itself. A coating that is too thick or resistant to digestion may limit the absorption of nutrients by the body. Coatings with optimal thickness, permeability, and digestibility can enhance nutrient absorption by allowing the nutrients to be released gradually during the digestive process (Bremenkamp & Sousa Gallagher, 2025).

Consumer Perception and Acceptance

Consumer perception and acceptance of nutrient-enriched edible coatings are important factors that influence the success of these coatings in the marketplace. The sensory characteristics of the coating, including taste, texture, color, and appearance, can impact how consumers perceive the product and whether they view it as beneficial or desirable (De Oliveira et al., 2023). Nutrient incorporation can sometimes alter the taste and texture of the edible coating. For example, the addition of minerals like calcium or magnesium may lead to a slightly bitter taste, while certain antioxidants and essential oils can impart strong flavors or aromas. These sensory changes must be carefully balanced to ensure that the coating remains palatable and does not detract from the overall eating experience (Bremenkamp & Sousa Gallagher, 2025). Alongside taste and texture, health perception plays a significant role in consumer acceptance. Consumers are increasingly concerned with the nutritional value of the foods they consume. As such, the addition of functional ingredients, such as vitamins, antioxidants, and fiber, can positively influence consumer perception of the product. However, transparency in labeling and clear communication about the health benefits of the coating are essential for consumer acceptance (Alba & Gabriel, 2018). The visual appeal of the coating also contributes to how it is perceived. The color and appearance of the edible coating can affect consumer acceptance, as natural pigments from carotenoids (e.g., beta-carotene) or anthocyanins can enhance the visual appeal of the coated food while also contributing to its nutritional value. However, the inclusion of certain nutrients may cause discoloration or undesirable changes in the appearance of the food, which may negatively affect consumer preference (Wan et al., 2007).

The incorporation of nutrients into edible coatings is a critical process for enhancing the nutritional value and preservation capabilities of food products. The ability to effectively integrate nutrients into the coating matrix depends on several factors, including the nature of the nutrients, the coating material, and the method of incorporation. Various mechanisms are used to incorporate nutrients into edible coatings, including techniques like encapsulation and microencapsulation, which help improve nutrient stability and bioavailability. However, challenges such as nutrient degradation, interaction with coating components, and maintaining functionality during storage must be addressed. Strategies to optimize nutrient retention and bioavailability continue to evolve, focusing on improving the compatibility between nutrients and coating materials while ensuring that the final product remains effective and appealing (Duncan, 2011).

Methods for the Incorporation of Nutrients into Edible Coatings/Films

Physical Methods

Physical incorporation involves the simple mixing or blending of nutrients into the edible coating material without altering their chemical structure. This process is typically used for water-soluble and fat-soluble nutrients, such as vitamins, minerals, and bioactive compounds, and it forms the basis of many edible coating formulations (Pirozzi et al., 2021).

Direct Blending

One of the most straightforward methods of incorporating nutrients into edible coatings is by directly blending them into the coating material. For example, vitamins such as vitamin C (ascorbic acid) or vitamin A (retinol) may be added to the polymer matrix during the preparation of the coating solution. This process usually involves dissolving the nutrients in a solvent (e.g., water or oil) that is compatible with the coating material. Direct blending is simple and cost-effective but may not always ensure optimal stability or distribution of the nutrients within the coating (Duncan, 2011).

Solvent Addition

Nutrients can be incorporated into edible coatings by dissolving them in a solvent that is compatible with both the nutrient and the coating material. This is particularly common for water-soluble nutrients, such as B vitamins and minerals, which can be dissolved in water-based solutions before being mixed with the coating polymers. For fat-soluble nutrients like vitamins A, D, E, and K, the nutrients are often dissolved in oils or emulsions to ensure uniform dispersion within the hydrophobic polymer matrix. The solvent serves to dissolve and distribute the nutrient throughout the coating material (Merzendorfer & Cohen, 2019).

Immersion

In some cases, food products coated with a base edible coating may be immersed in a nutrient-rich solution. This method is often used to enhance the nutrient content of foods such as fruits and vegetables by soaking them in a nutrient solution, followed by drying or setting. Immersion can be effective for incorporating nutrients into porous food products, although nutrient uptake may be limited by the permeability of the food's surface.

Chemical Incorporation of Nutrients

Chemical incorporation refers to the process by which nutrients are chemically bound or attached to the coating matrix through various reactions, creating a more stable interaction between the nutrient and the coating material. This method enhances nutrient stability and can improve the release of nutrients over time (Wang et al., 2023).

Covalent Bonding

Covalent bonding involves the formation of stable chemical bonds between the nutrients and the polymer matrix. This process typically requires the use of cross-linking agents or chemical modifiers that enable the nutrients to bond with the polymer. For example, mineral salts such as calcium can be covalently bonded to polysaccharides (e.g., chitosan or alginate) via ionic or covalent interactions, resulting in enhanced stability and controlled release. This method is often employed to incorporate minerals or antioxidants, ensuring their retention within the coating.

Cross-Linking

Cross-linking agents, such as glutaraldehyde, calcium ions, or other multifunctional chemicals, can be used to chemically link nutrients to the polymer matrix. Cross-linking creates a three-dimensional network within the edible coating, which can enhance the mechanical strength and barrier properties of the coating while stabilizing sensitive nutrients. For example, antioxidants like polyphenols may be cross-linked with the coating material to protect them from oxidative degradation while ensuring their effective distribution and release (Pirozzi et al., 2021).

Encapsulation Methods

Encapsulation involves enclosing the nutrient within a polymer matrix, where it is physically isolated from the external environment. The polymer matrix can be designed to slowly release the nutrient over time, providing controlled release and preventing nutrient degradation. Encapsulation can be achieved through various methods such as spray drying, coacervation, or solvent evaporation. For instance, water-soluble vitamins like vitamin C or bioactive compounds such as polyphenols are often encapsulated in polymers such as alginate, gelatin, or chitosan to improve their stability and bioavailability (Quirós-Sauceda et al., 2014).

Microencapsulation and nanoencapsulation are advanced techniques that involve the incorporation of nutrients within tiny carriers or capsules, typically ranging from micrometers to nanometers in size. These encapsulation methods are highly effective in protecting nutrients from degradation, controlling their release, and enhancing their functionality within edible coatings.

Microencapsulation

Microencapsulation involves the encapsulation of nutrients in a carrier material, such as polysaccharides, proteins, lipids, or synthetic polymers, to form microparticles or microcapsules. These capsules act as protective shells that shield the nutrients from environmental stressors such as heat, oxygen, light, and moisture. Microencapsulation also allows for controlled and sustained nutrient release, which is crucial for preserving the nutrients' effectiveness over time. For example, vitamins or essential oils can be encapsulated in a gel matrix or starch-based carrier, providing long-term protection and enhancing the coating's performance (Chaudhary et al., 2021).

Nanoencapsulation

Nanoencapsulation refers to the incorporation of nutrients within nanometer-sized carriers, such as liposomes, nanoparticles, or nanogels. The smaller size of the carrier particles allows for enhanced nutrient protection and a more efficient controlled release system. Nanoencapsulation is particularly useful for fat-soluble vitamins, antioxidants, and polyphenols, which may be sensitive to oxidation and environmental degradation. The incorporation of nanoencapsulated nutrients into edible coatings offers the potential for improved bioavailability, as the nanoparticles can penetrate the food surface more easily and be absorbed by the body (Chaudhary et al., 2021). Nutrients are shielded from oxidation, microbial degradation, and environmental conditions. The small size of nanoparticles allows for better absorption by the body, especially in the case of fat-soluble vitamins. Nanoencapsulation can provide controlled or targeted release of nutrients, ensuring that they remain effective during food storage and consumption. A variety of carrier systems are used for both microencapsulation and nanoencapsulation, including lipids, proteins, polysaccharides, and synthetic polymers (Quirós-Sauceda et al., 2014).

Emulsion-Based Methods

Emulsion-based methods are particularly effective for incorporating fat-soluble nutrients into edible coatings. In these methods, nutrients are emulsified within a carrier medium (typically oil- or lipid-based systems) and dispersed within the hydrophilic coating material. This process creates stable emulsions where the nutrients are encapsulated within lipid droplets, preventing direct exposure to water or oxygen and improving their stability (Liu & Zhai, 2022).

Oil-in-Water Emulsions

In oil-in-water emulsions, the nutrients are dissolved in the oil phase, and the oil droplets are dispersed in the water phase. This method is effective for incorporating fat-soluble vitamins and bioactive compounds, such as essential oils or polyphenols, into a coating matrix made of hydrophilic polymers. The emulsification process ensures that the hydrophobic nutrients are uniformly distributed throughout the coating without forming separate phases or causing instability (Quirós-Sauceda et al., 2014).

Water-in-Oil Emulsions

Water-in-oil emulsions are less commonly used in edible coatings but are sometimes employed to protect the hydrophilic nutrients (such as water-soluble vitamins or minerals) within a lipid-based coating. In this case, the nutrients are encapsulated within the water phase, which is dispersed in the oil phase, providing a barrier against moisture and oxygen (Liu & Zhai, 2022).

Challenges of Emulsion-Based Methods

Emulsion-based methods require careful control of emulsification parameters (e.g., mixing speed, temperature, and emulsifier concentration) to achieve a stable and homogenous dispersion of nutrients. Inadequate emulsification may lead to phase separation, nutrient leakage, or reduced stability of the final coating.

Self-Assembly and Bio-inspired Techniques

Self-assembly and bio-inspired techniques are emerging as innovative methods for nutrient incorporation into edible coatings. These methods utilize the principles of molecular self-organization and biomimicry to create complex nutrient delivery systems (Samyn et al., 2023).

Self-Assembly

Self-assembly refers to the spontaneous organization of molecules into well-defined structures without external intervention. In the context of edible coatings, self-assembling polymers can be used to incorporate nutrients in a way that allows for efficient distribution and controlled release. For example, natural polymers such as polysaccharides or proteins can be designed to self-assemble into structures that encapsulate nutrients and protect them from degradation (Liu & Zhai, 2022).

Bio-inspired Techniques

Bio-inspired techniques take inspiration from nature to design nutrient delivery systems. For example, the use of biodegradable nanoparticles mimicking the structure of natural cell membranes can be employed to encapsulate bioactive nutrients. This approach enhances the stability, bioavailability, and controlled release of the nutrients within the coating.

Impact of Nutrients on the Structural Properties of Edible Coatings

The incorporation of nutrients into edible coatings can significantly affect their structural properties, including film formation, mechanical strength, flexibility, permeability, and overall performance. While the primary goal of edible coatings is to enhance the preservation and nutritional value of food, understanding the impact of these nutrients on the physical and mechanical properties of the coatings is crucial for their successful application. This section discusses how different types of nutrients, such as vitamins, minerals, antioxidants, and other bioactive compounds, influence the structural properties of edible coatings (Günal-Köroğlu & Capanoglu, 2024).

Influence of Nutrients on Film Formation

Film formation is a fundamental process in the development of edible coatings. The ability of the coating to form a uniform, continuous, and coherent film determines its effectiveness in protecting food. The incorporation of nutrients into the coating matrix can either enhance or hinder this process, depending on the characteristics of the nutrients and the base material used for the coating.

Hydrophilic Nutrients

Nutrients such as vitamins and water-soluble minerals, including vitamin C, B vitamins, and calcium, are hydrophilic in nature. When these nutrients are incorporated into a coating made from hydrophilic materials, such as polysaccharides (e.g., chitosan, starch, or alginate), they can promote the formation of a more cohesive and dense film. However, excessive water-soluble nutrients may lead to increased water absorption by the coating, which can affect the integrity and stability of the film over time.

Hydrophobic Nutrients

Fat-soluble vitamins (e.g., vitamins A, D, E, and K) and bioactive compounds such as polyphenols and essential oils are hydrophobic in nature. The incorporation of these hydrophobic nutrients into the coating may result in phase separation if not properly emulsified or encapsulated. In such cases, the film may become less uniform and may exhibit lower mechanical strength. To address this, emulsifying agents or encapsulation technologies (e.g., liposomes and microcapsules) are often used to improve the dispersion of hydrophobic nutrients and promote smooth film formation.

Effect of Nutrient Concentration

The concentration of nutrients incorporated into the edible coating material is a key factor in determining film formation. Low-to-moderate concentrations of nutrients generally improve film formation by enhancing the interaction between the nutrient and the polymer matrix. However, high concentrations can lead to film defects, such as cracking or reduced cohesiveness, due to excessive water retention or phase separation of incompatible materials.

Mechanical Properties

The mechanical properties of edible coatings, such as tensile strength, elongation at break, and flexibility, are essential for their functionality in food preservation. These properties are influenced by the type and concentration of nutrients used in the formulation of the coating.

Tensile Strength

The tensile strength of an edible coating refers to its ability to resist breaking under tension. The addition of nutrients such as vitamins and minerals can either strengthen or weaken the coating's tensile properties. For instance, the incorporation of antioxidant-rich compounds such as polyphenols can reinforce the coating matrix due to their ability to interact with the coating material, resulting in improved tensile strength. However, the addition of large amounts of hydrophilic nutrients like vitamin C can decrease the tensile strength, as water-soluble compounds may weaken the film by increasing its water content.

Elongation and Flexibility

Elongation at break is a measure of the flexibility of the coating, and it is important for coatings that need to stretch or conform to the shape of the food. The addition of nutrients such as glycerol or other plasticizers can increase the flexibility and elongation of the edible coating. For example, the inclusion of lipophilic nutrients such as vitamin E may act as a plasticizer, increasing the flexibility of the coating and improving its ability to stretch without breaking. On the other hand, the inclusion of high concentrations of hydrophilic nutrients can lead to reduced flexibility and increased brittleness, as excessive water absorption may cause the film to become fragile.

Impact of Encapsulation

Nutrients that are encapsulated in microcapsules or other carrier systems often have a less pronounced impact on the mechanical properties of the coating. Encapsulation protects the nutrient from direct interaction with the polymer matrix, allowing for better integration of the coating material and improved mechanical properties. The use of encapsulated antioxidants and vitamins, for example, can help maintain or even improve the tensile strength and flexibility of the coating by reducing nutrient-induced film defects.

Permeability and Barrier Properties

One of the primary functions of edible coatings is to act as a barrier to moisture, gases, and volatile compounds, thereby preventing spoilage and extending the shelf life of food products. The incorporation of nutrients can have a direct impact on the permeability and barrier properties of the coating, influencing its effectiveness as a food preservation tool.

Moisture Permeability

The addition of nutrients, particularly hydrophilic ones like vitamins C and B, can increase the water absorption capacity of edible coatings, making them more susceptible to moisture permeability. In the case of water-soluble vitamins, their incorporation into the coating material can cause the film to become more hygroscopic, which means that it will absorb more moisture from the surrounding environment. This increased moisture content can reduce the barrier properties of the coating, leading to faster deterioration of the food product. To mitigate this effect, coating formulations can include plasticizers or hydrophobic compounds to reduce moisture absorption.

Gas Permeability

Nutrient incorporation can also affect the gas permeability of edible coatings, particularly in relation to oxygen and carbon dioxide. Oxygen permeability is a critical factor in determining the shelf life of perishable foods, as excessive exposure to oxygen accelerates oxidation and microbial growth. The incorporation of antioxidants like polyphenols or essential oils can reduce oxygen permeability by enhancing the coating's protective barrier. In some cases, the addition of lipophilic nutrients may result in a more impermeable barrier, as these nutrients can form a hydrophobic layer that reduces gas diffusion.

Volatile Compound Permeability

The incorporation of nutrient-rich compounds, such as essential oils and antioxidants, can also influence the permeability of the coating to volatile compounds, such as ethylene in fruits or sulfur compounds in vegetables. These volatile compounds are often responsible for the ripening process or the production of off-flavors. Nutrient-enriched coatings can help reduce the release of these volatile compounds, slowing down ripening and maintaining the freshness of the food product.

Thermal Stability and Durability

Thermal stability is an important property of edible coatings, particularly when they are subjected to heat during food processing or storage. The incorporation of nutrients, especially heat-sensitive vitamins and antioxidants, can influence the thermal stability and durability of the coating.

Heat Sensitivity of Nutrients

Certain nutrients, such as vitamin C and other water-soluble vitamins, are highly sensitive to heat and may degrade when exposed to high temperatures during the drying or baking processes of edible coatings. As these nutrients degrade, they can negatively affect the stability and performance of the coating. For example, the breakdown of vitamin C can weaken the coating structure and reduce its antioxidant capacity. In contrast, heat-stable nutrients like minerals and some antioxidants (e.g., vitamin E) are less affected by heat and may have minimal impact on the thermal stability of the coating.

Durability Under Heat Exposure

Edible coatings made from biopolymers such as starch or chitosan may experience changes in their mechanical properties under heat exposure, including softening, melting, or loss of cohesion. The incorporation of heat-stable nutrients or the use of encapsulation techniques can help mitigate these effects. Additionally, the use of plasticizers and stabilizers can enhance the thermal stability of the coating, making it more durable during processing and storage.

Impact of Nutrient Interactions on Coating Integrity

The interactions between nutrients and the coating material itself can influence the overall integrity and functionality of the edible coating. These interactions can either improve or impair the coating's performance, depending on the compatibility between the nutrient and the base material (Satchanska et al., 2024).

Nutrient–Polymer Interactions

The interaction between nutrients and polymer matrices (e.g., chitosan, alginate, or cellulose) can affect the structural properties of the coating. In some cases, nutrients can interact with the polymer chains to form complexes that enhance film strength and stability. For example, the chelation of metal ions (e.g., calcium) with polysaccharides can improve the mechanical properties of the coating. However, incompatible interactions between certain nutrients and polymers may cause phase separation, leading to weak points in the coating that can compromise its barrier properties and overall performance.

Nutrient Cross-Linking

Cross-linking agents, such as calcium ions or glutaraldehyde, can be used to enhance the structural integrity of edible coatings. The incorporation of nutrients that can participate in cross-linking reactions can further strengthen the coating matrix, improving its mechanical properties and resistance to degradation. For instance, the incorporation of polyphenolic compounds may facilitate cross-linking with the polymer matrix, leading to a more robust coating structure.

Nutrient Stability in Edible Coatings

The stability of nutrients in edible coatings is a critical factor in determining the effectiveness of these coatings for food preservation and nutrition enhancement. Edible coatings are designed to protect the nutrients within them from degradation during storage, processing, and exposure to environmental factors. However, the incorporation of nutrients into edible coatings introduces a challenge: the potential for nutrient loss due to chemical, physical, or microbial factors. This section explores the factors influencing nutrient stability in edible coatings, strategies for enhancing stability, and the potential impact on the functionality of nutrient-enriched coatings.

Factors Affecting Nutrient Stability

The stability of nutrients within edible coatings is influenced by several factors, including the nature of the nutrients themselves, the composition of the coating materials, storage conditions, and environmental factors like temperature, humidity, and light exposure. Understanding these factors is key to developing effective nutrient-enriched edible coatings.

Chemical Properties of Nutrients

Different nutrients exhibit varying degrees of stability under different conditions. For example, vitamins such as vitamin C, B vitamins, and are highly sensitive to heat, light, and oxygen. These vitamins are prone to oxidation, hydrolysis, and other forms of chemical degradation when exposed to unfavorable conditions. On the other hand, minerals such as calcium and magnesium are generally more stable than vitamins and can more easily be incorporated into coatings without significant degradation.

Temperature Sensitivity

The stability of heat-sensitive nutrients like vitamin C and carotenoids can be compromised when exposed to high temperatures during the production or storage of food products. Nutrient degradation can occur when the edible coating is exposed to temperatures above a certain threshold, such as during the drying process of the coating or while the food is stored under high-temperature conditions.

pH Sensitivity

The pH of the coating material can also influence nutrient stability. Nutrients such as vitamin C and folic acid are known to be unstable under acidic or alkaline conditions. The pH of the edible coating must be carefully controlled to ensure the preservation of these nutrients. For example, coatings based on acidic polysaccharides like alginate may cause the degradation of certain nutrients, while coatings made from neutral-pH materials like chitosan might provide better protection for some vitamins.

Oxygen and Light Exposure

Oxidative degradation of nutrients such as polyphenols and vitamins A, C, and E can be accelerated by exposure to oxygen and light. These nutrients are particularly vulnerable to oxidative processes, which can reduce their bioavailability and efficacy. When developing nutrient-enriched edible coatings, it is essential to consider the oxygen permeability of the coating material and the potential for light-induced degradation.

Water Activity and Humidity

Water activity plays a crucial role in nutrient stability, particularly for water-soluble nutrients. High water activity can promote the dissolution and leaching of nutrients from the coating, as well as accelerate microbial growth, which may further degrade the nutrients. A low-water-activity coating will help stabilize the nutrients, as it reduces the potential for nutrient leaching and microbial activity. However, the moisture content must be balanced to ensure that the coating remains functional.

Interactions with Coating Materials

The composition of the edible coating itself can impact the stability of the nutrients. Some natural polymers, such as chitosan, alginate, or starch, can interact with nutrients and influence their stability. For example, certain polysaccharides may chelate metal ions or form complexes with nutrients, which could either enhance or inhibit their stability. Understanding these interactions is important for optimizing nutrient preservation.

Strategies to Enhance Nutrient Stability in Edible Coatings

To improve the stability of nutrients in edible coatings, various strategies can be employed. These strategies aim to protect sensitive nutrients from degradation during coating production, storage, and application (Satchanska et al., 2024).

Use of Antioxidants and Stabilizers

The incorporation of natural antioxidants such as vitamin E, green tea extract, or rosemary extract into edible coatings can help protect sensitive nutrients from oxidative damage. These antioxidants scavenge free radicals and reactive oxygen species, preventing the breakdown of vitamins and other bioactive compounds. In addition to antioxidants, stabilizers like cyclodextrins can encapsulate sensitive nutrients and protect them from environmental stressors.

Encapsulation of Nutrients

Encapsulation is a technique where nutrients are enclosed within a protective matrix, such as liposomes, nanoparticles, or microcapsules, to prevent direct exposure to oxygen, light, or heat. This method provides a physical barrier that enhances the stability of sensitive nutrients. For example, the encapsulation of vitamin C or polyphenols can prevent their degradation and allow for controlled release, ensuring that the nutrients remain effective throughout the shelf life of the coated food.

Incorporation of Active Packaging Materials

Active packaging materials that can interact with the surrounding environment can be incorporated into edible coatings to enhance nutrient stability. These materials can include oxygen scavengers, moisture absorbers, or UV-blocking agents that protect both the coating and the nutrients within it. By reducing oxygen or light exposure, these materials help maintain the stability of sensitive nutrients over time.

Optimizing Coating Composition

The choice of edible coating material is key to enhancing nutrient stability. For example, polysaccharides like chitosan or xanthan gum can act as a barrier to oxygen, reducing nutrient oxidation. Similarly, the use of fats and lipids in the coating matrix can help encapsulate fat-soluble vitamins and prevent their exposure to light and oxygen. The composition of the coating should be tailored to match the specific needs of the nutrients being incorporated.

Incorporation of pH-Responsive Materials

For nutrients sensitive to pH fluctuations, incorporating pH-responsive materials into the edible coating can provide an additional layer of protection. These materials change their properties in response to changes in pH, creating microenvironments that help stabilize the nutrients. For instance, incorporating acidic pH– or alkaline pH–responsive polymers can help protect sensitive nutrients from extreme pH conditions during food storage and consumption.

Controlled Release Mechanisms

A controlled release system allows nutrients to be gradually released from the edible coating over time, which not only improves their stability but also ensures that they remain bioavailable for longer periods. This can be achieved using controlled release carriers, such as hydrogels, or by manipulating the coating thickness and permeability. Controlled release can enhance the shelf life of the nutrients and maintain their functional properties for a longer duration.

Impact of Nutrient Degradation on Food Quality and Functionality

The degradation of nutrients in edible coatings can have several consequences for both food quality and the overall functionality of the coating. Nutrient loss can impact the intended health benefits of the coating and may also affect the sensory and preservation properties of the food.

Loss of Health Benefits

If essential nutrients like vitamins, minerals, or antioxidants degrade, the intended health benefits of the nutrient-enriched coating may be compromised. For example, the breakdown of vitamin C or antioxidants can reduce the nutritional value of the food, rendering the coating less effective in promoting immune health or preventing oxidative damage.

Impact on Food Sensory Attributes

Nutrient degradation can also lead to changes in the sensory properties of the food. For instance, the oxidation of polyunsaturated fats or vitamins can result in rancid flavors, which can negatively affect consumer acceptance. Similarly, the breakdown of certain bioactive compounds may alter the aroma or color of the food, diminishing its visual and olfactory appeal.

Coating Integrity and Functionality

The loss of nutrients can affect the structural integrity of the edible coating itself. Some nutrients may act as plasticizers or stabilizing agents within the coating matrix. If these nutrients degrade, the coating's ability to retain moisture, protect against microorganisms, or maintain its physical properties may be compromised. This can reduce the overall effectiveness of the coating in preserving food.

Microbial Activity and Safety

The degradation of antimicrobial agents within the coating, such as essential oils or polyphenols, can result in reduced protection against microbial contamination. This can compromise the safety of the food product and increase the risk of spoilage or pathogenic contamination. Ensuring the stability of these bioactive components is essential to maintain the antimicrobial function of the coating.

Monitoring and Testing of Nutrient Stability

To ensure that nutrient-enriched edible coatings remain effective throughout the shelf life of the product, it is essential to monitor and test the stability of the nutrients under various conditions. Several methods can be employed to assess the stability of nutrients in coatings.

High-Performance Liquid Chromatography (HPLC)

HPLC is a commonly used technique to analyze the concentration of specific nutrients, such as vitamins and antioxidants, in edible coatings. This method allows for the precise measurement of nutrient degradation over time under different storage conditions.

Sensory Evaluation

Sensory tests, including taste, aroma, texture, and visual appearance assessments, can help determine the impact of nutrient degradation on the overall quality of the food. These tests provide valuable insights into how nutrient loss affects the sensory attributes of the coated food.

Microbial Analysis

Microbial testing is essential for assessing the effectiveness of the antimicrobial agents in the edible coating. By measuring microbial counts over time, researchers can determine whether the coating continues to provide sufficient protection against spoilage or pathogenic microorganisms.

Stability Testing Under Various Conditions

Nutrient stability should be evaluated under different temperature, humidity, and light conditions to simulate the storage and handling of food products in real-world scenarios. These tests can provide valuable information on the shelf life of the nutrient-enriched coatings and guide improvements in formulation.

Effects of Nutrients on Edible Coatings' Effectiveness in Food Preservation

Edible coatings and films are often used in the food industry to extend the shelf life of perishable products. The effectiveness of these coatings in food preservation is primarily determined by their ability to create barriers that protect food from environmental factors like moisture loss, oxygen exposure, and microbial contamination. The incorporation of nutrients into these coatings not only enhances their functional properties but also contributes to the preservation of food by mitigating spoilage processes. This section explores how various nutrients integrated into edible coatings impact food preservation, with a focus on their roles in preventing oxidation, moisture loss, and microbial growth.

Antioxidants and Their Role in Preventing Oxidation

One of the most significant challenges in food preservation is oxidation, which leads to the degradation of lipids, proteins, and other sensitive compounds in foods. Oxidation can result in rancidity, discoloration, flavor loss, and nutritional degradation. Antioxidants are key nutrients that play a pivotal role in preventing these oxidative processes. When incorporated into edible coatings, antioxidants can significantly extend the shelf life of food products by protecting them from oxidation-induced damage.

Vitamin C and Vitamin E

Both of these vitamins are widely used as antioxidants in food coatings. Vitamin C (ascorbic acid) acts as a potent reducing agent, neutralizing free radicals and inhibiting lipid oxidation, particularly in fruit and vegetable coatings. Vitamin E (α-tocopherol) is commonly used in coatings for meats and oils, as it protects against oxidative rancidity and preserves the flavor and texture of fatty foods.

Polyphenols

These plant-derived antioxidants, which include flavonoids, tannins, and phenolic acids, are effective at reducing oxidative damage in food. Polyphenol-rich coatings, such as those derived from green tea or grape seed extract, help extend the shelf life of foods by scavenging free radicals and inhibiting the formation of oxidation products.

Carotenoids

These compounds, found in carrots, tomatoes, and other colorful fruits and vegetables, are effective antioxidants with the additional benefit of contributing to the nutritional value of the coating. Carotenoids like beta-carotene help protect foods from oxidative spoilage, particularly in fruit and vegetable coatings, where they also provide natural color enhancement.

The inclusion of antioxidants in edible coatings not only enhances the preservation of sensory qualities (such as taste, aroma, and appearance) but also helps in retaining the nutritional value of food, especially by protecting vitamins and essential fatty acids from oxidative degradation.

Antimicrobials and Their Role in Inhibiting Microbial Growth

Microbial contamination is another major cause of spoilage and safety concerns in food products. Pathogenic and spoilage microorganisms, such as bacteria, molds, and yeasts, can degrade food quality and pose serious health risks. By incorporating antimicrobial nutrients and bioactive compounds into edible coatings, it is possible to slow down or inhibit microbial growth and thereby extend the shelf life of perishable foods (Rodríguez et al., 2006).

Essential Oils

Many essential oils, such as those derived from thyme, oregano, clove, and cinnamon, exhibit antimicrobial properties that can be incorporated into edible coatings to prevent the growth of bacteria and molds. These essential oils contain compounds like phenols and terpenes, which disrupt microbial cell membranes and inhibit their growth.

Chitosan

Chitosan, a natural polysaccharide derived from chitin, has inherent antimicrobial properties and is commonly used in edible coatings for its ability to inhibit the growth of a wide range of bacteria and fungi. It has been particularly effective in coatings for fresh fruits, vegetables, and meat products.

Nisin

A natural antimicrobial peptide produced by certain strains of bacteria, nisin is used in some food coatings to target gram-positive bacteria such as *Listeria* and *Staphylococcus*. The incorporation of nisin into edible coatings can provide an additional layer of protection against bacterial contamination in ready-to-eat meats and dairy products.

Zinc and Copper

Both minerals exhibit antimicrobial properties, with zinc having a broad-spectrum effect on bacteria and copper acting as an antibacterial agent. Zinc, in particular, is commonly included in coatings for fruits and vegetables to inhibit microbial growth. The antimicrobial action of these nutrients not only protects the food from spoilage but also reduces the risk of foodborne illnesses, making the coated food safer for consumption. The addition of antimicrobial agents to edible coatings is particularly beneficial in the preservation of high-risk foods such as meats, dairy products, and fresh produce.

Nutrients and Their Impact on the Sensory Qualities of Coated Food

In addition to preserving the shelf life of the food, the integration of nutrients into edible coatings also impacts the sensory qualities of the food. Nutrient-enriched coatings can preserve the appearance, texture, flavor, and overall sensory appeal of the food, which plays an important role in consumer acceptance and product success.

Color Preservation

Nutrients like antioxidants and carotenoids, which are rich in color (e.g., carotene and lycopene), can enhance the appearance of coated food products while also protecting them from discoloration caused by oxidation. For example, the use of beta-carotene in edible coatings helps maintain the vibrant orange color of carrots and other vegetables.

Flavor Retention

Certain nutrients, such as phenolic compounds and essential oils, have the ability to preserve the flavor of the food by protecting it from oxidation and rancidity. Additionally, antioxidants and antimicrobial agents can prevent the development of off-flavors, which often result from microbial spoilage or chemical degradation.

Texture Maintenance

The incorporation of nutrients like proteins and fibers into coatings can help maintain the texture of the food by preventing moisture loss, which is crucial for foods like fresh produce, meats, and bakery items. By regulating the water content, these nutrients contribute to the crispness or tenderness of the coated product.

Nutrient-enriched coatings enhance the sensory experience of the food product by preserving and improving its visual, taste, and textural qualities, ultimately contributing to consumer satisfaction.

Nutritional Aspects for Consumer Acceptance

The incorporation of nutrients into edible coatings has the potential to offer multiple benefits, not only in terms of food preservation but also in enhancing the nutritional profile of foods. However, the consumer acceptance of these nutrient-enriched edible coatings is influenced by a variety of factors, including the perception of health benefits, the sensory attributes of the food, and the familiarity with the ingredients used in the coatings. This section explores the key nutritional aspects that affect the consumer acceptance of edible coatings, focusing on health perceptions, the role of nutrient enrichment, and the challenges that may arise from such innovations in food technology (Köhler et al., 2019).

Perception of Health Benefits

Among the main factors influencing the consumer acceptance of nutrient-enriched edible coatings are the perceived health benefits. As consumers become more health-conscious and aware of the link between diet and chronic diseases, there is a growing demand for functional foods that provide more than just basic nutrition. Edible coatings that are fortified with essential vitamins, minerals, antioxidants, and dietary fibers can appeal to this demand by offering added health value without requiring significant changes to the existing dietary habits (Azmi et al., 2019).

Fortification with Essential Nutrients

Coatings that deliver nutrients such as vitamin C, vitamin A, calcium, iron, and fiber can be marketed as a way to increase the nutritional content of everyday foods. For example, a fruit or vegetable coated with a nutrient-enriched film could be perceived as providing additional immune support or contributing to bone health due to the presence of vitamin C or calcium. This functional aspect makes the coating particularly appealing to consumers who are looking for ways to boost their nutrient intake conveniently (Köhler et al., 2019).

Antioxidants for Disease Prevention

With increasing awareness of the role of oxidative stress in the development of diseases such as cancer, cardiovascular diseases, and diabetes, consumers are becoming more interested in antioxidant-rich foods. Edible coatings that include antioxidant-rich ingredients like polyphenols, flavonoids, or vitamins (A, C, and E) may be particularly appealing as part of a health-conscious diet. The inclusion of antioxidants can enhance the functional appeal of coatings, contributing to the overall perceived healthfulness of the food product.

Functional Foods and the Growing Market for "Healthier" Options

The rise of functional foods, which are marketed for their health-promoting properties beyond basic nutrition, has been a key trend in the food industry. Edible coatings that provide an additional health benefit by delivering nutrients that are often deficient in modern diets (such as fiber, vitamin D, or omega-3 fatty acids) can meet the needs of consumers who seek convenient, nutrient-dense options. Coatings that support digestive health (e.g., those with fiber) or cardiovascular health (e.g., those with polyphenols or omega-3s) can help position products as part of a healthy lifestyle (Sheoran et al., 2022).

While the potential health benefits are crucial in consumer decision-making, clear communication and education about the specific benefits of nutrient-enriched coatings are necessary to increase consumer acceptance. Consumers need to understand how the nutrients in the coatings contribute to overall health and wellness for them to perceive the added value.

Sensory Aspects of Nutrient-Enriched Coatings

The sensory properties of the food, such as the appearance, taste, texture, and aroma, are significant factors that influence consumer preferences. Even though a food coating may be nutrient-enriched, the coating must not negatively impact the sensory qualities of the food for it to be accepted by consumers. If the nutrient-enriched edible coating alters the food's flavor, color, or texture in an undesirable way, it may reduce the likelihood of consumer acceptance (Granato et al., 2022).

Consumer Awareness and Education

For consumers to accept and embrace nutrient-enriched edible coatings, there needs to be a clear understanding of the benefits these coatings provide. Consumer awareness and education are essential in fostering trust and acceptance of these innovative food products.

Labeling and Marketing

Transparent labeling that clearly communicates the nutritional benefits of the coating can help consumers make informed decisions. Claims such as "fortified with vitamin C" or "rich in antioxidants" can signal to consumers that the product offers enhanced health benefits. However, it is crucial that these claims are substantiated with scientific evidence to avoid skepticism and confusion. Providing information on how the nutrient-enriched coatings contribute to overall health—such as promoting immunity or preventing oxidative damage—can improve consumer confidence in the product.

Addressing Misconceptions

Consumers may be wary of food products with new technological advancements, particularly if they perceive these products as unnatural or overly processed. Educational campaigns that highlight the natural origin of many edible coatings (e.g., polysaccharides, proteins, and lipids from plant or animal sources) and the safety of nutrients used in the coatings can help overcome these concerns. It is also important to emphasize that the nutrients are integrated into the coatings in ways that preserve their bioavailability and ensure that they are safe for consumption.

Health Claim Regulations

Many countries have regulations governing the use of health claims in food labeling. To encourage consumer acceptance, nutrient-enriched edible coatings must comply with these regulations and provide clear, truthful information regarding the health benefits of the coatings. This can increase the credibility of the product and make consumers more likely to trust and adopt these innovations.

Consumer Research

Ongoing consumer studies and market research are vital in understanding the specific nutritional needs and concerns of target demographics. For example, parents may be more interested in the nutritional benefits of coatings for their children's foods, while older adults may be more interested in coatings that deliver nutrients that support bone health, such as calcium or vitamin D. Tailoring the marketing and communication strategies based on these insights can improve consumer acceptance.

Ethical and Cultural Considerations

Cultural preferences and ethical considerations also play a significant role in the acceptance of nutrient-enriched edible coatings. Some consumers may be hesitant to accept new technologies in food processing, especially if the coatings are perceived as unfamiliar or artificial.

Natural Versus Synthetic Nutrients

Some consumers may be more inclined to accept edible coatings that use nutrients derived from natural sources rather than synthetic additives. For instance, coatings enriched with natural antioxidants from plant sources (like green tea extract or grape seed extract) may be viewed more favorably than those using synthetic additives or fortifications.

Dietary Restrictions

Consumer acceptance also depends on how well the nutrient-enriched coatings cater to various dietary restrictions. For instance, coatings containing animal-derived ingredients (e.g., gelatin and whey) may not be acceptable to vegans or vegetarians. Similarly, coatings with certain nutrients, such as gluten or lactose, may be problematic for individuals with sensitivities or intolerances. To maximize consumer acceptance, edible coatings should ideally offer alternatives that cater to diverse dietary needs.

Sustainability and Transparency

Consumers are increasingly prioritizing sustainability and ethical practices in their food choices. The use of plant-based coatings and the incorporation of eco-friendly packaging can enhance the attractiveness of nutrient-enriched coatings, especially among environmentally conscious consumers. Ensuring that nutrient-enriched coatings are produced using sustainable practices and that the sourcing of ingredients is transparent will likely improve their acceptance among these consumer groups.

Conclusion

The integration of nutrient-enriched edible coatings represents a significant advancement in food preservation and nutritional enhancement. These coatings not only extend shelf life by regulating moisture, preventing microbial growth, and reducing oxidation but also offer added health benefits. The ability to maintain sensory attributes such as texture, flavor, and appearance while delivering functional benefits is critical to consumer acceptance and market success. Additionally, advancements in biomaterials science and nanotechnology enhance the controlled release of nutrients, ensuring their bioavailability and efficacy. Regulatory frameworks, consumer education, and transparent labeling will play a crucial role in fostering trust and promoting widespread adoption of these coatings. As the global food industry shifts toward sustainable and health-oriented solutions, nutrient-enriched edible coatings have the potential to transform food packaging and preservation strategies. Their successful implementation could lead to reduced food waste, improved dietary nutrition, and the development of next-generation functional foods that adapt to evolving consumer preferences and dietary needs.

References

Ahmed, W. (2021). The advances in the foliar application of micro and macronutrients and their effects on growth development and yield potential of tomato crops. *Journal Advance Nutrition Science Techechnoly, 1*, 19–89. https://doi.org/10.15228/ANST.2021.v01.i02.p04

Ahmed, W., Azmat, R., Khojah, E., Ahmed, R., Qayyum, A., Shah, A. N., Abbas, A., Moin, S., & Samra, B. N. (2022). The Development of a Green Innovative Bioactive Film for Industrial Application as a New Emerging Technology to Protect the Quality of Fruits. *Molecules., 27*(2), 486. https://doi.org/10.3390/molecules27020486

Ahmed, W., Azmat, R., Qayyum, A., Bibi, Y., Ali, S., Ahmed, R., & Ramadan, M. F. (2023). Application of HPLC, LC-Q-TOF–MS, UF-HPLC and LC–MS in the comparative analysis of flavonoids and sterols in the Citrus paradise cv. shamber (grapefruit). *Rendiconti Lincei: Scienze Fisiche E Naturali*. https://doi.org/10.1007/s12210-023-01188-8

Alba, D. V., & Gabriel, V. H. (2018). Effects of edible coatings on sensory quality of minimally processed carrot. *Advance Journal of Food Science and Technology, 16*(S), 41–46.

Azmi, N. a. N., Elgharbawy, A. a. M., Motlagh, S. R., Samsudin, N., & Salleh, H. M. (2019). Nanoemulsions: factory for food, pharmaceutical and cosmetics. *Processes, 7*(9), 617. https://doi.org/10.3390/pr7090617

Babalola, B. J., Li, J., Willing, C. E., Zheng, Y., Wang, Y., Gan, H., Li, X., Wang, C., Adams, C. A., Gao, C., & Guo, L. (2022). Nitrogen fertilisation disrupts the temporal dynamics of arbuscular mycorrhizal fungal hyphae but not spore density and community composition in a wheat field. *New Phytologist, 234*(6), 2057–2072. https://doi.org/10.1111/nph.18043

Bremenkamp, I., & Sousa Gallagher, M. J. (2025). Edible coatings for ready-to-eat products: Critical review of recent studies, sustainable packaging perspectives, challenges and emerging trends. *Polymers, 17*(3), 376.

Chaudhary, V., Thakur, N., Kajla, P., Thakur, S., & Punia, S. (2021). Application of encapsulation technology in edible films: carrier of bioactive compounds. *Frontiers in Sustainable Food Systems, 5*, 734921.

De Oliveira, A. C. S., Ribeiro, M. N., Ugucioni, J. C., Da Rocha, R. A., & Borges, S. V. (2023). Consumer perception of biodegradable packaging for food. *Polímeros, 33*(4). https://doi.org/10.1590/0104-1428.20230068

Dhumal, C. V., & Sarkar, P. (2018). Composite edible films and coatings from food-grade biopolymers. *Journal of Food Science and Technology, 55*(11), 4369–4383.

Díaz-Montes, E., & Castro-Muñoz, R. (2021). Edible films and coatings as food-quality preservers: An overview. *Foods, 10*(2), 249.

Duncan, T. V. (2011). Applications of nanotechnology in food packaging and food safety: Barrier materials, antimicrobials and sensors. *Journal of Colloid and Interface Science, 363*(1), 1–24. https://doi.org/10.1016/j.jcis.2011.07.017

Gallegos-Cedillo, V. M., Nájera, C., Signore, A., Ochoa, J., Gallegos, J., Egea-Gilabert, C., Gruda, N. S., & Fernández, J. A. (2024). Analysis of global research on vegetable seedlings and transplants and their impacts on product quality. *Journal of the Science of Food and Agriculture, 104*(9), 4950–4965. https://doi.org/10.1002/jsfa.13309

Granato, D., Carocho, M., Barros, L., Zabetakis, I., Mocan, A., Tsoupras, A., Cruz, A. G., & Pimentel, T. C. (2022). Implementation of Sustainable Development Goals in the dairy sector: Perspectives on the use of agro-industrial side-streams to design functional foods. *Trends in Food Science & Technology, 124*, 128–139. https://doi.org/10.1016/j.tifs.2022.04.009

Günal-Köroğlu, D., & Capanoglu, E. (2024). The effect of using natural plant-based waxes in coating/film materials on postharvest quality of fruits and vegetables. *Future Postharvest and Food, 1*(1), 166–172. https://doi.org/10.1002/fpf2.12004

Gupta, D., Lall, A., Kumar, S., Patil, T. D., & Gaikwad, K. K. (2024). Plant based edible films and coatings for food packaging applications: Recent advances, applications, and trends. *Sustainable Food Technology., 2*, 1428.

Hashemi, A., Jang, J., & Beheshti, J. (2023). A Machine Learning-Based surrogate finite element model for estimating dynamic response of mechanical systems. *IEEE Access, 11*, 54509–54525. https://doi.org/10.1109/access.2023.3282453

Hu, Q., Zhou, F., Ly, N. K., Ordyna, J., Peterson, T., Fan, Z., & Wang, S. (2023). Development of multifunctional nanoencapsulated trans-Resveratrol/Chitosan nutraceutical edible coating for strawberry preservation. *ACS Nano, 17*(9), 8586–8597. https://doi.org/10.1021/acsnano.3c01094

Joglar, J. A., Chung, M. K., Armbruster, A. L., Benjamin, E. J., Chyou, J. Y., Cronin, E. M., Deswal, A., Eckhardt, L. L., Goldberger, Z. D., Gopinathannair, R., Gorenek, B., Hess, P. L., Hlatky, M., Hogan, G., Ibeh, C., Indik, J. H., Kido, K., Kusumoto, F., Link, M. S., et al. (2023). 2023 ACC/AHA/ACCP/HRS Guideline for the Diagnosis and Management of Atrial Fibrillation: A report of the American College of Cardiology/American Heart Association Joint Committee on Clinical Practice Guidelines. *Circulation, 149*(1). https://doi.org/10.1161/cir.0000000000001193

Karnwal, A., Kumar, G., Singh, R., Selvaraj, M., Malik, T., & Al Tawaha, A. R. M. (2025). Natural biopolymers in edible coatings: Applications in food preservation. *Food Chemistry: X, 25*, 102171.

Köhler, J., Geels, F. W., Kern, F., Markard, J., Onsongo, E., Wieczorek, A., Alkemade, F., Avelino, F., Bergek, A., Boons, F., Fünfschilling, L., Hess, D., Holtz, G., Hyysalo, S., Jenkins, K., Kivimaa, P., Martiskainen, M., McMeekin, A., Mühlemeier, M. S., et al. (2019). An agenda for sustainability transitions research: State of the art and future directions. *Environmental Innovation and Societal Transitions, 31*, 1–32. https://doi.org/10.1016/j.eist.2019.01.004

Liu, Q., & Zhai, W. (2022). Hierarchical porous ceramics with distinctive microstructures by emulsion-based direct ink writing. *ACS Applied Materials & Interfaces, 14*(28), 32196–32205. https://doi.org/10.1021/acsami.2c03245

Matloob, A., Ayub, H., Mohsin, M., Ambreen, S., Khan, F. A., Oranab, S., et al. (2023). A review on edible coatings and films: Advances, composition, production methods, and safety concerns. *ACS Omega, 8*(32), 28932–28944.

Merzendorfer, H., & Cohen, E. (2019). Chitin/Chitosan: Versatile ecological, industrial, and biomedical applications. In *Biologically-inspired systems* (pp. 541–624). https://doi.org/10.1007/978-3-030-12919-4_14

Miteluţ, A. C., Popa, E. E., Drăghici, M. C., Popescu, P. A., Popa, V. I., Bujor, O. C., et al. (2021). Latest developments in edible coatings on minimally processed fruits and vegetables: A review. *Foods, 10*(11), 2821.

Perez-Vazquez, A., Barciela, P., Carpena, M., & Prieto, M. A. (2023). Edible coatings as a natural packaging system to improve fruit and vegetable shelf life and quality. *Foods, 12*(19), 3570. https://doi.org/10.3390/foods12193570

Pirozzi, A., Ferrari, G., & Donsì, F. (2021). The use of nanocellulose in edible coatings for the preservation of perishable fruits and vegetables. *Coatings, 11*(8), 990. https://doi.org/10.3390/coatings11080990

Priya, K., Thirunavookarasu, N., & Chidanand, D. V. (2023). Recent advances in edible coating of food products and its legislations: A review. *Journal of Agriculture and Food Research, 12*, 100623.

Qasim, U., Sirviö, J. A., Suopajärvi, T., Hu, L., Pratiwi, F. W., Lin, M. K. T. H., Anghelescu-Hakala, A., Ronkainen, V., Xu, C., & Liimatainen, H. (2024). A multifunctional biogenic films and coatings from synergistic aqueous dispersion of wood-derived suberin and cellulose nanofibers. *Carbohydrate Polymers, 338*, 122218. https://doi.org/10.1016/j.carbpol.2024.122218

Quirós-Sauceda, A. E., Ayala-Zavala, J. F., Olivas, G. I., & González-Aguilar, G. A. (2014). Edible coatings as encapsulating matrices for bioactive compounds: A review. *Journal of Food Science and Technology, 51*(9), 1674–1685. https://doi.org/10.1007/s13197-013-1246-x

Rodríguez, F., Alberdi, M., & García, R. (2006). Influence of edible coatings with antimicrobial agents on the shelf life of food. *Food Science and Technology International, 12*(4), 295–299. https://doi.org/10.1177/1082013206071292

Samyn, P., Rastogi, V. K., Baba, N. S. S., & Van Erps, J. (2023). Role of nanocellulose in light harvesting and artificial photosynthesis. *Catalysts, 13*(6), 986. https://doi.org/10.3390/catal13060986

Satchanska, G., Davidova, S., & Petrov, P. D. (2024). Natural and synthetic polymers for biomedical and environmental applications. *Polymers, 16*(8), 1159. https://doi.org/10.3390/polym16081159

Shah, Y. A., Bhatia, S., Al-Harrasi, A., Afzaal, M., Saeed, F., Anwer, M. K., et al. (2023). Mechanical properties of protein-based food packaging materials. *Polymers, 15*(7), 1724.

Sheoran, S., Kumar, S., Ramtekey, V., Kar, P., Meena, R. S., & Jangir, C. K. (2022). Current status and Potential of biofortification to Enhance crop Nutritional quality: An Overview. *Sustainability, 14*(6), 3301. https://doi.org/10.3390/su14063301

Shukla, P., Bhise, S., & Thind, S. S. (2019). Role of biodegradable edible films and coatings in food industry. *Acta Scientific Nutritional Health, 3*(5), 138–147.

Silue, Y., & Fawole, O. A. (2024). Global research network analysis of edible coatings and films for preserving perishable fruit crops: current status and future directions. *Foods, 13*(15), 2321.

Teixeira-Costa, B. E., & Andrade, C. T. (2021). Natural polymers used in edible food packaging—History, function and application trends as a sustainable alternative to synthetic plastic. *Polysaccharides, 3*(1), 32–58. https://doi.org/10.3390/polysaccharides3010002

Wan, V. C. H., Lee, C. M., & Lee, S. Y. (2007). Understanding consumer attitudes on edible films and coatings: Focus group findings. *Journal of Sensory Studies, 22*(3), 353–366.

Wang, Y., Liang, B., Bao, H., Chen, Q., Cao, Y., He, Y., & Li, L. (2023). Potential of crop straw incorporation for replacing chemical fertilizer and reducing nutrient loss in Sichuan Province, China. *Environmental Pollution, 320*, 121034. https://doi.org/10.1016/j.envpol.2023.121034

Wijesekara, T., & Xu, B. (2024). New insights into sources, bioavailability, health-promoting effects, and applications of chitin and chitosan. *Journal of Agricultural and Food Chemistry, 72*(31), 17138–17152. https://doi.org/10.1021/acs.jafc.4c02162. Epub 2024 Jul 23.

Xu, Y., Yan, X., Zheng, H., Li, J., Wu, X., Xu, J., et al. (2024). The application of encapsulation technology in the food Industry: Classifications, recent Advances, and perspectives. *Food Chemistry: X, 21*, 101240.

Chapter 14
Effects of Natural Additives/Bioactive Compounds on the Functional and Thermal Properties of Packaging

Himani, Himanshu, Muskan, and Nishant Kumar

Abstract Edible packaging is crucial for the preservation of food from chemical, mechanical, physical, and microbiological damage by providing an insulation toward moisture and oxidation and regulating enzyme activity. Beneficial organic compounds derived from plants, such as aromatic compounds, oils, and extracts from the plants, enable biopolymers to improve their mechanical, thermal, and functional properties, including antibacterial and antioxidant effects. Active packaging, including natural chemicals, like antioxidants and antibacterial agents, may provide a sustainable solution to this issue. These organically derived functional agents surpass synthetic alternatives because they are more versatile and flammable, considering the growing client desire for ecologically friendly packaging alternatives. This study compiles information on the integration of natural bioactives, including extracts from plants and aromatic compounds, and their impact upon the thermal as well as functional aspects of packaging.

Keywords Antioxidant · Antimicrobial · Bioactive compounds · Essential oils · Thermal properties · Different scanning colorimetry

---

Himani · Himanshu (✉)
Department of Life Sciences (Botany), Baba Mast Nath University, Rohtak, Haryana, India

Muskan
Department of Physics, Baba Mast Nath University, Rohtak, Haryana, India

N. Kumar
Amity Food and Agriculture Foundation (AFAF), Amity University Noida, Noida, Uttar Pradesh, India

N. Kumar et al. (eds.), *Trends in Edible Packaging: A sustainable approaches for postharvest management of fruits and vegetables*,
https://doi.org/10.1007/978-3-032-13199-7_14

Introduction

Food preservation is essential for protecting food from exposure to the environment and even contamination, hence ensuring durability, together with reducing spoilage and deterioration (Yao et al., 2024). Concurrently, improved ecological consciousness and increasing customer desire for reliable and quality food prompted us to investigate innovative food preservation methods, such as active food packaging (Jiménez Romanillos et al., 2024). Food packaging is essential for maintaining nutritional value and lapse period by safeguarding against mechanical, physical, atmospheric, and microbial deterioration (Baishya & Kumar, 2025). Biopolymer-driven packaging materials have garnered considerable interest in the field of food packaging as a viable alternative to petroleum-based plastics, which adversely impact human health and the environment (Agarwal et al., 2023). Biopolymers are recognized worldwide for their biodegradability, non-toxicity, plentiful renewability, and environmental friendliness (Dutta & Sit, 2024). Nonetheless, biopolymeric materials are deficient in barrier, mechanical, thermal, and optical characteristics (Baishya & Kumar, 2025). Active packaging, incorporating natural additives, offers a subsequent remediation for food deterioration during storage. This approach minimizes undesirable effects, extends shelf life, enhances quality, and may confer health benefits to consumers (Deshmukh & Gaikwad, 2024).

Food additives are substances that are either natural or synthetic and are incorporated into food for technical purposes, such as preventing spoilage and bacterial degradation, or for sensory purposes, such as improving taste or color (Mesías et al., 2021). Food additives are classified into six categories according to their functions: preservatives, nutritional additions, texturizers, coloring agents, flavorings, and miscellaneous additives (Novais et al., 2022). Depending on the type of food, some additives may be utilized concurrently (Mesías et al., 2021). The food is negatively impacted by chemical additives, and they are not economically feasible (Obahiagbon & Ogwu, 2024). Edible materials designed for packaging can serve as reservoirs for bioactive and nutraceutical substances, thereby fulfilling an active function in packaging. The purpose of micro- or nano-encapsulation of nutraceuticals and bioactive compounds is to make these substances more stable and useful (Trajkovska Petkoska et al., 2021).

Plant-derived extracts and essential oils (EOs) are examples of natural beneficial additives that can be added to biopolymers to make them better at protecting against chemicals, mechanical stress, and microbes, as well as making them stronger and more resistant to oxidation. These additives serve as environmentally friendly alternatives to synthetic nanofillers, which may pose toxicity risks and interact with food, especially in food packaging applications. Functional agents from natural sources are better than synthetic ones since they are biocompatible, biodegradable, and safer to use with food. This fits with the growing number of people who want eco-friendly and clean-label packaging options (Baishya & Kumar, 2025). Natural additions that stop browning, kill microorganisms, and protect from free radicals have improved the nutritional and structural qualities of edible films (Kaur et al., 2023).

The need for materials that are better for the environment and cause less pollution has forced all businesses to make changes. During this situation, edible coatings and films offer an appealing substitute to plastic replacements since they are safe, renewable, cheap, and long-lasting. Over the past few decades, there has been significant research conducted to find combinations of new coating materials to identify the optimal combinations of novel coating materials that can perform a variety of functions. Elderberry extracts have gained significant attention as they improve food quality. This gives the coating important properties, like antioxidant and antimicrobial activities. The present research concentrates on improving films and coatings to enable their extensive industrialization in an environmentally sustainable manner. The goal of this study is to summarize how natural bioactive chemicals affect the thermal and functional qualities of packaging materials. The research also investigates the techniques for integrating natural extracts into films and coatings. More research is needed, especially on a larger scale, to improve the quality of coatings and the conditions in which they work. It is also important to find out how well the packaging, the finished product, and the human body work together.

Natural Bioactive Compound–Based Packaging

The need for easy-to-use, clear, and long-lasting packaging has led to the creation of new packaging technologies. To make food safer, more stable, more useful, and longer-lasting, we need new ideas like active food packaging, which adds active substances to packaging materials. This will make consumers feel better (Yildirim et al., 2018). Bioactive chemicals derived from agricultural and culinary by-products exhibit considerable prospects regarding the development of edible films and coatings. The latest study asserts that the coatings should maintain or improve the quality of food, allow it to last longer, and help make packaging materials that are better for the environment (Gomes et al., 2020). Antimicrobial packaging involves the process in which antimicrobial substances, including chitosan (CHS), plant extracts, essential oils, and inorganic nanoparticles, are added to films to stop the growth of infectious germs and keep food from getting contaminated (Chawla et al., 2021). Blueberry fruit and leaf extracts (BLEs) were added to chitosan coatings to make blueberries last longer. BLE showed that it could kill bacteria, and the chitosan coatings that had BLE in them slowed down the decay process, maintained more phenolic content, and improved the ability of the coatings to scavenge radicals compared to the control samples. Research has shown that edible coatings have antioxidant and antibacterial properties, stop microbes from growing, slow down deterioration, and maintain good sensory qualities (Messinese et al., 2024).

The novel food packaging is played an important roles in the food preservation compared to traditional approaches. The inclusion of antioxidant compounds in food packaging system helps in the slow down lipid oxidation and keep food from diminishing in its taste and nutritional value (Siciliano et al., 2024). Studies have shown that beef samples that were packaged in active packaging were more durable and had a longer shelf life when it came to oxidative stability. The beneficial outcomes were because presence of bioactive compounds in the films acted as antioxidants (Grimaldi et al., 2022).

Potato remnants, such as peels, were used to make edible films. Peels were broken down in a mild acid solution and processed into gelatin using varied amounts of plasticizers (glycerol or polyglycerol-3), followed by casting and pressing to form films (Merino et al., 2021).The finding indicated that films made with polyglycerol-3 were better at resisting heat. The films that were plasticized with polyglycerol-3 had better mechanical characteristics. However, the oxygen permeability was not good enough to keep the food safe from oxidation for a long time. This implies that these materials are more suitable for packaging powdered food when oxygen-shielding properties are not required. This means that these materials are better as dry food packaging, in which oxygen-shielding properties are not required (Merino et al., 2021).

Incorporation Methods for Active Agents in Packaging

Pouches, tags, and pads are the most common ways in which active agents are added to or used on the surfaces of packaging materials. These compounds may emanate from nature or may be made in a lab, and they are used to achieve different goals, such as antibacterial or antioxidant properties. People are becoming more interested in using natural antioxidants and antibacterial agents in food because manufactured active additives and preservatives can have harmful effects. Also, the focus on natural substances is due to their bioactive parts and their possible economic value since they mostly come from plants that are not used much and food waste (Singh et al., 2022). Bioactive substances can be carried by edible films and coatings. This strategy may work better than putting them directly into food formulations, which usually only have short-term benefits. By allowing them to diffuse slowly through the food surface, adding bioactive chemicals to packaging materials may help them stay active for a longer time (Hamed et al., 2022).

Addition of Sachets into a Food Package

Incorporating a sachet into a food product has been determined to be effective for volatile antibacterial agents. The volatile disinfectant compounds are released into the region prior to the packaging, which stops microorganisms from growing without getting into proximity with the food (Rupérez et al., 2025). The inclusion of a sachet within the packing material for food is the best way to use an antimicrobial packaging solution in business (Ahmed et al., 2022). The carrier or emitting sachet is based on the idea that the antimicrobial agent exists within the container. The sachet is put inside the food packaging bags or containers, away from the external environment (Yadav & Dutta, 2024). Antimicrobial substances are either adhered to the interior of the package or are contained in loose packets. Many researchers have tried to make food last longer by using antimicrobial packaging and sachets composed of polymer and essential oil mixtures. This means that preservatives are not needed (Petchwattana et al., 2021). Plant cells produce essential oils, which are naturally occurring substances. They may be extracted from a variety of elements and plants and contain antibacterial, antiparasitic, and antiviral substances that can be effectively incorporated into antimicrobial packaging materials. The addition of *Bunium percicum* and *Mentha piperita* essential oils to polylactic acid (PLA) films demonstrated their exceptional antimicrobial activity against *Staphylococcus aureus*, *Enterobacteriaceae,* and *Pseudomonas*. Additionally, they were highly effective in the destruction of bacteria (Petchwattana et al., 2021).

Coating or Adsorption on Polymer Substrate

The use of an antibacterial agent, along with other substances, to cover fruits and vegetables is still in the early phases of development (Oyom et al., 2022). Waxes or coating preparations for treating fresh produce contain different active ingredients that work as antimicrobial agents to stop the growth of germs (Gupta et al., 2024). Products are often coated with antimicrobial active chemicals that cannot withstand the high temperatures encountered with polymer manufacturing. The principal contact layer in food packaging films is recommended to be composed of films that are edible and biodegradable as reservoirs for antimicrobial agents (Deshmukh & Gaikwad, 2024). Thin layers of edible ingredients make up edible coatings. Consumers can choose to eat them or take them off. The main benefits are that they are biodegradable and edible, and they can be made cheaply and can last a long time, which is good for the environment (Barrea et al., 2018). For instance, the application of essential oregano oil to mango surface reduces the risk of *Salmonella enterica* (Petchwattana et al., 2021), Also, the kiwi fruit has been coated with nanoemulsion coatings, which possess antibacterial properties (Manzoor et al., 2021), while gellan coatings have been added to strawberries, resulting in improved quality during the storage period (Tomadoni et al., 2018).

Immobilization of Antimicrobial Compounds

Previous research has examined the fixation of antimicrobial agents onto polymers. An immobilization process takes place at the functional groups of polymers and antimicrobials (Sofi et al., 2018). Functional groups within the domain of polymers include acrylic acid, polystyrene, vinyl acetate, and ionomers (Deshmukh &

Gaikwad, 2024). The sorts of bonds that hold together immobilized polymers or films have a big effect on their antibacterial activity, which makes it easier for active drugs to slowly leak into the food. Lysozyme and chitinase are effective against gram-positive bacteria. The covalent attachment of glucose oxidase to an insoluble substrate that works with packaging materials makes it easier for glucose and oxygen reaction mixtures to turn into antibacterial hydrogen peroxide (Deshmukh & Gaikwad, 2024).

Antimicrobial Agents' Incorporation into the Polymer Matrix

Heat-pressing, extrusion heating, or solution casting are all methods that are employed to incorporate antimicrobial compounds into the packaging matrix. The technique could be used in business for delivering drugs, insecticides, household appliances, fabrics, and medical gadgets. The main problem with this approach is that high-heat treatments like extrusion and heat-pressing make it less effective in killing germs (Iheaturu et al., 2018). Enzymes, such as lactoferrin, lysozymes, and lactoperoxidase, as well as peptides like magainins and cecropins, and natural phenols are widely used in packing, especially when making polymers. When chitosan–ZnO nanoparticles are mixed with clove essential oil (CEO), they can kill *P. aeruginosa*, *S. aureus*, and *E. coli* (Gasti et al., 2022). Incorporating *Eucalyptus globulus* essential oil (EGEO) into chitosan composite films exhibited antimicrobial properties by inhibiting microorganisms (Deshmukh & Gaikwad, 2024), as demonstrated by composite films of chitosan and thyme essential oil (Lian et al., 2020). Edible coatings are predominantly employed on dairy products, fruits and vegetables, and meat (Kumar et al., 2022a). Typically, packaging substances are composed of a cohesive, structured biopolymer, additive, and/or a solvent. They have the capacity to contain a diverse array of active elements, including colorants, flavor, nutrients, and antibacterial and antioxidant compounds. Adding these components to the final product can enhance its nutritional value, extend its shelf life, and enhance its flavor (Lara-Gómez et al., 2025). Health and environmental concerns have contributed to further examination of packaging that uses natural antioxidants and antibacterial sources, like natural extracts. These compounds are widely employed as active components in edible film and coating compositions (Kaur et al., 2023).

Effects of Natural Bioactive-Based Packaging on Different Properties of Packaging

In recent decades, considerable research has been conducted toward the advancement of sustainable packaging alternatives, primarily motivated by growing concerns regarding human health and environmental sustainability (Tandon et al., 2023). The incorporation and delivery of bioactive compounds, such as plant extracts and essential oils (EOs), have made biopolymer-based packaging materials intriguing candidates (Das et al., 2025). The incorporation of these natural additives has been demonstrated to enhance the bio-functional, barrier, and mechanical properties of packaging films, with a particular emphasis on their antioxidant and antimicrobial properties. However, these modifications frequently result in diminished thermal stability, thereby reducing heat resistance (Vieira et al., 2022). The impact of integrating natural materials on the thermal properties of food packaging materials has been the subject of a limited number of studies. This review examines the thermal properties of food packaging films with respect to the incorporation of natural materials, such as essential oils, extracts derived from plants, and nanomaterials (Kumarihami et al., 2023). The subsequent sections outline the effects of incorporating natural additives and bioactive compounds on thermal and functional properties.

Thermal Properties

The thermal characteristics of packaging materials are crucial because they significantly influence the material's ability to interact with the food it contains (Kokkuvayil Ramadas et al., 2024). Limitations in the thermal resistance of biopolymer-derived materials for packaging continue to be a significant obstacle because of their replacement of conventional nonbiodegradable plastics in the food industry. The preservation of structural integrity and the minimization of degradation under variable temperature conditions that are encountered during processing, transportation, and storage are contingent upon the maintenance of adequate thermal resistance in such films. This property is particularly important to ensure the quality and longevity of packaged food (Shah et al., 2024). The thermal resistance and ultimate functional efficiency of translucent κ-carrageenan films have been demonstrated to be improved by the addition of grapefruit essential oil (GFO). In this study, varying concentrations of grapefruit essential oil (0.1%, 0.2%, and 0.3%) were incorporated into κ-carrageenan-based packaging films. The outcomes suggested that the thermal stability of the developed films was considerably enhanced by the addition of grapefruit essential oil (GFO) (Kokkuvayil Ramadas et al., 2024). According to Gupta et al. (2024), the thermal stability of carrageenan films was enhanced by the incorporation of cellulose nanocrystals derived from amla pomace (Kokkuvayil Ramadas et al., 2024).

Functional Properties

Cellulose nanostructures derived from amla pomace were incorporated to improve the thermal endurance of carrageenan films. They have the potential to make the packaging better and add additional features, such as antibacterial effects, antioxidant characteristics, and sensors that can keep track of and detect various components in a container (Carvalho et al., 2025). The active ingredients that kill bacteria also help fight fungi, improve color, make food last longer, and show phytochemical potential (Kaur et al., 2023). Plant-derived antioxidants from sources such as herbs, spices, fruits, and vegetables are regarded as promising additives for incorporation into films and coatings. These bioactive molecules play a protective role by reducing food degradation caused by exposure to heat, oxygen, and light. Furthermore, the inclusion of antimicrobial agents, particularly essential oils and polyphenols, within biopolymer-based films and coatings effectively suppresses microbial proliferation, thereby enhancing product safety, extending the duration of storage, and maintaining overall quality (Kaur et al., 2023).

Phytochemicals in plants are synthesized through the metabolic breakdown of sugars, organic acids, and polysaccharides. These compounds exhibit a diverse array of bioactivities, like antiproliferative, antidiabetic, antioxidant, neuroprotective, and antihypertensive effects (Himanshu & Khangwal, 2025).

Numerous investigations have highlighted the antiviral and antibacterial properties of plants (Dutta & Sit, 2023). In recent years, advancements have been achieved in the regulation of the liberation and distribution of nutraceuticals and bioactive compounds and the extension of the shelf life of perishable food products by utilizing pomegranate fruit and its underutilized by-products. The application of pomegranate peel extract to edible film matrices has shown a high degree of compatibility between the peel components and the matrix. This has led to enhanced biochemical properties, mechanical strength, and structural integrity, as well as increased phenolic and antioxidant levels and antibacterial activity.

The incorporation of pomegranate rind extract into edible coatings that are applied to food items has been shown to delay pigment degradation, inhibit lipid oxidation, and suppress microbial growth, thereby extending product shelf life while preserving sensory attributes such as flavor and aroma.

A multitude of studies have emphasized the antiviral, antifungal, and antibacterial properties of plants (Table 14.1) (Dutta & Sit, 2023). This, in turn, preserves sensory attributes such as flavor and aroma, thereby extending the shelf life of the product (Kumar et al., 2022b).

Phoenix dactylifera seeds contain unique phytochemicals that exhibit a variety of health benefits, such as anti-inflammatory, antioxidant, and antimicrobial properties (Kumar et al., 2024). These seeds are specifically abundant in bioactive compounds, including flavonoids and polyphenols, which are responsible for their significant antioxidant and antibacterial properties. These attributes are advantageous in that they prolong the expiration life of food products and improve their safety (Al-Khalili et al., 2025).

Table 14.1 Applications of bioactive compounds in polymer-based films for extending shelf life and ensuring food safety

S. No.	Phytochemicals	Films	Applications	References
1.	Ferulic acid	The solvent casting method was employed to produce poly(butylene adipate-co-terephthalate) (PLA-PBAT) films that contained ferulic acid	Antibacterial activity against *Escherichia coli*	Sharma et al. (2020)
2.	Chitosan polymers grafted with vanillic acid (VA) and coumaric acid	Vanillic acid and coumaric acid were used to graft chitosan, and the resultant derivatives were evaluated for their antioxidant and antimicrobial properties in comparison to chitosan that was grafted with gallic acid and ferulic acid	*Staphylococcus aureus* proliferation was inhibited by chitosan grafted with coumaric acid, which demonstrated significant potential as a food preservative. Ferulic acid–modified chitosan exhibited the lowest minimum inhibitory concentration (MIC) against *S. aureus* among the tested derivatives, whereas gallic acid–grafted chitosan was the most efficacious against *Pseudomonas aeruginosa*	Chatterjee et al. (2015)
3.	Chitosan/polyvinyl alcohol (PVA) composites with vanillic acid	The solvent casting method, which is recognized for its cost-effectiveness, was employed to produce chitosan/polyvinyl alcohol films that were incorporated with vanillic acid	The synergistic effect against *S. aureus*, *E. coli*, and *C. albicans* infections resulted in an increased inhibition zone, demonstrating significant antimicrobial efficacy	Eelager et al. (2023)
4.	Vanillic acid–conjugated chitosan hydrogel films	A carbodiimide-mediated coupling method was employed to synthesize VA-g-CHS hydrogel films	The shelf life of cherry tomatoes was significantly enhanced compared to that of conventional storage when VA-g-CHS hydrogel films were employed, allowing them to remain fresh for up to 10 days at approximately 25 °C	Bakar et al. (2024)
5.	Polyvinyl alcohol/sodium alginate (SA) composite films were prepared with stigmasterol (STG) derived from corn silk	Stigmasterol (STG) was incorporated into polyvinyl alcohol/sodium alginate (PVA/SA) filaments that were derived from corn silk. The films' opacity increased, their solubility decreased, their water vapor permeability decreased, and their moisture content decreased as a consequence of the STG incorporation	The STG-loaded films demonstrated robust antioxidant and antimicrobial properties, which substantially extended the refrigerated shelf life of white button mushrooms	Khan et al. (2025)

(continued)

Table 14.1 (continued)

S. No.	Phytochemicals	Films	Applications	References
6.	β-cryptoxanthin encapsulated chitosome	Chitosan encapsulated with β-cryptoxanthin for meat preservation	The chitosan encapsulated with β-cryptoxanthin demonstrated antimicrobial activity against *Staphylococcus epidermidis*, *P. aeruginosa*, *E. coli*, *S. aureus*, and *Enterococcus faecalis*, highlighting its potential as an effective strategy for extending the freshness and shelf life of raw poultry meat	Dutta et al. (2025)
7.	Sweet whey/starch-based active films incorporated with chlorogenic acid (CGA)	Chlorogenic acid (CGA) was first embedded into a sweet whey/starch matrix as a functional compound for the development of biodegradable active packaging films	CGA demonstrates antioxidant potential and antibacterial activity against *E. coli*, as demonstrated by the ABTS (2,2′-azino-bis-(3-ethylbenzothiazoline-6-sulfonic acid) assay method. Additionally, it helps delay banana browning by reducing oxidative processes	Dai et al. (2024)
8.	Cellulose acetate–based packaging films enriched with natural pigments such as norbixin, lycopene, or zeaxanthin	Norbixin, zeaxanthin, or lycopene was incorporated into the formulation of cellulose acetate films	Films enriched with 0.1% norbixin or lycopene exhibited greater efficacy in preserving the stability of sunflower oil than those containing zeaxanthin	Assis et al. (2020)

In one study, varying concentrations of date seed powder (DSP) (5%, 10%, 15%, and 20%) were incorporated into a carboxymethyl cellulose (CMC) film-forming solution. The structural, mechanical, antioxidant, and antimicrobial properties of the resulting films were analyzed to ascertain the impact of DSP addition. The integration of DSP resulted in an increase in film thickness (from 0.11 to 0.15 mm), a decrease in solubility (from 95.31% to 77.23%), and the transmission of antimicrobial activity against foodborne pathogens, including *Escherichia coli, S. aureus,* and *Salmonella typhimurium.* It is important to note that the antimicrobial properties against *E. coli* and *S. typhimurium* were only observed at greater DSP concentrations (15% and 20%). The film's antioxidant potential, as determined by DPPH (2,2-diphenyl-1-picrylhydrazyl) and ABTS (2,2′-azino-bis-(3-ethylbenzothiazoline-6-sulfonic acid) radical scavenging assays, increased in proportion to its DSP content. Specifically, DPPH activity increased from 0% to 51.98% and ABTS activity from 0% to 84.58%. The film's extension at distortion increased from 79.03% to 85.95%, while its tensile strength decreased from 7.66 to 5.12 MPa despite its increased water vapor permeability. In summary, the results indicate that DSP has the potential to serve as both a bioactive and a reinforcing additive in biodegradable edible films based on CMC, thereby facilitating the creation of natural active food packaging systems (Lawal et al., 2023).

Latest advancements in edible coatings and films enriched with phenolic compounds have contributed significantly to enhancing packaging performance, similar to the developments observed in active packaging technologies. Research efforts have particularly emphasized the integration of phenolic compounds into edible matrices to strengthen their antioxidant and antimicrobial potential. Such functionalized coatings contribute significantly to the preservation of food quality and the extension of its expiring life. For example, coatings derived from chitosan that are augmented with green tea extract are rich in phenolics—such as catechin and epicatechin gallate—and have been applied to walnuts to prevent spoilage by inhibiting lipid oxidation and fungal contamination. The antioxidant capacity of green tea–derived phenolics is primarily attributed to their ability to suppress the initiation of radical chain reactions, chelate transition metal ions, and neutralize free radicals, thereby limiting lipid peroxidation (Singh et al., 2022).

Conclusion

Food packaging is essential in the modern food production chain, as it guarantees the longevity and safeguarding of food from refining to consumption. In recent years, packaging strategies have become increasingly innovative in response to consumer demands and global sustainability concerns. The design of packaging materials must consider not only manufacturing and marketing aspects but also consumer usability, environmental impact, and the incorporation of emerging technologies, including nanotechnology and biotechnology. Effective packaging is expected to extend shelf life, safeguard quality and safety, contribute to consumer health, and minimize ecological footprint.

One promising approach is the incorporation of natural plant-derived compounds into packaging systems, highlighting the intersection of phytochemistry and food technology. Bioactive molecules with antioxidant and antimicrobial properties are abundant in plants, which, when embedded in packaging materials, enhance food safety and preservation. This eco-friendly alternative to conventional packaging is particularly well-suited for food applications, as edible films and coatings are enriched with natural ingredients. Current developments emphasize the use of plant extracts, bioactive derivatives, and vitamins as functional additives in edible matrices. Commonly applied natural compounds include ascorbic acid, citric acid, tea polyphenols, α-tocopherol, rosemary extract, and murta extract. These bioactive substances provide multiple benefits, such as antimicrobial activity, the inhibition of enzymatic browning, and protection against oxidative stress, thereby enhancing both the functional and structural attributes of edible films and coatings.

References

Agarwal, A., Shaida, B., Rastogi, M., & Singh, N. B. (2023). Food packaging materials with special reference to biopolymers-properties and applications. *Chemistry Africa, 6*(1), 117–144.

Ahmed, S., Sameen, D. E., Lu, R., Li, R., Dai, J., Qin, W., & Liu, Y. (2022). Research progress on antimicrobial materials for food packaging. *Critical Reviews in Food Science and Nutrition, 62*(11), 3088–3102.

Al-Khalili, M., Rahman, S., & Al-Habsi, N. (2025). Date seed–Added biodegradable films and coatings for active food packaging applications: A review. *Packaging Technology and Science, 38*(6), 445–472.

Assis, R. Q., Rios, P. D., de Oliveira Rios, A., & Olivera, F. C. (2020). Biodegradable packaging of cellulose acetate incorporated with norbixin, lycopene or zeaxanthin. *Industrial Crops and Products, 147*, 112212.

Baishya, H., & Kumar, S. (2025). Natural functional agents reinforced biopolymer composites for active food packaging applications: A review. *Trends in Food Science & Technology, 161*, 105042.

Bakar, B., Pekdemir, S. S., Birhanlı, E., Ulu, A., Pekdemir, M. E., & Ateş, B. (2024). Unveiling the effect of molecular weight of vanillic acid grafted chitosan hydrogel films on physical, antioxidant, and antimicrobial properties for application in food packaging. *International Journal of Biological Macromolecules, 256*, 128397.

Barrea, L., Di Somma, C., Muscogiuri, G., Tarantino, G., Tenore, G. C., Orio, F., Colao, A., & Savastano, S. (2018). Nutrition, inflammation and liver-spleen axis. *Critical Reviews in Food Science and Nutrition, 58*(18), 3141–3158.

Carvalho, L. G., de Carvalho, A. P. A., & Conte-Junior, C. A. (2025). Bioactive compounds, antimicrobial properties of Açaí (Euterpe oleracea mart.) and its processing residues–insights for smart packaging and green chemistry. Biocatalysis and agricultural. *Biotechnology, 65*, 103540.

Chatterjee, N. S., Panda, S. K., Navitha, M., Asha, K. K., Anandan, R., & Mathew, S. (2015). Vanillic acid and coumaric acid grafted chitosan derivatives: Improved grafting ratio and potential application in functional food. *Journal of Food Science and Technology, 52*(11), 7153–7162.

Chawla, R., Sivakumar, S., & Kaur, H. (2021). Antimicrobial edible films in food packaging: Current scenario and recent nanotechnological advancements-a review. *Carbohydrate Polymer Technologies and Applications, 2*, 100024.

Dai, X., Dong, F., Dong, Z., Bai, Z., & Mao, L. (2024). Enhanced antibacterial and antioxidant activities of chlorogenic acid loaded sweet whey/starch active films for edible food packaging. *LWT, 199*, 116118.

Das, A., Toki, G. F. I., & Mia, R. (2025). Antimicrobial biopolymers for food preservation: Recent developments and applications based on essential oils and nanomaterials. *Green Materials for Active Food Packaging*, 93–130.

Deshmukh, R. K., & Gaikwad, K. K. (2024). Natural antimicrobial and antioxidant compounds for active food packaging applications. *Biomass Conversion and Biorefinery, 14*(4), 4419–4440.

Dutta, D., Shrivastava, S., Gupta, B. D., & Dutta, D. (2025). β-cryptoxanthin encapsulated chitosome: An innovative coating for raw chicken meat preservation. *Food Packaging and Shelf Life, 50*, 101542.

Dutta, D., & Sit, N. (2023). Application of natural extracts as active ingredient in biopolymer based packaging systems. *Journal of Food Science and Technology, 60*(7), 1888–1902.

Dutta, D., & Sit, N. (2024). A comprehensive review on types and properties of biopolymers as sustainable bio-based alternatives for packaging. *Food Biomacromolecules, 1*(2), 58–87.

Eelager, M. P., Masti, S. P., Chougale, R. B., Hiremani, V. D., Narasgoudar, S. S., Dalbanjan, N. P., & SK, P. K. (2023). Evaluation of mechanical, antimicrobial, and antioxidant properties of vanillic acid induced chitosan/poly (vinyl alcohol) active films to prolong the shelf life of green chilli. *International Journal of Biological Macromolecules, 232*, 123499.

Gasti, T., Dixit, S., Hiremani, V. D., Chougale, R. B., Masti, S. P., Vootla, S. K., & Mudigoudra, B. S. (2022). Chitosan/pullulan based films incorporated with clove essential oil loaded chitosan-ZnO hybrid nanoparticles for active food packaging. *Carbohydrate Polymers, 277*, 118866.

Gomes, A. C. A., da Costa Lima, M., de Oliveira, K. Á. R., dos Santos Lima, M., Magnani, M., Câmara, M. P. S., & de Souza, E. L. (2020). Coatings with chitosan and phenolic-rich extract from acerola (Malpighia emarginata DC) or jabuticaba (Plinia jaboticaba (Vell.) Berg) processing by-product to control rot caused by Lasiodiplodia spp. in papaya (Carica papaya L.) fruit. *International Journal of Food Microbiology, 331*, 108694.

Grimaldi, M., Pitirollo, O., Ornaghi, P., Corradini, C., & Cavazza, A. (2022). Valorization of agro-industrial byproducts: Extraction and analytical characterization of valuable compounds for potential edible active packaging formulation. *Food Packaging and Shelf Life, 33*, 100900.

Gupta, S., Ghoshal, G., & Arya, R. K. (2024). Antimicrobial coatings for fruits and vegetables. *Functional Coatings for Biomedical, Energy, and Environmental Applications*, 543–569.

Hamed, I., Jakobsen, A. N., & Lerfall, J. (2022). Sustainable edible packaging systems based on active compounds from food processing byproducts: A review. *Comprehensive Reviews in Food Science and Food Safety, 21*(1), 198–226.

Himanshu, N. K., & Khangwal, I. (2025). Functional attributes and bio-prospects of fruit peel waste. *Foods and Raw Materials*.

Iheaturu, N. C., Nwakaudu, A. A., Nwakaudu, M. S., & Owuamanam, C. I. (2018). The use of natural antioxidant active polymer packaging for food preservation: A review. *Futo Journal Series, 4*, 94–112.

Jiménez Romanillos, E., Williams, H., & Wever, R. (2024). Unpacking behaviours: A literature study and research agenda on consumer behaviour in packaging-free systems. *Packaging Technology and Science, 37*(9), 821–840.

Kaur, J., Singh, J., Rasane, P., Gupta, P., Kaur, S., Sharma, N., & Sowdhanya, D. (2023). Natural additives as active components in edible films and coatings. *Food Bioscience, 53*, 102689.

Khan, U., Shao, Y., Abdalmegeed, D., Sedjoah, R.-C. A.-A., Wu, Z., Hayat, F., Li, L., & Xin, Z. (2025). Characterization of antibacterial polyvinyl alcohol/sodium alginate films loaded with stigmasterol isolated from corn silk (stigma maydis) for food packaging. *Progress in Organic Coatings, 200*, 109004.

Kokkuvayil Ramadas, B., Rhim, J.-W., & Roy, S. (2024). Recent progress of carrageenan-based composite films in active and intelligent food packaging applications. *Polymers, 16*(7), 1001.

Kumar, L., Ramakanth, D., Akhila, K., & Gaikwad, K. K. (2022a). Edible films and coatings for food packaging applications: A review. *Environmental Chemistry Letters, 20*(1), 875–900.

Kumar, N., Daniloski, D., D'cunha, N. M., Naumovski, N., & Petkoska, A. T. (2022b). Pomegranate peel extract–A natural bioactive addition to novel active edible packaging. *Food Research International, 156*, 111378.

Kumar, N., Khangwal, I., & Upadhyay, A. (2024). Assessment of nutritional composition, phytochemical screening, antioxidant, and antibacterial activities of date palm (Phoenix dactylifera) seeds. *Discover Food, 4*(1), 1–12.

Kumarihami, H. M. P. C., Kumar, N., Pratibha, P., & A. T., & Neeraj. (2023). Effect of natural materials on thermal properties of food packaging film: An overview. In *Natural materials for food packaging application* (pp. 255–274).

Lara-Gómez, A. B., Castro-Rosas, J., Hernández-Hernández, E., Cadena-Ramírez, A., Hernández-Cruz, M. Á., & Gómez-Aldapa, C. A. (2025). Mechanical and barrier properties of films containing bioactive compounds with antimicrobial and antioxidant activity. *Food and Bioprocess Technology*, 1–31.

Lawal, K. G., Riaz, A., Mostafa, H., Stathopoulos, C., Manikas, I., & Maqsood, S. (2023). Development of carboxymethylcellulose based active and edible food packaging films using date seed components as reinforcing agent: Physical, biological, and mechanical properties. *Food Biophysics, 18*(4), 497–509.

Lian, H., Shi, J., Zhang, X., & Peng, Y. (2020). Effect of the added polysaccharide on the release of thyme essential oil and structure properties of chitosan based film. *Food Packaging and Shelf Life, 23*, 100467.

Manzoor, S., Gull, A., Wani, S. M., Ganaie, T. A., Masoodi, F. A., Bashir, K., Malik, A. R., & Dar, B. N. (2021). Improving the shelf life of fresh cut kiwi using nanoemulsion coatings with antioxidant and antimicrobial agents. *Food Bioscience, 41*, 101015.

Merino, D., Paul, U. C., & Athanassiou, A. (2021). Bio-based plastic films prepared from potato peels using mild acid hydrolysis followed by plasticization with a polyglycerol. *Food Packaging and Shelf Life, 29*, 100707.

Mesías, F. J., Martín, A., & Hernández, A. (2021). Consumers' growing appetite for natural foods: Perceptions towards the use of natural preservatives in fresh fruit. *Food Research International, 150*, 110749.

Messinese, E., Pitirollo, O., Grimaldi, M., Milanese, D., Sciancalepore, C., & Cavazza, A. (2024). By-products as sustainable source of bioactive compounds for potential application in the field of food and new materials for packaging development. *Food and Bioprocess Technology, 17*(3), 606–627.

Novais, C., Molina, A. K., Abreu, R. M. V., Santo-Buelga, C., Ferreira, I. C. F. R., Pereira, C., & Barros, L. (2022). Natural food colorants and preservatives: A review, a demand, and a challenge. *Journal of Agricultural and Food Chemistry, 70*(9), 2789–2805.

Obahiagbon, E. G., & Ogwu, M. C. (2024). Organic food preservatives: The shift towards natural alternatives and sustainability in the global south's markets. In *Food safety and quality in the global south* (pp. 299–329). Springer.

Oyom, W., Zhang, Z., Bi, Y., & Tahergorabi, R. (2022). Application of starch-based coatings incorporated with antimicrobial agents for preservation of fruits and vegetables: A review. *Progress in Organic Coatings, 166*, 106800.

Petchwattana, N., Naknaen, P., Cha-Aim, K., Suksri, C., & Sanetuntikul, J. (2021). Controlled release antimicrobial sachet prepared from poly (butylene succinate)/geraniol and ethylene vinyl alcohol coated paper for bread shelf-life extension application. *International Journal of Biological Macromolecules, 189*, 251–261.

Rupérez, D., Nerín, C., & Silva, F. (2025). Controlling beef microbial spoilage with diacetyl-based active packaging sachet. *Food Research International, 209*, 116176.

Shah, Y. A., Bhatia, S., Al-Harrasi, A., Oz, F., Khan, M. H., Roy, S., Esatbeyoglu, T., & Pratap-Singh, A. (2024). Thermal properties of biopolymer films: Insights for sustainable food packaging applications. *Food Engineering Reviews, 16*(4), 497–512.

Sharma, S., Jaiswal, A. K., Duffy, B., & Jaiswal, S. (2020). Ferulic acid incorporated active films based on poly (lactide)/poly (butylene adipate-co-terephthalate) blend for food packaging. *Food Packaging and Shelf Life, 24*, 100491.

Siciliano, S., Lopresto, C. G., & Lamonaca, F. (2024). From traditional packaging to smart bio-packaging for food safety: A review. *Euro-Mediterranean Journal for Environmental Integration, 9*(4), 1971–1986.

Singh, A. K., Kim, J. Y., & Lee, Y. S. (2022). Phenolic compounds in active packaging and edible films/coatings: Natural bioactive molecules and novel packaging ingredients. *Molecules, 27*(21), 7513.

Sofi, S. A., Singh, J., Rafiq, S., Ashraf, U., Dar, B. N., & Nayik, G. A. (2018). A comprehensive review on antimicrobial packaging and its use in food packaging. *Current Nutrition & Food Science, 14*(4), 305–312.

Tandon, R., Nikhanj, P., & Kaur, M. (2023). Green packaging, an imminent eco-friendly approach. In *Encyclopedia of green materials* (pp. 1–11). Springer.

Tomadoni, B., del Moreira, M. R., Pereda, M., & Ponce, A. G. (2018). Gellan-based coatings incorporated with natural antimicrobials in fresh-cut strawberries: Microbiological and sensory evaluation through refrigerated storage. *LWT, 97*, 384–389.

Trajkovska Petkoska, A., Daniloski, D., Kumar, N., Pratibha, & Broach, A. T. (2021). Active edible packaging: A sustainable way to deliver functional bioactive compounds and nutraceuticals. In *Sustainable Packaging* (pp. 225–264). Springer.

Vieira, I. R. S., de Carvalho, A. P. A., & de, & Conte-Junior, C. A. (2022). Recent advances in biobased and biodegradable polymer nanocomposites, nanoparticles, and natural antioxidants for antibacterial and antioxidant food packaging applications. *Comprehensive Reviews in Food Science and Food Safety, 21*(4), 3673–3716.

Yadav, S., & Dutta, P. K. (2024). Moisture-absorbent food packaging systems and the role of chitosan. In *Smart food packaging systems: Innovations and technology applications* (pp. 169–193). Wiley.

Yao, Q., Huang, F., Lu, Y., Huang, J., Ali, M., Jia, X.-Z., Zeng, X.-A., & Huang, Y. (2024). Polysaccharide-based food packaging and intelligent packaging applications: A comprehensive review. *Trends in Food Science & Technology, 147*, 104390.

Yildirim, S., Röcker, B., Pettersen, M. K., Nilsen-Nygaard, J., Ayhan, Z., Rutkaite, R., Radusin, T., Suminska, P., Marcos, B., & Coma, V. (2018). Active packaging applications for food. *Comprehensive Reviews in Food Science and Food Safety, 17*(1), 165–199.

Chapter 15
Effects of the Deposition Methods of Edible Coatings on Fruits and Vegetables

Moufida Chaari, Khaoula Elhadef, Sarra Akermi, and Slim Smaoui

Abstract Edible coatings have emerged as an innovative postharvest approach to preserving fruits and vegetables, addressing both food security challenges and agricultural sustainability. While edible films and coatings serve similar functions, they differ in processing methods and physical–mechanical properties. These coatings are specifically designed to enhance food shelf life, act as protective barriers, and improve nutritional value. With sustainability gaining importance, biomacromolecules—classified as Generally Recognized as Safe (GRAS) substances—are widely utilized in the development of edible coatings. Key biopolymer sources, such as polysaccharides, proteins, and lipids, or composite materials contribute to the formation of stable colloidal dispersions, facilitating the ease of processing through various techniques like blending, spraying, dipping, and electrospinning. However, certain structural and barrier limitations necessitate improvements in edible coatings. To address these challenges, researchers are developing composite materials, incorporating targeted functional molecules, and creating multilayered structures to enhance performance. These advancements help extend food preservation and shelf life while maintaining the texture, flavor, and nutritional integrity of fruits and vegetables.

Keywords Biopolymers · Bio-coating · Edible coatings · Fruits and vegetables · Deposition methods

M. Chaari · K. Elhadef · S. Akermi · S. Smaoui (✉)
Laboratory of Microbial and Enzymatic Biotechnologies and Biomolecules (LBMEB), Center of Biotechnology of Sfax (CBS), University of Sfax, Sfax, Tunisia
e-mail: slim.smaoui@cbs.rnrt.tn

N. Kumar et al. (eds.), *Trends in Edible Packaging: A sustainable approaches for postharvest management of fruits and vegetables*,
https://doi.org/10.1007/978-3-032-13199-7_15

Introduction

Fruits' and vegetables' quality is influenced by continuous changes in nutritional composition, sensory attributes, and microbiological properties throughout processing, storage, and distribution (Suhag et al., 2020). These variations result from interactions between the food and its surrounding environment, leading to significant alterations in texture, appearance, and safety. Several factors, including oxidative reactions, respiration, microbial spoilage, and enzymatic activity, can accelerate food deterioration, ultimately shortening its shelf life (Meitha et al., 2020). Given the increasing consumer demand for fresh and minimally processed foods, implementing effective preservation strategies is essential to maintaining food safety, extending shelf life, and reducing fruit and vegetable waste.

Fresh-cut fruits and vegetables have gained immense popularity due to their health benefits and convenience (Wallace et al., 2020). These ready-to-eat products provide essential nutrients while eliminating the need for extensive preparation. However, unlike canned, freeze-dried, or spray-dried products, fresh-cut produce is highly perishable, with a significantly shorter shelf life. The high moisture content and numerous exposed cut surfaces generated during processing increase susceptibility to texture degradation, enzymatic browning, moisture loss, and microbial contamination (Giannakourou & Tsironi, 2021). As a result, maintaining the freshness and quality of fresh-cut fruits and vegetables remains a significant challenge in the food industry. Several preservation techniques have been developed to address these issues, including modified atmosphere packaging, sterile water washing, heat treatment, irradiation, and the application of edible films and coatings (Giannakourou & Tsironi, 2021). Among these, edible coatings have garnered significant attention due to their environmentally friendly nature, cost-effectiveness, and ability to improve food safety and quality without altering the product's natural attributes. In fact, edible coatings serve as protective layers that regulate water vapor permeability, create a barrier against lipid oxidation, inhibit microbial growth, prevent enzymatic browning, minimize weight loss, slow down respiration rates, and control enzymatic activity (Zhang et al., 2020). Additionally, they help maintain the food's natural color, texture, and flavor, which are critical factors in consumer acceptance (Pham et al., 2023). The incorporation of natural antimicrobial agents and antioxidants further enhances their preservative properties, providing an added layer of protection against spoilage and foodborne pathogens (Ammar et al., 2025; Sowmyashree et al., 2021; Deng et al., 2024). These coatings can be formulated using a variety of biopolymers, including polysaccharides, proteins, and lipids, each offering distinct functional properties. Polysaccharide-based coatings, such as those derived from chitosan, alginate, and cellulose, are widely used due to their excellent film-forming abilities, biodegradability, and ability to provide effective moisture and gas barriers (Shiekh et al., 2022). Protein-based coatings, including those made from whey, soy, or gelatin, offer superior flexibility, while lipid-based coatings, such as beeswax (BW) and carnauba wax, provide enhanced water resistance, reducing moisture loss in high-humidity environments (Bari & Giannouli, 2022; Miranda et al., 2022; Chetia et al., 2023).

Extensive research has been conducted on developing edible coatings using various film-forming materials and active ingredients. These coatings have been applied to a wide range of fruits and vegetables, including apples (Najafi Marghmaleki et al., 2021), kiwis (Alav et al., 2024; Manzoor et al., 2021), strawberries (Yan et al., 2019), cantaloupes (Sami et al., 2021), tomatoes (Wardak et al., 2024), and cucumbers (Fan et al., 2021; Kumar et al., 2021). These studies have demonstrated that edible coatings can effectively delay ripening, reduce microbial spoilage, and maintain firmness, thereby extending the postharvest life of fresh produce. The incorporation of antimicrobial agents, antioxidants, and other functional ingredients has further improved the efficacy of edible coatings (Medina-Jaramillo et al., 2020). Essential oils (EOs) (Gutiérrez-Jara et al., 2021) and plant extracts (Esmaeili et al., 2022; Carbone et al., 2021; Ebrahimi & Rastegar, 2020), such as *Humulus lupulus*, *Eremurus,* and *Spirulina* extracts, possess strong antimicrobial and antifungal properties, making them effective additives in edible coatings. These natural compounds inhibit the growth of spoilage microorganisms and foodborne pathogens while enhancing the sensory attributes of coated foods. Similarly, the addition of antioxidants such as ascorbic acid and tocopherols helps prevent oxidative degradation, thereby preserving the color and nutritional quality of fruits (Najafi Marghmaleki et al., 2021; Keshari et al., 2022). Despite their numerous advantages, edible coatings face several challenges that limit their widespread industrial application. One of the primary limitations is the instability of bioactive compounds, which can reduce the effectiveness of edible coatings over time. To overcome these challenges, researchers have explored innovative approaches, such as biopolymer blends, composite materials, and multilayered structures, to enhance the performance of edible coatings (Sowmyashree et al., 2021; Yan et al., 2019; Zhang et al., 2025). Combining different biopolymers, such as polysaccharides and proteins, can improve the film strength, flexibility, and barrier properties. Additionally, surface modifications, cross-linking, and functionalization using eco-friendly techniques have gained attention for their ability to enhance the physical, chemical, and biological properties of edible coatings (Qin et al., 2020). While challenges remain, ongoing research and technological advancements in the field of edible coatings hold great promise for the future of food preservation.

Methods of Coating Application in Fresh Fruits and Vegetables

Dipping Method

The dipping method is the simplest technique, involving three key steps (Fig. 15.1): (i) immersion and dwelling, (ii) deposition, and (iii) evaporation of solvents (Ammar et al., 2025). After the excess solution is drained, the fruits or the vegetables are usually dried under ambient conditions or treated with a dryer. Previous studies have indicated that the density and morphology of coatings formed by dipping are greatly influenced by several factors, including immersion time; withdrawal speed; number of dip-coating cycles; coating solution properties, such as density, viscosity, and surface tension; substrate surface characteristics; and drying conditions (Tang & Yan, 2017). However, the dipping method has several disadvantages. It often results in a thick coating layer, which can significantly reduce fruit respiration, cause surface damage, and impair functionality. Additionally, microorganisms and dirt from the fruit surface can contaminate the coating solution, making industrial scaling difficult. Another limitation of the dipping method is the large amount of solution required per unit mass of product to ensure optimal dipping conditions (Pham et al., 2023).

Spraying Method

Spraying is the process of using a set of nozzles to distribute small droplets on the fruit surface (Fig. 15.1). Spraying methods are used in three different ways, namely air spray atomization, pressure atomization, and air-assisted airless atomization (Priya et al., 2023). The spraying technology allows multilayered applications such as interlayer solutions, as well as consistent coating with homogeneous thickness. Additionally, the coating thickness is greater than that of the dipping method due to

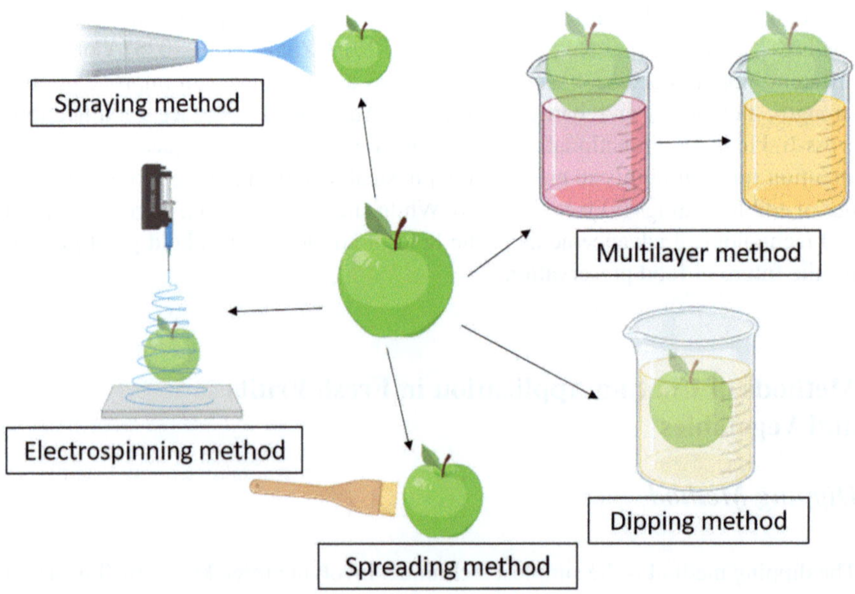

Fig. 15.1 Examples of bio-coating methods of fruits

the low viscosity of the solution (Ammar et al., 2025). Spraying is fast, efficient, and scalable for commercial production.

Spreading Method

The spreading technique is particularly effective for coating solutions with high viscosity (Pham et al., 2023). Several parameters influence the efficiency of coating deposition through spreading, including the quality of the substrate, drying conditions, liquid properties, and surface geometry (Priya et al., 2023). Brushing is usually performed by specialized operators, meaning that the human factor plays a significant role in determining the quality of the coating and the uniformity of its thickness (Shiekh et al., 2022).

Multilayered Coating (Layer-by-Layer)

The layer-by-layer (LBL) deposition method relies on the electrostatic interactions between the food surface and charged polyelectrolytes (Zhang et al., 2025). These interactions enhance the adhesion of the coating to the food surface and can be used to create coatings with two or more thin layers that are chemically or physically bonded to each other (Pham et al., 2023). This multilayered coating approach improves the effectiveness of the coating compared to traditional edible coatings. The use of multilayered coatings to increase the compactness of coating layers during postharvest storage of fruits has been documented, particularly for polysaccharides and charged polyelectrolytes that can form hydrogen and covalent bonds (Sowmyashree et al., 2021; Li et al., 2022a).

Electrospinning Coating Method

Electrospinning technology utilizes polymers as base materials, where liquid droplets are electrically charged to form a jet that undergoes stretching and elongation to produce fibers (Fig. 15.1). The key components of an electrospinning setup include a high-voltage power supply, a syringe pump, a spinneret (typically a blunt-tip hypodermic needle), and a conductive collector. During the process, the increasing potential difference between the spinneret and the collector enhances the surface charge density on the droplet, facilitating fiber formation (Xue et al., 2019). Due to their ultrafine structure, electrospun fibers can penetrate the surface pores of fruits and legumes, ensuring a more uniform and thorough coating compared to conventional methods (Shi et al., 2023). A key advantage of electrospinning is its ability to produce coatings of varying thicknesses, ranging from just a few nanometers to several microns, depending on the number of layers applied (Xue et al., 2019).

Cross-Linked Coating Method

This technique involves linking polymer chains using covalent and non-covalent bonds to form a three-dimensional network. Cross-linked coatings are usually applied by spraying, dipping, or spreading the coating solution onto the food surface, followed by the addition of a cross-linking agent to create a more compact and stable coating (Pham et al., 2023). Cross-linking is particularly effective for biopolymer materials made from proteins or polysaccharides, with proteins being more commonly used than polysaccharides due to their higher number of functional groups (Dai et al., 2020). Cross-linked coatings can reduce gas permeability and moisture loss, slowing down the ripening process in fruits and vegetables (Deng et al., 2024). These coatings can be used to encapsulate and release active agents like antimicrobial and antioxidant compounds, further extending the shelf life and safety of the produce (Wardak et al., 2024; Yıldırım-Yalçın et al., 2022). For example, edible coatings developed by Yıldırım-Yalçın et al. (Yıldırım-Yalçın et al., 2022) from grape juice, cross-linked starch, and glycerol were applied to strawberries.

Effects of Bio-coatings on the Physicochemical Quality of Fruits and Vegetables

Chitosan-Based Coatings

Chitosan, a biopolymer derived from chitin found in crab shells and insect exoskeletons, is gaining recognition in the food industry for its potential as a natural and sustainable coating material (Chakravarty & Edwards, 2022). It is biodegradable and environmentally friendly, making it a safe option for food applications. Regulatory authorities generally classify it as a safe substance, allowing its use as an edible coating for different food items. Due to its excellent film-forming capabilities, chitosan has drawn significant interest in the food sector as a functional coating material (Basumatary et al., 2022; Alizadeh-Sani et al., 2021; Amalraj et al., 2020). Additionally, chitosan can produce flexible, transparent films that act as protective barriers against moisture and gases, which are crucial for preserving food quality (Pires et al., 2021; Yilmaz et al., 2020). To further enhance its functionality, chitosan coatings can be enriched with bioactive compounds such as antioxidants, minerals, nanoparticles, and flavoring agents, improving both the nutritional profile and sensory appeal of food products (Chetia et al., 2023; Amor et al., 2021). For instance, the findings of Basumatary et al. (2022) revealed that incorporating eugenol oil nanoemulsion and *Aloe vera* gel into the chitosan-based coatings significantly improved their physicochemical and antimicrobial activity against foodborne bacterial pathogens *(Escherichia coli, Alcaligenes faecalis, Staphylococcus aureus,* and *Bacillus subtilis*). These enhanced coatings not only maintained the freshness of pineapples but also preserved their quality, extending their shelf life by up to 3 weeks under ambient storage conditions (Basumatary et al., 2022).

Findings from several studies on the impact of chitosan composite coatings on the physicochemical properties of fruits and vegetables are summarized in Table 15.1. For example, Zhang et al. (2020) demonstrated that a composite coating made of chitosan, zein, and α-tocopherol effectively preserved *Agaricus bisporus* mushrooms at low temperatures (4 °C). Mushrooms treated with this coating exhibited reduced weight loss, lower electrolyte leakage, lower browning index, decreased respiration rate, and lower polyphenol oxidase (PPO), peroxidase, and malondialdehyde levels compared to the control group. By the end of 12 days of storage, untreated mushrooms had a peroxidase activity level 4.96 times higher than those coated with the chitosan–zein–α-tocopherol (C/Z/T) film. Additionally, mushrooms coated with the C/Z/T film showed improved firmness, higher levels of catalase and SOD activity, increased total phenolic content, and more significant 2,2-diphenyl-1-picrylhydrazyl (DPPH) radical scavenging activity. The C/Z/T coating also extended the peak catalase activity period to the ninth day—longer than the sixth day observed in other treatments. Similarly, SOD activity peaked on the ninth day for coated mushrooms, compared to the third day in untreated samples. These findings suggest that the C/Z/T film effectively enhanced antioxidant capacity and delayed quality deterioration at 4 °C. In addition, a chitosan-based coating solution was formulated by incorporating varying concentrations of lemon essential oil (0.25%, 0.5%, 0.75%, and 1% v/v) (Chetia et al., 2023). This coating was applied to fresh grapes to assess its impact on their physicochemical properties and storage quality. The results indicated that grapes coated with 1% essential oil maintained the highest quality for up to 15 days at room temperature (25 ± 2 °C). This study suggests that a lemon essential oil–chitosan coating could serve as a natural preservative to enhance the storage quality of grapes (Chetia et al., 2023). Furthermore, Sultan et al. (2021) coated pears with chitosan–beeswax/pollen grain and exhibited a noticeable reduction in weight loss, decay, and softening rate. Thus, the chitosan–beeswax/pollen grain coating proves to be a safe and effective solution for fruit preservation.

Alginate-Based Coatings

Sodium alginate, a water-soluble polysaccharide, has a molecular weight ranging from 12,000 to 180,000 Daltons and exhibits colloidal behavior when dissolved in water. It is a linear copolymer composed of β-D-mannuronic acid and α-L-guluronic acid units linked by 1,4-glycosidic bonds. Sodium alginate serves as a thickening

Table 15.1 Bio-coatings for preserving fruits and vegetables

Deposition method	Coating material	Target	Main effects on quality	References
Dipping	Sodium alginate (SA)–α-tocopherol acetate	Carrot slice	Two percent sodium alginate and 1% α-tocopherol exhibited minimal variation in weight loss, total soluble solids (TSS), pH, whiteness index, reducing sugar, ascorbic acid content, and bacterial growth	Keshari et al. (2022)
Dipping	Sodium alginate–citric acid–ascorbic acid	Apples	Changes in eating quality parameters, appearance acceptance, and textural properties	Najafi Marghmaleki et al. (2021)
Dipping	Sodium alginate–carvacrol	Blueberries	Delaying postharvest spoilage	Medina-Jaramillo et al. (2020)
Dipping	Chitosan–ascorbic acid	Papaya	Maintaining the postharvest storage quality	Zhou et al. (2022)
Dipping	Chitosan–*Aloe vera*–*Moringa oleifera*	Banana	Increasing storage life	Odetayo et al. (2022)
Dipping	Dextrin–carboxymethyl cellulose (CMC), chitosan, and starch	Zucchini	Dextrin-based coating was the most effective at extending postharvest life and enhancing the overall quality	Castro-Cegrí et al. (2023)
Dipping	Sodium alginate–rhubarb extract	Peach	Peach fruits treated with rhubarb–SA coating maintained good physiological quality and sensory quality compared with the control	Li et al. (2019)
Dipping	Sodium alginate–gelatin–Ag nanocomposite	Tangerine	The coating solution could slow down the loss of nutrients and prolong the shelf life of tangerines	Li et al. (2022b)
Dipping	Carnauba wax–shellac wax and carvacrol nanoemulsion	Mandarin	Mandarin coated with 2% carvacrol nanoemulsion exhibited significantly lower weight loss while retaining firmness and total soluble solids over 30 days of ambient storage	Devi et al. (2025)

(continued)

Table 15.1 (continued)

Deposition method	Coating material	Target	Main effects on quality	References
Dipping	Chitosan–beeswax–pollen grain	Pears	Decrease in weight loss, decay, and rate of softening	Sultan et al. (2021)
Dipping	Beeswax with sodium caseinate and guar gum	Strawberries	Reducing the respiration and transpiration rates of strawberries	Miele et al. (2022)
Dipping	Paraffin wax–eucalyptus oil–Bavistin	Mandarin	Prolongation of shelf life, as well as the quality of mandarin fruit	Joshi et al. (2020)
Dipping	Carnauba wax–oleic acid	Tomatoes	Increased fruit gloss and improved appearance of tomatoes	Miranda et al. (2022)
Dipping	Arabic gum	Persimmon fruits	Higher total phenolics, ascorbic acid, antioxidant activity, and titratable acidity	Saleem et al. (2020)
Dipping	Tragacanth gum with *Eremurus* extract	Sweet cherries	Reducing postharvest losses and increasing the shelf life	Esmaeili et al. (2022)
Dipping	Guar gum–ethanolic extract of *Spirulina platensis*	Mango	Increasing the shelf life of mango fruit	Ebrahimi and Rastegar (2020)
Dipping	Starch–lemongrass essential oil	Guava fruit	Coating material effectively prolonged the storage of guava fruit while preserving its postharvest quality for up to 9 days	Yadav et al. (2022)
Dipping	Chitosan–nanocrystal cellulose	Okra	The weight loss and decay rate of treated fruits were lower than those of the control during storage	Wang et al. (2023)
Dipping	Alginate–*Humulus lupulus* extract	Kiwi	The coating solution helped maintain the quality and nutraceutical properties of fresh-cut kiwi during cold storage	Carbone et al. (2021)
Dipping	Chitosan–lemon peel essential oil (EO)	Grape	The coating solution demonstrated that grapes treated with a 1% EO coating maintained optimal quality for up to 15 days at room temperature	Chetia et al. (2023)

(continued)

Table 15.1 (continued)

Deposition method	Coating material	Target	Main effects on quality	References
Dipping	Locust bean gum–carboxy cellulose nanocrystal	Strawberries and cherry tomatoes	Extending the shelf life of strawberries and cherry tomatoes	Li et al. (2021)
Layer-by-layer	Chitosan–alginate	Pear	Minimized fruit respiration and ethylene production rates, inhibited flesh firmness loss, and prevented peel color change	Hira et al. (2022)
Layer-by-layer	Chitosan–pectin	Strawberries	The coated strawberries also maintained their color, firmness, and pH during this period	Adhikari et al. (2023)
Layer-by-layer	Chitosan–carboxymethyl cellulose	Strawberries	The LBL coating is effective in preserving the quality of strawberries by maintaining lower levels of metabolites after 8 days of storage	Yan et al. (2019)
Layer-by-layer	Alginate–chitosan	Mandarin fruits	The preservation of bioactive compounds and organic acids in fruits	Jurić et al. (2023)
Electrospinning	Polylactic acid–carbon nanotube–chitosan composite fibers	Strawberry	An effective preservation for strawberries, with the fiber containing 7 wt% chitosan, provided the most optimal results	Liu et al. (2019)
Electrospinning	Jujube extract–electrospun polyvinyl alcohol nanofiber	Strawberry	Treated strawberries had acceptable overall appearance for marketing, as no obvious sign of decay was observed on the 12th day of storage	Zeinali et al. (2021)
Electrospinning	Polyvinyl butyral–camellia oil–ZnO–TiO_2	Strawberry and tomato	Increasing the storage duration of strawberries by 4 days while notably delaying the ripening process of small tomatoes	Cheng et al. (2024)
Electrospraying	Zein–*Artemisia sphaerocephala* Krasch.	Banana	A lower browning rate, higher hardness, and longer shelf life	Fan et al. (2022)
Spraying and dipping	*Aloe vera*	Grapes	Prolonging the storage life of table grapes and maintaining their quality	Ehtesham Nia et al. (2022)
Spreading	Chitosan	Mandarin fruit	Maintaining fruit quality parameters	Baswal et al. (2020)

(continued)

Table 15.1 (continued)

Deposition method	Coating material	Target	Main effects on quality	References
Spraying	Alginate, zein, potato starch, and essential oil	Potato tubers	Most of the edible coatings used for potatoes have limited effects on quality properties: weight loss and firmness	Emragi et al. (2022)
Spraying	Chitosan + cinnamon essential oil	Mango	The coating solution slowed the physicochemical parameters Coating at a ratio of 1:0.5 could extend the shelf life to 18 days during ambient storage	Deshi et al. (2024)
Spraying	Nanoemulsion of orange essential oil	Avocado	Fifty percent and 25% nanoemulsions yielded the best results in terms of firmness and weight loss, showing a reduction in polyphenol oxidase activity and a noticeable delay in fruit darkening	de Cenobio-Galindo et al. (2019)
Spraying	Clove essential oil + three starches (corn starch, mango kernel, and litchi starch)	Khasi mandarins	The use of various starch-based edible coatings infused with clove essential oil proved effective in extending the shelf life of Khasi mandarins	Kumar et al. (2024)
Spraying	Chitosan–eugenol–*Aloe vera*	Pineapple	Preserving its quality and prolonging its shelf life	Basumatary et al. (2022)
Cross-linked coating	Alginate–oil nanoemulsion	Sweet cherries	Increased cracking tolerance by 53% and increased firmness. Exhibited a 10% increase in water loss	Gutiérrez-Jara et al. (2021)
Cross-linked coating	Sodium alginate + whey protein isolate	Banana	The composite coating effectively reduced water loss, slowed down the ripening and browning processes, and extended the shelf life of bananas for 6 days at 25 °C and 50% relative humidity	Deng et al. (2024)

(continued)

Table 15.1 (continued)

Deposition method	Coating material	Target	Main effects on quality	References
Cross-linked coating	Corn starch–grape seed extract	Strawberry	Coating strawberries with grape derivatives and cross-linked starch can extend the shelf life of fresh fruit	Yıldırım-Yalçın et al. (2022)
Cross-linked coating	Citric acid–cross-linked starch and cellulose nanofiber	Tomato	The coated tomato exhibited a significant reduction in weight loss while retaining the firmness and color of the fruit	Wardak et al. (2024)

agent and emulsifier (Jadach et al., 2022). Due to its excellent film-forming properties, it is commonly used in the production of edible films. To enhance the physical and antimicrobial properties of alginate-based coatings, various substances have been incorporated, including plant extracts, essential oils, and antioxidants. For instance, Li et al. (2019) investigated the effectiveness of a rhubarb extract and sodium alginate (SA) coating in preserving peaches. After 7 days, peaches treated with the rhubarb–SA coating retained a firmness of 5.6 N, significantly higher than the control group, which softened to 1.8 N. At the end of the storage period, the peaches treated with rhubarb–SA exhibited the highest concentration of soluble solids and the lowest respiration rate throughout the entire duration. Additionally, the malondialdehyde content in uncoated peaches reached 1.78 μmol/L, 37.1% higher than in rhubarb–SA-treated samples. Moreover, PPO enzyme activity in untreated peaches was 20.8% higher than in those coated with rhubarb–SA. Sensory evaluations also confirmed that rhubarb–SA-treated peaches received the highest scores in terms of quality and freshness. Meanwhile, Keshari et al. (2022) examined the impact of sodium alginate–based edible coatings (1%, 2%, and 3% w/v) combined with α-tocopherol acetate (0.5% and 1% w/v) on the quality of minimally processed carrot slices stored for 15 days at 10 °C and 65% relative humidity. Among them, treatment T4 (2% sodium alginate +1% α-tocopherol acetate) proved to be the most effective in maintaining the overall quality and nutritional value. Carrot slices coated with T4 exhibited minimal changes in weight loss, total soluble solids (TSS), pH, whiteness index, reducing sugars, ascorbic acid, antioxidant activity, total carotenoids, and microbial counts throughout storage.

In another study, Najafi Marghmaleki et al. (2021) found that after 7 days of storage at 4 °C, the sodium alginate–based coating enriched with both citric acid and ascorbic acid demonstrated the best preservation effects, showing the lowest weight loss (0.33%), microbial load (2.95 log CFU/g), and electrolyte leakage (35.5%). Additionally, it maintained the highest lightness value (96.11) and antioxidant capacity (11.03 mmol Fe^{2+}/kg).

When carvacrol (0%, 0.03%, 0.06%, or 0.09%) was added to blends of sodium alginate (2% w/v) to coat blueberries, the coating solution maintained better quality over 21 days of storage than that of the control due to reduced respiration rates and lower water loss compared to uncoated berries (Medina-Jaramillo et al., 2020). Additionally, the coating improved the fruit's appearance and gloss. While the control group experienced a significant decline in firmness, the coated blueberries retained this essential postharvest attribute throughout the storage period (Medina-Jaramillo et al., 2020). Furthermore, Duong et al. (2022) found that the weight loss of rose apple (*Syzygium samarangense*) coated with alginate (7%) was significantly lower ($p < 0.05$) compared to uncoated fruits. In another study, grapes (*Vitis vinifera*) coated with a composite of alginate, galactomannans, cashew gum, and gelatin showed a lower weight/mass loss of 20% compared to uncoated fruits (30%) (de Souza et al., 2021). Similarly, peaches (*Prunus persica*) coated with an alginate-based coating enriched with rhubarb (*Rheum rhaponticum* L.) extract exhibited 7% weight/mass loss compared to 13% in the control treatment (Li et al., 2019).

Cellulose-Based Coatings

Cellulose-based edible coatings are typically transparent, odorless, tasteless, nontoxic, nonallergenic, water-soluble, and resistant to oils and fats (Shoaib et al., 2020). These coatings can also act as carriers for functional additives such as antimicrobial agents, antioxidants, anti-browning compounds, and nutrients, making them highly suitable for horticultural products (Panahirad et al., 2021). For example, carboxymethyl cellulose (CMC) is an anionic polysaccharide produced by incorporating $-CH_2COOH$ functional groups into cellulose chains, making it a promising material for food packaging applications. According to the study of Ali et al. (2021), the effectiveness of a 1% carboxymethyl cellulose (CMC) coating in preserving Kinnow mandarins stored at 5 °C was demonstrated after 30 days. In fact, the weight loss of CMC-coated mandarins was significantly lower (2.93%) compared to the control group (7.38%). Additionally, electrolyte leakage and hydrogen peroxide accumulation in coated fruits were substantially reduced. The hydrogen peroxide concentration in the peel of coated mandarins was 1.46 times lower than that of the control, while the malondialdehyde levels increased 1.43 times less in coated samples than in untreated ones. Enzyme activity analysis revealed that coating treatment significantly prevented the decline of ascorbate peroxidase (APX), catalase (CAT), and superoxide dismutase (SOD). After 30 days, the total phenol content of coated mandarins was 1.53 times higher than that of the control, and their DPPH radical scavenging activity was 1.47 times higher. Furthermore, the rise in total soluble solids (TSS) concentration was 1.39 times lower in coated mandarins than in uncoated ones.

The application of a CMC coating can help maintain the quality of Valencia oranges during storage, according to the study of Abbasi et al. (2023). Among the primary postharvest diseases impacting citrus fruits are green mold and blue mold, both caused by *Penicillium digitatum* and *Penicillium italicum*, leading to considerable economic losses each year. In a study conducted by Abbasi et al. (2023), the antimicrobial properties of thyme essential oil (TEO) and CMC coating were assessed both in vitro and in vivo against these pathogens. The findings revealed that TEO and CMC effectively suppressed the mycelial growth and spore germination of *P. digitatum* and *P. italicum* while also significantly lowering the occurrence of green and blue mold in the fruit (Abbasi et al., 2023).

Additionally, Panahirad et al. (2020) investigated the impact of CMC and pectin coatings on preserving plum fruit and discovered that CMC coating played a crucial role in maintaining its nutritional quality after harvest. The coating application led to a reduction in polygalacturonase activity, which helped retain the firmness of the plums. Baswal et al. (2020) developed a double-layer edible coating using CMC and chitosan, with CMC as the inner layer and chitosan as the outer layer, achieving optimal results in maintaining the physiological quality of citrus fruits. Additionally, a comparative study on the effects of chitosan, CMC, and beeswax coatings on Kinnow mandarins demonstrated that CMC was the most effective in preserving fruit quality. Sousa et al. (2021) investigated the use of hydroxypropyl methylcellulose (HPMC) and beeswax (BW) coatings to extend the shelf life of Palmer mangoes at 21 °C. Throughout a 15-day storage period, the respiration rate of uncoated mangoes was approximately twice as high as that of coated ones. Alcohol dehydrogenase activity, associated with fruit ripening, was significantly lower in coated mangoes than in the control group. Coated fruits maintained a color closer to their harvest state, indicating a delayed ripening process. Additionally, coated mangoes exhibited 19% less weight loss than uncoated ones. At the end of storage, the firmness of uncoated mangoes was only one-fourth of that of coated fruits. The titratable acidity of mangoes treated with HPMC +10% BW was 40% higher than that of the control group. While the control group had higher concentrations of soluble and reducing sugars, mangoes coated with HPMC +20% BW retained 20% more ascorbic acid. Furthermore, at the end of storage, coated mangoes contained significantly lower levels of beta-carotene and malondialdehyde compared to uncoated fruits, highlighting the protective effect of the coating against oxidative degradation.

Starch-Based Coatings

Starch, a naturally occurring polysaccharide sourced from plants like corn, potatoes, rice, and wheat, has attracted considerable interest as an eco-friendly material for food coating applications. Starch is a widely studied material for biodegradable film production due to its affordability and abundance (Dutta & Sit, 2024). Starch-based coatings help reduce weight loss, maintain firmness and color, and slow down the degradation of essential nutrients and flavor compounds (Sapper & Chiralt, 2018). Additionally, these coatings lower the respiration rate and inhibit the activity of certain enzymes, thereby extending the shelf life of coated produce (Oyom et al., 2022). Additionally, the incorporation of natural extracts, essential oils, and bioactive compounds can strengthen the antibacterial and antioxidant properties of these coatings, further enhancing food preservation (Bari & Giannouli, 2022; Dai et al.,

2020; Sapper & Chiralt, 2018). A distinct study demonstrated that cross-linked cassava starch (CCS) films exhibit desirable properties, including good mechanical strength and resistance to mild and controlled water vapor permeability, making them excellent film-forming substrates. Starch nanocrystals (SNCs) are commonly utilized as organic nanoparticle enhancers (Dai et al., 2020).

Sweet potato starch has limitations, such as low transparency, poor viscosity, and weak thickening ability. However, physical modification using moist heat treatment significantly improves its air permeability, film-forming ability, and mechanical characteristics. While modified sweet potato starch coatings have limited antimicrobial effectiveness in preserving fruits and vegetables, adding antimicrobial essential oils enhances their antimicrobial properties (Oyom et al., 2022). *Spondias purpurea* L. fruit contains high levels of starch and possesses strong antioxidant properties due to its flavonoid and phenolic content (Rodrigues et al., 2021).

One challenge with starch-based coatings is their incompatibility with polylactic acid (Niu et al., 2018). However, surface hydrophobic modification can resolve this issue. For instance, modifying starch with octenyl succinic anhydride (OSA) results in a hierarchical structure containing both hydrophilic and hydrophobic functional groups, allowing it to act as an intermediate layer to enhance compatibility between the two materials. A composite film with a layered structure can be created using pea starch, OSA-modified pea starch, and polylactic acid (Zhou et al., 2021). Oyom et al. (Dai et al., 2020) investigated the use of cross-linked cassava starch and starch nanocrystals (SNCs) for preserving Huangguan pears. Their findings demonstrated that the starch-based coating significantly inhibited weight loss and browning. Furthermore, the coating effectively reduced the respiration rate of pears while suppressing the activity of key oxidative enzymes such as peroxidase (POD) and polyphenol oxidase (PPO), which are responsible for enzymatic browning. Rodrigues et al. (2021) examined the application of starch combined with phenolics extracted from *Spondias purpurea* L. for mango preservation. The results showed that this coating effectively reduced moisture transfer, respiration rate, and oxidative processes, leading to an extended storage life. Additionally, the soluble solids content in coated mangoes was lower compared to the control group, suggesting a slower ripening process and better retention of fruit quality during storage. In a different study, Oyom et al. (2022) observed that a starch-based antifungal coating effectively minimized black spot rot caused by *Alternaria alternata* in Asian pears stored for 21 days. This protective effect was attributed to the presence of antimicrobial agents in the coating formulation.

Gelatin-Based Coating

Gelatin, a protein extracted from collagen found in animal bones, skin, and connective tissues, has gained prominence in the food industry due to its versatility and sustainability as a coating material (Toniciolli Rigueto et al., 2022). Gelatin forms strong, flexible, and transparent films that serve as effective barriers against moisture and gases, both crucial for maintaining food integrity (Lu et al., 2022). Its excellent gelation properties and compatibility with other biopolymers make it suitable for coating a variety of foods, including fruits and vegetables (Bari & Giannouli, 2022; Lu et al., 2022). For instance, the study of Dehghani et al. (2022) suggest that fish gelatin (FG) conjugated with bitter almond gum (BAG) could serve as an effective edible coating material. Coating dispersions were formulated using BAG, FG, or their conjugates (1:2, 1:1, 2:1) and assessed for their impact on the physicochemical and quality parameters of tomatoes stored at 20 °C for 28 days. The conjugate-based coating demonstrated superior preservation of tomato quality compared to single-component coatings, maintaining better sensory attributes and overall acceptability. While pH, titratable acidity, and most color indices remained unaffected, the BAG: FG (2:1) conjugate significantly delayed changes in firmness and lycopene content while minimizing weight loss and decay. These findings suggest that a higher BAG ratio in FG-based coatings could be a promising approach for extending fruit shelf life and maintaining quality during storage. In addition, edible coating solutions were formulated by Sekarina et al. (2023) using chitosan, gelatin, glycerol, and black tea at varying concentrations (0%, 5%, 10%, and 15%). These coatings were applied to minimally processed papaya and stored for 1, 4, 7, and 10 days to evaluate their effects on the fruit's physical, chemical, and microbiological characteristics. The results showed that the edible coating effectively reduced weight loss and helped maintain papaya texture throughout the storage period.

Moreira Pereira et al. (2022) evaluated the preservation of 'Pedro Sato' red guavas using chitosan and gelatin coatings applied through the layer-by-layer technique. Fruits were immersed in four treatments: (1) control, (2) chitosan, (3) gelatin, and (4) chitosan with glycerol under gelatin. After drying, the guavas were stored for 8 days at 21 ± 0.3 °C and 77% Relative Humidity (RH). Various parameters were analyzed, including coating composition, weight loss, color, firmness, soluble solids, titratable acidity, ascorbic acid, chlorophylls, lycopene, total phenolics, antioxidant activity, lipid oxidation, and enzyme activity (catalase and peroxidase). The gelatin/chitosan coating effectively delayed ripening, protected against diseases, reduced weight loss and lipid oxidation, preserved color and firmness, and enhanced catalase activity.

In fact, due to its affordability, ease of application, and excellent functional properties, gelatin is considered a valuable biopolymer for food coatings. The addition of antioxidants and other bioactive compounds can further extend the shelf life of highly perishable items by delaying oxidation. For example, after 7 days of storage, Frassinetti et al. (2021) mentioned that tomatoes coated with gelatin and blueberry juice exhibited higher phenolic and flavonoid content, along with enhanced antioxidant activity, while the carotenoid levels remained unchanged. Additionally, bacterial growth was significantly suppressed throughout the storage period. These findings suggest that the coating effectively preserved the nutritional quality of the tomatoes.

Fatty Acid and Wax Coatings

Fatty acids and waxes are naturally occurring substances that have long been utilized as coatings to enhance the shelf life of various food products (Liyanapathiranage et al., 2023). As key components of lipids, fatty acids play a crucial role in forming protective layers that minimize moisture loss, regulate gas exchange, and inhibit microbial growth, thereby maintaining the freshness and quality of food (de Freitas et al., 2019; Prudnikov et al., 2023). Furthermore, many fatty acids and waxes are recognized as safe for consumption and have been approved as food additives by regulatory authorities such as the Food and Drug Administration (FDA) and European Food Safety Agency (EFSA). Fatty acid and wax coatings serve as an effective and sustainable method for extending the shelf life of food products while preserving their quality. Their natural abundance, biodegradability, and excellent barrier properties make them ideal for food coating applications.

In the study of Devi et al. (2023), sustainable active coatings based on carnauba wax were formulated, analyzed, and applied to extend the postharvest shelf life of Khasi mandarin and sweet lemon. In fact, neem oil nanoemulsion was also incorporated as an active agent to enhance the antimicrobial properties of the composite coatings. These coatings effectively minimized weight loss, preserved firmness, maintained total soluble solids and titratable acidity, and slowed the deterioration of coated Khasi mandarins and sweet lemons under ambient storage conditions. According to another study, Joshi et al. (2020) mentioned that the postharvest treatment of *Citrus reticulate* using 10% wax combined with 0.1% Bavistin resulted in the lowest physiological weight loss (6.61%) and the highest juice recovery (43.72%), which was statistically comparable to fruits treated with 10% wax, 0.1% Bavistin, 2% cinnamon oil, 2% eucalyptus oil, and 1% calcium chloride. By the end of the storage period, the highest total soluble solids (TSS) content (15.45 °Brix) was observed in fruits treated with 1% calcium chloride, with no significant difference compared to the control, 2% eucalyptus oil, or 0.1% Bavistin treatments. The pH remained stable throughout storage, while the longest postharvest life (73 days) was recorded in fruits treated with 10% wax and 0.1% Bavistin, compared to only 46 days in the untreated control (Table 15.1).

Various essential oils, such as oregano (Gutiérrez-Pacheco et al., 2020), clove (Rajaei et al., 2017), and lemon (Chetia et al., 2023), along with different combinations, were formulated to enhance their in vitro performance when incorporated into edible coatings made of polymers and other materials. These formulations significantly improve antibacterial and antifungal properties. In the food industry, essential oils demonstrate greater efficiency when integrated into an appropriate delivery system, which also helps prevent interactions with other food components. To minimize their impact on sensory attributes, essential oils are often delivered using nanotechnology, allowing better interaction with cell molecules. Edible coatings play a crucial role in extending the shelf life of fruits by reducing respiration rates and water loss, thereby preventing postharvest deterioration (Hasan et al., 2020; Pandey et al., 2022).

In the study of Gutiérrez-Pacheco et al. (2020), the authors aim to assess the impact of individual and combined coatings of chitosan (0.008 g/mL) and carnauba wax (0.1 g/mL) enriched with oregano essential oil (OEO, 0.08 g/mL) on minimizing dehydration and microbial spoilage in fresh cucumbers stored at 10 °C. The chitosan–OEO–wax films exhibited the lowest water vapor transmission rate (0.141 g/m^2·h), outperforming single chitosan films (0.257 g/m^2·h). Chitosan–OEO films effectively inhibited the in vitro growth of *Alternaria alternata* and reduced the populations of *Salmonella typhimurium* and *Escherichia coli* O157:H7 associated with decayed cucumbers. Carnauba wax and wax–OEO coatings were the most effective in preventing weight loss over a 15-day storage period at 10 °C, while chitosan and chitosan–OEO coatings demonstrated the highest antimicrobial efficacy. In addition, Hashemi et al. (2021) made edible coatings from gum arabic at concentrations of 6% and 8%, combined with *Zataria multiflora* essential oil at 0.3% and 0.5%, to evaluate the quality of fresh in-hull pistachios stored at 85% relative humidity and 2 °C for up to 36 days. The results demonstrated that the 6% gum arabic coating, when combined with thyme essential oil at both 0.3% and 0.5%, effectively reduced color change and polyphenol oxidase activity in fresh pistachios compared to the untreated control fruits.

The incorporation of oils into nanoemulsion coatings helps stabilize their bioactivity, enhances cellular absorption through passive mechanisms, provides antimicrobial benefits, and preserves key sensory attributes such as flavor, aroma, and taste (Hasan et al., 2020; Pandey et al., 2022). For instance, Gutiérrez-Jara et al. (2021) coated sweet cherries using electrospraying with an edible nanoemulsion composed of alginate and soybean oil. Tabassum and Khan (2020) explored the effectiveness of an alginate-based (2% w/v) edible coating infused with nanoemulsions of *Thymus vulgaris* and *Origanum vulgare* essential oils (0.5, 1.0, and 2.0%) in extending the shelf life of papaya by preventing mold and yeast growth. Their study, conducted over 12 days at 4 °C, revealed that uncoated papayas exhibited yeast and mold levels ranging from 5.90 to 8.40 CFU/g, surpassing the acceptable consumption threshold. Chu et al. (2020) developed a pullulan-based coating (2% w/v) containing a nanoemulsion of cinnamon essential oil to assess its impact on mold and yeast contamination in strawberries. Their findings demonstrated that the coating effectively reduced fruit decay over a six-day storage period at 20 °C. While decay in both the control and pure pullulan–coated strawberries began after 2 days and reached 65% by day six, the essential oil–infused pullulan coating lowered the decay rate by 30%. This antifungal effect was attributed to cinnamaldehyde, the primary active compound, which disrupted fungal cell membranes. Additionally, the nanoemulsion coating significantly reduced fungal growth from 10^4 to 1.958 CFU/g, a result linked to the nanometric size of the essential oil droplets, which enhanced surface area and membrane penetration. A further study by Gago et al. (2020) examined the effectiveness of an alginate-based (2% w/w) nanoemulsion containing 1.25% w/w lemongrass essential oil in preserving Rocha pears stored at 22 °C with 70% relative humidity for 7 days. Their study found that the coating helped maintain key quality attributes, such as color, soluble solids content, titratable acidity, and firmness. Additionally, the nanoemulsion slowed down the ripening process, contributing to overall fruit quality preservation.

Conclusion and Future Prospects

Edible coatings play a crucial role in food preservation, especially in the face of global hunger and supply chain disruptions. Those derived from renewable biomacromolecules have gained significant attention due to their availability, cost-effectiveness, biocompatibility, non-toxicity, and biodegradability. The colloidal chemistry of biomacromolecules such as polysaccharides, proteins, and lipids indicates their processing capabilities, film-forming potential, and casting and lamination properties. Additionally, biopolymer-based hydrocolloids offer excellent air and gas barrier properties, making them ideal candidates for developing edible coatings and films. Sustainable production methods, the use of renewable materials, green synthesis, and environmental considerations are key aspects of modern edible packaging. While biomacromolecules hold promise for advancing food packaging, edible films face challenges limiting their industrial applications, such as structural limitations, aroma, and flavor retention issues. Some films also exhibit instability of bioactive compounds, weak film-forming abilities, and limited surface adhesion. The incorporation of essential oils to enhance antimicrobial properties introduces challenges such as poor water solubility, strong odors, and high volatility. At the industrial level, edible coating materials from bio-based sources often struggle with moisture barrier inefficiency, temperature resistance, and humidity control. Furthermore, the absence of standardized regulations and safety guidelines presents additional hurdles to the widespread adoption of edible films and coatings. To address these challenges, innovative techniques such as biopolymer blends, composites, and multilayered structures have emerged for edible film and coating development. Cross-linking, functionalization, and surface modifications using eco-friendly methods have gained attention for improving the physical, chemical, and biological properties of these films. Active edible coating, enriched with antimicrobial and antifungal agents, essential oils, and plasticizers, has been developed to enhance food preservation. Additionally, the integration of nanoparticles in edible coatings improves their functionality and extends the shelf life of perishable foods. To promote sustainability, consumers, food producers, and regulatory agencies should encourage the adoption of edible coatings derived from bio-based materials. This approach can help mitigate food losses, aligning with circular economy principles.

References

Abbasi, M., Dastjerdi, A. M., Seyahooei, M. A., et al. (2023). Postharvest control of green and blue molds on Mexican lime fruit caused by Penicillium species using Thymus vulgaris essential oil and carboxy methyl cellulose. *Journal of Plant Diseases and Protection, 130*, 1017–1026. https://doi.org/10.1007/s41348-023-00773-1

Adhikari, M., Koirala, S., & Anal, A. K. (2023). Edible multilayer coating using electrostatic layer-by-layer deposition of chitosan and pectin enhances shelf life of fresh strawberries. *International Journal of Food Science and Technology, 58*, 871–879. https://doi.org/10.1111/ijfs.15704

Alav, A., Kutlu, N., Kına, E., & Meral, R. (2024). A novel green tea extract-loaded nanofiber coating for kiwi fruit: Improved microbial stability and nutritional quality. *Food Bioscience, 62*, 105043. https://doi.org/10.1016/j.fbio.2024.105043

Ali, S., Anjum, M. A., Ejaz, S., et al. (2021). Carboxymethyl cellulose coating delays chilling injury development and maintains eating quality of 'Kinnow' mandarin fruits during low temperature storage. *International Journal of Biological Macromolecules, 168*, 77–85. https://doi.org/10.1016/j.ijbiomac.2020.12.028

Alizadeh-Sani, M., Tavassoli, M., Mohammadian, E., et al. (2021). pH-responsive color indicator films based on methylcellulose/chitosan nanofiber and barberry anthocyanins for real-time monitoring of meat freshness. *International Journal of Biological Macromolecules, 166*, 741–750. https://doi.org/10.1016/j.ijbiomac.2020.10.231

Amalraj, A., Raj, K. K. J., Haponiuk, J. T., et al. (2020). Preparation, characterization, and antimicrobial activity of chitosan/gum arabic/polyethylene glycol composite films incorporated with black pepper essential oil and ginger essential oil as potential packaging and wound dressing materials. *Advanced Composites and Hybrid Materials, 3*, 485–497. https://doi.org/10.1007/s42114-020-00178-w

Ammar, E. E., Zou, X., Ghosh, S., et al. (2025). Fresh futures: Cutting-edge eco-friendly coating techniques for fruits. *Journal of Food Processing and Preservation, 2025*, 5201632. https://doi.org/10.1155/jfpp/5201632

Amor, G., Sabbah, M., Caputo, L., et al. (2021). Basil essential oil: Composition, antimicrobial properties, and microencapsulation to produce active chitosan films for food packaging. *Foods, 10*, 121. https://doi.org/10.3390/foods10010121

Bari, A., & Giannouli, P. (2022). Evaluation of biodegradable gelatin and gelatin–rice starch coatings to fresh cut Zucchini Slices. *Horticulturae, 8*, 1031. https://doi.org/10.3390/horticulturae8111031

Basumatary, I. B., Mukherjee, A., Katiyar, V., et al. (2022). Chitosan-based active coating for pineapple preservation: Evaluation of antimicrobial efficacy and shelf-life extension. *LWT, 168*, 113940. https://doi.org/10.1016/j.lwt.2022.113940

Baswal, A. K., Dhaliwal, H. S., Singh, Z., et al. (2020). Influence of carboxy methylcellulose, chitosan and beeswax coatings on cold storage life and quality of Kinnow mandarin fruit. *Scientia Horticulturae, 260*, 108887. https://doi.org/10.1016/j.scienta.2019.108887

Carbone, K., Macchioni, V., Petrella, G., et al. (2021). Humulus lupulus cone extract efficacy in alginate-based edible coatings on the quality and nutraceutical traits of fresh-cut kiwifruit. *Antioxidants, 10*, 1395. https://doi.org/10.3390/antiox10091395

Castro-Cegrí, A., Ortega-Muñoz, M., Sierra, S., et al. (2023). Application of polysaccharide-based edible coatings to improve the quality of zucchini fruit during postharvest cold storage. *Scientia Horticulturae, 314*, 111941. https://doi.org/10.1016/j.scienta.2023.111941

Chakravarty, J., & Edwards, T. A. (2022). Innovation from waste with biomass-derived chitin and chitosan as green and sustainable polymer: A review. *Energy Nexus, 8*, 100149. https://doi.org/10.1016/j.nexus.2022.100149

Cheng, J., Li, D., Mai, Z., et al. (2024). In-situ electrospinning PVB/camellia oil/ZnO-TiO2 nanofibrous membranes with synergistic antibacterial and degradation of ethylene applied in fruit preservation. *Food Chemistry, 460*, 140629. https://doi.org/10.1016/j.foodchem.2024.140629

Chetia, J., Adhikary, P., Devi, L. M., & Badwaik, L. S. (2023). Extraction of essential oil from Assam lemon peels and its incorporation in chitosan based coating for maintaining grape quality. *Sustainable Chemistry and Pharmacy, 32*, 101034. https://doi.org/10.1016/j.scp.2023.101034

Chu, Y., Gao, C., Liu, X., et al. (2020). Improvement of storage quality of strawberries by pullulan coatings incorporated with cinnamon essential oil nanoemulsion. *LWT, 122*, 109054. https://doi.org/10.1016/j.lwt.2020.109054

Dai, L., Zhang, J., & Cheng, F. (2020). Cross-linked starch-based edible coating reinforced by starch nanocrystals and its preservation effect on graded Huangguan pears. *Food Chemistry, 311*, 125891. https://doi.org/10.1016/j.foodchem.2019.125891

de Cenobio-Galindo, A. J., Ocampo-López, J., Reyes-Munguía, A., et al. (2019). Influence of bioactive compounds incorporated in a nanoemulsion as coating on avocado fruits (Persea americana) during postharvest storage: Antioxidant activity, physicochemical changes and structural evaluation. *Antioxidants (Basel), 8*, 500. https://doi.org/10.3390/antiox8100500

de Freitas, C. A. S., de Sousa, P. H. M., Soares, D. J., et al. (2019). Carnauba wax uses in food—A review. *Food Chemistry, 291*, 38–48. https://doi.org/10.1016/j.foodchem.2019.03.133

de Souza, W. F. C., de Lucena, F. A., da Silva, K. G., et al. (2021). Influence of edible coatings composed of alginate, galactomannans, cashew gum, and gelatin on the shelf- life of grape cultivar '*Italia*': Physicochemical and bioactive properties. *LWT, 152*, 112315. https://doi.org/10.1016/j.lwt.2021.112315

Dehghani, S., Hosseini, E., & Rousta, E. (2022). Shelf-life extension of tomato (*Solanum lycopersicum* L.) using an edible coating of bitter almond gum-fish gelatin conjugates. *Progress in Organic Coatings, 170*, 106980. https://doi.org/10.1016/j.porgcoat.2022.106980

Deng, P., Zhang, Y., Deng, Q., et al. (2024). Multifunctional sodium alginate-based self-healing edible cross-linked coating for banana preservation. *Food Hydrocolloids, 151*, 109753. https://doi.org/10.1016/j.foodhyd.2024.109753

Deshi, V. V., Awati, M. G., Terdal, D., et al. (2024). Cinnamon essential oil incorporated chitosan submicron emulsion as a sustainable alternative for extension of mango shelf life. *Sustainable Chemistry and Pharmacy, 41*, 101736. https://doi.org/10.1016/j.scp.2024.101736

Devi, L., Mukherjee, A., Dutta, D., & Kumar, S. (2023). Carnauba wax-based sustainable coatings for prolonging postharvest shelf-life of citrus fruits. *Sustainable Food Technology, 1*. https://doi.org/10.1039/D2FB00049K

Devi, L. S., Mukherjee, A., Sharma, S., et al. (2025). Natural wax-based edible coatings for preserving postharvest quality of mandarin orange. *Food Chemistry: X, 26*, 102302. https://doi.org/10.1016/j.fochx.2025.102302

Duong, N. T. C., Uthairatanakij, A., Laohakunjit, N., et al. (2022). An innovative single step of cross-linked alginate-based edible coating for maintaining postharvest quality and reducing chilling injury in rose apple cv. "Tabtimchan" (Syzygium samarangenese). *Scientia Horticulturae, 292*, 110648. https://doi.org/10.1016/j.scienta.2021.110648

Dutta, D., & Sit, N. (2024). Comprehensive review on developments in starch-based films along with active ingredients for sustainable food packaging. *Sustainable Chemistry and Pharmacy, 39*, 101534. https://doi.org/10.1016/j.scp.2024.101534

Ebrahimi, F., & Rastegar, S. (2020). Preservation of mango fruit with guar-based edible coatings enriched with *Spirulina platensis* and *Aloe vera* extract during storage at ambient temperature. *Scientia Horticulturae, 265*, 109258. https://doi.org/10.1016/j.scienta.2020.109258

Ehtesham Nia, A., Taghipour, S., & Siahmansour, S. (2022). Effects of salicylic acid preharvest and *Aloe vera* gel postharvest treatments on quality maintenance of table grapes during storage. *South African Journal of Botany, 147*, 1136–1145. https://doi.org/10.1016/j.sajb.2022.05.010

Emragi, E., Kalita, D., & Jayanty, S. S. (2022). Effect of edible coating on physical and chemical properties of potato tubers under different storage conditions. *LWT, 153*, 112580. https://doi.org/10.1016/j.lwt.2021.112580

Esmaeili, A., Jafari, A., Ghasemi, A., & Gholamnejad, J. (2022). Improving postharvest quality of sweet cherry fruit by using Tragacanth and Eremurus. *International Journal of Fruit Science, 22*, 370–382. https://doi.org/10.1080/15538362.2022.2043221

Fan, K., Zhang, M., Guo, C., et al. (2021). Laser-induced microporous modified atmosphere packaging and chitosan carbon-dot coating as a novel combined preservation method for fresh-cut cucumber. *Food and Bioprocess Technology, 14*, 968–983. https://doi.org/10.1007/s11947-021-02617-y

Fan, X., Rong, L., Li, Y., et al. (2022). Fabrication of bio-based hierarchically structured ethylene scavenger films via electrospraying for fruit preservation. *Food Hydrocolloids, 133*, 107837. https://doi.org/10.1016/j.foodhyd.2022.107837

Frassinetti, S., Antonella, C., Marco, S., et al. (2021). Gelatin-based coating enriched with blueberry juice preserves the nutraceutical quality and reduces the microbial contamination of tomato fruit. *Natural Product Research, 35*, 6088–6092. https://doi.org/10.1080/14786419.2020.1824224

Gago, C., Antão, R., Dores, C., et al. (2020). The effect of nanocoatings enriched with essential oils on 'Rocha' pear long storage. *Foods, 9*, 240. https://doi.org/10.3390/foods9020240

Giannakourou, M. C., & Tsironi, T. N. (2021). Application of processing and packaging hurdles for fresh-cut fruits and vegetables preservation. *Food, 10*, 830. https://doi.org/10.3390/foods10040830

Gutiérrez-Jara, C., Bilbao-Sainz, C., McHugh, T., et al. (2021). Effect of cross-linked alginate/oil nanoemulsion coating on cracking and quality parameters of sweet cherries. *Foods, 10*, 449. https://doi.org/10.3390/foods10020449

Gutiérrez-Pacheco, M. M., Ortega-Ramírez, L. A., Silva-Espinoza, B. A., et al. (2020). Individual and combined coatings of chitosan and carnauba wax with oregano essential oil to avoid water loss and microbial decay of fresh cucumber. *Coatings, 10*, 614. https://doi.org/10.3390/coatings10070614

Hasan, S. M. K., Ferrentino, G., & Scampicchio, M. (2020). Nanoemulsion as advanced edible coatings to preserve the quality of fresh-cut fruits and vegetables: A review. *International Journal of Food Science and Technology, 55*, 1–10. https://doi.org/10.1111/ijfs.14273

Hashemi, M., Dastjerdi, A. M., Mirdehghan, S. H., et al. (2021). Incorporation of *Zataria multiflora* Boiss essential oil into gum Arabic edible coating to maintain the quality properties of fresh in-hull pistachio (*Pistacia vera* L.). *Food Packaging and Shelf Life, 30*, 100724. https://doi.org/10.1016/j.fpsl.2021.100724

Hira, N., Mitalo, O. W., Okada, R., et al. (2022). The effect of layer-by-layer edible coating on the shelf life and transcriptome of 'Kosui' Japanese pear fruit. *Postharvest Biology and Technology, 185*, 111787. https://doi.org/10.1016/j.postharvbio.2021.111787

Jadach, B., Świetlik, W., & Froelich, A. (2022). Sodium alginate as a pharmaceutical excipient: Novel applications of a well-known polymer. *Journal of Pharmaceutical Sciences, 111*, 1250–1261. https://doi.org/10.1016/j.xphs.2021.12.024

Joshi, P., Ojha, B. R., & Kafle, A. (2020). Effect of different postharvest treatments on prolonging shelf life of Citrus reticulata Blanco. *Nepalese Horticulture, 14*, 1–8. https://doi.org/10.3126/nh.v14i1.30603

Jurić, S., Bureš, M. S., Vlahoviček-Kahlina, K., et al. (2023). Chitosan-based layer-by-layer edible coatings application for the preservation of mandarin fruit bioactive compounds and organic acids. *Food Chemistry: X, 17*, 100575. https://doi.org/10.1016/j.fochx.2023.100575

Keshari, D., Tripathi, A. D., Agarwal, A., et al. (2022). Effect of α-dl tocopherol acetate (antioxidant) enriched edible coating on the physicochemical, functional properties and shelf life of minimally processed carrots (*Daucus carota* subsp. *sativus*). *Future Foods, 5*, 100116. https://doi.org/10.1016/j.fufo.2022.100116

Kumar, R., Ghoshal, G., & Goyal, M. (2021). Biodegradable composite films/coatings of modified corn starch/gelatin for shelf life improvement of cucumber. *Journal of Food Science and Technology, 58*, 1227–1237. https://doi.org/10.1007/s13197-020-04685-y

Kumar, N., Khan, A. A., Pyngrope, D., et al. (2024). Development and characterization of novel starch (mango kernel and litchi seed) based active edible coatings and films using ultrasonication: Effects on postharvest shelf life of *Khasi* mandarins. *Sustainable Chemistry and Pharmacy, 39*, 101610. https://doi.org/10.1016/j.scp.2024.101610

Li, X., Du, X., Liu, Y., et al. (2019). Rhubarb extract incorporated into an alginate-based edible coating for peach preservation. *Scientia Horticulturae, 257*, 108685. https://doi.org/10.1016/j.scienta.2019.108685

Li, T., Liu, R., Zhang, C., et al. (2021). Developing a green film from locust bean gum/carboxycellulose nanocrystal for fruit preservation. *Future Foods, 4*, 100072. https://doi.org/10.1016/j.fufo.2021.100072

Li, H., Huang, Z., Addo, K. A., & Yu, Y. (2022a). Evaluation of postharvest quality of plum (*Prunus salicina* L. cv. 'French') treated with layer-by-layer edible coating during storage. *Scientia Horticulturae, 304*, 111310. https://doi.org/10.1016/j.scienta.2022.111310

Li, H., Li, W., Zhang, J., et al. (2022b). Preparation and characterization of sodium alginate/gelatin/Ag nanocomposite antibacterial film and its application in the preservation of tangerine. *Food Packaging and Shelf Life, 33*, 100928. https://doi.org/10.1016/j.fpsl.2022.100928

Liu, Y., Wang, S., Lan, W., & Qin, W. (2019). Fabrication of polylactic acid/carbon nanotubes/chitosan composite fibers by electrospinning for strawberry preservation. *International Journal of Biological Macromolecules, 121*, 1329–1336. https://doi.org/10.1016/j.ijbiomac.2018.09.042

Liyanapathiranage, A., Dassanayake, R. S., Gamage, A., et al. (2023). Recent developments in edible films and coatings for fruits and vegetables. *Coatings, 13*, 1177. https://doi.org/10.3390/coatings13071177

Lu, Y., Luo, Q., Chu, Y., et al. (2022). Application of gelatin in food packaging: A review. *Polymers, 14*, 436. https://doi.org/10.3390/polym14030436

Manzoor, S., Gull, A., Wani, S. M., et al. (2021). Improving the shelf life of fresh cut kiwi using nanoemulsion coatings with antioxidant and antimicrobial agents. *Food Bioscience, 41*, 101015. https://doi.org/10.1016/j.fbio.2021.101015

Medina-Jaramillo, C., Quintero-Pimiento, C., Díaz-Díaz, D., et al. (2020). Improvement of Andean blueberries postharvest preservation using carvacrol/alginate-edible coatings. *Polymers, 12*, 2352. https://doi.org/10.3390/polym12102352

Meitha, K., Pramesti, Y., & Suhandono, S. (2020). Reactive oxygen species and antioxidants in postharvest vegetables and fruits. *International Journal of Food Science, 2020*, 8817778. https://doi.org/10.1155/2020/8817778

Miele, N. A., Volpe, S., Torrieri, E., & Cavella, S. (2022). Improving physical properties of sodium caseinate based coating with the optimal formulation: Effect on strawberries' respiration and transpiration rates. *Journal of Food Engineering, 331*, 111123. https://doi.org/10.1016/j.jfoodeng.2022.111123

Miranda, M., Ribeiro, M. D. M. M., Spricigo, P. C., et al. (2022). Carnauba wax nanoemulsion applied as an edible coating on fresh tomato for postharvest quality evaluation. *Heliyon, 8*. https://doi.org/10.1016/j.heliyon.2022.e09803

Moreira Pereira, E., Dellinghausen Borges, C., dos Santos Formiga, A., et al. (2022). Conservation of red guava "Pedro Sato" using chitosan and gelatin-based coatings produced by the layer-by-layer technique. *Process Biochemistry, 121*, 35–44. https://doi.org/10.1016/j.procbio.2022.06.020

Najafi Marghmaleki, S., Mortazavi, S. M. H., Saei, H., & Mostaan, A. (2021). The effect of alginate-based edible coating enriched with citric acid and ascorbic acid on texture, appearance and eating quality of apple fresh-cut. *International Journal of Fruit Science, 21*, 40–51. https://doi.org/10.1080/15538362.2020.1856018

Niu, X., Liu, Y., Song, Y., et al. (2018). Rosin modified cellulose nanofiber as a reinforcing and co-antimicrobial agents in polylactic acid/chitosan composite film for food packaging. *Carbohydrate Polymers, 183*, 102–109. https://doi.org/10.1016/j.carbpol.2017.11.079

Odetayo, T., Sithole, L., Shezi, S., et al. (2022). Effect of nanoparticle-enriched coatings on the shelf life of Cavendish bananas. *Scientia Horticulturae, 304*, 111312. https://doi.org/10.1016/j.scienta.2022.111312

Oyom, W., Xu, H., Liu, Z., et al. (2022). Effects of modified sweet potato starch edible coating incorporated with cumin essential oil on storage quality of 'early crisp.'. *LWT - Food Science and Technology, 153*, 112475. https://doi.org/10.1016/j.lwt.2021.112475

Panahirad, S., Naghshiband-Hassani, R., Bergin, S., et al. (2020). Improvement of postharvest quality of plum (Prunus domestica L.) using polysaccharide-based edible coatings. *Plants, 9*, 1148. https://doi.org/10.3390/plants9091148

Panahirad, S., Dadpour, M., Peighambardoust, S. H., et al. (2021). Applications of carboxymethyl cellulose- and pectin-based active edible coatings in preservation of fruits and vegetables: A review. *Trends in Food Science & Technology, 110*, 663–673. https://doi.org/10.1016/j.tifs.2021.02.025

Pandey, V. K., Islam, R. U., Shams, R., & Dar, A. H. (2022). A comprehensive review on the application of essential oils as bioactive compounds in Nano-emulsion based edible coatings of fruits and vegetables. *Applied Food Research, 2*, 100042. https://doi.org/10.1016/j.afres.2022.100042

Pham, T. T., Nguyen, L. L. P., Dam, M. S., & Baranyai, L. (2023). Application of edible coating in extension of fruit shelf life: Review. *AgriEngineering, 5*, 520–536. https://doi.org/10.3390/agriengineering5010034

Pires, J., de Paula, C. D., Souza, V. G. L., et al. (2021). Understanding the barrier and mechanical behavior of different nanofillers in chitosan films for food packaging. *Polymers, 13*, 721. https://doi.org/10.3390/polym13050721

Priya, K., Thirunavookarasu, N., & Chidanand, D. V. (2023). Recent advances in edible coating of food products and its legislations: A review. *Journal of Agriculture and Food Research, 12*, 100623. https://doi.org/10.1016/j.jafr.2023.100623

Prudnikov, E., Polishchuk, I., Sand, A., et al. (2023). Self-assembled fatty acid crystalline coatings display superhydrophobic antimicrobial properties. *Materials Today Bio, 18*, 100516. https://doi.org/10.1016/j.mtbio.2022.100516

Qin, Z., Jia, X., Liu, Q., et al. (2020). Enhancing physical properties of chitosan/pullulan electrospinning nanofibers via green crosslinking strategies. *Carbohydrate Polymers, 247*, 116734. https://doi.org/10.1016/j.carbpol.2020.116734

Rajaei, A., Hadian, M., Mohsenifar, A., et al. (2017). A coating based on clove essential oils encapsulated by chitosan-myristic acid nanogel efficiently enhanced the shelf-life of beef cutlets. *Food Packaging and Shelf Life, 14*, 137–145. https://doi.org/10.1016/j.fpsl.2017.10.005

Rodrigues, F. A. M., dos Santos, S. B. F., de Lopes, M. M. A., et al. (2021). Antioxidant films and coatings based on starch and phenolics from *Spondias purpurea* L. *International Journal of Biological Macromolecules, 182*, 354–365. https://doi.org/10.1016/j.ijbiomac.2021.04.012

Saleem, M. S., Ejaz, S., Anjum, M. A., et al. (2020). Postharvest application of gum arabic edible coating delays ripening and maintains quality of persimmon fruits during storage. *Journal of Food Processing and Preservation, 44*, e14583. https://doi.org/10.1111/jfpp.14583

Sami, R., Almatrafi, M., Elhakem, A., et al. (2021). Effect of nano silicon dioxide coating films on the quality characteristics of fresh-cut cantaloupe. *Membranes, 11*, 140. https://doi.org/10.3390/membranes11020140

Sapper, M., & Chiralt, A. (2018). Starch-based coatings for preservation of fruits and vegetables. *Coatings, 8*, 152. https://doi.org/10.3390/coatings8050152

Sekarina, A. S., Supriyadi, S., Munawaroh, H. S. H., et al. (2023). Effects of edible coatings of chitosan - fish skin gelatine containing black tea extract on quality of minimally processed papaya during refrigerated storage. *Carbohydrate Polymer Technologies and Applications, 5*, 1–9. https://doi.org/10.1016/j.carpta.2023.100287

Shi, C., Fang, D., Huang, C., et al. (2023). Electrospun biopolymer material for antimicrobial function of fresh fruit and vegetables: Application perspective and challenges. *LWT, 174*, 114374. https://doi.org/10.1016/j.lwt.2022.114374

Shiekh, K. A., Ngiwngam, K., & Tongdeesoontorn, W. (2022). Polysaccharide-based active coatings incorporated with bioactive compounds for reducing postharvest losses of fresh fruits. *Coatings, 12*, 8. https://doi.org/10.3390/coatings12010008

Shoaib, M. H., Yousuf, R. I., Ahmed, F. R., et al. (2020). Polymer coatings for pharmaceutical applications. In *Polymer coatings* (pp. 275–317). Wiley.

Sousa, F. F., Pinsetta Junior, J. S., Oliveira, K. T. E. F., et al. (2021). Conservation of "palmer" mango with an edible coating of hydroxypropyl methylcellulose and beeswax. *Food Chemistry, 346*, 128925. https://doi.org/10.1016/j.foodchem.2020.128925

Sowmyashree, A., Sharma, R. R., Rudra, S. G., & Grover, M. (2021). Layer-by-layer coating of hydrocolloids and mixed plant extract reduces fruit decay and improves postharvest life of nectarine fruits during cold storage. *Acta Physiologiae Plantarum, 43*, 112. https://doi.org/10.1007/s11738-021-03256-8

Suhag, R., Kumar, N., Petkoska, A. T., & Upadhyay, A. (2020). Film formation and deposition methods of edible coating on food products: A review. *Food Research International, 136*, 109582. https://doi.org/10.1016/j.foodres.2020.109582

Sultan, M., M. Hafez, O., A. Saleh, M., & M. Youssef, A. (2021). Smart edible coating films based on chitosan and beeswax–pollen grains for the postharvest preservation of Le Conte pear. *RSC Advances, 11*, 9572–9585. https://doi.org/10.1039/D0RA10671B

Tabassum, N., & Khan, M. A. (2020). Modified atmosphere packaging of fresh-cut papaya using alginate based edible coating: Quality evaluation and shelf life study. *Scientia Horticulturae, 259*, 108853. https://doi.org/10.1016/j.scienta.2019.108853

Tang, X., & Yan, X. (2017). Dip-coating for fibrous materials: Mechanism, methods and applications. *Journal of Sol-Gel Science and Technology, 81*, 378–404. https://doi.org/10.1007/s10971-016-4197-7

Toniciolli Rigueto, C. V., Rosseto, M., Alessandretti, I., et al. (2022). Gelatin films from wastes: A review of production, characterization, and application trends in food preservation and agriculture. *Food Research International, 162*, 112114. https://doi.org/10.1016/j.foodres.2022.112114

Wallace, T. C., Bailey, Regan, L., Blumberg, Jeffrey, B., et al. (2020). Fruits, vegetables, and health: A comprehensive narrative, umbrella review of the science and recommendations for enhanced public policy to improve intake. *Critical Reviews in Food Science and Nutrition, 60*, 2174–2211. https://doi.org/10.1080/10408398.2019.1632258

Wang, J., Wang, H., Liu, Z., et al. (2023). Chitosan-nanocrystal cellulose composite coating could inhibit fruit decay rate and maintain the texture parameters of fruit in okra (Abelmoschus esculentus [L.] Moench). *Quality Assurance and Safety Crops & Foods, 15*, 20–31. https://doi.org/10.15586/qas.v15i2.1208

Wardak, M. H., Nkede, F. N., Van, T. T., et al. (2024). Development of edible films and partial coating, a novel coating technique for tomato fruits, using citric acid-crosslinked starch and cellulose nanofiber. *Progress in Organic Coatings, 187*, 108127. https://doi.org/10.1016/j.porgcoat.2023.108127

Xue, J., Wu, T., Dai, Y., & Xia, Y. (2019). Electrospinning and electrospun nanofibers: Methods, materials, and applications. *Chemical Reviews, 119*, 5298–5415. https://doi.org/10.1021/acs.chemrev.8b00593

Yadav, A., Kumar, N., Upadhyay, A., et al. (2022). Effect of mango kernel seed starch-based active edible coating functionalized with lemongrass essential oil on the shelf-life of guava fruit. *Quality Assurance and Safety Crops & Foods, 14*, 103–115. https://doi.org/10.15586/qas.v14i3.1094

Yan, J., Luo, Z., Ban, Z., et al. (2019). The effect of the layer-by-layer (LBL) edible coating on strawberry quality and metabolites during storage. *Postharvest Biology and Technology, 147*, 29–38. https://doi.org/10.1016/j.postharvbio.2018.09.002

Yıldırım-Yalçın, M., Şeker, M., & Sadıkoğlu, H. (2022). Effect of grape derivatives and cross-linked maize starch coatings on the storage life of strawberry fruit. *Progress in Organic Coatings, 167*, 106850. https://doi.org/10.1016/j.porgcoat.2022.106850

Yilmaz, M. T., Taylan, O., Karakas, C. Y., & Dertli, E. (2020). An alternative way to encapsulate probiotics within electrospun alginate nanofibers as monitored under simulated gastrointestinal conditions and in kefir. *Carbohydrate Polymers, 244*, 116447. https://doi.org/10.1016/j.carbpol.2020.116447

Zeinali, T., Alemzadeh, E., Zarban, A., et al. (2021). Fabrication and characterization of jujube extract-loaded electrospun polyvinyl alcohol nanofiber for strawberry preservation. *Food Science & Nutrition, 9*, 6353–6361. https://doi.org/10.1002/fsn3.2601

Zhang, L., Liu, Z., Sun, Y., et al. (2020). Combined antioxidant and sensory effects of active chitosan/zein film containing α-tocopherol on *Agaricus* bisporus. *Food Packaging and Shelf Life, 24*, 100470. https://doi.org/10.1016/j.fpsl.2020.100470

Zhang, L., Shao, G., Jin, Y., et al. (2025). Applying layer-by-layer deposition to enhance stability and rheological behavior of emulsions: A review. *Food Hydrocolloids, 158*, 110537. https://doi.org/10.1016/j.foodhyd.2024.110537

Zhou, X., Cheng, R., Wang, B., et al. (2021). Biodegradable sandwich-architectured films derived from pea starch and polylactic acid with enhanced shelf-life for fruit preservation. *Carbohydrate Polymers, 251*, 117117. https://doi.org/10.1016/j.carbpol.2020.117117

Zhou, Y., Hu, L., Chen, Y., et al. (2022). The combined effect of ascorbic acid and chitosan coating on postharvest quality and cell wall metabolism of papaya fruits. *LWT, 171*, 114134. https://doi.org/10.1016/j.lwt.2022.114134

Chapter 16
Effect of Edible Coating on Post-harvest Characteristics of Fruits and Vegetables

Nishant Kumar, Ajay Yadav, Pratibha, Rushali Jaiswal, and Nutan Kaushik

Abstract Edible coatings are thin layers applied to fruits and vegetables to preserve their quality characteristics during post-harvest handling and storage. Fruits and vegetables are widely consumed globally due to their inherent nutrients, but they have a very short shelf life. Furthermore, approximately 40% of the world's fruits and vegetables are discarded as waste due to post-harvest losses. Edible coatings are environmentally friendly and can be consumed along with the produce, aiding in shelf life extension by maintaining an appropriate gaseous composition and controlling moisture migration on the surface of fruits and vegetables. A substantial amount of literature exists on the application of edible coatings for enhancing the shelf life of fruits and vegetables. This chapter systematically presents the available literature on how edible coatings affect the quality characteristics of fruits and vegetables. The primary quality characteristics influenced by edible coatings include physiological weight loss, firmness, respiration rate, ethylene production, colour, pH, titratable acidity, antioxidant properties, and sensory attributes.

Keywords Edible · Coatings · Fruits · Vegetables · Post-harvest · Quality

---

N. Kumar · N. Kaushik
Amity Food and Agriculture Foundation (AFAF), Amity University Noida, Noida, Uttar Pradesh, India

A. Yadav (✉)
ICAR National Dairy Research Institute, Karnal, Haryana, India

Pratibha
Centre of Management and Humanities (CMH), Punjab Engineering College (Deemed to be University), Chandigarh, India

R. Jaiswal
Department of FST, National Institute of Food Technology Entrepreneurship and Management, Kundli, Sonipat, India

N. Kumar et al. (eds.), *Trends in Edible Packaging: A sustainable approaches for postharvest management of fruits and vegetables*,
https://doi.org/10.1007/978-3-032-13199-7_16

Introduction

The consumption of plastic and synthetic materials has rapidly increased from 5 million tonnes to 413.8 million tonnes between 1950 and 2023 globally. In 2022, China, Japan, and the rest of Asia produced 216 million tonnes of plastic, which accounted for 54% of the total global production. The packaging industry has accounted for 37% of all plastic production. Natural, biodegradable, edible, functional, and economical packaging is becoming increasingly popular (Han & Gennadios, 2005; Prasad et al., 2025). Food manufacturers favoured edible and food-grade packaging because it increased the food products' stability, safety, and shelf life. Edible packaging made of natural biopolymers, like edible coatings and films, has a lot of potential for both the environment and business. Compared to chemical and synthetic packaging materials, edible packaging has drawn a lot of attention in recent decades because of its advantages and environmentally friendly methods (Teixeira-Costa & Andrade, 2021). It has been established that edible packaging (coating/film) is safe to eat and can deliver active compounds and gradients, including flavour, colourants, antioxidants, antimicrobials, anti-browning agents, and functional components (Kumar & Neeraj, 2019; Thamarsha et al., 2024). It controls the moisture loss, gas exchange, colour, aroma, ethylene production, and microbiological contamination of food products, such as fruits, vegetables, dairy, and meat products, by acting as the primary packaging and coming into direct contact with the food surface (Kumar et al., 2024a). The term edible coatings is used interchangeably to indicate that the surface of a food is covered by a relatively thin layer of material of a certain composition using immersion, dipping, spraying, and drenching (coating) methods. However, the edible film is different from coating materials, and it is used as a wrapping and packaging material to laminate the foodstuff (Kumar et al., 2023). Edible coatings and films are eco-friendly, biodegradable, and safe for consumption and an alternative to chemical and synthetic packaging (Kumar et al., 2024b). The various types of biopolymers, such as polysaccharides, proteins, lipids & waxes, have been used for developing edible packaging materials for food products. The biopolymer formulations used for developing edible coating and film should be capable of forming a cohesive three-dimensional matrix (Petkoska et al., 2025). The properties and effectiveness of the developed coating/film depend on the properties of the used biopolymers (Vanaraj et al., 2024). Various researchers have successfully applied different types of biopolymer-based (polysaccharide, proteins, and lipids) based edible coatings and films to improve the stability, shelf life and organoleptic properties of the fruits, vegetables, fresh-cut, meat and other food products i.e. mango (Fonseca et al., 2004), banana (Suseno et al., 2014), papaya (Prasad et al., 2024), avocado (Le et al., 2021), Sapota (Chettri et al., 2023), Pomegranate (Anean et al., 2023), blueberries (Davidović et al., 2021), apple (Aayush et al., 2022), pear (Hira et al., 2022), plum (Kaur et al., 2023), cherry (Zhang et al., 2021), strawberries (de Albuquerque Sousa et al., 2024), tomato (Yadav et al., 2022), cucumber (Rux et al., 2023), fresh cut (Yousuf et al., 2021), fish and fish products (Bremenkamp & Sousa-Gallagher, 2024), and meat, beef and chicken (Zhang et al., 2022; Yang et al., 2023; Zhao et al., 2022) etc. The present book chapter extensively reviews edible coatings and films and their impacts on the quality and post-harvest shelf life of fruits and vegetables. In addition, the mechanism and concept of edible coatings and the physiological parameters of fruits and vegetables have also been explored.

Edible Packaging

Historical Background of Edible Packaging

Edible films and coatings materials have traditionally been used for the preservation of food products against moisture loss and microbial contamination. The term edible coatings is used interchangeably to indicate that the surface of a food is covered by a relatively thin layer of material of a certain composition. However, edible films are different from coating materials, and they are used as a wrapping and packaging material to cover the food surface. Edible coatings and films are eco-friendly, biodegradable, and safe for consumption and are an alternative to chemical and synthetic packaging. The edible films and coatings are made with a film-forming biopolymer and additives such as colour agents, flavour, antimicrobial, antioxidant, anti-browning, and functional ingredients. The incorporation of bioactive compounds helps in the prevention (physicochemical and biological deterioration) of the food product (Kumar & Neeraj, 2019). Edible packaging on food products has a long history. In the twelfth and thirteenth centuries, China applied a wax coating on citrus fruits to prolong the shelf life during the storage period. Japan developed a type of edible coating and film, "Yuba", using soybean in the fifteenth century to maintain the quality attributes and improve the shelf life of food products (Tural et al., 2017; Ulusoy et al., 2018). In the nineteenth century, the United States issued a patent for gelatin-based edible packaging for meat and meat products. The commercial use of the edible coatings and films to improve the shelf life and preservation of fruits & vegetables (apple, orange, grape, cherries, tomato, capsicum, and strawberry), dairy, confectionery, and meat & meat products has been used for a long time (Oğuzhan & Yangilar, 2016; Hassan et al., 2018). In the past, in 1967, edible packaging had very little commercial use for food preservation. In 1986, more than 10 industries applied edible coatings and films to products, whereas in 1996, the number of industries grew and reached up to 600. In the last decade, consumer awareness and consumption of edible films have rapidly expanded, and the total revenue has exceeded $100 million per annum. In India, three types of waxes, i.e. wax, carnauba wax, and shellac wax, have been permitted to be applied to fruits and vegetables by FSSAI, 2011 (2.3.6). Commercially, the first waxing was invented and applied to fruits and vegetables in 1922 (Raghav et al., 2016). Previously, various researchers applied different types of edible coating/film, i.e. polysaccharide, protein, lipid and wax, to improve the quality and prolong the shelf life of food products, i.e. fruit and vegetables, dairy, and meat products (Tarkhasi, 2016; Kharchoufi et al., 2018a, b; Licciardello et al., 2018; Taberner et al., 2018; Jaishankar et al., 2025).

Edible Coating/Film

At present, consumers are aware of eco-friendly and biopolymer-based edible packaging and have a demand for natural products with high quality and food safety (Aguirre-Joya et al., 2018). Furthermore, industries and researchers have also focused on developing natural packaging materials for food products to ensure food quality and consumers' safety. Numerous types of natural biopolymers, such as polysaccharides, proteins, lipids, and wax-based components, have been used as film-forming materials to develop edible coatings and films. At present, numerous industries have been using edible coatings and films on a commercial scale for the preservation of food products (Table 16.1).

These components of edible packaging are non-toxic, eco-friendly, and biodegradable in nature, and are obtained from plant and animal derivatives (Kumar & Neeraj, 2019). The functional and specific properties of edible films are responsible for the deceleration of organic vapour, such as solvent, aroma, water vapour, solute, such as salts, lipids, food additives and pigments, and gases, such as carbon dioxide, oxygen, and nitrogen. The water barrier properties of films are responsible for retarding the surface dehydration of fruits, vegetables, meat, seafood and other frozen food products (Dhall, 2013; Dehghani et al., 2018). Edible coatings are a barrier to gas exchange, mostly oxygen and control the ripening of fruits, and help to minimise oxidation of oxygen-sensitive food products. Edible film and coating are a controller of mass transfer between the food products and the environment. It also helps to minimise radical air reaction (UV light effect) in foods. It also improves the mechanical properties of food and food products during handling and storage. The organoleptic properties, such as transparency, colour, roughness, shininess, and stickiness, can be improved by edible films and packaging. Table 16.2 shows the reasons for considering edible coatings and films in the food processing sector.

Functional Properties of Edible Coatings

The edible coating should provide the following functions:

(i) An edible coating provides a stable quality for food products with market safety, nutritional value, and economic cost of production in the food processing sector (Bhardwaj et al., 2019).

Table 16.1 Commercially available edible films and coatings for food preservation

Edible coating/film	Name	Matrix/formulations	Applications
Film	BioEnvelop®	Cellulose and starch	Frying fish, potatoes, and other vegetables
	COGIN®	Carrageenan	
	ENAK®	HPMC, alginate, and carrageenan	
	GREENSOL®	Hydroxypropylmethylcellulose (HPMC)	
Coatings	Chris-Kraft Polymer Inc.	Hydroxypropylmethylcellulose	Fruits and vegetables
	Freshseel™	Polysaccharide	Melon and pome fruits
	Semperfresh™		
	Fry Shield™	Calcium pectinate	Frying fish, potatoes, and other vegetables
	Nature Seal™	Calcium ascorbate	Apples, avocado, carrot, and other vegetables
	Sta-Fresh 2103a	Carnauba	Citrus, pome fruits, pomegranate, pineapple, and tropical fruits
	Sta-Fresh 2104a		
	Sta-Fresh 2109a		
	Sta-Fresh 900a		
	Sta-Fresh 965a		
	Sta-Fresh 975a		
	Sta-Fresh 925a		
	Sta-Fresh 2505a		
	Sta-Fresh 2119a		
	MPa		
	Natural Shine™ 9000b		
	Natural Shine™ 960b		
	Natural Shine™ 990b		
	Natural Shine™ TFC 210b		

(continued)

Table 16.1 (continued)

Edible coating/film	Name	Matrix/formulations	Applications
	Sta-Fresh 223HSa	Shellac	Citrus, pome fruits, and pomegranate
	Sta-Fresh 2210a		
	Sta-Fresh 227a		
	Sta-Fresh 819Fa		
	Sta-Fresh 123a		
	CitraShine™ 585 Ftb		
	CitraShine™ 595 Ftb		
	Lustre Dryb		
	Lustre Dry w/TBZb		
	Sta-Fresh 216a	Shellac/resin	Citrus, pome fruits, pomegranate, and pineapple
	Sta-Fresh 710 (810)a		
	Sta-Fresh 2210a	Shellac/carnauba	Citrus
	Sta-Fresh 890SSa		
	Sta-Fresh 4201a	Poly	Citrus, pineapple
	Sta-Fresh 7100a		
	Sta-Fresh 7055a		
	Sta-Fresh 711(811)a	Oil	Lime, stone fruits, potato, melon, vegetables, pineapple, and citrus
	Sta-Fresh 151a		
	Sta-Fresh 7051a	Paraffin	Pineapple
	PacRite® Durafreshb	Petroleum wax	Stone fruits and vegetables
	PrimaFresh® 200b		
	Natural Shine™ 320Orb,	Vegetable wax	Citrus, pear, and lemon
	Natural Shine™ 4000b,		
	Natural Shine™ 4000b		
	Natural Shine™ 2015b	Vegetable wax/shellac	Citrus
	PacRite® 425b	Vegetable, resin, shellac, and waxes	

	PrimaFresh® 45b	Carnauba	Stone fruits
	PrimaFresh® 50-Vb	Oil	Stone fruits and vegetables
	PrimaFresh® FDb	Carnauba	Citrus
	PrimaFresh® Ultrab	Vegetable and shellac waxes	Pome fruit
	PrimaFresh® Goldb	Carnauba	Apple
	Shield-Brite® AP-40b	Shellac	Apples
	Shield-Brite® AP-431FDb	Shellac	Apples
	Shield-Brite® AP-747b	Shellac	Apples
	Shield-Brite®	Carnauba	Melon
	Decco Lustr® 202	Wax	Citrus, cantaloupe, cherry, mango, avocado, and stone fruits
	Vegetable Lustr® 227Fc	Oil, petrolatum, and paraffin	Vegetables
	Peach, Nectarine Plum Lustr® 251c	Carnauba wax/hydrocarbons/paraffin	Stone fruits, mango, cherry, vegetables, and potato
	Peach, Nectarine Plum Lustr® 255c	Oils	Stone fruits, pineapple, cherry, avocado, and cantaloupe
	Citrus Lustr 267c	Resins/wax	Citrus
	Peach, Nectarine Plum Lustr® 282c	Oils	Stone fruits, vegetables, cherry, mango, and potato
	Citrus Lustr® 402c	Resins/fatty acids/silicons	Citrus
	Decco Muli Lustr® 295c	Hydrocarbons/minerals	Stone fruits
	Decco Lustr®	Carnauba	Citrus, vegetables, apple, pear, mango, papaya, and sweet potato
	Decco Lustr® 231c	Carnauba	Citrus, pome fruits, stone fruits, vegetables, pomegranate, apple, roots, mango, and papaya
	Premium Apple Lustr®—Fast Dryc	Shellac	Apple

(continued)

Table 16.1 (continued)

Edible coating/film	Name	Matrix/formulations	Applications
	Decco Pearl Lustr®c	Rsins/fatty acid/propylene glycol/minerals	Citrus
	SemperFreshd	Sodium CMC	Fruits, vegetables, nectarine
	NatureSeald	Polysaccharide	Apple, pear, avocado, and banana
	Fresh Seald	Cellulose	Avocado, cantaloupe, mango, and papaya
	Fruit Coating Resinsd	Resins	Apple and citrus
	Nutrasave™	Chitosan	Avocado
	Opta Glaze™	Wheat gluten	Prevent microbial growth
	Seal gum, Spray gum	Calcium acetate	Potato
	SHELLAC (E904)	Resin from lac insect	Confectioneries glaze
	Z-Coat™	Zein-corn proteins	Nut meats, pecan, and chocolate covered peanut

Source: Olivas et al. (2008) and Erkmen and Barazi (2018)

Table 16.2 Reasons for consideration of edible coating/film in the food processing sector

1.	Chemical safety	As per WHO and FAO all food ingredients and additives, the safety is a fundamental requirement, so edible coating are free from chemicals and can be directly consumed by consumers
2.	Food quality	The incorporation of pomegranate peel in an edible coating prolongs the shelf life of fruits and vegetables during storage period and improves their organoleptic properties such as appearance, flavour and mouthfeel
3.	Nutritive value	Edible coating with pomegranate peel tends to increase physiochemical properties of food products
4.	Barrier properties	The developed edible coating may help to provide barrier properties for water, gas and UV light between food and the atmosphere
5.	Environmental	It is a biodegradable and eco-friendly packaging, maybe an alternative of plastics packaging and helps to reduce environmental problem
6.	Waste utilisation	It uses the phenolic and antioxidant agents from the food waste
7.	Cost	It is a biologically safe and low-cost material prepared from pomegranate waste and seaweed waste to reduce the post-harvest losses of fruits and vegetables during storage conditions

(ii) Edible coating materials and films protect food products such as fruits, vegetables, meat, fish, and other products from physical and mechanical factors of damage, with the control of microbial spoilage (Rooney, 2005).

(iii) The edible coating layer helps to minimise lipid oxidation, maintain quality parameters, appearance, and enhance the shelf life of the food products (Lacroix & Ouattara, 2000).

(iv) Natural antioxidant and antimicrobial agents have also been incorporated into the edible film to retard autoxidation of high-fat-rich food products and increase the oil resistance of fried food products (Daraei et al., 2012).

(v) Edible coating and packaging materials provide the barrier properties against moisture and gases of fresh produce during storage to retard the enzymatic oxidation and provide protection of food from browning discolouration, and texture softening. It also has the ability to prevent the loss of natural volatile flavour compounds and colour components (Dehghani et al., 2018). Figures 16.1 and 16.2 schematically represent the functional properties of edible packaging and reciprocal relations between the packaging materials and food product quality attributes.

Types of Edible Coating

Polysaccharide

Polysaccharide-based components have unique physical, chemical, and biological properties. They are polymers of simple sugars and are linked with glycosidic bonds. The term polysaccharide stands for "poly", which is derived from "many"

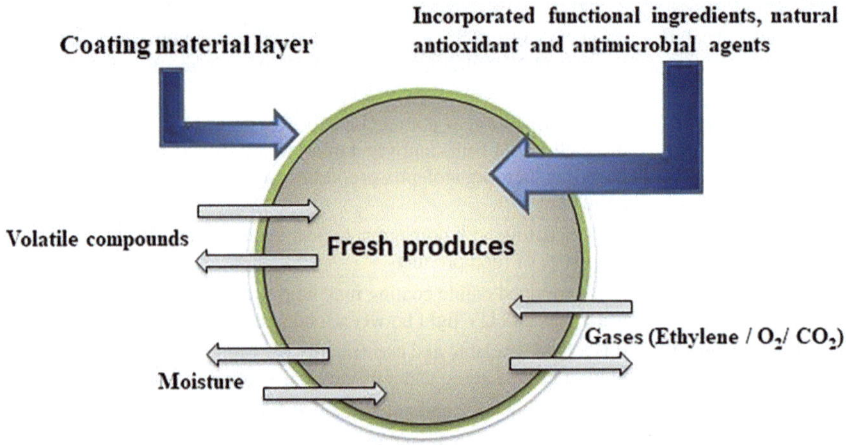

Fig. 16.1 Schematic representation of edible packaging and its functions on food products. (Kumar & Neeraj, 2019)

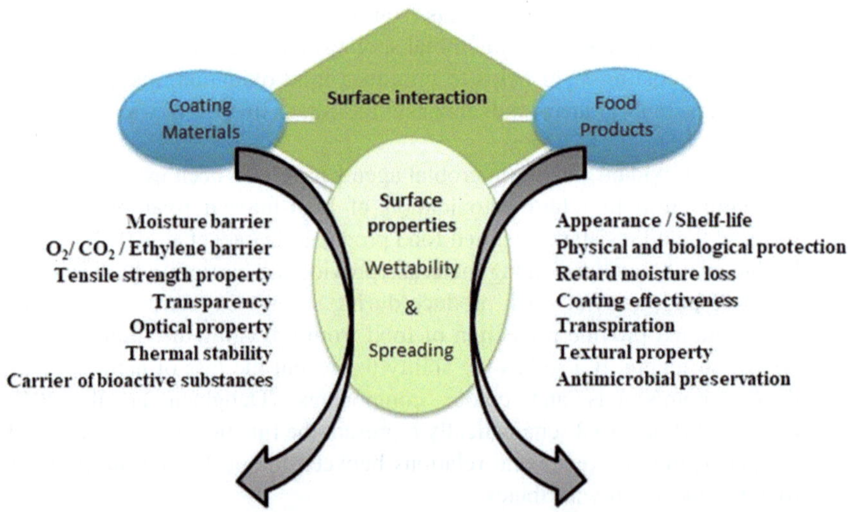

Fig. 16.2 Relationship between edible coating and food product quality attributes. (Kumar & Neeraj, 2019)

and "saccharide", which is derived from "sugar or sweet". They are known as the most abundant polymers in nature (Benalaya et al., 2024). The polysaccharide-based structural polymers have originated from plants, animals, bacteria, and marine origins. Numerous types of polysaccharide-based components have been used for developing edible packaging (coating/film) to prevent food products from microbial degradation. The polysaccharide-based components, such as starch, cellulose, hemicellulose, pectin, guar gum, gum arabic, glycogen, chitin, chitosan, pullulan, alginate, carragenan, bacterial cellulose/gellan, xylan, agar, and xanthan gum, have been potentially used in the formation of eco-friendly and edible packaging for food application due to their hydrocolloid and gelling nature and also exhibit gas and water barrier properties (Kumar et al., 2019; Suhag et al., 2020; Prasad et al., 2025). These components are non-toxic, flexible, biodegradable, biocompatible, highly chemically reactive, chiral, polyfunctional, and absorptive in nature. They also have high hydrophobicity due to the presence of primary amino and hydroxyl groups of the glucose unit (Crini, 2005). The characteristics of the polysaccharide-based structural components serve to prolong the shelf life of food products, as well as fruits and vegetables (Pan et al., 2024). Silva (2018) developed a chitosan-based edible film under different concentrations and studied its effect on guava. The application of the film suppresses the guava respiration rate, delays firmness, weight loss, and ethylene peak, and reduces the rate of degradation of skin pigment and controls the oxidation process of physiological compounds. It also helps in the prevention of phenylalanine ammonia-lyase (PAL) activity and increases in the antioxidant process.

Protein-Based Edible Coating

Proteins are biomolecules composed of long chains of amino acids linked by peptide bonds. They exhibit diverse functionalities depending on their application. Recently, proteins have garnered significant interest as biomolecules for food packaging applications due to their abundance, high nutritional value, and fibrous and globular nature. Proteins for coating applications can be sourced from either plants or animals. Plant sources include proteins from grains (such as wheat, rice, and sorghum), pulse proteins (such as soybean and pea), and other sources like millet proteins. Animal proteins include milk proteins (such as casein, sodium caseinate, and whey), collagen, and gelatin (Lacroix & Vu, 2014).

Coatings made from proteins demonstrate good forming properties and are generally transparent. Protein-based edible coatings exhibit tensile strength comparable to polysaccharide-based coatings. Additionally, protein-based edible coatings possess excellent barrier properties against gaseous transmission, particularly oxygen and carbon dioxide. A study by Bourtoom (2009) reported that protein-based edible coatings made from soy protein exhibited lower gaseous transmission compared to polysaccharide-based coatings. However, a limitation of protein-based coatings is their poor barrier to water vapour due to the presence of hydrophilic amino acids in their structure. Tien (2001) studied the effect of whey and calcium caseinate protein coating solutions on apple and potato slices. During this study, it was observed that the whey protein showed high antioxidant activity as compared to calcium caseinate. These coating materials help to delay the browning of coated apple and potato slices due to their oxygen barrier properties. Feng (2018) developed a whey protein isolate nano-fibril-based material film and coating material with the addition of different concentrations of glycerol. The study was conducted for comparison between WPI, WPNF, and WPNF (whey protein) based material and their effect on fresh-cut apples. The formation of whey protein (WPNFs) with glycerol and trehalose was found to be the best combination to protect fresh-cut apples. It exhibits a more well-ordered structure of β-sheet, excellent antioxidant and hydrophobic activity, as compared to other films. It helps to increase surface continuity, transparency, homogeneity, and water solubility of the developed film. The WPNF-based coating material with the incorporation of glycerol and trehalose protects fresh-cut apple slices and exhibits excellent protective action to retain phenolic activity and retard browning and moisture loss during storage of apple slices at 4 °C.

Lipid/Wax

Lipid-based edible coatings consist of lipid moieties or waxes as structural ingredients. The lipids used for fabricating edible coatings include neutral and polar lipids, fatty acids, mineral oil, and waxes (Milani &Nemati, 2022). The advantage of lipid-based edible coatings is their hydrophobic nature, which prevents water migration through their structure, thereby providing excellent barrier properties against water vapour transmission. In addition to neutral and polar lipids, waxes used for lipid-based coatings include carnauba, beeswax, and paraffin wax (Islam et al., 2024). Recently, essential oils have been incorporated into lipid-based edible coatings to enhance hydrophobic and antimicrobial properties. Essential oils such as clove, lemongrass, cinnamon, basil, and oregano are widely used in the fabrication of edible coatings for fruits and vegetables. Hashemi (2017) demonstrated the effect of basil seed gum-based edible coating with the addition of oregano essential oil (1–6%) on fresh-cut apricots during cold storage at 4°C for 8 days. The incorporation of oregano essential oil in edible film-forming solutions improves WVP and the mechanical properties of the film. The results indicate that the incorporation of 6% oregano essential oil is found to be most effective in reducing pathogen load to maintain the quality of fresh-cut apricots. It also helps to increase phenolic and antioxidant activity and the shelf life of stored fresh-cut apricots for 8 days. Oregel-Zamudio (2017) studied the effect of candelilla wax-based edible coating alone or with the incorporation of B. subtilis biocontrol bacteria on the quality of strawberries during a 6-day storage period at 25 °C. The present research indicates that the addition of biocontrol bacteria helps to reduce post-harvest deterioration of strawberries. Both applied edible wax coatings preserved the strawberry fruit's quality, such as total soluble solids (TSS), moisture, weight loss, pH, and shelf life. It reduces the risk of fruit decay and protects against the R. stolonifer pathogen.

Composite Edible Coating

Polysaccharide-based edible coatings offer good mechanical strength but lack water barrier properties, while lipid-based coatings provide excellent water vapour barriers but are rigid and inelastic. Proteins are effective in gaseous transmission but lack water vapour permeability. To address the limitations of individual polymers, two or more polymers are blended to form composite films with desired functionalities. In fruits and vegetables, composite films are widely used for coating applications to maintain the appropriate gaseous atmosphere and water permeability across the food surface and coating material. Composite films can be further categorised into binary films, including protein-protein coatings, protein-lipid coatings, lipid-lipid coatings, carbohydrate-protein coatings, carbohydrate-lipid coatings, and polysaccharide-polysaccharide coatings (Prasad et al., 2025; Jaishankar et al., 2025). Gamboa-Santos and Campañone (2019) developed an alginate-lactate-based edible film and studied its effect on strawberries during osmotic dehydration. It was found that the coating material had a positive impact on strawberries, water evaporation, and drying rate through subsequent microwave drying. Escamilla-García et al. (2018) developed an edible film with a blend of chitosan-oxidised starch. The developed material was applied to papaya at room temperature. During the 15-day storage study, it was observed that there was 10-times greater butyric acid production in uncoated papaya compared with coated papaya. The coating material helps to delay fermentation and reduces surface microbial contamination of papaya. Kharchoufi et al. (2018a, b) developed a chitosan and locust bean gum-based antifungal edible coating with the incorporation of pomegranate peel extract (aqueous and methanolic) and biocontrol agents (Wickerhamomyces anomalus) to reduce the physiological and post-harvest decay of orange fruits. The control used chitosan and locust bean gum without the incorporation of PGP extract and biocontrol agents. The results revealed that the incorporation of 0.361 g of the dry powder pomegranate extract and biocontrol in both coating materials reduced disease incidence in oranges and showed excellent antifungal activity against *Penicillium digitatum.*

Effect of Edible Coating on Fruits and Vegetables

The application of an edible coating on fruits and vegetables could potentially extend the postharvest shelf life of fruits and vegetables with their visual appearance and minimum losses of nutrients and other postharvest characteristics, i.e. physicochemical and sensory characteristics, due to minimising the gas and water transpiration from the product surface. The potential effect of different biopolymer-based edible coatings on fruits and vegetables' physiochemical, textural, microbiological, physiological and sensory characteristics has been discussed below.

Physiological Loss in Weight

Weight loss in horticulture is mainly caused by the transpiration of water during storage. It may affect the appearance of the fruits and vegetables (Caleb et al., 2012; Aloui et al., 2014). Therefore, the edible coating helps to prevent moisture loss and enhances the post-harvest characteristics of fruits and vegetables. The edible coating acts as a physical barrier against water loss of horticultural produce during storage (Shamloo et al., 2013; Maftoonazad & Ramaswamy, 2019). The edible coating creates a barrier on the surface of fruits to restrict water vapour loss (Poverenova et al., 2014). Various studies have reported that the edible coating is able to prevent weight loss as compared to untreated fruits and vegetables during storage (Vanoli et al., 2015; Zam, 2019). Divya (2018) reported on the effect of chitosan-based nanocoating on brinjal and chilli. During the study, it was observed that the chitosan-based edible coating was an effective coating agent for increasing water vapour transmission resistance and reducing water loss. The 4% and 5% nanoparticles of chitosan concentration help prevent weight loss and shelf life of the coated brinjals and chillies. Adetunji et al. (2019), Kumar et al. (2020, 2021a, b) extended the shelf life of litchi fruits, tomatoes, and green bell peppers using chitosan–pullulan composite edible coating with pomegranate peel extract by reducing the physiological weight loss of fruits and vegetables during the storage period at room and cold temperatures. Furthermore, various researchers applied different biopolymer-based edible coatings to extend the shelf life of fruits and vegetables with minimum losses of colour, flavour, and appearance of the produce (Rattanapanone et al., 2007; Gol & Rao, 2014; Kumari et al., 2015; Tripathi et al., 2016; Sucharitha et al., 2018; Zahedi et al., 2019) due to the barrier properties of the edible coating against water evaporation and gas transpiration.

Titratable Acidity

The titratable acidity of horticultural produce gradually decreases with increasing time of storage (Eshetu et al., 2019). The accumulation of organic acids has usually been seen during fruit development and is converted into respiratory substrates in glycolysis and the TCA cycle (Famiani et al., 2015). The rise in acidity could be attributed to the production of organic acids during ripening (Batista-Silva et al., 2018), which could be due to higher malic acid accumulation caused by the activity of the malate dehydrogenase enzyme. Another aspect contributing to the increasing acidity during storage could be an indication of the initiation of fermentative pathways (Muñoz et al., 2004). Another aspect related to the increase in acidity of stone fruits and vegetables, as suggested in the literature, was that the acidity in the head of mangoes increased with storage at 13 °C and 5 °C, which may be attributed to the accumulation of citric acid. The accumulation of citric acid may be due to the blocking effect of aconitase activity, which interferes with the conversion of citrate to aconitic acid (Degu et al., 2011). Alternatively, there is also a possibility that a lower temperature could reduce the activity of citrate synthase, which synthesises citric acid (Książek, 2023).

The edible coating helps to maintain and control the titratable acidity of horticultural produce during storage as compared to untreated horticultural produce. The edible coating controls the titratable acidity of horticultural produce, probably due to the retarded respiration rate (Eshetu et al., 2019). The post-harvest management practice, such as edible coating, helps to delay ripening and respiration rate, and maintain the acidity of horticultural produce during storage (Priya et al., 2023). Eshetu et al. (2019) investigated the effect of edible coating on mango post-harvest characteristics during the storage period and reported that the post-harvest characteristics of mango fruit were improved. The edible coating controls the enzymatic reaction of organic acids and helps to maintain the value of titratable acidity of horticultural products (Yaman & Bayoindirli, 2002; Mani et al., 2017). Doreyappa and Huddar (2001) reported similar patterns of results of mango fruit during storage at 18–34 °C. Kumar et al. (2020) also maintained the acidity and pH of litchi fruits and green bell peppers using chitosan–pullulan composite edible coating with pomegranate peel extract by controlling the water loss, gas exchange, and retarded enzymatic reactions throughout the 18-day storage period at room and cold temperatures.

Colour

Colour attributes of fruits and vegetables are the most important factor for consumers. The colour influences consumer preferences by ensuring the freshness of the produce. The colour difference of the product during the storage period has been used to determine the degradation of lycopene and chlorophyll structures in fruits and vegetables. The oxidative degradation of chlorophyll in fruits and vegetables happens due to the high metabolic activity during storage, which is responsible for the degradation of colour in fruits and vegetables (Artes & Gomez, 2003; Mohammadi et al., 2016). The influence of biopolymer-based edible coating slow down the respiration and ethylene production process (Ali et al., 2019). The edible coating may lead to a reduction in the colour difference of fruits and vegetables during the storage period. This is probably due to the control of enzymatic browning and delay in pigmentation (Dilmacunal et al., 2011; Sharma et al., 2019). The complete colour difference (ΔE) has been widely used to determine maturation owing to the degradation of the chlorophyll and lycopene structures. It is evident that the fruits and vegetables show distinction in colour at different maturity stages (Escamilla-García et al., 2018; Mandal et al., 2018). Ali et al. (2011) observed a delay in colour growth in papaya fruit covered with greater chitosan levels, owing to a slow respiration rate and decreased ethylene production. The treatment resulted in a delay in fruit maturation and senescence, leading to a decreased shift in colour development by controlling enzymatic browning and delaying ethylene synthesis and pigmentation (Sharma et al., 2019; Kumar et al., 2024a). The previous researchers Lin and Zhao (2007), Lin et al. (2011), de Oliveira et al. (2014), and Kumar et al. (2020, 2021a) controlled the colour degradation of fruits and vegetables during the storage period using active and non-active edible coatings due to the reduction of polyphenol oxidase activity, rate of respiration, ethylene production, control of enzymatic browning, and delaying the pigmentation process.

TSS

Total soluble solids (TSS) is the most important factor that is responsible for consumer acceptance. The TSS of the horticultural produce increases during the storage period (Abebe et al., 2017). TSS of the fruits and vegetables during storage was found to increase rapidly. The treated fruits and vegetables with edible coating exhibited a slow increase as compared to untreated fruits and vegetables due to controlled ripening and maturity of the horticultural produce (Maftoonazad & Ramaswamy, 2019; Nair et al., 2018). Various researchers have investigated the effect of different types of edible coatings on fruits and vegetables, resulting in maintaining TSS during the storage period by controlling the conversion of complex polysaccharides to simple sugars (Kumar et al., 2019; Yin et al., 2019).

Firmness

Firmness (softening) is the most important parameter of the horticultural produce (Batu, 2004). The firmness of fruits and vegetables is directly associated with the process of solubilisation of pectin substances and breakdown of the soluble sugars (Mebratie et al., 2015). The edible coating helps to protect the firmness of horticultural produce while controlling moisture loss. The edible coating-treated horticulture produce retains a better firmness level as compared to untreated fruits. The edible coating has been found effective in delaying ripening and preventing produce from early softening (Eshetu et al., 2019). Previous researchers, Lin et al. (2011), Jorge et al. (2014), Poverenova et al. (2014), and Yin et al. (2019) have reported that the application of edible coatings retains the firmness of litchi fruits, tomatoes, bell peppers, and mango fruits during storage. This might also be attributed to the effect of edible coating, which could lead to reduced enzymatic activity (PPO/POD) and might create barrier properties against water vapour loss (Zam, 2019). The results of this study are in line with previous work conducted by Lin et al. (2011), who reported that the application of chitosan-based edible coating improved the firmness of litchi fruit during the storage period. Various researchers and scientists reported that the application of different types of biopolymer-based edible coatings can control the degradation of fruits and vegetables' textural properties (firmness) by creating a semi-permeable barrier against water vapour loss from the coated surface

(Mandal et al., 2018; Eshetu et al., 2019; Pillai et al., 2024). The barrier properties of edible coatings could lead to the control of lipid oxidation, water transpiration, and hydrolysis of pectin substances (Poverenova et al., 2014; Prasad et al., 2025).

pH

The pH represents the nature (acid/base) of horticultural produce. The pH of the horticultural produce has been reported by research workers to show a gradual increase during the storage time due to the high respiration rates and delayed utilisation of organic compounds (Abebe et al., 2017). The edible coating helps to control the respiration rate and helps to reduce the pH in treated fruits and vegetables as compared to untreated ones during storage conditions such as ambient and cold storage (Duan et al., 2011; Sanaa et al., 2017). Various researchers have reported that the application of edible coatings can maintain the pH of horticultural produce during storage conditions by delaying the utilisation of organic compounds, controlling anthracnose, and activating defence enzymes (Bill et al., 2014; Miranda et al., 2024).

Respiration Rate

Some fruits and vegetables, being climacteric fruits, continue to undergo respiration post-harvest. The respiration rate escalates to a stage known as the climacteric peak, which signifies the onset of senescence (Baseer et al., 2024). Edible coatings are designed to mitigate the respiration rate, thereby delaying or preventing the occurrence of the climacteric peak by altering the gaseous environment surrounding the entire fresh tomato (Zhang et al., 2019). Pagno et al. (2018) demonstrated that a 1.5% chitosan coating effectively reduced the respiration rate of tomatoes by 21% compared to uncoated samples stored at 20 °C and 85% relative humidity for 14 days. Torres-León (2018) utilised mango by-products and developed an edible film and coating from mango seed (by-products) extract with the addition of glycerol plasticiser. The developed edible coating material was applied to a peach to investigate the coating effect. The developed film showed good properties of colour, antioxidant, hydrophobicity, permeability, and surface properties. During this investigation, it was found that the mango seed extract-based coating material helped in the reduction of O_2, CO_2 consumption, and ethylene production of the coated peach. Nair (2018) developed chitosan- and alginate-based edible coating material with the incorporation of pomegranate peel extract to improve the quality of guava. The incorporation of pomegranate peel extract showed a positive impact on guava characteristics. It helps to reduce the respiration rate and increase total phenolic, total flavonoids, and antioxidant activities of the guava fruit.

Ethylene Production

Ethylene, commonly referred to as the ripening hormone, plays a crucial role in expediting the ripening process by enhancing the activities of ripening enzymes such as ACC synthase and oxidase (Paul et al., 2010). Oxygen is regarded as a key substrate for the activity of these ripening enzymes (Leszczuk et al., 2024). In the context of edible-coated fruit and vegetables, the edible coating acts as a barrier to oxygen, thereby decelerating enzyme activity and delaying ripening-associated changes. Various studies indicate that edible coatings are effective in reducing ethylene production in fruit and vegetables.

Microbial Decay

In fruit and vegetables, microbial decay is attributed to both fungal and bacterial rots. These rots not only compromise the quality of fruit and vegetables but also raise concerns regarding food safety. Numerous studies have demonstrated the efficacy of edible coatings in mitigating microbial decay in fruit and vegetables. The enhanced antimicrobial activity resulting from the incorporation of antimicrobial compounds, such as plant extracts and essential oils, has also been documented. Hashemi et al. (2021) have applied an alginate-based edible coating enriched with Shirazi thyme essential oil to pistachio and cashew. Who reported that the application of the edible coating reduced the risk of microbial contamination and controlled the peroxidation of pistachio during the storage period. Grande-Tovar (2018) concluded various research studies in his review and reported the antifungal properties of chitosan-based edible film and emulsion, and their impact on fungal decay and preservation of fruits. Tayel et al. (2016) developed a bioactive edible coating to control decay of citrus fruits during storage conditions. The bioactive edible coating materials were formulated with chitosan and various plant (pomegranate peel, cress seeds, olive leaves, and senna pods) extracts. Results of this study revealed that the incorporation of each plant extract in the chitosan-based coating individually showed great potential against fungal growth on citrus fruits during storage. The most effective combination with cress and pomegranate peel extract of the coating material prevented citrus fruits from decaying and blue and green mould for a 14-day storage period. The author recommended pomegranate peel extract and cress seed extract, natural derivatives, as a powerful alternative source to retard fungal growth of fruits and vegetables.

Jaiswal et al. (2018) developed a herbal coating material by using aloe vera and neem extract with the incorporation of salicylic acid and citric acid to reduce the loss of post-harvest quality in tomatoes at 4 and 25 °C temperatures. The neem and aloe vera gel help in the extension of the shelf life of tomatoes, and help to retard microorganism contamination and improve the quality, firmness, colour, and marketability of tomatoes. Sanchís et al. (2016) increased the shelf life and sensory quality of fresh-cut Rojo Brillante slices during a 9-day storage period at 5 °C. The pectin-based edible coating was developed with the incorporation of anti-browning agents such as citric acid and calcium chloride. The results indicated that the incorporation of anti-browning agents in the pectin-based edible coating helps to extend shelf life, control microbial growth, firmness, colour, polyphenol oxidase activity, and organoleptic properties of fresh-cut Rojo Brillante slices during storage. The developed edible coating was found to be effective against *E. coli* and *L. monocytogenes* microbes. Mannozzi et al. (2018) developed an edible coating material with the combination of chitosan and Procyanidin. The developed edible coating material showed a positive impact on maintaining firmness and increasing antioxidant activities of the coated blueberries. The coating material also exhibits antifungal properties. Al-Naamani et al. (2018) demonstrated a comparison between a nano composite and a chitosan film alone. Nanocomposite-based coating helps to reduce of fungal growth on okra during a 12-day storage period. The treated film of polyethene showed better antifungal and antimicrobial properties as compared to the chitosan-based film.

Sensory

Previous researchers have reported that the application of edible coatings with or without the addition of active agents may be helpful to prevent the losses of quality and sensory attributes (Nair et al., 2018; Kumar et al., 2024b). Krasniewska et al. (2014, 2015) incorporated a plant extract of *Satureja hortensis* L. with pullulan-based edible coating to maintain the quality and sensory characteristics of apples and peppers during the storage period and obtained favourable results. De Leon-Zapata et al. (2015) also applied candelilla wax coating enriched with tarbush (*Flourensiacernua*) extract on apples and reported that the applied material on apples did not cause undesirable changes in the sensorial aspects and appearance during the 8-week storage period. Guerreiro et al. (2015) reported that the incorporation of natural antioxidant agents in the edible coating could favourably influence the sensory profiling of the fruits and vegetables. In addition, the results of Minh et al. (2019) also indicated that the application of an edible coating (gum arabic) on litchi fruits was effective in maintaining a higher sensory score for overall acceptability during the storage period.

Phenolic, Flavonoid, and Antioxidant Activity

The biochemical properties, such as phenolic content, flavonoid content, and antioxidant activities of fruits and vegetables, are very important for reducing free radical activity. It represents the presence of the most important bioactive compounds, which are responsible for antioxidant activity and free radical scavenging activity

(Toor & Savage, 2005). Edible coatings are able to enhance the phenolic activity of coated fruits and vegetables as compared to uncoated fruits and vegetables (Lopez-Mata et al., 2013). The addition of pomegranate peel extract in the edible coating has been found effective in enhancing the phenolic activity of the coating material as well as treated horticultural produce (Kharchoufi et al., 2018a, b; Taberner et al., 2018). The obtained results indicated that the edible coating improved the phenolic content activity of litchi fruits, tomatoes, bell peppers, and mango fruits during storage conditions (Lin et al., 2011; Jorge et al., 2014; Poverenova et al., 2014; Yin et al., 2019). As per the previous research, the incorporation of pomegranate peel extract in edible coating enhanced the phenolic activities of the material as well as that of treated fruits and vegetables (Kharchoufi et al., 2018a, b; Nair et al., 2018; Taberner et al., 2018).

In support of the results of this study, previous studies indicated that the incorporation of pomegranate peel extract in edible coating was found to enhance the phenolic activities of the prepared coating material (Kharchoufi et al., 2018a, b; Taberner et al., 2018; Kumar et al., 2019). The present study also revealed that the applied coating treatment was found to be effective in controlling the degradation of phenolic content in mango fruits. This might be possible due to the barrier properties and control of enzymatic metabolic activities (Wang & Gao, 2013; Zam, 2019).

Ali et al. (2013) reported that the lower values of flavonoid content in fruits are related to a higher rate of respiration during shelf life. The incorporation of pomegranate peel extract in the edible coating helps in better retention of flavonoid content in fruits during the storage period, resulting in lower rates of respiration (Nair et al., 2018; Zam, 2019). The work conducted by Kumari et al. (2015) reported that the application of chitosan and salicylic acid-based edible coating retained 1.29-fold higher flavonoid content of litchi fruit as compared to the control.

The applied edible coating was found to be effective in retaining higher antioxidant activities in treated mangoes under both storage conditions (room and cold). The results of this study are supported by the previous work of Ma et al. (2011), Palafox-Carlos et al. (2012), and Jorge (2014), as they also reported that the application of edible coating is protective against the reduction of antioxidant agents from fruits and vegetables. Tayel et al. (2016) reported that the addition of a natural plant extract to the chitosan edible coating can enhance the antioxidant activity of fruits during the storage period.

The results demonstrated that the use of chitosan and pullulan-blended edible coating was effective in preventing loss of phenolic content in litchi fruit during the storage period at both room temperature and 4 °C storage conditions. This might also be attributed to the effect of edible coating, which could lead to control of moisture loss and compartmentation between the PPO and POD enzymes (Awad & De Jager, 2000). The results of this study are in line with previous work conducted by Sun et al. (2010), who reported that the application of a chitosan edible coating on litchi fruits controlled the loss of phenolic compounds during storage.

The use of chitosan- and pullulan-blended edible coating was effective in preventive flavonoid content in tomatoes during the storage period at both room temperature and 4°C storage conditions. The work of Ali et al. (2010) and Ghasemnezhad et al. (2010) reported that an edible coating prepared with arabic gum and chitosan significantly controlled the reduction of flavonoid content in tomatoes and apricots. Howard et al. (2000) reported that the reduction of flavonoid content in fruits and vegetables may occur due to the higher respiration rate and degradation of phenolic compounds.

Shelf Life

Various researchers investigated the effects of different types of edible coating treatments on the shelf life and post-harvest characteristics of fruits and vegetables throughout the storage periods at different temperature conditions. They maintained the overall quality attributes, i.e. physical properties, biochemical properties, textural properties, appearance, colour, flavour, and other organoleptic properties of fruits and vegetables during the storage period. Details of the investigations conducted by some previous researchers on fruits and vegetables have been described below.

Wang and Gao (2013) studied the effect of chitosan-based coating solution on storing strawberries at 5 °C or 10 °C, respectively. The researcher used chitosan solution at different concentrations (0.5, 1.0 and 1.5 g/100 mL). The chitosan-based edible coating helps in maintaining the quality of treated strawberries. It also helps to increase the level of the phenolic content activity, anthocyanins, the flavonoid content, and antioxidant activity. The treatment retards the decrease in ascorbic acid, glutathione (GSH) content and b1,3-glucanase activity as compared to the control. Therefore, the chitosan-based edible material could be favourable for extending shelf life and controlling strawberry decay during storage at 5 °C or 10 °C.

Nadim (2015) also reported the effect of methylcellulose-based edible film on strawberry physicochemical characteristics. The methylcellulose-based coating material helps in reducing firmness, mould infections, weight loss, colour change, and delay in softening and textural properties of strawberries.

Radi (2017) developed an edible coating from gelatin-based material with the addition of green and black tea extract (5, 10%), and aloe vera (50, 100%) gel. The developed edible coating material was applied to fresh-cut oranges and stored for 17 days at 4 °C to evaluate its physical and chemical characterisation. Significant results were observed between coated and control samples of fresh-cut oranges. The developed gelatin-based coating material with 100% aloe vera gel and 10% black tea extract showed minimal effect on the quality of fresh-cut oranges. It helps to retain sensory quality, reduce weight loss, and retard microbial growth of the fresh-cut orange during storage at 4 °C. The formulation of black tea extract with aloe vera gel was found to be most effective as compared to green tea extract and aloe vera gel combination in gelatin-based edible coating to protect the shelf life and organoleptic properties of stored fresh-cut oranges.

Moreira (2017) investigated the effect of fibre-rich pectin-based edible coating and pulse light (PL) treatments on fresh-cut apple physicochemical properties. The pulse light (PL) treatment is an alternative source to extend the shelf life of fresh-cut fruits. The results indicated that the pectin-based edible coating and PL treatment help to minimise browning and softening of coated fresh-cut apples fruits during storage. The additionally incorporated fibre in the edible film did not show any impact on the microbial load of coated fresh-cut apples. The combination of both pectin and PL treatment showed good antioxidant activity, antimicrobial activity, and maintained organoleptic characteristics and shelf life of the coated products.

Divya (2018) studied the effect of chitosan-based nanocoating on brinjal and chilli. During the study, was observed that the chitosan-based edible coating is an effective coating agent for increasing water vapour transmission resistance and reducing water loss. The 4% and 5% nanoparticles of chitosan concentration helped in the prevention of weight loss and the extension of the shelf life of the coated brinjals and chillies.

Saberi et al. (2018) developed a lipid-based edible coating with the incorporation of guar gum and pea starch, shellac, and oleic acid. During this study, it was found that the shellac and oleic acid-based edible coating had a positive impact as compared to the pea starch and guar gum on the quality of the coated oranges. It helped in the reduction of the respiration rate, ethylene production, weight and firmness loss, peel pitting, and fruit decay index rate of oranges.

Aitboulahsen Mohamed et al. (2018) developed bioactive edible packaging with gelatine and essential oil (Mentha pulegium) to improve the physicochemical properties of strawberries. During this study, the developed edible bioactive material was applied to strawberries and their sensory and organoleptic properties were characterised. The coated strawberries exceeded their physicochemical and organoleptic properties, such as weight loss, TA, TSS, firmness, colour, shelf life and phenolic activity. The combinations of both materials showed excellent properties as compared to gelatin alone.

Tomadoni et al. (2018a, b) developed a gellan gum-based edible coating material with the incorporation of geraniol and pomegranate extract at different concentrations. The results were shown, and incorporation of pomegranate extract in gellan gum-based edible coating material could not control microbial contamination, nor could it improve the firmness of fresh-cut strawberries. The applied coating of fresh-cut strawberries failed to extend their shelf life and sensory properties.

Dong and Wang (2018) demonstrated the guar-gum- and ginseng-extract-blended edible coating to apply to sweet cherries. The developed edible coating showed a positive response in reducing weight loss, respiration rate, and a reduction in the change of TSS, TA, AA, total anthocyanins, phenolic properties, and improved firmness of the coated sweet cherries.

López-Palestina (2018) developed a gelatin-based coating with the incorporation of oily tomato extract. The developed coating material was applied to the garambullo fruit under storage conditions at 5 °C. The edible coating with the incorporation of oily tomato extract helped in the reduction of weight loss, TA, colour, Bx, and pH of the coated garambullo fruit during storage. It also extended the biological properties such as higher antioxidant activity, betalains, ascorbic acid, phenolic, and flavonoid content. The researcher also reported that the garambullo fruits are an attractive source of antioxidants and can be preserved using tomato oil extract in an edible coating.

Amiri et al. (2018) studied the effect of a gelatin-based edible coating on fresh-cut apples. The citric acid, ascorbic acid, $CaCO_3$ and gelatin were used to obtain the edible coating with the incorporation of green tea extract and aloe vera at different concentrations. The effect of the edible coating was demonstrated on fresh-cut apples for 16 days at 4 °C. The incorporation of green tea extract and aloe vera in the gelatin-based edible coating helps to maintain freshness, colour and quality attributes of apple slices during the storage period. The higher concentrations of green tea extract and aloe vera have shown good antimicrobial properties and significantly decreased TSS at the end of storage. The highest sensory score was achieved by the BF2 sample with the 150% aloe vera during the storage time of apple slices at 4 °C.

Taberner (2016) studied the effect of a pectin-based edible coating with incorporated pomegranate peel extract and anti-browning agents (calcium ascorbate and citric acid) on Fresh-cut "Rojo Brillante" fruits during 5 days of storage at 5 °C. The coating with pomegranate peel extract helps in maintaining the L* value, which indicates the anti-browning activity on coated fruit during storage. The formulation of pectin and pomegranate peel extract or calcium ascorbate significantly reduced the firmness of the coated fruits as compared to the control samples. Overall, all the combinations of pectin-based edible coating retained excellent flavour. The marketability of the fruits was evaluated after 7 days of storage.

Nair et al. (2018) evaluated antifungal properties and characterisation of coated capsicum during the storage period at 10 °C. The polysaccharide-based edible coating was developed by using chitosan and alginate with incorporated pomegranate peel extract to improve capsicum quality during storage. The results indicated that the incorporation of pomegranate peel extract helped to reduce weight loss, firmness, colour, ascorbic acid, and total chlorophyll content, showing antifungal properties against Colletotrichum gloeosporioides in storage capsicum.

Tomadoni et al. (2018a, b) investigated a comparative study between the incorporation of the pomegranate peel extract and geraniol extract in gellan-based coatings. The coating was applied to a fresh-cut strawberry. As per the results of this study, the high positive impact to improve quality, sensory and antimicrobial properties was highly exhibited by the geraniol extract-based coating as compared to the pomegranate peel extract incorporated edible coating.

Settier-Ramírez (2019) developed an edible film by using five different film-forming solutions (polyvinyl alcohol, gelatin, sodium caseinate, hydrolysates of casein and gelatin, and yeast extract) with the incorporation of lactic acid bacteria. The viability of bacteria was assessed after 4 weeks of storage at 20 °C and 43.2% RH. The antimicrobial activity was assessed against L. monocytogenes, and the PVA-based film was used as a control with the addition of bacteria. During the stationary phase, L. lactis embedded within films could reach a maximum level. Incorporation of hydrolysates showed good antimicrobial properties as compared to other materials.

Yan (2019) investigated the qualitative effect of a layer-by-layer edible coating consisting of chitosan and carboxymethyl cellulose and chitosan alone, on stored strawberries. The layer-by-layer coating was most effective compared to the chitosan-based edible coating for the retained moisture loss, firmness, and aroma of stored strawberries at 0 °C. Both chitosan and layer-by-layer coating helped to delay the volatile composition as well as primary and secondary metabolism during the storage of strawberries at 0 °C. The researcher suggested that layer-by-layer coating is a good approach to improve the shelf life and other post-harvest quality of strawberries.

Conclusion

Edible coatings have been shown to significantly extend the shelf life of fresh fruits and vegetables by forming a semipermeable membrane on their surfaces, thereby allowing selective gas permeation similar to modified atmospheric packaging. Additionally, these coatings act as a barrier to moisture diffusion from the interior of the fruit surface to the external environment, thereby aiding in the maintenance of firmness and reducing physiological weight loss. The literature highlighted that the shelf life of fruits and vegetables can be increased by approximately 1.5–2 times at both ambient and cold temperatures. Most studies on the edible coating of fruits and vegetables have been conducted on a small scale, primarily at laboratory or pilot levels. There is a critical need to scale up the application of these developed coatings for large-scale use. Furthermore, additional research is necessary to understand the potential risks associated with the use of nanoparticles in edible coatings. The incorporation of essential oils in edible coatings poses challenges to the sensory properties of fruits and vegetables. Further studies are required to mitigate the strong aroma of essential oils without compromising the antimicrobial properties of the edible coatings.

References

Aayush, K., McClements, D. J., Sharma, S., Sharma, R., Singh, G. P., Sharma, K., & Oberoi, K. (2022). Innovations in the development and application of edible coatings for fresh and minimally processed Apple. *Food Control, 141*, 109188.

Abebe, Z., Tola, Y. B., & Mohammed, A. (2017). Effects of edible coating materials and stages of maturity at harvest on storage life and quality of tomato (Lycopersicon Esculentum Mill.) fruits. *African Journal of Agricultural Research, 12*(8), 550–565.

Adetunji, O. C., Ojediran, O. J., Adetunji, B. J., & Owa, O. S. (2019). Influence of chitosan edible coating on postharvest qualities of Capsicum annum L. during storage in evaporative cooling system. Croatian. *Journal of Food Science and Technology, 11*(1), 59–66.

Aguirre-Joya, J. A., De Leon-Zapata, M. A., Alvarez-Perez, O. B., Torres-León, C., Nieto-Oropeza, D. E., Ventura-Sobrevilla, J. M., et al. (2018). Basic and applied concepts of edible packaging for foods. In *Food packaging and preservation* (pp. 1–61). Academic Press.

Aitboulahsen Mohamed, Z., Said, L., Amin, C., Hicham, A., Abdelhay, B., Mohammed, Z. H., & Mounir. (2018). Gelatin-based edible coating combined with mentha pulegium essential oil as bioactive packaging for strawberries. *Journal of Food Quality, 2018*, 1–7.

Ali, A., Maqbool, M., Ramachandran, S., & Alderson, P. G. (2010). Gum arabic as a novel edible coating for enhancing shelf-life and improving postharvest quality of tomato (Solanum lycopersicum L.) fruit. *Postharvest Biology and Technology, 58*(1), 42–47.

Ali, A., Muhammad, M. T. M., Sijam, K., & Siddiqui, Y. (2011). Effect of chitosan coatings on the physicochemical characteristics of Eksotika II papaya (Carica papaya L.) fruit during cold storage. *Food Chemistry, 124*(2), 620–626.

Ali, A., Maqbool, M., Alderson, G. P., & Zahid, N. (2013). Effect of gum arabic as an edible coating on antioxidant capacity of tomato (Solanum lycopersicum L.) fruit during storage. *Postharvest Biology and Technology, 76*, 119–124.

Ali, A., Chen, Y., Liu, H., Yu, L., Baloch, Z., Khalid, S., et al. (2019). Starch-based antimicrobial films functionalized by pomegranate peel. *International Journal of Biological Macromolecules, 129*, 1120–1126.

Al-Naamani, L., Dutta, J., & Dobretsov, S. (2018). Nanocomposite zinc oxide-chitosan coatings on polyethylene films for extending storage life of okra (Abelmoschus esculentus). *Nanomaterials, 8*(7), 479.

Aloui, H., Khwaldia, K., Sánchez-González, L., Muneret, L., Jeandel, C., Hamdi, M., & Desobry, S. (2014). Alginate coatings containing grapefruit essential oil or grapefruit seed extract for grapes preservation. *International Journal of Food Science & Technology, 49*(4), 952–959.

Amiri, S., Akhavan, H. R., Zare, N., & Radi, M. (2018). Effect of gelatin-based edible coatings incorporated with aloe vera and green tea extracts on the shelf-life of fresh-cut apple. *Italian Journal of Food Science, 30*(2018), 61–74.

Anean, H. A., Mallasiy, L. O., Bader, D. M., & Shaat, H. A. (2023). Nano edible coatings and films combined with zinc oxide and pomegranate peel active phenol compounds has been to extend the shelf life of minimally processed pomegranates. *Materials, 16*(4), 1569.

Artes, F., & Gomez, P. (2003). Packaging and color control: the case of fruit and vegetables. In R. Avenhainen (Ed.), *Novel Food Packaging Techniques* (p. 416). CRC Press and Woodhead Publishing.

Awad, M. A., & De Jager, A. (2000). Flavonoid and chlorogenic acidconcentrations in skin of 'Jonagold' and 'Elstar' apples during and after regular and ultra low oxygen storage. *Postharvest Biology and Technology, 20*, 15–24.

Baseer, A. Q., Niazi, P., Monib, A. W., Hassand, M. H., Hejran, A. B., Sarwari, A., et al. (2024). Lifecycle transitions in plant development: Ripening, senescence, & cell death. *Journal of Pharma Insights and Research, 2*(2), 169–179.

Batista-Silva, W., Nascimento, V. L., Medeiros, D. B., Nunes-Nesi, A., Ribeiro, D. M., Zsögön, A., & Araújo, W. L. (2018). Modifications in organic acid profiles during fruit development and ripening: correlation or causation? *Frontiers in Plant Science, 9*, 1689.

Batu, A. (2004). Determination of acceptable firmness and colour values of tomatoes. *Journal of Food Engineering, 61*(3), 471–475.

Benalaya, I., Alves, G., Lopes, J., & Silva, L. R. (2024). A review of natural polysaccharides: sources, characteristics, properties, food, and pharmaceutical applications. *International Journal of Molecular Sciences, 25*(2), 1322.

Bhardwaj, A., Alam, T., & Talwar, N. (2019). Recent advances in active packaging of agri-food products: a review. *Journal of Postharvest Technology, 7*(1), 33–62.

Bill, M., Sivakumar, D., Korsten, L., & Thompson, A. K. (2014). The efficacy of combined application of edible coatings and thyme oil in inducing resistance components in avocado (Persea americana Mill.) against anthracnose during post-harvest storage. *Crop Protection, 64*, 159–167.

Bourtoom, T. J. I. F. R. J. (2009). Edible protein films: Properties enhancement. *International Food Research Journal, 16*(1), 1–9.

Bremenkamp, I., & Sousa-Gallagher, M. J. (2024). Design and development of an edible coating for a ready-to-eat fish product. *Polymers, 16*(3), 346.

Caleb, O. J., Opara, U. L., & Witthuhn, C. R. (2012). Modified atmosphere packaging of pomegranate fruit and arils: a review. *Food and Bioprocess Technology, 5*(1), 15–30.

Chettri, S., Sharma, N., & Mohite, A. M. (2023). Utilization of lima bean starch as an edible coating base material for sapota fruit shelf-life enhancement. *Journal of Agriculture and Food Research, 12*, 100615.

Crini, G. (2005). Recent developments in polysaccharide-based materials used as adsorbents in wastewater treatment. *Progress in polymer science, 30*(1), 38–70.

Daraei, G. A., Mirzaei, H. O., Maghsudlo, Y., Kashaninejad, M., & Jafari, S. M. (2012). Production of low fat french-fries with single and multi-layer hydrocolloid coatings. *Journal of Food Science and Technology, 1*(8), 660–669.

Davidović, S., Miljković, M., Gordic, M., Cabrera-Barjas, G., Nesic, A., & Dimitrijević-Branković, S. (2021). Dextran-based edible coatings to prolong the shelf life of blueberries. *Polymers, 13*(23), 4252.

de Albuquerque Sousa, T. C., de Lima Costa, I. H., Gandra, E. A., & Meinhart, A. D. (2024). Use of edible coatings as a new sustainable alternative to extend the shelf life of strawberries (Fragaria ananassa): A review. *Journal of Stored Products Research, 108*, 102375.

De Leon-Zapata, M. A., Saenz-Galindo, A., Rojas-Molina, R., Rodriguez-Herrera, R., Jasso-Cantu, D., & Aguilar, C. N. (2015). Edible candelilla wax coating with fermented extract of tarbush improves the shelf life and quality of apples. *Food Packaging and Shelf Life, 3*, 70–75.

de Oliveira, C. E. V., Magnani, M., de Sales, C. V., de Pontes, A. L. S., Campos-Takaki, G. M., Stamford, T. C. M., et al. (2014). Effects of chitosan from Cunninghamella elegans on virulence of post-harvest pathogenic fungi in table grapes (Vitis labrusca L.). *International Journal of Food Microbiology, 171*, 54–61.

Degu, A., Hatew, B., Nunes-Nesi, A., Shlizerman, L., Zur, N., Katz, E., et al. (2011). Inhibition of aconitase in citrus fruit callus results in a metabolic shift towards amino acid biosynthesis. *Planta, 234*(3), 501–513.

Dehghani, S., Seyed, H. V., & Joe, R. M. (2018). Edible films and coatings in seafood preservation: A review. *Food Chemistry, 240*, 505–513.

Dhall, R. K. (2013). Advances in edible coatings for fresh fruits and vegetables: A review. *Critical Reviews in Food Science and Nutrition, 53*, 435–450.

Dilmacunal, T., Koyunchu, M. A., Aktas, H., & Bayindir, D. (2011). The effect of several post-harvest treatments on shelf life, quality of bunch tomatoes. *Notulae Botanicae Horti Agrobotanici Cluj-Napoca, 39*, 209–213.

Divya, K., Vijayan, S., & Jisha, M. S. (2018). Antifungal, antioxidant and cytotoxic activities of chitosan nanoparticles and its use as an edible coating on vegetables. *International Journal of Biological Macromolecules, 114*, 572–577.

Dong, F., & Wang, X. (2018). Guar gum and ginseng extract coatings maintain the quality of sweet cherry. *LWT-Food Science and Technology, 89*, 117–122.

Doreyappa, G., & Huddar, A. G. (2001). Studies on ripening changes in mango (Mangifera indica L.) fruits. *Journal of Food Science and Technology, 38*(2), 135–137.

Duan, J., Wu, R., Strik, C. B., & Zhao, Y. (2011). Effect of edible coatings on the quality of fresh blueberries (Duke and Elliott) under commercial storage conditions: *Postharvest Biology and Technology, 59*(1), 71–79.

Erkmen, O., & Barazi, A. O. (2018). General characteristics of edible films. *Journal of Food Biotechnology Research, 2*(1–3), 1–4.

Escamilla-García, M., Rodríguez-Hernández, M. J., Hernández-Hernández, H. M., Delgado-Sánchez, L. F., García-Almendárez, B. E., Amaro-Reyes, A., & Regalado-González, C. (2018). Effect of an edible coating based on chitosan and oxidized starch on shelf life of Carica papaya L., and its physicochemical and antimicrobial properties. *Coatings, 8*(9), 318.

Eshetu, A., Ibrahim, M. A., Forsido, F. S., & Kuyu, G. C. (2019). Effect of beeswax and chitosan treatments on quality and shelf life of selected mango (Mangifera indica L.) cultivars. *Heliyon, 5*(1), 1–22.

Famiani, F., Battistelli, A., Moscatello, S., Cruz-Castillo, J. G., & Walker, R. P. (2015). The organic acids that are accumulated in the flesh of fruits: occurrence, metabolism and factors affecting their contents – A review. *Revista Chapingo Serie Horticultura, 21*(2), 97–128.

Feng, Z., Guangxin, W., Chunhong, L., Dongmei, L., Bin, J., & Xiaosong, Z. (2018). Edible coating based on whey protein isolate nanofibrils for antioxidation and inhibition of product browning. *Food Hydrocolloids, 79*, 179–188.

Fonseca, M. D. O., Cecon, P. R., Salomão, L. C. C., & Pushmann, R. (2004). Pulp and skin pigments in mango'Haden'treated with fungicides and wax.

Gamboa-Santos, J., & Campañone, L. A. (2019). Application of osmotic dehydration and microwave drying to strawberries coated with edible films. *Drying Technology, 37*(8), 1002–1012.

Ghasemnezhad, M., Shiri, M. A., & Sanavi, M. (2010). Effect of chitosan coatings on some quality indices of apricot (Prunus armeniaca L.) during cold storage. Caspian. *Journal of Environmental Sciences, 8*(1), 25–33.

Gol, B. N., & Rao, T. V. R. (2014). Influence of zein and gelatin coatings on the postharvest quality and shelf life extension of mango (Mangifera indica L.). *Fruits, 69*, 101–115.

Grande-Tovar, D. C., Chaves-Lopezb, C., Annalisa, S., Chiara, R., & Antonello, P. (2018). Chitosan coatings enriched with essential oils: Effects on fungi involved in fruit decay and mechanisms of action. *Trends in Food Science & Technology, 78*, 61–71.

Guerreiro, A. C., Gago, C. M., Faleiro, M. L., Miguel, M. G., & Antunes, M. D. (2015). The effect of alginate-based edible coatings enriched with essential oils constituents on Arbutus unedo L. fresh fruit storage. *Postharvest Biology and Technology, 100*, 226–233.

Han, J. H., & Gennadios, A. (2005). Edible films and coatings: A review. In U. K. Oxford (Ed.), *Innovations in Food Packaging* (pp. 239–262). Elsevier Academic Press.

Hashemi, S. M. B., Amin, K. M., Maryam, G. G., & Ismail, E. (2017). Basil-seed gum containing Origanum vulgare subsp. Viride essential oil as edible coating for fresh cut apricots. *Postharvest Biology and Technology, 125*(2017), 26–34.

Hashemi, M., Dastjerdi, A. M., Shakerardekani, A., & Mirdehghan, S. H. (2021). Effect of alginate coating enriched with Shirazi thyme essential oil on quality of the fresh pistachio (Pistacia vera L.). *Journal of Food Science and Technology, 58*(1), 34–43.

Hassan, B., Chath, S. A. S., Hussain, A. L., Zia, K. M., & Akhtar, N. (2018). Recent advances on polysaccharides, lipids and protein based edible films and coatings: A review. *International Journal of Biological Macromolecules, 109*, 1095–1107.

Hira, N., Mitalo, O. W., Okada, R., Sangawa, M., Masuda, K., Fujita, N., et al. (2022). The effect of layer-by-layer edible coating on the shelf life and transcriptome of 'Kosui'Japanese pear fruit. *Postharvest Biology and Technology, 185*, 111787.

Howard, L. R., Talcott, S. T., Brenes, C. H., & Villalon, B. (2000). Changes in phytochemical and antioxidant activity of selected pepper cultivars (Capsicum species) as influenced by maturity. *Journal of Agricultural and Food Chemistry, 48*, 1713–1720.

Islam, R. U., Khan, T., Junaid, P. M., & Naseem, S. (2024). Recent developments in sustainable bio-based wax nanoemulsions as edible coatings for food commodities: A comprehensive review. *Packaging Technology and Science, 37*(12), 1175–1194.

Jaishankar, P., Nishant, K., & Sujata, P. S. (2025). Shelf-life extension of guava and eggplant fruits using chitosan-sodium alginate based antimicrobial composite coating functionalized with clove-soy lecithin nanoemulsion. *Plant Science Today, 12*(3), 1–11.

Jaiswal, A. K., Kumar, S., & Bhatnagar, T. (2018). Studies to enhance the shelf life of tomato using aloe vera and neem based herbal coating. *Journal of Postharvest Technology, 6*(2), 21–28.

Jorge, E. D. A., José, A. V. R., Mónica, A. V. O., Orlando, T. O., Guadalupe, I. O., Ayala-Zavala, J. F., & González-Aguilar, G. A. (2014). Effect of edible coatings on bioactive compounds and antioxidant capacity of tomatoes at different maturity stages. *Journal of Food Science and Technology, 51*(10), 2706–2712.

Kaur, P., Sandhu, K. S., Bangar, S. P., & Whiteside, W. S. (2023). Effect of edible coating from octenyl succinic anhydride modified rye starch to extend shelf life of plums. *International Journal of Food Science & Technology, 58*(2), 987–994.
Kharchoufi, S., Parafati, L., Licciardello, F., Muratore, G., Hamdi, M., Cirvilleri, G., & Restuccia, C. (2018a). Edible coatings incorporating pomegranate peel extract and biocontrol yeast to reduce Penicillium digitatum postharvest decay of oranges. *Food Microbiology, 74*, 107–112.
Kharchoufi, S., Parafati, L., Licciardello, F., Muratore, G., Hamdi, M., Cirvilleri, G., & Cristina, R. (2018b). Edible coatings incorporating pomegranate peel extract and biocontrol yeast to reduce Penicillium digitatum postharvest decay of oranges. *Food Microbiology, 74*, 107–112.
Krasniewska, K., Gniewosz, M., Synowiec, A., Przybyl, J. L., Baczek, K., & Weglarz, Z. (2014). The use of pullulan coating enriched with plant extracts from Satureja hortensis L. to maintain pepper and apple quality and safety. *Postharvest Biology and Technology, 90*, 63–72.
Krasniewska, K., Gniewosz, M., Synowiec, A., Przybyl, J. L., Baczek, K., & Weglarz, Z. (2015). The application of pullulan coating enriched with extracts from Bergenia crassifolia to control the growth of food microorganisms and improve the quality of peppers and apples. *Food and Bioproducts Processing, 94*, 422–433.
Książek, E. (2023). Citric acid: properties, microbial production, and applications in industries. *Molecules, 29*(1), 22.
Kumar, N., & Neeraj. (2019). Polysaccharide-based component and their relevance in edible film/coating: A review. *Nutrition & Food Science, 49*(5), 793–823.
Kumar, N., Neeraj, Ojha, A., & Singh, R. (2019). Preparation and characterization of chitosan - pullulan blended edible films enrich with pomegranate peel extract. *Reactive and Functional Polymers, 144*, 104350.
Kumar, N., Neeraj, P., & Singla, M. (2020). Enhancement of storage life and quality maintenance of litchi (Litchi chinensis Sonn.) fruit using chitosan: Pullulan blend antimicrobial edible coating. *International Journal of Fruit Science, 20*(sup3), S1662–S1680.
Kumar, N., Neeraj, P., & Trajkovska Petkoska, A. (2021a). Improved shelf life and quality of tomato (Solanum lycopersicum L.) by using chitosan-pullulan composite edible coating enriched with pomegranate peel extract. ACS. *Food Science & Technology, 1*(4), 500–510.
Kumar, N., Ojha, A., Upadhyay, A., Singh, R., & Kumar, S. (2021b). Effect of active chitosan-pullulan composite edible coating enrich with pomegranate peel extract on the storage quality of green bell pepper. *LWT, 138*, 110435.
Kumar, N., Pratibha, Prasad, J., Yadav, A., Upadhyay, A., Neeraj, et al. (2023). Recent trends in edible packaging for food applications—Perspective for the future. *Food Engineering Reviews, 15*(4), 718–747.
Kumar, N., Khan, A. A., Pyngrope, D., Alanazi, A. M., Upadhyay, A., & Shukla, S. (2024a). Development and characterization of novel starch (mango kernel and litchi seed) based active edible coatings and films using ultrasonication: Effects on postharvest shelf life of Khasi mandarins. *Sustainable Chemistry and Pharmacy, 39*, 101610.
Kumar, N., Upadhyay, A., Shukla, S., Bajpai, V. K., Kieliszek, M., Yadav, A., & Kumaravel, V. (2024b). Next generation edible nanoformulations for improving post-harvest shelf-life of citrus fruits. *Journal of Food Measurement and Characterization, 18*(3), 1825–1856.
Kumari, P., Barman, K., Patel, V. B., Siddiqui, W. M., & Kole, B. (2015). Reducing postharvest pericarp browning and preserving health promoting compounds of litchi fruit by combination treatment of salicylic acid and chitosan. *Scientia Horticulturae, 197*, 555–563.
Lacroix, M., & Ouattara, B. (2000). Combined industrial processes with irradiation to assure innocuity and preservation of food products – A review. *Food Research International, 33*(2), 719–724.
Lacroix, M., & Vu, K. D. (2014). Edible coating and film materials: Proteins. In *Innovations in food packaging* (pp. 277–304). Academic.
Le, K. H., Nguyen, M. D. B., Dai Tran, L., Thi, H. P. N., Van Tran, C., Van Tran, K., et al. (2021). A novel antimicrobial ZnO nanoparticles-added polysaccharide edible coating for the

preservation of postharvest avocado under ambient conditions. *Progress in Organic Coatings, 158*, 106339.

Leszczuk, A., Kutyrieva-Nowak, N., Nowak, A., Nosalewicz, A., & Zdunek, A. (2024). Low oxygen environment effect on the tomato cell wall composition during the fruit ripening process. *BMC Plant Biology, 24*(1), 503.

Licciardello, F., Kharchoufi, S., Muratore, G., & Restuccia, C. (2018). Effect of edible coating combined with pomegranate peel extract on the quality maintenance of white shrimps (Parapenaeus longirostris) during refrigerated storage. *Food Packaging and Shelf Life, 17*, 114–119.

Lin, D., & Zhao, Y. (2007). Innovations in the development and application of edible coatings for fresh and minimally processed fruits and vegetables. *Comprehensive Reviews in Food Science and Food Safety, 6*(3), 60–75.

Lin, B., Dua, Y., Liang, X., Wang, X., Wang, X., & Yanga, J. (2011). Effect of chitosan coating on respiratory behavior and quality of stored litchi under ambient temperature. *Journal of Food Engineering, 102*(1), 94–99.

Lopez-Mata, M. A., Ruiz-Cruz, S., Silva-Beltran, N. P., Ornelas-Paz, J. D., Zamudio-Flores, P. B., & Burruel-Ibarra, S. E. (2013). Physicochemical, antimicrobial and antioxidant properties of chitosan films incorporated with carvacrol. *Molecules, 18*, 13735–13753.

López-Palestina, C. U., Aguirre-Mancilla, C. L., Raya-Pérez, J. C., Ramírez-Pimentel, J. G., Gutiérrez-Tlahque, J., & Hernández-Fuentes, A. D. (2018). The effect of an edible coating with tomato oily extract on the physicochemical and antioxidant properties of garambullo (Myrtillocactus geometrizans) fruits. *Agronomy, 8*(11), 248.

Ma, X., Wu, H., Liu, L., Yao, Q., Wang, S., & Zhan, R. (2011). Polyphenolic compounds and antioxidant properties in mango fruits. *Scientia Horticulturae, 129*, 102–107.

Maftoonazad, N., & Ramaswamy, S. H. (2019). Application and evaluation of a pectin-based edible coating process for quality change kinetics and shelf-life extension of Lime Fruit (Citrus aurantifolium). *Coatings, 9*(285), 1–13.

Mandal, D., Lalrinpuii, S., Hazarika, K. T., & Amritesh, S. C. (2018). Effect of edible coating on shelf life and quality of local mango cv. Rangkuai of Mizoram. *Research on Crops, 19*(3), 419–424.

Mani, A., Jain, N., Singh, K., & Arun, S. M. (2017). Effects of aloevera edible coating on quality and postharvest physiology of Ber (Zizyphus mauritiana Lamk.) under ambient storage conditions. *International Journal of Pure & Applied Bioscience, 5*(6), 43–53.

Mannozzi, C., Tylewicz, U., Chinnici, F., Siroli, L., Rocculi, P., Dalla Rosa, M., & Romani, S. (2018). Effects of chitosan based coatings enriched with procyanidin by-product on quality of fresh blueberries during storage. *Food Chemistry, 251*, 18–24.

Mebratie, M. A., Woldetsadik, K., Ayalew, A., & Haji, J. (2015). Comparative study of different banana ripening methods. *Science Technology and Arts Research Journal, 4*(2), 32–38.

Milani, J. M., & Nemati, A. (2022). Lipid-based edible films and coatings: A review of recent advances and applications. *Journal of Packaging Technology and Research, 6*(1), 11–22.

Minh, P. N., Nhi, T. Y. T., Vien, T. B. L., Ha, T. T. T., & Yen, T. K. N. (2019). Effect of arabic gum coating on postharvest quality of Litchi (Litchi chinensis) Fruits. *Journal of Pharmaceutical Sciences and Research, 11*(4), 1464–1468.

Miranda, M., Bai, J., Pilon, L., Torres, R., Casals, C., Solsona, C., & Teixidó, N. (2024). Fundamentals of edible coatings and combination with biocontrol agents: A strategy to improve postharvest fruit preservation. *Foods, 13*(18), 2980.

Mohammadi, A., Hashemi, M., & Hosseini, S. M. (2016). Postharvest treatment of nanochitosan-based coating loaded with Zataria multiflora essential oil improves antioxidant activity and extends shelf-life of cucumber. *Innovative Food Science & Emerging Technologies, 33*, 580–588.

Moreira, M. R., Álvarez, M. V., Martín-Belloso, O., & Soliva-Fortuny, R. (2017). Effects of pulsed light treatments and pectin edible coatings on the quality of fresh-cut apples: A hurdle technology approach. *Journal of the Science of Food and Agriculture, 97*(1), 261–268.

Muñoz, T., Sanchez-Ballesta, M. T., Ruiz-Cabello, J., Escribano, M. I., & Merodio, C. (2004). The acid metabolism of Annona fruit during ripening. *The Journal of Horticultural Science and Biotechnology, 79*(3), 472–478.

Nadim, Z., Ahmadi, E., Sarikhani, H., & Amiri Chayjan, R. (2015). Effect of methylcellulose-based edible coating on strawberry fruit's quality maintenance during storage. *Journal of Food processing and Preservation, 39*(1), 80–90.

Nair, M. S., Alok, S., & Kaur, C. (2018). Characterization and antifungal activity of pomegranate peel extract and its use in polysaccharide-based edible coatings to extend the shelf-life of Capsicum (Capsicum annuum L.). *Food and Bioprocess Technology, 11*(7), 1317–1327.

Oğuzhan Yıldız, P., & Yangilar, F. (2016). The use of edible films and coatings in food industry: Compilation. *Bitlis Eren University Journal of Science and Technology, 5*(1), 27–35.

Olivas, G. I., Dávila-Aviña, J. E., Salas-Salazar, N. A., & Molina, F. J. (2008). Use of edible coatings to preserve the quality of fruits and vegetables during storage. *Stewart Postharvest Review, 3*(6), 1–10.

Oregel-Zamudio, E., Angoa-Pérez, V. M., Guadalupe, O. S., Cristóbal, A. G. N., & Hortencia, M. V. G. (2017). Effect of candelilla wax edible coatings combined with biocontrol bacteria on strawberry quality during the shelf-life. *Scientia Horticulturae, 214*(2017), 273–279.

Pagno, C. H., Castagna, A., Trivellini, A., Mensuali-Sodi, A., Ranieri, A., Ferreira, E. A., et al. (2018). The nutraceutical quality of tomato fruit during domestic storage is affected by chitosan coating. *Journal of Food Processing and Preservation, 42*(1), e13326.

Palafox-Carlos, H., Yahia, E. M., & Gonzalez-Aguilar, G. A. (2012). Identification and quantification of major phenolic compounds from mango (Mangifera indica, cv. Ataulfo) fruit by HPLC-DAD-MS/MS-ESI and their individual contribution to the antioxidant activity during ripening. *Food Chemistry, 135*, 105–111.

Pan, J., Li, C., Liu, J., Jiao, Z., Zhang, Q., Lv, Z., et al. (2024). Polysaccharide-based packaging coatings and films with phenolic compounds in preservation of fruits and vegetables—A review. *Foods, 13*(23), 3896.

Paul, V., Singh, A., & Pandey, R. (2010). Estimation of total soluble solids (TSS). In *Laboratory manual on post-harvest physiology of fruits and flowers* (pp. 41–43).

Petkoska, A. T., Kumar, N., Koraqi, H., Ali-Haliti, S., & Broach, A. T. (2025). Bionanocomposites in food packaging and preservation. In *Sustainable materials for food packaging and preservation* (pp. 149–170). Elsevier.

Pillai, A. R., Eapen, A. S., Zhang, W., & Roy, S. (2024). Polysaccharide-based edible biopolymer-based coatings for fruit preservation: A review. *Foods, 13*(10), 1529.

Poverenova, E., Zaitsev, Y., Arnon, H., Granit, R., Alkalai-Tuvia, S., Perzelan, Y., Weinberg, T., & Fallik, E. (2014). Effects of a composite chitosan–gelatin edible coating on postharvest quality and storability of red bell peppers. *Postharvest Biology and Technology, 96*, 106–109.

Prasad, K., Akshatha, H., Pradhan, J., Singh, S. K., Udit, K., Saroj, N., et al. (2024). Eco-safe composite edible coating of hydrocolloids with papaya leaf extract improves postharvest quality and shelf life of papaya fruit under ambient storage. *Journal of Food Science, 89*(2), 1114–1126.

Prasad, J., Kumar, N., Jaiswal, R., Yadav, A., Sharma, S. P., Fawole, O. A., & Kaushik, N. (2025). Biopolymer based composite packaging: A sustainable approach for fruits and vegetables preservation. *Applied Food Research, 5*, 101211.

Priya, K., Thirunavookarasu, N., & Chidanand, D. V. (2023). Recent advances in edible coating of food products and its legislations: A review. *Journal of Agriculture and Food Research, 12*, 100623.

Radi, M., Firouzi, E., Akhavan, H., & Amiri, S. (2017). Effect of gelatin-based edible coatings incorporated with Aloe vera and black and green tea extracts on the shelf life of fresh-cut oranges. *Journal of Food Quality, 2017*(1), 9764650.

Raghav, P. K., Agarwal, N., & Saini, M. (2016). Edible coating of fruits and vegetables: A review. *Education, 1*(2), 188–204.

Rattanapanone, N., Plotto, A., & Baldwin, E. (2007, December). Effect of edible coatings and other surface treatments on pericarp color of Thai Lychee cultivars. In *Proceedings of the Florida State Horticultural Society* (Vol. 120, pp. 222–227).

Rooney, M. L. (2005). Introduction to active food packaging technologies. *Innovations in Food Packaging*, 63–69.

Rux, G., Labude, C., Herppich, W. B., & Geyer, M. (2023). Investigation on the potential of applying bio-based edible coatings for horticultural products exemplified with cucumbers. *Current Research in Food Science, 6*, 100407.

Saberi, B., Golding, J. B., Marques, J. R., Pristijono, P., Chockchaisawasdee, S., Scarlett, C. J., & Stathopoulos, C. E. (2018). Application of biocomposite edible coatings based on pea starch and guar gum on quality, storability and shelf life of 'Valencia'oranges. *Postharvest biology and technology, 137*, 9–20.

Sanaa, S. H., Aly, E. N. M., & Entsar, S. A. (2017). Effect of edible coating on extending the shelf life and quality of fresh cut taro. *American Journal of Food Technology, 12*, 124–131.

Sanchís, E., González, S., Ghidelli, C., Sheth, C. C., Mateos, M., Palou, L., & Pérez-Gago, M. B. (2016). Browning inhibition and microbial control in fresh-cut persimmon (Diospyros kaki Thunb. cv. Rojo Brillante) by apple pectin-based edible coatings. *Postharvest Biology and Technology, 112*, 186–193.

Settier-Ramírez, L., López-Carballo, G., Gavara, R., & Hernández-Muñoz, P. (2019). Antilisterial properties of PVOH-based films embedded with Lactococcus lactis subsp. lactis. *Food Hydrocolloids, 87*, 214–220.

Shamloo, M., Sharifani, M., Gramakhany, A. D., & Seifie, E. (2013). Alternation of flavonoid compounds in Valencia Orange fruit (Citrus sinensis) peel as a function of storage period and edible covers. *Minerva Biotecnologica, 25*, 191–197.

Sharma, L., Saini, S., Sharma, C., & Harish, K. (2019). Biocomposite edible coatings based on cross linked-sesame protein and mango puree for the shelf life stability of fresh-cut mango fruit. *Journal of Food Process Engineering, 42*(1), 1–9.

Silva, W. B., Glaucia, S. M. C., Diederson, S. B., et al. (2018). Chitosan delays ripening and ROS production in guava (Psidium guajava L.) Fruit. *Food Chemistry, 242*, 232–238.

Sucharitha, K. V., Beulah, A. M., & Ravikiran, K. (2018). Effect of chitosan coating on storage stability of tomatoes (Lycopersicon esculentum Mill). *International Food Research Journal, 25*(1), 93–99.

Suhag, R., Kumar, N., Petkoska, A. T., & Upadhyay, A. (2020). Film formation and deposition methods of edible coating on food products: A review. *Food Research International, 136*, 109582.

Sun, D., Liang, G., Xie, J., Lei, X., & Mo, Y. (2010). Improved preservation effects of litchi fruit by combining chitosan coating with ascorbic acid treatment during postharvest storage. *African Journal of Biotechnology, 922*, 3272–3279.

Suseno, N., Savitri, E., Sapei, L., & Padmawijaya, K. S. (2014). Improving shelf-life of cavendish banana using chitosan edible coating. *Procedia Chemistry, 9*, 113–120.

Taberner, V., SanchÃs, E., Mateos, M., Palou, L., & Pérez-Gago, M. B. (2016). Pectin-based edible coatings formulated with pomegranate peel extracts and other antibrowning agents to extend shelf life of fresh-cut'Rojo Brillante'persimmon. In *VIII international postharvest symposium: Enhancing Supply chain and consumer benefits-ethical and technological issues* 1194 (pp. 887–894).

Taberner, V., Sanchís, E., Mateos, M., Palou, L., & Pérez-Gago, M. B. (2018). Pectin-based edible coatings formulated with pomegranate peel extracts and other antibrowning agents to extend shelf life of fresh-cut 'Rojo Brillante. *Acta Horticulturae, 1194*, 887–894.

Tarkhasi, A. (2016). Effect of edible coating containing pomegranate peel extract on quality and shelf life of Silver Carp (Hypophthalmichthys molitrix) fillet during refrigerated storage. *Journal of Food and Industrial Microbiology, 2*(2), 1000112.

Tayel, A. A., Moussa, S. H., Salem, M. F., Mazrou, K. E., & El-Tras, W. F. (2016). Control of citrus molds using bioactive coatings incorporated with fungal chitosan/plant extracts composite. *Journal of the Science of Food and Agriculture, 96*(4), 1306–1312.

Teixeira-Costa, B. E., & Andrade, C. T. (2021). Natural polymers used in edible food packaging—History, function and application trends as a sustainable alternative to synthetic plastic. *Polysaccharides, 3*(1), 32–58.

Thamarsha, A. K. A. N. W. M. R. K., Kumar, N., Pratibha, K. M., Upadhyay, A., Sudhani, H. P., & Pamu, V. S. (2024). A review of functional properties and applications of legume-based edible coatings. *Legume Science, 6*(4), e70004.

Tien, C. L., Vachon, C., Mateescu, M. A., & Lacroix, M. (2001). Milk protein coatings prevent oxidative browning of apples and potatoes. *Journal of Food Science, 66*(4), 512–516.

Tomadoni, B., Moreira, R. M., Pereda, M., & Ponce, G. A. (2018a). Gellan-based coatings incorporated with natural antimicrobials in fresh-cut strawberries: Microbiological and sensory evaluation through refrigerated storage. *LWT – Food Science and Technology, 97*(2018), 384–389.

Tomadoni, B., Moreira, M. R., Pereda, M., & Ponce, A. G. (2018b). Gellan-based coatings incorporated with natural antimicrobials in fresh-cut strawberries: Microbiological and sensory evaluation through refrigerated storage. *LWT, 97*, 384–389.

Toor, R. K., & Savage, G. P. (2005). Antioxidant activity in different fractions of tomatoes. *Food Research International, 38*(5), 487–494.

Torres-León, C., Vicente, A. A., Flores-López, M. L., Rojas, R., Serna-Cock, L., Alvarez-Pérez, O. B., & Aguilar, C. N. (2018). Edible films and coatings based on mango (var. Ataulfo) by-products to improve gas transfer rate of peach. *LWT-Food Science and Technology, 97*, 624–631.

Tripathi, A., Mishra, D. S., Nimbolkar, P. K., & Chander, S. (2016). Physico-chemical properties and post-harvest life of Litchi: A review. *Journal of Agricultural Engineering and Food Technology, 3*(3), 189–193.

Tural, S., Sarıcaoğlu, F. T., & Turhan, S. (2017). Edible films and coatings: their production, application methods, functions and uses in muscle foods. 84–94.

Ulusoy, B. H., Yildirim, F. K., & Hecer, C. (2018). Edible films and coatings: A good idea from past to future technology. *Journal of Food Technology Research, 5*(1), 28–33.

Vanaraj, R., Kumar, S. M. S., Mayakrishnan, G., Rathinam, B., & Kim, S. C. (2024). A current trend in efficient biopolymer coatings for edible fruits to enhance shelf life. *Polymers, 16*(18), 2639.

Vanoli, M., Grassi, M., Buccheri, M., & Rizzolo, A. (2015). Influence of edible coatings on postharvest physiology and quality of Honeydew melon fruit (Cucumis melo L. inodorus). *Advances in Horticultural Science, 29*(2-3), 65–74.

Wang, S. Y., & Gao, H. (2013). Effect of chitosan-based edible coating on antioxidants, antioxidant enzyme system, and postharvest fruit quality of strawberries (Fragaria x aranassa Duch.). *LWT-Food sScience and Technology, 52*(2), 71–79.

Yadav, A., Kumar, N., Upadhyay, A., Sethi, S., & Singh, A. (2022). Edible coating as postharvest management strategy for shelf-life extension of fresh tomato (Solanum lycopersicum L.): An overview. *Journal of Food Science, 87*(6), 2256–2290.

Yaman, Ö., & Bayoindirli, L. (2002). Effects of an edible coating and cold storage on shelf-life and quality of cherries. *LWT- Food Science and Technology, 35*, 146–150.

Yan, J., Luo, Z., Ban, Z., Lu, H., Li, D., Yang, D., et al. (2019). The effect of the layer-by-layer (LBL) edible coating on strawberry quality and metabolites during storage. *Postharvest Biology and Technology, 147*, 29–38.

Yang, X., Zhao, D., Ge, S., Bian, P., Xue, H., & Lang, Y. (2023). Alginate-based edible coating with oregano essential oil/β-cyclodextrin inclusion complex for chicken breast preservation. *International Journal of Biological Macromolecules, 251*, 126126.

Yin, C., Huang, C., Wang, J., Liu, Y., Lu, P., & Huang, L. (2019). Effect of chitosan- and alginate-based coatings enriched with cinnamon essential oil microcapsules to improve the postharvest quality of mangoes. *Materials, 12*(13), 2039.

Yousuf, B., Wu, S., & Siddiqui, M. W. (2021). Incorporating essential oils or compounds derived thereof into edible coatings: Effect on quality and shelf life of fresh/fresh-cut produce. *Trends in Food Science & Technology, 108*, 245–257.

Zahedi, M. S., Hosseini, S. M., Karimi, M., & Ebrahimzadeh, A. (2019). Effects of postharvest polyamine application and edible coating on maintaining quality of mango (Mangifera indica L.) cv. Langra during cold storage. *Food Science & Nutrition, 7*(2), 433–441.

Zam, W. (2019). Effect of alginate and chitosan edible coating enriched with olive leaves extract on the shelf life of sweet cherries (Prunus avium L.). *Journal of Food Quality, 8192964*, 1–7.

Zhang, H., Hou, H., Liu, P., Wang, W., & Dong, H. (2019). Effects of acid hydrolysis on the physicochemical properties of pea starch and its film forming capacity. *Food Hydrocolloids, 87*, 173–179.

Zhang, Y. L., Cui, Q. L., Wang, Y., Shi, F., Liu, Y. P., Liu, J. L., & Nie, G. W. (2021). Effect of carboxymethyl chitosan-gelatin-based edible coatings on the quality and antioxidant properties of sweet cherry during postharvest storage. *Scientia Horticulturae, 289*, 110462.

Zhang, M., Luo, W., Yang, K., & Li, C. (2022). Effects of sodium alginate edible coating with cinnamon essential oil nanocapsules and Nisin on quality and shelf life of beef slices during refrigeration. *Journal of Food Protection, 85*(6), 896–905.

Zhao, R., Zhang, Y., Chen, H., Song, R., & Li, Y. (2022). Performance of eugenol emulsion/chitosan edible coating and application in fresh meat preservation. *Journal of Food Processing and Preservation, 46*(3), e16407.

Index

N. Kumar et al. (eds.), *Trends in Edible Packaging: A sustainable approaches for postharvest management of fruits and vegetables*,
https://doi.org/10.1007/978-3-032-13199-7

MIX
Papier aus verantwortungsvollen Quellen
Paper from responsible sources
FSC® C105338

If you have any concerns about our products,
you can contact us on
ProductSafety@springernature.com

In case Publisher is established outside the EU,
the EU authorized representative is:
Springer Nature Customer Service Center GmbH
Europaplatz 3, 69115 Heidelberg, Germany

Printed by Libri Plureos GmbH
in Hamburg, Germany